Electromagnetic Constants

Coulomb constant	k	$8.99 \times 10^9\ \mathrm{N \cdot m^2/C^2}$
	$4\pi k/c$	$377\ \Omega = 377\ \mathrm{N \cdot m \cdot s/C^2}$
Permittivity constant	ε_0	$8.85 \times 10^{-12}\ \mathrm{C^2 \cdot N^{-1} \cdot m^{-2}} = (4\pi k)^{-1}$
Permeability constant	μ_0	$1.26 \times 10^{-6}\ \mathrm{N \cdot s^2/C^2} = 4\pi k/c^2$
Conductivity:	σ_c	
Silver		$6.3 \times 10^7\ (\Omega \cdot \mathrm{m})^{-1}$
Copper		$5.9 \times 10^7\ (\Omega \cdot \mathrm{m})^{-1}$
Nichrome		$6.7 \times 10^5\ (\Omega \cdot \mathrm{m})^{-1}$

Electromagnetic Units and Conversion Factors

1 C = 1 coulomb = total charge of 6.242×10^{18} protons
1 N/C = units for the electric and magnetic fields = V/m
1 T = 1 tesla = $1\ \mathrm{N \cdot s \cdot C^{-1} \cdot m^{-1}}$, a magnetic field unit equivalent to 299.792458 MN/C
1 G = 1 gauss = another unit for the magnetic field, equivalent to 10^{-4} T or about 30 kN/C
1 V = 1 volt = 1 J/C = $1\ \mathrm{N \cdot m/C} = 1\ \mathrm{kg \cdot m^2 \cdot s^{-2} \cdot C^{-1}}$ = unit of energy per unit charge
1 A = 1 ampere = 1 C/s = unit of current
$1\ \Omega$ = 1 ohm = 1 V/A = $1\ \mathrm{J \cdot s/C^{-2}} = 1\ \mathrm{N \cdot m \cdot s/C^{-2}} = 1\ \mathrm{kg \cdot m^2 \cdot s^{-1} \cdot C^{-2}}$ = unit of resistance
1 W = 1 watt = 1 J/s = $1\ \mathrm{V \cdot A} = 1\ \mathrm{kg \cdot m^2/s^3}$ = unit of power
1 F = 1 farad = 1 C/V = $1\ \mathrm{C^2/J} = \mathrm{C^2 \cdot s^2 \cdot m^{-2} \cdot kg^{-1}}$ = unit of capacitance
1 H = 1 henry = $1\ \Omega \cdot \mathrm{s} = 1\ \mathrm{kg \cdot m^2/C^2}$ = unit of inductance
Units of conductivity = $\mathrm{C^2 \cdot s \cdot m^{-3} \cdot kg^{-1}} = \mathrm{A \cdot m^2 \cdot (N/C)^{-1}} = (\Omega \cdot \mathrm{m})^{-1}$
Units of current density = $\mathrm{A/m^2} = (\mathrm{C/m^3})(\mathrm{m/s})$
$\vec{B}$ is magnetic field measured in teslas, $\vec{\mathbb{B}} \equiv c\vec{B}$ is magnetic field measured in newtons per coulomb

Useful Integrals

$$\int \frac{dx}{(x^2+u^2)^{1/2}} = \ln[x + (x^2+a^2)^{1/2}]$$

$$\int \frac{dx}{x^2+a^2} = \frac{1}{a}\tan^{-1}\left(\frac{x}{a}\right)$$

$$\int \frac{dx}{(x^2+a^2)^{3/2}} = \frac{1}{a^2}\frac{x}{(x^2+a^2)^{1/2}}$$

$$\int \frac{x\,dx}{(x^2+a^2)^{1/2}} = (x^2+a^2)^{1/2}$$

$$\int \frac{x\,dx}{(x^2+a^2)^{3/2}} = -\frac{1}{(x^2+a^2)^{1/2}}$$

$$\int \frac{x^2\,dx}{(x^2+a^2)^{3/2}} = -\frac{x}{(x^2+a^2)^{1/2}} + \ln[x + (x^2+a^2)^{1/2}]$$

Maxwell's Equations

$$\oint \vec{E} \cdot d\vec{A} = 4\pi k \rho$$

$$\oint \vec{\mathbb{B}} \cdot d\vec{A} = 0$$

$$\oint \vec{\mathbb{B}} \cdot d\vec{S} - \frac{1}{c}\int \frac{\partial \vec{E}}{\partial t} \cdot d\vec{A} = \frac{4\pi k}{c}\vec{J}$$

$$\oint \vec{E} \cdot d\vec{S} + \frac{1}{c}\int \frac{\partial \vec{\mathbb{B}}}{\partial t} \cdot d\vec{A} = 0$$

Six Ideas That Shaped Physics

Unit E: Electric and Magnetic Fields Are Unified

Second Edition

Thomas A. Moore

Boston Burr Ridge, IL Dubuque, IA Madison, WI New York San Francisco St. Louis
Bangkok Bogotá Caracas Kuala Lumpur Lisbon London Madrid Mexico City
Milan Montreal New Delhi Santiago Seoul Singapore Sydney Taipei Toronto

McGraw-Hill Higher Education

A Division of The **McGraw-Hill** *Companies*

SIX IDEAS THAT SHAPED PHYSICS, UNIT E: ELECTRIC AND MAGNETIC FIELDS ARE UNIFIED, SECOND EDITION

Published by McGraw-Hill, a business unit of The McGraw-Hill Companies, Inc., 1221 Avenue of the Americas, New York, NY 10020.

This book is printed on recycled, acid-free paper containing 10% postconsumer waste.

1 2 3 4 5 6 7 8 9 0 QPD/QPD 0 9 8 7 6 5 4 3 2

ISBN 0–07–239711–X

Publisher: *Kent A. Peterson*
Sponsoring editor: *Daryl Bruflodt*
Developmental editor: *Spencer J. Cotkin, Ph.D.*
Marketing manager: *Debra B. Hash*
Senior project manager: *Susan J. Brusch*
Senior production supervisor: *Sandy Ludovissy*
Media project manager: *Sandra M. Schnee*
Lead media technology producer: *Judi David*
Designer: *David W. Hash*
Cover/interior designer: *Rokusek Design*
Cover image: *© Corbis Images*
Senior photo research coordinator: *Lori Hancock*
Photo research: *Chris Hammond/PhotoFind LLC*
Supplement producer: *Brenda A. Ernzen*
Compositor: *Interactive Composition Corporation*
Typeface: *10/12 Palatino*
Printer: *Quebecor World Dubuque, IA*

Credit List: **Chapter 1** E1.3: © McGraw-Hill Higher Education/ Photo by Chris Hammond; E1.10: © Kip Peticolas/Fundamental Photographs, New York; E1.12b: © McGraw-Hill Higher Education/ Photo by Chris Hammond; E1.14: © Charles D Winters/Photo Researchers. **Chapter 3** E3.7: © Corbis /R-F Website. **Chapter 4** E4.5: © McGraw-Hill Higher Education/ Photo by Chris Hammond; E4.10: © PhotoDisc/Vol. #31; E4.11: © Adam Hart-Davis/SPL/Photo Researchers. **Chapter 5** E5.3: Courtesy of PASCO scientific; E5.9, E5.11, E5.12: © McGraw-Hill Higher Education/Photo by Chris Hammond. **Chapter 6** E6.5, E6.20: © McGraw-Hill Higher Education/ Photo by Chris Hammond. **Chapter 7** E7.3: © Phil Degginger/Color-Pic; E7.9b: Courtesy of PASCO scientific; E7.11 "Used with permission of the Stanford Linear Accelerator Center, Stanford University." E7.12a: Photo by Lee Snyder, Geophysical Institute, University of Alaska Fairbanks; E7.12b: © L. A. Frank, University of Iowa; E7.14: © Tom Pantages. **Chapter 8** E8.3a: © Corbis/Vol. #160; E8.4b: ©1990 Richard Megna Fundamental Photographs, NYC; E8.11a: Courtesy Dr.R.D.Gomez; 165: © Morton Beebe/Corbis Images; E8.16: Courtesy of Pasco Scientific. **Chapter 11** E11.9: © Corbis/R-F Website; E11.15b: Photo Courtesy of Gepco Wire and Cable. **Chapter 14** E14.1: © Richard Menga/Fundamental Photographs; E14.6, E14.8b: © McGraw-Hill Higher Education/ Photo by Chris Hammond; p. 293: Courtesy Sargent-Welch/Central Scientific Company. **Chapter 15** E15.1a: © D. Boone/Corbis Images; E15.1b: © Tom Pantages; E15.1c: NASA Dryden Flight Research Center, photo by Dr. Leonard Weinstein; E15.1d: Kirk Borne (ST ScI), and NASA; E15.9: Courtesy of Pasco Scientific. **Chapter 16** E16.9a-b: © Corbis/R-F Website; E16.9c: © McGraw-Hill Higher Education/ Photo by Chris Hammond; E16.9d: © David Young-Wolff/PhotoEdit.

Library of Congress Cataloging-in-Publication Data

Moore, Thomas A. (Thomas Andrew)
Six ideas that shaped physics. Unit E: Electric and magnetic fields are unified. / Thomas A. Moore. — 2nd ed.
p. cm.
Contents: [1] Unit C: Conservation laws constrain interactions — [2] Unit N: The laws of physics are universal — [3] Unit R: The laws of physics are frame-independent — [4] Unit E: Electric and magnetic fields are unified — [5] Unit Q: Particles behave like waves — [6] Unit T: Some processes are irreversible.
Includes bibliographical references and index.
ISBN 0–07–239711–X (acid-free paper)
1. Physics—Study and teaching (Higher) 2. Physics—Problems, exercise, etc. I. Title: Electric and magnetic fields are unified. II. Title.

QC32 .M647 2003 2002032552
530—dc21 CIP

www.mhhe.com

Dedication

To Joyce
The light of my life.

Table of Contents for *Six Ideas That Shaped Physics*

Unit C
Conservation Laws Constrain Interactions

C1 Introduction to Interactions
C2 Vectors
C3 Interactions Transfer Momentum
C4 Particles and Systems
C5 Applying Momentum Conservation
C6 Introduction to Energy
C7 Some Potential Energy Functions
C8 Force and Energy
C9 Rotational Energy
C10 Thermal Energy
C11 Energy in Bonds
C12 Power, Collisions, and Impacts
C13 Angular Momentum
C14 Conservation of Angular Momentum

Unit N
The Laws of Physics Are Universal

N1 Newton's Laws
N2 Vector Calculus
N3 Forces from Motion
N4 Motion from Forces
N5 Statics
N6 Linearly Constrained Motion
N7 Coupled Objects
N8 Circularly Constrained Motion
N9 Noninertial Reference Frames
N10 Projectile Motion
N11 Oscillatory Motion
N12 Introduction to Orbits
N13 Planetary Motion

Unit R
The Laws of Physics Are Frame-Independent

R1 The Principle of Relativity
R2 Synchronizing Clocks
R3 The Nature of Time
R4 The Metric Equation
R5 Proper Time
R6 Coordinate Transformations
R7 Lorentz Contraction
R8 The Cosmic Speed Limit
R9 Four-Momentum
R10 Conservation of Four-Momentum

Unit E
Electric and Magnetic Fields Are Unified

E1 Electrostatics
E2 Electric Fields
E3 Electric Potential
E4 Conductors
E5 Driving Currents
E6 Analyzing Circuits
E7 Magnetic Fields
E8 Currents and Magnets
E9 Symmetry and Flux
E10 Gauss's Law
E11 Ampere's Law
E12 The Electromagnetic Field
E13 Maxwell's Equations
E14 Induction
E15 Introduction to Waves
E16 Electromagnetic Waves

Unit Q
Particles Behave Like Waves

Q1 Standing Waves
Q2 The Wave Nature of Light
Q3 The Particle Nature of Light
Q4 The Wave Nature of Matter
Q5 The Quantum Facts of Life
Q6 The Wavefunction
Q7 Bound Systems
Q8 Spectra
Q9 Understanding Atoms
Q10 The Schrödinger Equation
Q11 Energy Eigenfunctions
Q12 Introduction to Nuclei
Q13 Stable and Unstable Nuclei
Q14 Radioactivity
Q15 Nuclear Technology

Unit T
Some Processes Are Irreversible

T1 Temperature
T2 Ideal Gases
T3 Gas Processes
T4 Macrostates and Microstates
T5 The Second Law
T6 Temperature and Entropy
T7 Some Mysteries Resolved
T8 Calculating Entropy Changes
T9 Heat Engines

Contents: Unit E

Electric and Magnetic Fields Are Unified

About the Author xiii
Preface xv
Introduction for Students xx

Chapter E1
Electrostatics 2

	Chapter Overview	2
E1.1	Introduction to the Unit	4
E1.2	What Is the Nature of Charge?	5
E1.3	How Objects Become Charged	7
E1.4	Conservation of Charge	9
E1.5	Coulomb's Law	10
E1.6	Conductors and Insulators	13
E1.7	Electrostatic Polarization	15
	Two-Minute Problems	18
	Homework Problems	18
	Answers to Exercises	20

Chapter E2
Electric Fields 22

	Chapter Overview	22
E2.1	The Field Concept	24
E2.2	An Operational Definition of $\vec{E}$	25
E2.3	The Field of a Point Charge	27
E2.4	The Superposition Principle	29
E2.5	The Field of a Dipole	31
E2.6	Handling Charge Distributions	33
	Two-Minute Problems	36
	Homework Problems	37
	Answers to Exercises	41

Chapter E3
Electric Potential 42

	Chapter Overview	42
E3.1	The Electric Potential	44
E3.2	Calculating ϕ from $\vec{E}$	47
E3.3	Calculating $\vec{E}$ from ϕ	49
E3.4	Useful Theorems About Charge Distributions	53
E3.5	The Electric Field as Dispersed Energy	55
	Two-Minute Problems	57
	Homework Problems	58
	Answers to Exercises	62

Chapter E4
Conductors 64

	Chapter Overview	64
E4.1	Introduction to Current	66
E4.2	A Microscopic Model	67
E4.3	Current Density	71
E4.4	Static Charges on Conductors	74
E4.5	Capacitance	77
	Two-Minute Problems	79
	Homework Problems	79
	Answers to Exercises	83

Chapter E5
Driving Currents 84

	Chapter Overview	84
E5.1	A Mechanical Model of a Battery	86
E5.2	Surface Charges Direct Currents	87
E5.3	The Emf of a Battery	90
E5.4	Resistance	93
E5.5	The Power Dissipated in a Conductor	95
E5.6	Discharging a Capacitor	97
	Two-Minute Problems	99
	Homework Problems	101
	Answers to Exercises	105

Chapter E6
Analyzing Circuits 106

	Chapter Overview	106
E6.1	Two Wires in Series	108
E6.2	Circuit Elements in Series	110

E6.3	Circuit Diagrams	112
E6.4	Circuit Elements in Parallel	114
E6.5	Analyzing Complex Circuits	116
E6.6	Realistic Batteries	119
E6.7	Electrical Safety Issues	119
	Two-Minute Problems	120
	Homework Problems	121
	Answers to Exercises	125

Chapter E7
Magnetic Fields 126

	Chapter Overview	126
E7.1	The Phenomenon of Magnetism	128
E7.2	The Definition of the Magnetic Field	128
E7.3	Magnetic Forces on Moving Charges	130
E7.4	A Free Particle in a Uniform Magnetic Field	135
E7.5	The Magnetic Force on a Wire	138
	Two-Minute Problems	143
	Homework Problems	144
	Answers to Exercises	149

Chapter E8
Currents and Magnets 150

	Chapter Overview	150
E8.1	Creating Currents in Moving Conductors	152
E8.2	The Magnetic Field of a Moving Charge	154
E8.3	The Magnetic Field of a Wire Segment	158
E8.4	The Magnetic Field of a Long, Straight Wire	159
E8.5	The Magnetic Field of a Circular Loop	161
E8.6	*All* Magnets Involve Circulating Charges	163
	Two-Minute Problems	165
	Homework Problems	166
	Answers to Exercises	170

Chapter E9
Symmetry and Flux 172

	Chapter Overview	172
E9.1	Symmetry Arguments	174
E9.2	The Electric Fields of Symmetric Objects	174
E9.3	The Mirror Rule for Magnetic Fields	178
E9.4	The Magnetic Fields of Symmetric Current Distributions	179
E9.5	The Flux Through a Surface	183
	Two-Minute Problems	186
	Homework Problems	188
	Answers to Exercises	190

Chapter E10
Gauss's Law 192

	Chapter Overview	192
E10.1	Gauss's Law	194
E10.2	Using Gauss's Law to Calculate Electric Fields	195
E10.3	The Field of a Spherical Charge Distribution	197
E10.4	The Field of an Infinite Cylindrical Distribution	199
E10.5	The Field of an Infinite Planar Slab	201
E10.6	Gauss's Law for the Magnetic Field	204
	Two-Minute Problems	206
	Homework Problems	208
	Answers to Exercises	212

Chapter E11
Ampere's Law 214

	Chapter Overview	214
E11.1	Ampere's Law	216
E11.2	Using Ampere's Law	218
E11.3	The Magnetic Field of an Axial Current Distribution	219
E11.4	The Magnetic Field of an Infinite Planar Slab	221
E11.5	The Magnetic Field of an Infinite Solenoid	223
E11.6	Ampere's Law for the Electric Field	226
	Two-Minute Problems	228
	Homework Problems	230
	Answers to Exercises	234

Chapter E12
The Electromagnetic Field 236

	Chapter Overview	236
E12.1	Why $\vec{E}$ and $\vec{B}$ Must Be Related	238
E12.2	How the Fields Transform	242
E12.3	A Bar Moving in a Magnetic Field	248
E12.4	Motion in a Velocity Selector	249
	Two-Minute Problems	250
	Homework Problems	251
	Answers to Exercises	255

Chapter E13
Maxwell's Equations 256

	Chapter Overview	256
E13.1	Introduction to Dynamic Fields	258
E13.2	Correcting Ampere's Law	258

E13.3	Faraday's Law	262
E13.4	Gauss's Laws Need No Correction	267
E13.5	Maxwell's Equations	268
E13.6	Local Field Equations (optional)	270
	Two-Minute Problems	272
	Homework Problems	274
	Answers to Exercises	278

Chapter E14

Induction 280

	Chapter Overview	280
E14.1	Magnetic Flux and Induced EMF	282
E14.2	Lenz's Law	284
E14.3	Self-induction	285
E14.4	"Discharging" an Inductor	287
E14.5	The Energy in a Magnetic Field	289
E14.6	Transformers	291
	Two-Minute Problems	294
	Homework Problems	295
	Answers to Exercises	298

Chapter E15

Introduction to Waves 300

	Chapter Overview	300
E15.1	What Is a Wave?	302
E15.2	A Sinusoidal Wave	304
E15.3	The Phase Velocity of a Wave	307
E15.4	The Wave Equation	309
	Two-Minute Problems	315
	Homework Problems	316
	Answers to Exercises	319

Chapter E16

Electromagnetic Waves 320

	Chapter Overview	320
E16.1	Electromagnetic Waves	322
E16.2	Characteristics of Electromagnetic Waves	324
E16.3	The Energy in an Electromagnetic Wave	328
E16.4	The Power Radiated by a Charge	329
E16.5	Why the Sky Is Blue	331
E16.6	Maxwell's Rainbow	332
	Two-Minute Problems	335
	Homework Problems	335
	Answers to Exercises	338

Glossary 341

Index 351

About the Author

Thomas A. Moore graduated from Carleton College (magna cum laude with Distinction in Physics) in 1976. He won a Danforth Fellowship that year that supported his graduate education at Yale University, where he earned a Ph.D. in 1981. He taught at Carleton College and Luther College before taking his current position at Pomona College in 1987, where he won a Wig Award for Distinguished Teaching in 1991. He served as an active member of the steering committee for the national Introductory University Physics Project (IUPP) from 1987 through 1995. This textbook grew out of a model curriculum that he developed for that project in 1989, which was one of only four selected for further development and testing by IUPP.

He has published a number of articles about astrophysical sources of gravitational waves, detection of gravitational waves, and new approaches to teaching physics, as well as a book on special relativity entitled *A Traveler's Guide to Spacetime* (McGraw-Hill, 1995). He has also served as a reviewer and an associate editor for American Journal of Physics. He currently lives in Claremont, California, with his wife Joyce and two college-aged daughters. When he is not teaching, doing research in relativistic astrophysics, or writing, he enjoys reading, hiking, scuba diving, teaching adult church-school classes on the Hebrew Bible, calling contradances, and playing traditional Irish fiddle music.

Preface

Introduction

Opening comments about *Six Ideas That Shaped Physics*

This volume is one of six that together comprise the text materials for *Six Ideas That Shaped Physics*, a fundamentally new approach to the two- or three-semester calculus-based introductory physics course. *Six Ideas That Shaped Physics* was created in response to a call for innovative curricula offered by the Introductory University Physics Project (IUPP), which subsequently supported its early development. In its present form, the course represents the culmination of more than a decade of development, testing, and evaluation at a number of colleges and universities nationwide.

This course is based on the premise that innovative approaches to the presentation of topics and to classroom activities can help students learn more effectively. I have completely rethought from the ground up the presentation of every topic, taking advantage of research into physics education wherever possible, and have done nothing just because "that is the way it has always been done." Recognizing that physics education research has consistently underlined the importance of active learning, I have also provided tools supporting multiple opportunities for active learning both inside and outside the classroom. This text also strongly emphasizes the process of building and critiquing physical models and using them in realistic settings. Finally, I have sought to emphasize contemporary physics and view even classical topics from a thoroughly contemporary perspective.

I have not sought to "dumb down" the course to make it more accessible. Rather, my goal has been to help students become *smarter.* I intentionally set higher than-usual standards for sophistication in physical thinking, and have then used a range of innovative approaches and classroom structures to help even average students reach this standard. I don't believe that the mathematical level required by these books is significantly different from that in most university physics texts, but I do ask students to step beyond rote thinking patterns to develop flexible, powerful conceptual reasoning and model-building skills. My experience and that of other users are that normal students in a wide range of institutional settings can, with appropriate support and practice, meet these standards.

The six volumes of the *Six Ideas* text

The six volumes that comprise the complete *Six Ideas* course are:

Unit C (**C**onservation Laws):	Conservation Laws Constrain Interactions
Unit N (**N**ewtonian Mechanics):	The Laws of Physics Are Universal
Unit R (**R**elativity):	The Laws of Physics Are Frame-Independent
Unit E (**E**lectricity and Magnetism):	Electricity and Magnetism Are Unified
Unit Q (**Q**uantum Physics):	Matter Behaves Like Waves
Unit T (**T**hermal Physics):	Some Processes Are Irreversible

I have listed these units in the order that I recommend they be taught, though other orderings are possible. At Pomona, we teach the first three units during the first semester and the last three during the second semester of a year-long course, but one can easily teach the six units in three quarters or even over three semesters if one wants a slower pace. The chapters of all these texts have been designed to correspond to what one might realistically discuss in a single 50-minute class session at the *highest possible pace:* while one might design a syllabus that covers chapters at a slower rate, one should *not* try to discuss more than one chapter in a 50-minute class.

For more information than I can include in this short preface about the goals of the *Six Ideas* course, its organizational structure (and the rationale behind that structure), the evidence for its success, and information about how to cut and/or rearrange material, as well as many other resources for both teachers and students, please visit the *Six Ideas* website (see the next section).

Important Resources

Instructions about how to use this text

I have summarized important information about how to read and use this text in an *Introduction for Students* immediately preceding chapter E1. Please look this over, particularly if you have not seen other volumes of this text.

The *Six Ideas* website

The *Six Ideas* website contains a wealth of up-to-date information about the course that I think both instructors and students will find very useful. The URL is

www.physics.pomona.edu/sixideas/

Essential computer programs

One of the most important resources available at this site are a number of computer applets that illustrate important concepts and aid in difficult calculations. In several places, this unit draws on some of these programs, and past experience indicates that students learn the ideas much more effectively when these programs are used both in the classroom and for homework. These applets are freeware and are available for both the Mac (Classic) and Windows operating systems.

Some Notes Specifically About Unit E

The goal of this unit

The fundamental goal of unit E is to provide a compact and modern introduction to electromagnetic field theory that takes students from Coulomb's law to electromagnetic waves in as short a time as possible. I have written this unit partly because I have been profoundly dissatisfied with most traditional treatments of this material, which tend to treat topics as essentially unrelated bits of "stuff to know" instead of helping students perceive the underlying shape and beauty of the theory. Therefore, I have sought to provide a carefully logical and sequential development of this material whose ultimate goal is to have students appreciate why electricity and magnetism *must be unified* if their field theories are to make sense in the context of the principle of relativity. The climax of the unit is therefore in chapters E12 and E13, where we see how relativity and logical necessity link the electric and magnetic fields and constrain the form of Maxwell's equations.

This is an admittedly 21st-century perspective on the topic, not a development that parallels the historical development of the discipline. However, I consider this a strength: I strongly believe that 21st-century hindsight in this case makes the structure of electromagnetic field theory much clearer and easier for students to understand. Moreover, I think that it provides a more interesting and exciting storyline than history does.

Because time in this unit is so tight (to make room for contemporary physics in other units), I have tried to streamline it as much as possible, keeping the focus on what moves the main argument forward rather than on interesting but tangential material. Someone comparing this unit to a traditional text may find it somewhat sparse: I intentionally ignore topics such as dielectrics, AC circuits, *LRC* circuits, mutual induction, diamagnetism, paramagnetism, superconductivity, and so on. Rather than "cover" such a laundry list of (admittedly fascinating) topics, I have tried to focus on the logic and interconnectedness of the material I do present as well as to take the time to explicitly teach important problem-solving skills in circuit analysis and field calculations.

This unit is a perennially difficult one for students. In this revision particularly, my goal has also been to make the mathematics as simple and as straightforward as possible, and to make liberal use of concrete visual and mechanical models to make the abstractions easier to understand. Even so, it is probably inevitable that students will find this unit to be the most difficult of the six. At Pomona, we flesh out the material on electric circuits by using a sequence of laboratory exercises designed to challenge misconceptions and help students practice circuit analysis skills.

Nontraditional aspects of the unit

Instructors will note several nontraditional aspects about the way that I approach the material. First, I avoid the use of field line diagrams. Published articles have recently discussed why two-dimensional field line diagrams can be inaccurate and misleading and how students routinely misunderstand these diagrams (for example, by assuming that charges will move along the field lines). I have been convinced by my colleagues Bruce Sherwood and Ruth Chabay that these diagrams are not really helpful to students, and that it is more important to spend time developing the field arrow representation of electric and magnetic fields. This representation can be awkward to draw, but its interpretation is straightforward and less prone to misunderstanding. Moreover, it focuses students on, rather than distracting them from, the basic mathematical model of the fields.

I also use the Coulomb constant k instead of ε_0 and μ_0 in electric and magnetic field equations. There are multiple reasons for doing this. The Coulomb constant has an easily remembered magnitude in SI units, and it is easier to understand its conceptual role (especially if one is not spending much time with dielectrics or magnetic materials). However, the most important reason is that using the Coulomb constant helps clarify the links between electric fields, magnetic fields, and gravitational fields, and so it serves the goal of enabling students to develop a more powerful understanding of the field concept. It will make some equations look odd to some instructors, but I believe that the benefits of this notation for helping students with the difficult task of making connections more than makes up for the minor adjustment required of instructors (whose training should give them more than enough flexibility to handle the shift in notation).

Similarly, starting in chapter E7, I use the symbol $\vec{\mathbb{B}} = c\vec{B}$ (read "B-bar") to stand for the magnetic field expressed in newtons per coulomb (N/C). This notation gives one all the advantages of gaussian units in simplifying equations and clearly displaying the link between electric and magnetic fields *without* abandoning any of the more familiar and useful SI electrical units. I still describe $\vec{B}$ as being measured in teslas and describe the definition of that unit (along with the gauss), but my premise here is that the tesla is the electromagnetic unit that is both least familiar and least useful to students, so if it is effectively replaced by the newton per coulomb, the cost is again more than made up by the gain.

Instructors will also find that I take unconventional approaches to the use of symmetry in magnetic fields, to developing the time-dependent terms in Maxwell's equations, and to explaining the wave equation and how electromagnetic fields satisfy the wave equation. I hope that these approaches make these perennially difficult topics more straightforward for students.

The instructor can play an important and positive role in helping all these approaches succeed pedagogically by spending the time required to get used to them and embracing them (and the value of their pedagogical goals) in front of the class.

How this unit is related to the other units

This unit is based on the foundation of newtonian mechanics constructed in units C and N, and so must follow these units in any projected course sequence. However, this unit does not depend on these particular units in any detailed way, so any calculus-based introduction to mechanics would be suitable preparation for this unit.

In the recommended unit sequence, this unit also follows unit R, which discusses the theory of relativity. This unit, because it looks at electricity and magnetism from a 21st-century perspective, does make use of some relativity. Indeed, the unit's main point is to show that the principle of relativity *requires* that electric and magnetic fields be linked! However, I have been deliberate about arranging things so that students need to know only three simple things about relativity: (1) the principle of relativity, which states that the laws of physics are the same in all inertial reference frames, (2) the idea that the speed of light is the ultimate speed limit, and (3) the fact that moving objects are Lorentz-contracted. No other knowledge about relativity is required. Therefore, if it is impossible at your institution to discuss unit R in full before getting into this unit, you can prepare students adequately by spending a single class day exploring these three ideas. I strongly believe that using these simple relativistic ideas pays off handsomely in terms of helping students understand the deep link between electricity and magnetism.

The material in this unit is not needed for any of the other units, except for chapter E15, which helps prepare students for unit Q.

How to make cuts if absolutely necessary

The unit's stated goal is essentially achieved by chapter E13, so chapters E14 through E16 could be dropped if necessary. Of the three, I think that chapter E14, which introduces an important implication of Faraday's law, is most valuable in supporting the main goal of the unit. However, omitting even chapters E15 and E16 means that students would miss out on seeing how electromagnetic waves emerge from Maxwell's field theory, an important secondary goal for this unit. Moreover, chapter E15 really is required for unit Q, if that unit follows this one.

If more severe cuts are absolutely necessary, chapters E1 through E8 provide an essentially irreducible basic introduction to electric and magnetic fields. Such a truncation would essentially lobotomize the unit, putting its basic goal (and stated "great idea") out of reach, but would still have some value as an first treatment of electromagnetic fields. Adding chapters E9 through E11 would also provide a basic introduction to Gauss's and Ampere's laws.

Please see the Instructor's Manual for more detailed comments about this unit and suggestions about how to teach it effectively.

Appreciation

Thanks!

A project of this magnitude cannot be accomplished alone. I would first like to thank the others who served on the IUPP development team for this project: Edwin Taylor, Dan Schroeder, Randy Knight, John Mallinckrodt, Alma Zook, Bob Hilborn, and Don Holcomb. I'd like to thank John Rigden and other members of the IUPP steering committee for their support of the project in its early stages, which came ultimately from an NSF grant and the special efforts of Duncan McBride. Users of the texts, especially Bill Titus, Richard Noer, Woods Halley, Paul Ellis, Doreen Weinberger, Nalini Easwar, Brian Watson, Jon Eggert, Catherine Mader, Paul De Young, Alma Zook, Dan Schroeder, David Tanenbaum, Alfred Kwok, and Dave Dobson, have offered invaluable feedback and encouragement. I'd also like to thank Alan Macdonald, Roseanne Di Stefano, Ruth Chabay, Bruce Sherwood, and Tony French for ideas, support, and useful suggestions. Thanks also to Robs Muir for helping with several of the indexes. My editors Jim Smith, Denise Schanck, Jack Shira, Karen Allanson, Lloyd Black, J. P. Lenney, and Daryl Bruflodt as well as Spencer Cotkin, Donata Dettbarn, David Dietz, Larry Goldberg, Sheila Frank, Jonathan Alpert, Zanae Roderigo, Mary Haas, Janice Hancock, Lisa Gottschalk, Debra Hash, David Hash, Patti Scott, Chris Hammond, Rick Hecker, and Susan Brusch have all worked very hard to make this text happen, and I deeply appreciate their efforts. I'd like to thank all the reviewers, including Edwin Carlson, David Dobson, Irene Nunes, Miles Dressler, O. Romulo Ochoa, Qichang Su, Brian Watson, and Laurent Hodges, for taking the time to do a careful reading of various units and offering valuable suggestions.

I also wish to thank the following panel of reviewers for providing careful and insightful comments on Unit E:

Soumya Chakravarti, *Cal State University—Pomona*
Nathanael Fortune, *Smith College*
Hugh Gallagher, *State University of New York—Oneonta*
Doug Harper, *Western Kentucky University*
Andrzej Herczynski, *Boston College*
Paul Lee, *California State University—Northridge*
David Lynch, *Iowa State University*
Allen Miller, *Syracuse University*
Kimberly Shaw, *Southern Illinois University—Edwardsville*
William Smith, *Boise State University*
Shubha Tewari, *Mount Holyoke College*
Joel Weisberg, *Carleton College*

Thanks to Connie Wilson, Hilda Dinolfo, Connie Inman, and special student assistants Michael Wanke, Paul Feng, and Mara Harrell, Jennifer Lauer, Tony Galuhn, Eric Pan, and all the Physics 51 mentors for supporting (in various ways) the development and teaching of this course at Pomona College. Thanks also to my Physics 51 students, and especially Win Yin, Peter Leth, Eddie Abarca, Boyer Naito, Arvin Tseng, Rebecca Washenfelder, Mary Donovan, Austin Ferris, Laura Siegfried, and Miriam Krause, who have offered many suggestions and have together found many hundreds of typos and other errors. Eric and Brian Daub and Ryan McLaughlin were indispensable in helping me put this edition together. Finally, very special thanks to my wife, Joyce, and to my daughters, Brittany and Allison, who contributed with their support and patience during this long and demanding project. Heartfelt thanks to all!

Thomas A. Moore
Claremont, California

Introduction for Students

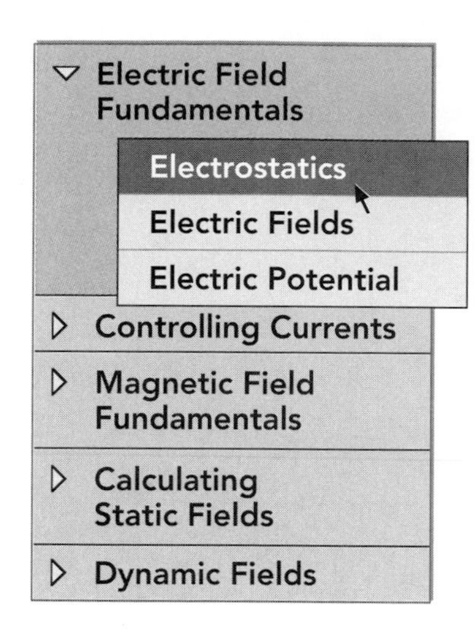

Introduction

Welcome to *Six Ideas That Shaped Physics!* This text has been designed using insights from recent research into physics learning to help you learn physics as effectively as possible. It thus has many features that may be different from science texts you have probably encountered. This section discusses these features and how to use them effectively.

Why Is This Text Different?

Research consistently shows that people learn physics most effectively if they participate in *activities* that help them *practice* applying physical reasoning in realistic situations. This is so because physics is not a collection of facts to absorb, but rather is a set of *thinking skills* requiring practice to master. You cannot learn such skills by going to factual lectures any more than you can learn to play the piano by going to concerts!

This text is designed, therefore, to support *active learning* both inside and outside the classroom by providing (1) resources for various kinds of learning activities, (2) features that encourage active reading, and (3) features that make it easier for the text (as opposed to lectures) to serve as the primary source of information, so that more class time is available for active learning.

The Text as Primary Source

Features that help the text serve as the primary source of information

To serve the last goal, I have adopted a conversational style that I hope will be easy to read, and I tried to be concise without being so terse that you need a lecture to fill in the gaps. There are also many text features designed to help you keep track of the big picture. The unit's **central idea** is summarized on the front cover where you can see it daily. Each chapter is designed to correspond to one 50-minute class session, so that each session is a logically complete unit. The two-page **chapter overview** at the beginning of each chapter provides a compact summary of that chapter's contents to consider before you are submerged by the details (it also provides a useful summary when you review for exams). An accompanying **chapter-location diagram** uses a computer menu metaphor to display how the current chapter fits into the unit (see the example at the upper left). Major unit subdivisions appear as gray boxes, with the current subdivision highlighted in color. Chapters in the current subdivision appear in a submenu with the current chapter highlighted in black and indicated by an arrow.

All technical terms are highlighted using a **bold** type when they first appear, and a **Glossary** at the end of the text summarizes their definitions. Please also note the tables of useful information, including definitions of common symbols, that appear inside the front cover.

A physics *formula* is both a mathematical equation and a *context* that gives the equation meaning. Every important formula in this text appears in a **formula box.** Each contains the equation, a ***purpose*** (describing the formula's meaning and utility), a definition of the ***symbols*** used in the equation, a description of any ***limitations*** on the formula's applicability, and possibly some other useful ***notes.*** Treat everything in such a box as an *indivisible unit* to be remembered and used together.

Active Reading

What it means to be an *active reader*

Like passively listening to a lecture, passively scanning a text does not really help you learn. *Active* reading is a crucial study skill for effectively learning from this text (and other types of technical literature as well). An active reader stops frequently to pose internal questions such as these: *Does this make sense? Is this consistent with my experience? Am I following the logic here? Do I see how I might use this idea in realistic situations?* This text provides two important tools to make this easier.

Tools to help you become an active reader

Use the **wide margins** to (1) record *questions* that occur to you as you read (so that you can remember to get them answered), (2) record *answers* when you receive them, (3) flag important passages, (4) fill in missing mathematics steps, and (5) record insights. Doing these things helps keep you actively engaged as you read, and your marginal comments are also generally helpful as you review. Note that I have provided some marginal notes that summarize the points of crucial paragraphs and help you find things quickly.

The single most important thing you can do

The **in-text exercises** help you develop the habits of (1) filling in missing mathematics steps and (2) posing questions that help you *practice* using the chapter's ideas. Also, though this text has many examples of worked problems similar to homework or exam problems, *some* of these appear in the form of in-text exercises (as you are more likely to *learn* from an example if you work on it some yourself instead of just scanning someone else's solution). Answers to *all* exercises appear at the end of each chapter so you can get immediate feedback on how you are doing. Doing at least some of the exercises as you read is probably the *single most important thing you can do* to become an active reader.

Active reading does take effort. *Scanning* the 5200 words of a typical chapter might take 45 minutes, but active reading could take several times as long. I personally tend to "blow a fuse" in my head after about 20 minutes of active reading, so I take short breaks to do something else to keep alert. Pausing to fill in missing math also helps me to stay focused longer.

Class Activities and Homework

End-of-chapter problems support active learning

The problems appearing at the end of each chapter are organized into categories that reflect somewhat different active-learning purposes. **Two-minute problems** are short, concept-oriented, multiple-choice problems that are primarily meant to be used *in* class as a way of practicing the ideas and/or exposing conceptual problems for further discussion. (The letters on the back cover make it possible to display responses to your instructor.) The other types of problems are primarily meant for use as homework *outside* class. **Basic** problems are simple, drill-type problems that help you practice in straightforward applications of a single formula or technique. **Synthetic** problems are more challenging and realistic questions that require you to bring together multiple formulas and/or techniques (maybe from different chapters) and to think carefully about physical principles. These problems define the level of sophistication that you should strive to achieve. **Rich-context** problems are yet more challenging problems that are often written in a narrative format and ask you to answer a practical, real-life question rather than explicitly asking for a numerical result. Like situations you will encounter in real life, many provide too little information and/or too much information, requiring you to make estimates and/or discard irrelevant data (this is true of some *synthetic* problems as well). Rich-context problems are generally too difficult for most students to solve alone; they are designed for *group* problem-solving sessions. **Advanced** problems are very sophisticated problems that provide supplemental discussion of subtle or advanced issues related to the material discussed in the chapter. These problems are for instructors and truly exceptional students.

Read the Text *Before* Class!

Class time works best if you are prepared

You will be able to participate in the kinds of activities that promote real learning *only* if you come to each class having already read and thought about the assigned chapter. This is likely to be *much* more important in a class using this text than in science courses you may have taken before! Class time can also (*if* you are prepared) provide a great opportunity to get your *particular* questions about the material answered.

E1 Electrostatics

▽ Electric Field Fundamentals
- Electrostatics
- Electric Fields
- Electric Potential

▷ Controlling Currents
▷ Magnetic Field Fundamentals
▷ Calculating Static Fields
▷ Dynamic Fields

Chapter Overview

Introduction

This chapter begins our study of the rich and fascinating topic of electricity and magnetism. Historically, the work done by a handful of 19th-century physicists not only set the tone and agenda for contemporary physics but also ignited a technological revolution that has utterly changed our lives. Much of what is distinctive about contemporary civilization (electric power, telecommunications, computers, television, movies, and so on) would not exist if not for their efforts.

Section E1.1: Introduction to the Unit

To physicists before the 19th century, **electricity, magnetism,** and **light** seemed like entirely distinct phenomena. Perhaps the greatest triumph of 19th-century physics was James Clerk Maxwell's beautiful and comprehensive theory linking these phenomena. In this theory, electricity and magnetism are two different manifestations of an **electromagnetic field** whose structure and time evolution is described by **Maxwell's equations.** The final surprise was that light is an oscillating electromagnetic field.

This unit explores the great idea that *electricity and magnetism are unified* in five subdivisions. The first three explore the fundamentals of electric fields, creating and controlling electric currents, and using currents to create magnetic fields. The fourth introduces Maxwell's equations as powerful tools for calculating static (time-independent) electric and magnetic fields. The last shows that special relativity *requires* that electric and magnetic fields be parts of a unified electromagnetic field, and shows how we have to modify Maxwell's equations to handle **dynamic** (time-dependent) fields.

This chapter launches the first subdivision by exploring the nature of electric charge and the forces that charged particles at rest exert on each other.

Section E1.2: What Is the Nature of Charge?

Only particles having nonzero **electric charge** can interact electromagnetically. Experiments show that there are exactly two types of charge, which we call *positive* and *negative* because we can model them using positive and negative numbers. We observe that the interaction between two charged particles is repulsive if the particle charges have like signs and attractive if they have opposite signs. However, which of the two kinds of charge we identify as *negative* is purely a matter of convention.

Section E1.3: How Objects Become Charged

Rubbing surfaces can become charged because rubbing causes the electron clouds of molecules on the surfaces to become intermingled, and some molecules are more prone than others to snatch an extra electron when the clouds separate. If only one molecule in a million captures such an extra electron, the surface can become noticeably charged.

The SI unit of charge is the **coulomb** (abbreviation: C), which we will define to be a charge equivalent to that on 6.242×10^{18} protons. This means that the charge of a proton is $1.602 \times 10^{-19}\,\text{C} \equiv +e$, and that of an electron is $-e$.

Section E1.4: Conservation of Charge

The electric charge of an isolated system is *always* conserved. Most physical processes merely shuffle around protons and electrons without creating or destroying them (and thus trivially conserve charge), but even exotic nuclear interactions that *can* create or destroy protons or electrons still conserve charge.

Section E1.5: Coulomb's Law

We find experimentally that the electrostatic interaction between two charged particles exerts on each particle a force whose magnitude depends on their charge and separation as follows:

$$F_e = \frac{k|q_1 q_2|}{r^2} \quad \text{where } k = 8.99 \times 10^9\ \text{N} \cdot \text{m}^2/\text{C}^2 \qquad \text{(E1.3)}$$

Purpose: This equation specifies the magnitude F_e of the electrostatic force exerted on each of two interacting charged particles separated by a distance r.

Symbols: q_1 and q_2 are the charges of the two particles; k is a constant of proportionality called the **Coulomb constant.**

Limitations: This equation applies only to *point particles* that are at *rest*.

Notes: The absolute value signs are necessary because q_1 and q_2 might be negative but $F_e \equiv \text{mag}(\vec{F}_e)$ must be positive.

The force on each particle points *away from* the other if the two charges have the same sign, or *toward* the particle if the charges have opposite signs. This form of **Coulomb's law** is analogous to Newton's law of universal gravitation $F_g = Gm_1 m_2/r^2$.

The **superposition principle** states that the force on any point charge q_1 due to its interactions with a *set* of point charges $q_2, q_3, \ldots$ is simply the *vector sum* of the forces that $q_2, q_3, \ldots$ would exert if each interacted with q_1 alone. Even though Coulomb's law applies only to point particles, the superposition principle implies that we can simply sum over all the pairs of interacting charges in a pair of macroscopic objects to find the net force that each exerts on the other.

Section E1.6: Conductors and Insulators

We call a substance a **conductor** if charges can readily move through it and an **insulator** if they cannot. Metals and salty water are good conductors; glass, rubber, and most plastics are good insulators. We will define these terms quantitatively in chapter E3.

Rubbing a conducting object can deposit charge on the conductor's surface, but these charges usually disperse (since they repel each other and are free to move) by flowing through the conductor and out through the person holding it. Therefore it is usually easier to charge insulators.

Section E1.7: Electrostatic Polarization

Forces exerted by an external charge can distort an atom's electron cloud, turning the atom into an **electric dipole,** which we can model by two point charges of opposite sign separated by a small distance. This process of **electrostatic polarization** explains how a charged object can attract an electrically neutral object.

E1.1 Introduction to the Unit

Electricity, magnetism, and light are linked

Since the beginning of history, humanity has pondered three great mysteries that no one originally imagined were related. Our ancestors experienced **electricity** in the form of lightning and the tiny sparks that amber produces when rubbed with cloth. Certain rocks (including those from the region called Magnesia by the ancient Greeks) exhibited an attractive effect that came to be called **magnetism.** The greatest mystery of all was that of **light,** essential to life and sight, which according to the ancient Hebrews was the first thing created by God, and which in many times and places has served as a metaphor for divinity and knowledge.

One of the greatest intellectual triumphs of the late 19th century was the recognition that electricity, magnetism, and light are not separate and distinct phenomena, but instead are but different aspects of an **electromagnetic field.** The creators of this model were surprised to find that the model required electromagnetic fields to have a reality somewhat apart from the electric charges that create them. The model also required that **dynamic** (time-dependent) fields obey "equations of motion" (called **Maxwell's equations**) in much the same way that material objects obey Newton's second law. The final surprise was that light turns out to be dynamic electromagnetic field. All these conclusions spring from this unit's fundamental idea: *Electricity and magnetism are unified.*

Our technological civilization is founded on this great idea. The generation and transportation of electric power, radio, electronics, radar, television, and computers are all progeny of this idea, and these technologies have changed the direction of history. The power of this idea also opened the floodgates of science, not only providing scientists with a host of new ways to probe and measure the universe, but also uncovering perplexing new mysteries that rapidly led to the development of relativity and quantum mechanics in the early 20th century.

Our purpose in this unit is to explore the concept of the electromagnetic field and the meaning of its equations of motion, so as to better understand the consequences of the idea that *electricity and magnetism are unified.* Rather than follow the historical path of this idea's development, which is full of twists and blind alleys, we will take a retrospective look from a contemporary perspective, since we now understand the full power and breadth of this idea.

The first of this unit's five subdivisions introduces the concept of a *field* and explores electric fields created by stationary charged particles. The second discusses how we can use electric fields to create and control *electric currents.* The third explores how such currents both react to and create magnetic fields. The fourth introduces Maxwell's equations as powerful tools for calculating **static** (time-independent) electric and magnetic fields.

The final subdivision builds on the background provided by the first four to explore the unit's core idea and its consequences. That subdivision's first chapter argues that special relativity *requires* that electric and magnetic fields be inseparable aspects of a more general electromagnetic field. In chapter E13, we will use these ideas to extend Maxwell's equations so that they can handle dynamic fields. The remaining chapters explore important applications and implications of these equations. *Induction* is a time-dependent electromagnetic phenomenon predicted by Maxwell's equations that makes it possible to *create* currents with time-dependent magnetic fields. This phenomenon has abundant applications in modern technology. Chapters E15 and E16 discuss electromagnetic waves and their relationship to light. The structure of the entire unit is summarized in figure E1.1.

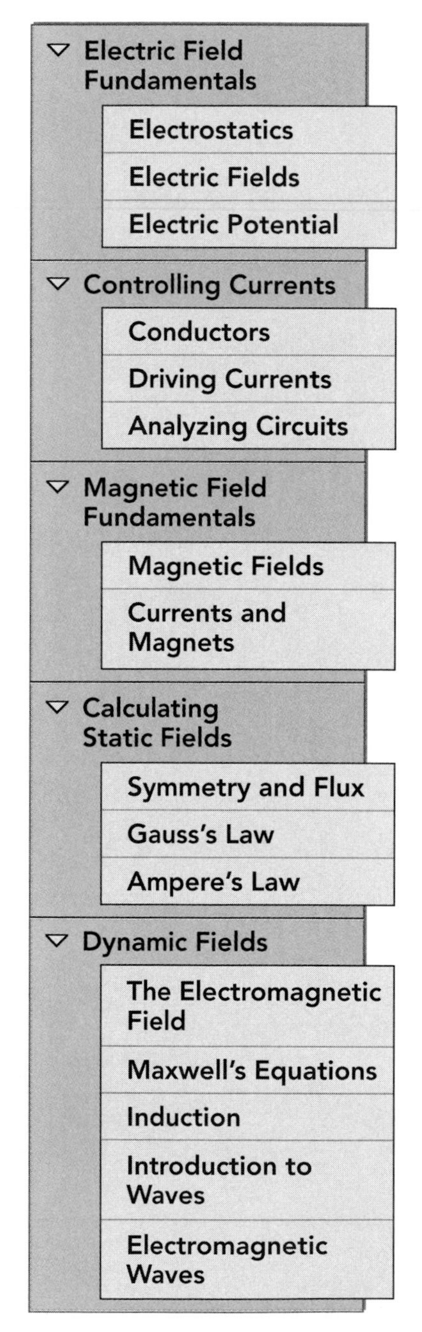

Figure E1.1
An overview of the structure of this unit.

In this subdivision, we will focus on **electrostatics,** the aspect of the electromagnetic interaction that acts between charges at rest. We begin in this chapter by exploring the most basic questions about such interactions: What is the nature of electric charge? Why do we describe it as being positive and negative? How can we describe electrostatic forces mathematically?

E1.2 What Is the Nature of Charge?

Historical notes

At least 2500 years ago, Greek scholars knew that rubbing amber (which they called *elektron*) changed it in a way that enabled it to attract light objects such as straw or feathers, and that sometimes little sparks could be seen jumping between the amber and other objects. The Greeks thought that this was a unique property of amber, but in 1600 William Gilbert showed that many objects could be similarly **electrically charged.** (This was not easy to show, because few *natural* substances other than amber are easily charged.)

Objects made of plastic, rubber, or synthetic fibers make the phenomenon of *static electricity* a common experience nowadays (particularly on days when the humidity is low). Nearly everyone has experienced delivering an electric spark to something after rubbing his or her shoes on a rug. After running a rubber or plastic comb through your hair, you will find that the comb attracts dust and bits of paper, and you may even hear the crackle of sparks jumping if you bring it near to something. When you pull clothes out of the drier (particularly clothes made of synthetic fibers), you can feel how they attract each other and hear the crackle of sparks as you pull them apart (you can actually see the sparks in the dark).

Learning about charge with sticky tape experiments

We can learn much about the nature of electric charge by doing some simple experiments with ordinary transparent adhesive (Scotch) tape. The characteristics of the adhesive make it easy to electrify a piece of such tape (as you may already know from experience!), and the cellophane or plastic used as a backing holds charges well. Find a dispenser of transparent tape: cheap, shiny cellophane tape is fine (maybe actually better than some brands of expensive "magic" tape, which are actually treated to reduce electrification). Tear off two pieces of tape about 20 cm (8 in.) long. Fold about one-half inch of the tape over on itself to make a nonsticky handle, as shown in figure E1.2a. Label the handle of one tape T and the handle of the other U. Then (this is important!) apply the tacky side of the T tape to the nontacky side of the U tape along the length of both, as shown in figure E1.2b (the T and U labels refer to whether it is the tacky or untacky side of the tape in each case that is in contact with the other). Grip the handles, and then quickly pull the two tapes apart. Hang one piece of tape handle down from the edge

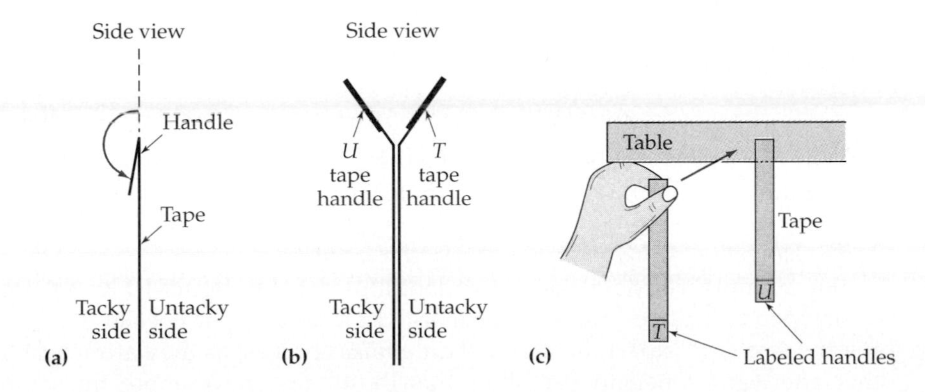

Figure E1.2
(a) Give a piece of tape a nonsticky handle by folding about a 1/2 in. of the tape down on itself. (b) To electrify the tapes, place the tacky side of the T tape in contact with the untacky side of the U tape, and pull the tapes apart suddenly. (c) Hang one tape from the edge of a table with its handle down, and bring the other near it. What happens?

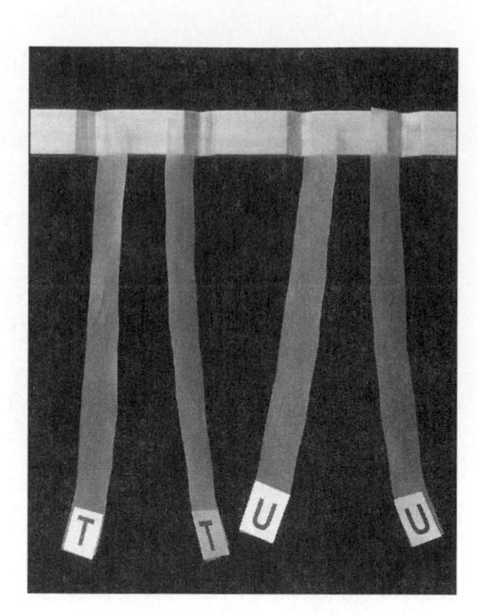

Figure E1.3
A photograph of charged tapes interacting.

of a table or chair, and bring the other tape near to it, as shown in figure E1.2c. You should find that the two pieces of tape strongly attract each other. Tearing the tapes apart therefore somehow charges both.

The way that electrically charged objects attract each other is similar to the way that massive objects attract each other gravitationally. From this point of view, the surprising thing is that electrically charged objects can also *repel* each other. This was first documented in 1733 by the King of France's gardener, Charles François de Cisternay Dufay. You can demonstrate this as follows. Prepare and label *another* two pieces of tape as shown in figure E1.2. You should find that the new *U* tape *repels* the first *U* tape, the new *T* tape repels the first *T* tape, and the *T* and *U* tapes attract each other (see figure E1.3).

There are exactly *two* kinds of electric charge

These observations imply that the *T* and *U* tapes must have different kinds of electric charge: if there were only one kind of electric charge, all the charged tapes would either attract or repel each other, not attract some tapes and repel others. In a series of papers published in the late 1740s and early 1750s, Benjamin Franklin and several collaborators proposed a simple model for understanding electrostatic phenomena. Franklin argued that only *two* kinds of charge, which he called **positive** and **negative,** were needed, since *all* charges (by experiment) either attract negative charges and repel positive charges or vice versa. Franklin also asserted that objects having *like* charges repel each other, while objects having *unlike* charges always attract.

Exercise E1X.1

How are we sure that it is *like* charges that repel each other? Explain how the tape experiments described above support Franklin's conclusion. (What makes us think that the two *U* tapes have the same type of charge?)

Charges combine like positive and negative numbers

Why did Franklin choose to describe these two types of charges as being positive and negative (and not, for example, *U* and *T*, or type A and type B)? These designations are a metaphorical way of saying that in certain ways, *electric charges behave like positive and negative numbers.*

For example, consider the experiment illustrated in figure E1.2b, where we stripped apart two originally uncharged pieces of tape. The act of stripping the two tapes apart must give them *unlike* charges, as they attract each other afterward. One can also show that these tapes have about the same *magnitude* of charge (for example, by showing that the magnitude of the attractive or repulsive force exerted by each on a third tape is the same). When the tapes are brought together again, they behave as if they were again uncharged. This suggests, as Franklin first argued, that the charges on these objects can be described by positive and negative numbers. The net charge on the uncharged tapes is originally zero. When the tapes are separated, they are found to have unlike charges with the same magnitudes, just like positive and negative numbers that have the same absolute value. When the tapes are brought together, they become uncharged again, just as a positive and a negative number having the same absolute value add to zero. The behavior of signed numbers is thus a good metaphor for the behavior of charges.

The choice of which kind of charge is positive and which is negative, however, is completely arbitrary. The accepted convention is based on Franklin's historical choice: a glass rod rubbed with silk cloth becomes *positively* charged. A more modern way of describing the same convention is to say that if you comb your hair with a plastic comb, the comb becomes *negatively* charged.

Which of the two types of charge we call *positive* is a matter of convention

Exercise E1X.2

According to this convention, what is the sign of the charge on a U tape? What is the sign of the charge on a T tape? Describe and perform an experiment that answers this question.

E1.3 How Objects Become Charged

Neither Gilbert nor Franklin was able to offer anything more than speculation about *how* objects such as amber, glass, and rubber could become electrically charged. (Franklin speculated that electric charge was some kind of continuous fluid, and that a surplus of this fluid gave an object a positive charge while a deficit of the fluid gave it a negative charge.) Since the early decades of the 20th century, scientists have known with certainty that material objects are made up of *atoms*, and we now know a great deal about the internal structure of the atoms themselves. This atomic model of matter enables us to offer a better explanation of how macroscopic objects can become electrically charged.

The atomic model of matter explains how objects can become charged

A macroscopic object is constructed from a huge number (on the order of magnitude of 10^{23}) of tiny atoms (roughly 0.2 nm in diameter). Each atom in turn consists of a *very* tiny central nucleus (comprised of protons and neutrons) surrounded by a cloud of electrons. The number of electrons in a normal atom is the same as the number of protons in its nucleus, and the number of neutrons in the nucleus is usually a bit larger than the number of protons. A schematic diagram of a copper atom is shown in figure E1.4.

Avogadro's number (6.02×10^{23}) of either protons or neutrons has a mass of very nearly 1.0 g (this is essentially how Avogadro's number was defined). The mass of an electron is more than 1800 times smaller than the mass of either. This means that the mass of an atom is essentially determined by the number of protons and neutrons it has: the total mass of the electrons in the atom is negligible in comparison. On the other hand, the size of this massive nucleus is tiny compared to the atom as a whole (the radius of the nucleus is roughly 20,000 times smaller than the atom's radius): most of the *volume* of an atom is occupied by the electron clouds.

Avogadro's number of protons or neutrons has a mass of very nearly 1 g

Protons are always observed to have *positive* charge (according to Franklin's convention), while electrons are always negatively charged and

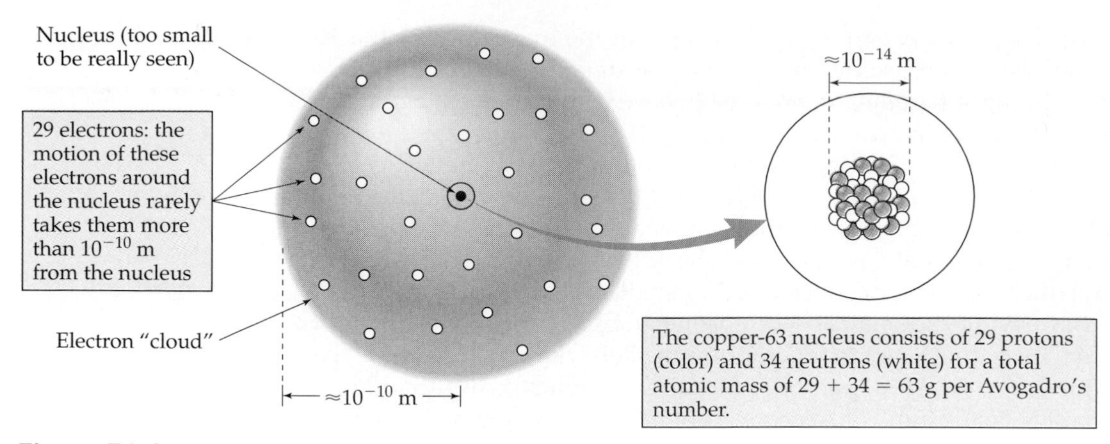

Figure E1.4
A schematic representation of a copper-63 atom. (Quantum mechanics implies that the charge of the 29 electrons behaves as if it were distributed smoothly throughout the gray "cloud" instead of being concentrated at 29 points.)

neutrons are always uncharged. It is precisely the strong electrostatic attraction between the electrons and protons that keeps the electrons attached to the atom's nucleus. Experiments also show that an atom with an equal number of protons and electrons behaves as if it were exactly uncharged (to many decimal places), implying that a proton and electron must have electric charges that are exactly equal in magnitude and opposite in sign. The numerical magnitude of this basic atomic unit of charge is designated by the letter e: an electron is said to have a charge of $-e$ and the proton a charge of $+e$.

The SI unit of charge

The SI unit of electric charge in is the **coulomb** (abbreviation: C), after Charles-Augustin de Coulomb, who quantitatively investigated the forces between charged objects in the late 1780s. This unit was defined long before the structure of the atom was understood, and so it is not related to the fundamental unit of charge e in any simple manner. We will define the coulomb as follows (although the *technical* SI definition is somewhat different for reasons beyond our scope here):

$$1\text{ C} \equiv \text{charge of } 6.242 \times 10^{18} \text{ protons} = (6.242 \times 10^{18})e \qquad \text{(E1.1}a\text{)}$$

$$\Rightarrow \qquad e = 1.602 \times 10^{-19}\text{ C} \qquad \text{(E1.1}b\text{)}$$

Any uncharged macroscopic object therefore contains an enormous amount of positive charge (in the form of protons) exactly balanced by an enormous amount of negative charge (in the form of electrons). If we define the net charge of an object to be the sum of the total positive and total negative charges on the object, an uncharged object has a net charge of zero, since the positive charges exactly cancel the negative charges. When two originally uncharged objects come into close physical contact (as when rubbed, or when two pieces of tape are stuck together), the electron clouds of the atoms on the surfaces of the two objects become somewhat intermingled. When the surfaces are separated, the atoms on one surface sometimes pull away some extra electrons, leaving the other with a deficit of electrons. The surface with the excess of electrons will thus have a small negative net charge, while the other has a small positive net charge.

Rubbing objects together can transfer electrons

Certain kinds of materials are more prone to lose electrons while other materials are more prone to capture extra electrons. For example, rubber molecules are slightly more able to capture and hold electrons than fur or cloth molecules are; so when a rubber balloon is rubbed with cloth or fur, the balloon ends up with a net negative charge. The imbalance between these materials does not have to be very large: even if only about one in every million surface molecules gets an extra electron, the balloon will accumulate an obvious net charge.

Table E1.1 lists a number of common substances in decreasing order of their empirically measured tendency to lose electrons when rubbed by another substance. When two substances are rubbed, whichever is higher on the list will lose electrons to the one lower on the list: the higher substance will therefore become positive and the lower negative. Physicists call such a list a **triboelectric series** (*tribein* is a Greek verb meaning "to rub"). One has to be careful in taking this list too literally, however, because surface finishes or contamination can greatly influence where an actual sample of a substance appears in the series.

Table E1.1 A triboelectric series

Air
Human skin (if very dry)
Rabbit fur
Glass
Human hair
Nylon
Wool
Lead
Silk
Aluminum
Paper
Cotton
Steel
Wood
Amber
Hard rubber
Nickel, copper
Brass, silver
Synthetic rubber
Gold, platinum
Polyester
Styrene (Styrofoam)
Saran Wrap
Polyethylene (used in Scotch tape)
Polypropylene
Vinyl (used in PVC pipes)
Silicon
Teflon
Silicone rubber

(Adapted from J. M. Zavista, *Understanding Static Electricity;* www.howstuffworks.com; and other sources)

Exercise E1X.3

A copper penny has a mass of about 5 g. The protons in this penny represent roughly how much total positive charge, in coulombs? (For comparison, a typical charged comb or balloon has a net charge on the order of 10 nC to 100 nC, that is, 10^{-8} C to 10^{-7} C.)

E1.4 Conservation of Charge

One of the most important assertions made by Franklin and his collaborators was that *electric charge is conserved:* that is, the net charge of any isolated system of objects remains constant in time. We now can see that this follows from the atomic model considered in section E1.3. Everyday physical processes do not destroy or create either protons or electrons, so an isolated system has a fixed number of protons, a fixed number of electrons, and thus a fixed net charge. Electrons (or very rarely, protons) can be shuffled around between objects in the system, but the net number of protons and electrons will not change. Therefore the net charge of any isolated system will remain constant.

Since about the 1930s, physicists have known of physical processes that *do* destroy and/or create electrons and protons. What happens to charge conservation in this case? In all physical processes known to date, the net charge in an isolated system is conserved even when the particles carrying that charge are created or destroyed. For example, an isolated neutron (zero charge) spontaneously decays after about 10 minutes, creating a proton (charge $+e$), an electron (charge $-e$), and a third particle called an *antineutrino* (zero charge): the net charge of the system is zero both before and after the decay. No physical process has *ever* been discovered that violates charge conservation at either the macroscopic or microscopic level.

An isolated system's total charge is conserved, even in processes that create or destroy particles.

Exercise E1X.4

Electrons have the smallest mass of any known *charged* particle, although a variety of less massive *neutral* particles (such as photons and neutrinos) exist. Argue that the principles of conservation of relativistic energy and charge imply that an electron *cannot* decay into something else.

E1.5 Coulomb's Law

How can we mathematically describe the forces that motionless charged objects exert on each other? The force exerted by a charged *point* particle on another charged *point* particle is described by **Coulomb's law,** a law very similar in form to Newton's law of universal gravitation (which was discussed in unit N). Newton's law of universal gravitation asserts that the gravitational force exerted by a point particle of mass m_1 on a point particle of mass m_2 has a magnitude given by

$$F_g = \frac{Gm_1m_2}{r^2} \quad \text{where } G = 6.67 \times 10^{-11}\ \text{N} \cdot \text{m}^2/\text{kg}^2 \qquad \text{(E1.2)}$$

In this equation, r is the distance between the two particles, and G is a constant of proportionality having the stated value. The force acting on either particle is directed *toward* the other particle along the line connecting them. Note that the force acting on either particle is equal in magnitude and opposite in direction to the force acting on the other particle, so the law of universal gravitation is consistent with Newton's third law.

Similarly, Coulomb's law asserts that the electrostatic force of attraction or repulsion exerted by two point particles on each other has the magnitude

Coulomb's law

$$F_e = \frac{k|q_1q_2|}{r^2} \quad \text{where } k = 8.99 \times 10^9\ \text{N} \cdot \text{m}^2/\text{C}^2 \qquad \text{(E1.3)}$$

Purpose: This equation specifies the magnitude F_e of the electrostatic force exerted on each of two interacting charged particles separated by a distance r.

Symbols: q_1 and q_2 are the charges of the two particles; k is a constant of proportionality called the **Coulomb constant.**

Limitation: This equation applies only to *point* particles that are at rest.

Notes: The absolute value signs are necessary because q_1 and q_2 might be negative, but $F_e \equiv \text{mag}(\vec{F}_e)$ must be positive.

The force acting on one particle points either *toward* the other particle (if the particles have opposite signs) or *away from* the other particle (if the particles have like signs). In either case (see figure E1.5) the forces exerted by the

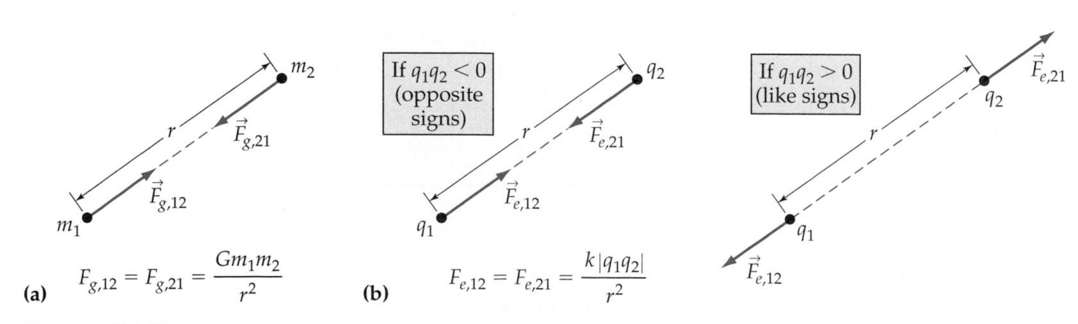

Figure E1.5
(a) The gravitational force exerted by one point particle on another points directly toward the other particle. (b) The electrostatic force exerted by one particle on the other points either directly *toward* or directly *away* from the other particle.

particles on each other have equal magnitudes and opposite directions, consistent with Newton's third law.

Limitations of Coulomb's law

Coulomb's law has two important limitations that you must understand to use it properly (the same restrictions actually apply to the law of universal gravitation). First, it strictly applies only to point particles (for example, electrons). It gives the *approximate* force between two macroscopic objects if the sizes of both objects are very small compared to their separation. Second, Coulomb's law technically only applies to particles *at rest*. We'll see later how we have to generalize this law in dynamic cases.

Exercise E1X.5

Consider the copper penny discussed in exercise E1X.3. Imagine that we were miraculously able to remove 1 out of every 1000 electrons on two such coins. If we were to hold the coins 1.0 m apart, what would be the magnitude of the force that each coin exerted on the other? Compare to your weight. (Pennies are small enough that at a separation of 1.0 m they can be considered to be approximately pointlike.)

How to compute electrostatic forces on macroscopic objects

How can we compute the electrostatic force exerted by one macroscopic object on another if the objects are not small enough to be considered pointlike? While a macroscopic object may not itself be pointlike, it is constructed of charged particles (quarks and electrons) that *are*.

The superposition principle

The *net* force exerted on any point charge q_1 by a *set* of point charges $q_2, q_3, \ldots$ is simply the *vector sum* of the forces exerted by each individual charge $q_2, q_3, \ldots$ on the charge q_1. This is one way to state what is called the **superposition principle** for the electrostatic interaction. Experiments strongly support the superposition principle as an accurate model of electrostatic interactions between particles in a vacuum, even when the charges are very strong.

Therefore (at least theoretically), we can calculate the net force exerted on a macroscopic charged object A by another such object B by determining the net force on *each* particle in A due to *all* the particles in B (using the superposition principle) and then summing over all the particles in A.

Example E1.1

Problem Consider three point charges q_1, q_2, and q_3 equally spaced along the x axis, as shown in figure E1.6. If the net force on q_3 is zero, what can we say about the magnitudes and signs of the other charges?

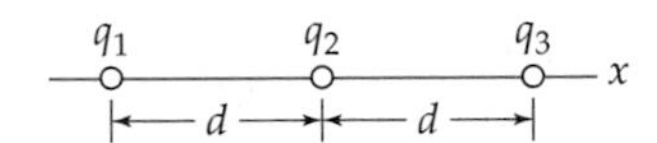

Figure E1.6
Three point charges equally spaced along the x axis.

Model Since all three charges lie along the x axis, the vectors describing the electrostatic forces exerted by q_1 and q_2 on q_3 must be parallel to the x direction. According to the superposition principle, the net force on q_3 is the vector sum of the forces exerted on it by q_1 and q_2 individually.

Solution The only way that these force vectors can add to zero is for the vectors to be equal in magnitude but opposite in direction. Therefore, q_1 and q_2 must have opposite signs to exert opposite forces on q_3. For the forces to have the same *magnitude*, we must have

$$\frac{k|q_1q_3|}{(2d)^2} = \frac{k|q_2q_3|}{d^2} \quad \text{according to Coulomb's law} \tag{E1.4}$$

since q_1 is a distance $2d$ from q_3, while q_2 is a distance d from q_3. Dividing both sides of this equation by $k|q_3|/d^2$, we find that $|q_1|/4 = |q_2|$. So from the information given, we can determine that q_1 and q_2 must have opposite signs and that the magnitude of q_1 must be 4 times that of q_2.

Evaluation We cannot determine anything more about the signs of the charges, and since $|q_3|$ divides out of equation E1.4, we can say nothing about the sign or magnitude of q_3.

Example E1.2

Problem Point charges q_1, q_2, and q_3 are arranged in an equilateral triangle whose sides have length d, as shown in figure E1.7a. All three charges have the same magnitude; and q_2 is negative, and the other two charges are positive. Qualitatively, what is the direction of the force on charge q_1?

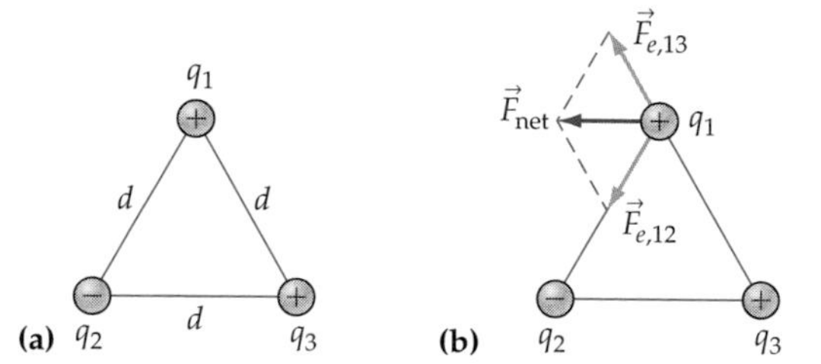

Figure E1.7
(a) Three charges arranged in the form of an equilateral triangle.
(b) The net force on q_1 is the vector sum of the forces exerted on that charge by q_2 and q_3 individually.

Solution Since $|q_2| = |q_3|$, the magnitudes of the forces that q_2 and q_3 exert on q_1 will be the same. But because q_1 and q_2 have opposite signs, the force that q_2 exerts on q_1 points *toward* q_2; but since q_1 and q_3 have the same sign, the force that q_3 exerts on q_1 points *away* from q_3, as shown in figure E1.7b. The vector sum of these two vectors points leftward in the diagram. (Note that $\vec{F}_{e,12}$ here stands for the electrostatic force *on* q_1 due to its interaction with q_2.)

Exercise E1X.6

Consider an object constructed of two point charges separated by a small distance. Assume that the charges have equal magnitudes but opposite signs: the net charge on the object is thus zero. Imagine that we place two such objects on the x axis, with the positive side of each object oriented in the $+x$ direction, as shown in figure E1.8. Will the objects attract each other, repel each other, or have no effect on each other? Explain.

Object A Object B x

Figure E1.8
Interaction of two neutral objects.

Example E1.3

Problem In the situation shown in figure E1.7, calculate the magnitude and direction of $\vec{F}_{net}$ in terms of $q \equiv |q_1| = |q_2| = |q_3|$ and d.

Translation Let us define a coordinate system for this problem as shown in figure E1.9. Since the triangle is equilateral, the $\theta = 60°$. Note that this means that both $\vec{F}_{e,13}$ and $\vec{F}_{e,12}$ make an angle of θ with respect to the x axis.

Model Assuming that these charges are small enough to be pointlike, we can use Coulomb's law to calculate the forces $\vec{F}_{e,13}$ and $\vec{F}_{e,12}$ separately. The superposition principle implies that the net force $\vec{F}_{net}$ on q_1 is the vector sum of these forces.

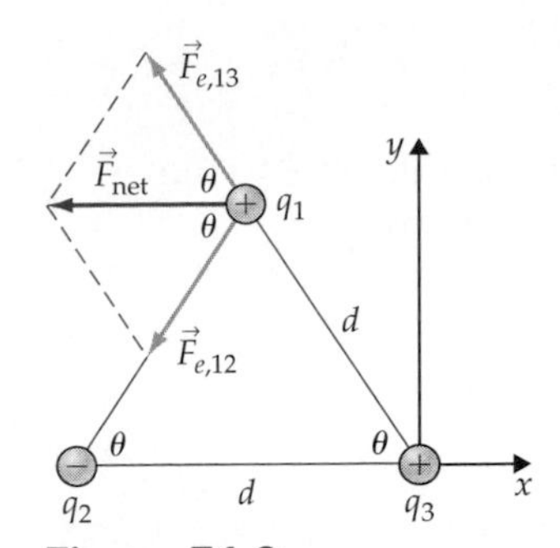

Figure E1.9
A diagram of the situation discussed in example E1.3.

Solution Since charges q_1 and q_3 are separated by the distance d, the magnitude of the force on q_1 will be

$$F_{e,13} \equiv \text{mag}(\vec{F}_{e,13}) = k\frac{|q_1q_3|}{r^2} = k\frac{q^2}{d^2} \tag{E1.5}$$

The magnitude of $\vec{F}_{e,12}$ will be the same. Looking at the diagram, we see that the components of these two force *vectors* are

$$\vec{F}_{e,13} = \begin{bmatrix} -F_{e,13}\cos\theta \\ F_{e,13}\sin\theta \\ 0 \end{bmatrix} = \frac{kq^2}{d^2}\begin{bmatrix} -\cos\theta \\ \sin\theta \\ 0 \end{bmatrix} \tag{E1.6a}$$

$$\vec{F}_{e,12} = \begin{bmatrix} -F_{e,12}\cos\theta \\ -F_{e,12}\sin\theta \\ 0 \end{bmatrix} = \frac{kq^2}{d^2}\begin{bmatrix} -\cos\theta \\ -\sin\theta \\ 0 \end{bmatrix} \tag{E1.6b}$$

Adding, we find that

$$\vec{F}_{net} = \vec{F}_{e,13} + \vec{F}_{e,12} = \frac{kq^2}{d^2}\begin{bmatrix} -\cos\theta \\ \sin\theta \\ 0 \end{bmatrix} + \frac{kq^2}{d^2}\begin{bmatrix} -\cos\theta \\ -\sin\theta \\ 0 \end{bmatrix}$$

$$= \frac{kq^2}{d^2}\begin{bmatrix} -2\cos\theta \\ 0 \\ 0 \end{bmatrix} = \frac{kq^2}{d^2}\begin{bmatrix} -1 \\ 0 \\ 0 \end{bmatrix} \tag{E1.7}$$

since $\cos 60° = \frac{1}{2}$. We see that $\vec{F}_{net}$ points entirely in the $-x$ direction and has a magnitude of kq^2/d^2.

Evaluation This direction is consistent with the result of example E1.2.

E1.6 Conductors and Insulators

Why do materials such as amber, plastic, rubber, and glass become easily charged when rubbed, while most other materials do not? For example, you can easily charge a plastic cup by rubbing it with a towel, but you do not become appreciably charged when you dry yourself after a shower. While it is easy to charge a rubber balloon by rubbing it on your sweater, you can rub a piece of silverware on your sweater all day, and it will never gain a discernible charge. Is it just because materials such as amber, plastic and rubber are that much more prone than flesh or metal to gain or lose electrons in the rubbing process?

Charges are not free to move in an insulator

It is true that some materials more effectively lose or gain electrons than others, but this is not the most important issue. Materials such as amber, plastic, and so on are easy to charge mostly because electrons in these substances (even the extra electrons created by rubbing) cannot move around on the surface of the substance or into its interior. So if rubbing a plastic comb through your hair, for example, deposits extra electrons on the surface of the comb, those extra electrons *stay* on the comb where they were deposited. A material whose protons and electrons are both essentially immobile is called an **insulator.** Glass, amber, most plastics, rubber, and dry cloth are all reasonably good insulators.

Charges *are* free to move in a conductor

At the other extreme are materials in which charges move very easily from place to place: such materials are called **conductors.** Salty water is a fairly good conductor, and most metals are excellent conductors. Copper and silver are especially good conductors: electrons move through either of these metals about 10^{27} times more easily than they move through polystyrene (a common plastic). In a metal, the outermost electron in each atom essentially becomes detached from its atom and becomes free to roam around inside the metal (although such electrons cannot easily *leave* the metal). Charge can flow through salty water because the salt disassociates into ions that are free to move through the water (chemically pure water with no dissolved ions is an excellent insulator).

Why insulators are relatively easy to charge

Materials that are exceptionally good insulators are easy to charge: any charges created by rubbing and deposited on the surface of an insulator remain fixed in place, in spite of the fact that these charges strongly repel each other. On the other hand, extra electrons, for example, deposited on the surface of a non-isolated conductor almost immediately disperse in response to the repulsive forces that they exert on each other. For example, if you rub a spoon on your sweater, any excess electrons deposited on the spoon, in their rush to get away from each other, disperse themselves by moving through the spoon into your hand and then through your body (which is mostly salt water) to the ground, where they can disperse themselves very widely. So it is not so much that rubbing a spoon does not create excess charge as it is that this excess charge is almost instantly dissipated because the electrons are free to move away from each other.

Because rubbing typically generates only a tiny amount of excess charge (on the order of 10^{-8} C), a material must be an exceptionally good insulator if this tiny charge is not to escape. Most natural (non-metallic) materials (even such things as wood or rock) are pretty bad conductors, but they are not terrific insulators either. Almost anything that is the least bit moist will conduct electrons well enough to quickly dissipate the small amount of

Figure E1.10
This person, who is insulated from the ground, has been highly charged by the Van de Graaff generator.

charge generated by rubbing: this is why it is difficult to do experiments involving static electricity on a humid day.

How to charge a conducting object

It is possible to charge a conducting object if that object is separated from any other conductor by an exceptionally good insulator. A metal sphere on an insulating plastic or glass stand can retain an electric charge very well. A piece of silverware can become nicely charged if you hold it with plastic gloves. Even a human body can become highly charged if it is well insulated from its surroundings. A classic demonstration of static electricity involves a person who holds on to one end of a Van de Graaff generator (a device that mechanically transports electrons from one end to the other, making both ends strongly charged) while standing on an insulating stool. Assuming the generator produces a negative charge, then electrons from the generator will disperse themselves through the person's body, including in the person's hair. Eventually the person's hair begins to stand up as the electrons in the hair strain to get as far away from each other as possible (see figure E1.10)!

E1.7 Electrostatic Polarization

A charged comb will pick up small bits of paper. A charged balloon will be attracted to a wall. Yet the paper and wall in these cases are *electrically neutral* (uncharged). Coulomb's law implies that a charged object should exert *zero* force on an uncharged object. So why do we observe attractive forces in these cases?

The charge of a normal atom's positive nucleus is exactly canceled by the charge of its negative electrons. Moreover, an *isolated* atom's electron cloud is normally exactly centered on its nucleus (see figure E1.11a). If an atom were to remain this way, it could neither attract nor be attracted by an external charge.

The definition of an *electric dipole*

However, if we place an atom close to an external negatively charged object, the charged object will pull the atom's positively charged nucleus toward it while pushing the negatively charged electron cloud in the opposite direction. This typically displaces the center of the electron cloud a small amount relative to the nucleus, as shown in figure E1.11b. As we will discuss

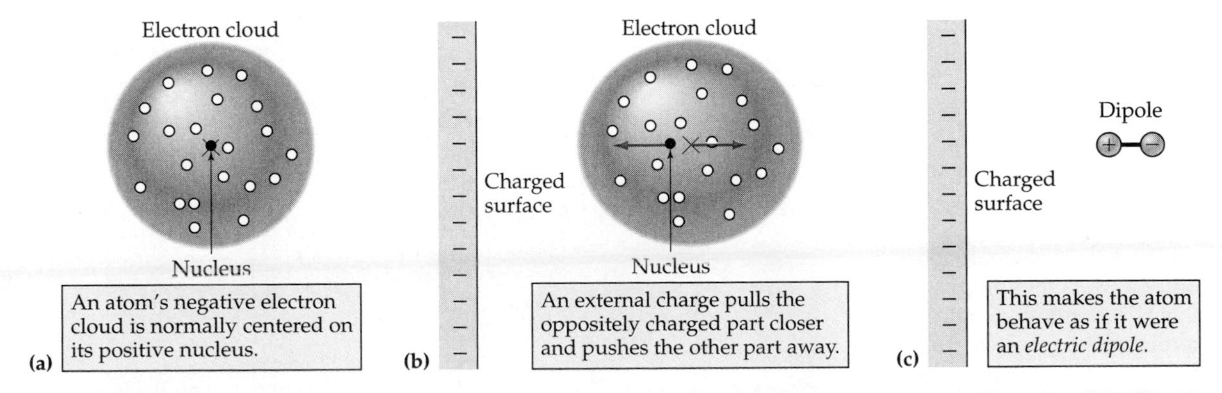

Figure E1.11
(a) An atom far from any charged object. (b) An atom in an insulator can become *polarized* by an external charge: the nucleus is pulled toward a negative external charge, for example, and the electrons are collectively pushed the other way. (c) A schematic representation of an electric dipole (not drawn to scale).

in the chapter E3, the essentially spherical electron cloud will interact with an external charge as if all the cloud's charge were located at its center. Therefore, to an excellent degree of approximation we can treat the atom's nucleus and displaced electron cloud as if they were two point charges of equal magnitude but opposite charge separated by a small distance. We call such a pair of point charges an **electric dipole** (see figure E1.11c).

How a charged object can attract a neutral object

Note that this process (which we call **electrostatic polarization**) will always pull the oppositely charged part of the atom closer, independent of the sign of the external charge. Since the Coulomb force exerted by two charges on each other decreases as the separation between the charges increases, the external charge will therefore attract the closer charge in the dipole a little bit more strongly than it repels the more distant charge. This means that the external charge will exert a *net* force on the dipole that is weakly attractive. By Newton's third law, the dipole will also exert a weak attractive force on the external charge.

This explains how a charged object can attract a neutral object. Imagine, for example, that we bring a negatively charged balloon close to an electrically neutral wall. The Coulomb forces that the balloon's charge exerts on the positively and negatively charged parts of the wall's atoms polarize those atoms, making them behave as dipoles (see figure E1.12). These dipoles are weakly attracted to (and in turn weakly attract) the balloon. Each individual wall atom is only *slightly* polarized by the balloon's charge, and the attractive force between each atom and the balloon is correspondingly tiny. But because there are a huge number of atoms in the wall, the sum of these tiny attractive forces can be enough to hold the balloon to the wall.

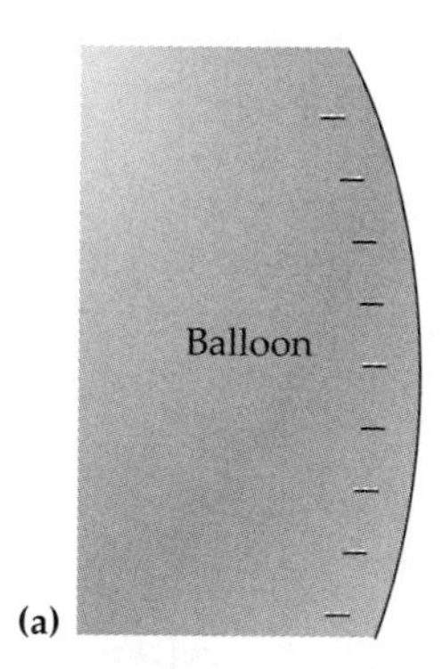

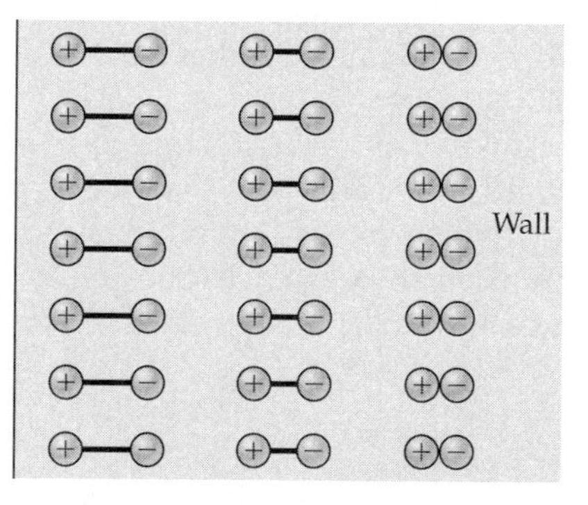

Figure E1.12
(a) A charged balloon polarizes atoms on a nearby wall. (Note that because the Coulomb forces are stronger on the atoms closer to the balloon, they will be polarized more strongly than atoms farther away: the diagram shows this.) Each atomic dipole then exerts a small attractive force on the balloon, since the dipole's positive end is a bit closer to the balloon than its negative end is. (b) The net effect is that the balloon is attracted to the wall.

The same effect explains why bits of paper are attracted to a charged comb, and so on. Note that the force between a charged object and a polarizable neutral object is *always* attractive. We will also see in chapter E2 that the force between a small charged object and a small but polarizable neutral object decreases as $1/r^5$(!), so the charged object has to be pretty close to the neutral object to have an appreciable affect.

Some molecules have permanent dipoles

Certain kinds of molecules behave as natural dipoles even when there is no charged object nearby. For example, the oxygen atom in a water molecule grips the molecule's electrons more tightly than the two hydrogen atoms do: this makes the oxygen atom somewhat negatively charged and the hydrogen atoms positively charged (see figure E1.13). The fact that a water molecule is a natural dipole is part of the reason that water is such a good solvent: the charged ends of the water molecule grab onto the oppositely charged ends of the solute molecules, enabling the water molecules to pull the solute molecules into solution.

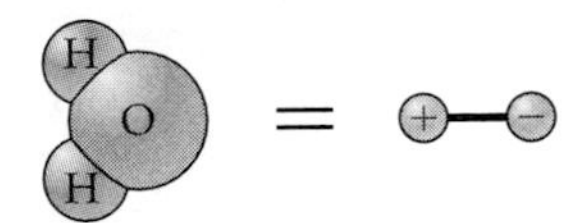

Figure E1.13
Because oxygen atoms tend to grab electrons more strongly than hydrogen atoms do, a water molecule has a permanent natural dipole.

If we bring a charged object near a sample of water, one end of any given water molecule will be attracted toward the object while the other end will be repelled. The molecule will thus try to pivot to align its oppositely charged end toward the object, and in that orientation, it is attracted to the object. Thermal collisions with adjacent molecules continually knock the molecule *out* of alignment, but even so, the forces exerted by the charged object cause the molecule's *average* orientation to be slightly weighted toward the attractive orientation. Therefore, at the macroscopic level, water gets polarized by an external charge pretty much as an insulating solid does, although the mechanism is different.

The natural charge separation in a water molecule is about 100,000 times larger than the charge separation a typical atom gets even in the presence of very strong external charges. So although the dealigning thermal collisions substantially reduce the average polarization of water molecules, empirically, water still responds to an external charge from 10 to 50 times as strongly as an equivalent volume of a typical solid (see figure E1.14).

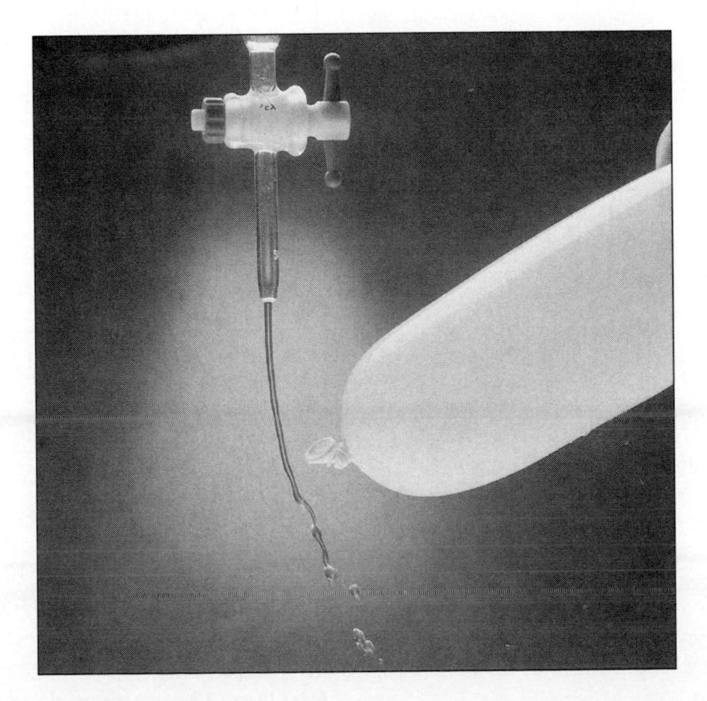

Figure E1.14
An electrically neutral stream of water is attracted by a charged balloon.

TWO-MINUTE PROBLEMS

E1T.1 We can certify that an object is definitely positively charged by using which of the following tests?
- A. Showing that the object is attracted by a negative charge
- B. Showing that the object is repelled by a positive charge
- C. Doing either of the tests above

E1T.2 All uncharged macroscopic objects contain a large amount of positive charge exactly balanced by a large amount of negative charge. The positive charge in 1 cm^3 of water (mass $= 1$ g) is closest to what value?
- A. 10^{-19} C
- B. 10^{-4} C
- C. 1 C
- D. 10^4 C
- E. 10^{23} C
- F. Other (specify)

E1T.3 If you rub Saran Wrap on your cotton shirt, which becomes positively charged?
- A. The cotton shirt
- B. The Saran Wrap
- C. Neither: both are insulators, and so the charges remain where they are

E1T.4 Two nickels are each given a charge of $+100\mu$C. If these nickels are placed 1.0 m apart, the magnitude of the force that each exerts on the other is closest to
- A. 0
- B. 10^{-8} N
- C. 100 N
- D. 10^7 N
- E. None of the above, because Coulomb's law doesn't apply (even approximately)

E1T.5 Consider two point charges q_1 and q_2 lying on the x axis. Consider a third charge q_3 located somewhere near q_1 and q_2. Assume that the net force exerted by these charges on q_3 is zero. This implies that q_3 *must* lie somewhere along the x axis, true (T) or false (F)?

E1T.6 Consider two point particles lying along the x axis separated by a distance d. These particles have charges that are equal in magnitude but opposite in sign. Where can a third point charge be placed so the net force on it is zero?
- A. Halfway between the two charges.
- B. A distance d from both charges (so the three charges form an equilateral triangle).
- C. Some point on the x axis but not between the two charges (I could compute the exact location if I had time).
- D. There is *no* point where the force would be zero.
- E. The answer depends on the charges' signs and/or magnitudes.
- F. Other (explain).

E1T.7 A point charge of $+1\ \mu$C is placed at the exact center of a plastic ring whose radius is 10 cm and that has a total charge of $+1\ \mu$C uniformly distributed about the ring. The magnitude of the net force on the point charge is most nearly
- A. 0
- B. +1 N
- C. −1 N
- D. 1000 N
- E. Other (specify)

E1T.8 Three identical charged particles are arranged in an equilateral triangle. If the force that each *individual* particle exerts on another is 1.0 N, the magnitude of the net force F_e exerted on any one of the three particles is
- A. $F_e = 0$
- B. $0 < F_e < 1.0$ N
- C. $F_e = 1.0$ N
- D. $1.0\text{ N} < F_e < 2.0$ N
- E. $F_e = 2.0$ N
- F. $F_e > 2.0$ N

HOMEWORK PROBLEMS

Basic Skills

E1B.1 Two point particles separated by 8.0 cm have charges $q_1 = +12$ nC and $q_2 = -42$ nC, respectively. Find the magnitude of the force that each exerts on the other.

E1B.2 What is the magnitude of the force exerted by a proton on an electron when the two are 0.05 nm apart? (This is roughly the typical separation of these particles in a hydrogen atom.)

E1B.3 What is the total charge (in coulombs) of the protons in 1 g of hydrogen gas?

E1B.4 Avogadro's number of water molecules has a mass of 18 g. If we were somehow able to remove an electron from one out of every 10^9 molecules in 1 g of water, what would be its charge?

E1B.5 Two point charges ($q_1 = +20$ nC and $q_2 = -40$ nC) are placed 10.0 cm apart along the x axis. Where can a third charge ($q_3 = +20$ nC) be placed so that the net electrostatic force on that charge is zero?

E1B.6 Two point charges ($q_1 = +25$ nC and $q_2 = +100$ nC) are placed 12 cm apart along the x axis. Where can a third charge ($q_3 = -15$ nC) be placed so that the net electrostatic force on that charge is zero?

E1B.7 Two point particles (with charges $q_1 = +30$ nC and $q_2 = -30$ nC) are placed on the x axis 20 cm apart, with the negative particle to the left of the positive particle. A third point particle (with charge $q_3 = +20$ nC) is placed on the x axis halfway between the first two particles. What is the force (magnitude and direction) on the third particle?

E1B.8 Two point particles (with charges $q_1 = +120$ nC and $q_2 = +30$ nC) are placed on the x axis 20 cm apart, with the latter to the left of the first. A third particle (with $q_3 = -60$ nC) is placed on the x axis halfway between the first two particles. What is the force (magnitude and direction) on the third particle?

E1B.9 Compute the ratio between the magnitudes of the gravitational force and the electrostatic force between two electrons separated by 1.0 m. How does this ratio change if we bring the electrons closer?

Synthetic

E1S.1 Two small steel balls (mass ≈ 5.6 g each) are separated by a distance of 5.0 cm. If one in every billion electrons could be removed from each ball, what would be the approximate magnitude of the electrostatic force exerted by each ball on the other? (*Hint.* Steel is essentially iron, and an iron nucleus contains 26 protons and 30 neutrons.)

E1S.2 Two small steel balls (mass ≈ 5.6 g each) separated by 10 cm are given the same positive charge and are found to repel each other with a force of 0.1 N (somewhat less than an ounce). What fraction of the electrons in each steel ball has been removed? (*Hint:* Steel is essentially iron, and the iron nucleus contains 26 protons and 30 neutrons.)

E1S.3 If the density of charge on a surface exceeds about 5×10^{-5} C/m^2, the electrostatic forces exerted by the charge will be strong enough to ionize air near the surface, making the air a conductor and thus draining the charge away from the surface. What is the maximum charge that can be put on the surface of a sphere 1.0 cm in diameter? (This effect is what in practice defines the upper limit of the charge that can be put on an object by rubbing.)

E1S.4 Three particles, each having a charge of +10 nC, are arranged in an equilateral triangle 10 cm on a side. What is the magnitude of the force acting on any one of the particles?

E1S.5 Four identical particles, each having a charge of +120 nC, are arranged in a square 6.0 cm on a side. Find the magnitude of the force acting on any one of the charges, and describe the direction of this force in words.

E1S.6 Four particles are arranged in a square 5.0 cm on a side. The magnitude of each of the charges is 33 nC, but the signs of the charges alternate as you go around the square. Find the magnitude of the force acting on any one of the particles, and describe in words the direction of the force vector. Does your description of the direction of the force depend on the sign of the charge you chose?

E1S.7 A particle with charge $q_1 = +30$ nC is placed on the x axis of a certain coordinate system at $x = 10$ cm. A particle with charge $q_2 = -60$ nC is placed on the y axis at $y = 15$ cm. Determine the magnitude and direction of the force acting on a particle with charge $q_3 = +10$ nC placed at the origin.

E1S.8 A particle with charge $q_1 = +1.0\ \mu$C is placed on the x axis of a certain coordinate system at $x = 10$ cm. A particle with charge $q_2 = -0.50\ \mu$C is placed on the y axis at $y = -20$ cm. Determine the magnitude and direction of the force acting on a particle with charge $q_3 = +0.10\ \mu$C at the origin.

E1S.9 Two point charges, one with charge $+q$ and the other with charge $+9q$, are placed a distance d apart. It is possible to place a third charge so that the net electrostatic force on *all three* charges is zero. What are the sign, magnitude, and position of this third charge (in terms of q and d)?

E1S.10 Two small spheres with mass m are suspended with insulating threads of length L from a common point. When the spheres are uncharged, they hang so they touch each other; but when they are both given the same charge q, they repel each other and hang a distance d apart, as shown in figure E1.15. Assume that d is pretty small compared to L but large compared to the diameter of the balls.

(a) Explain, using a net force diagram, why the magnitude of the electrostatic force $\vec{F}_e$ acting on either ball must be

$$F_e = mg \tan\theta \tag{E1.8}$$

(more)

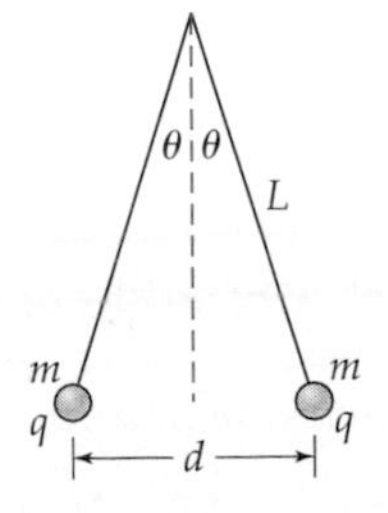

Figure E1.15 The situation discussed in problem E1S.10.

(b) Explain why in this situation we can approximate

$$\tan\theta \approx \frac{d}{2L} \tag{E1.9}$$

By what percentage is this expression in error if $d = L/10$?

(c) Combine the answers to parts (a) and (b) with Coulomb's law to show that

$$|q| = \sqrt{\frac{mgd^3}{2kL}} \tag{E1.10}$$

(d) If this situation is demonstrated in your class, use the measured values of L, m, and d to estimate q. Otherwise, calculate q, assuming that $L = 70$ cm, $d = 4.0$ cm, and $m = 0.4$ g.

Rich-Context

E1R.1 Two small 0.1-g Styrofoam balls are strung like beads on a vertical insulating thread. The lower ball is glued to the thread, but the upper ball is free to move. Imagine that both are given an equal amount of negative charge so that the upper ball is suspended above the lower ball and their centers are 4.0 cm apart. (Assume that each ball repels the other as if it were a point charge located at its center.) Roughly what fraction of the electrons originally on each ball have been added to each ball? (*Hint:* Styrofoam is made mostly of low-mass atoms having roughly equal numbers of protons and neutrons.)

E1R.2 Imagine that you rub about a 10-cm length of a rubber rod 1.0 cm in diameter with fur, giving it a charge of −100 nC (this is close to the maximum charge one can put on such a rod). Rubber (which has a density about equal to water) can be considered to be made of building blocks of C_5H_8. Using this information, very roughly estimate what fraction of these building blocks on the surface of the rubber have gained an electron in the rubbing process.

Advanced

E1A.1 Imagine that we place two particles, both with charge $+q$, on the y axis at $y = +d$ and $y = -d$, respectively. Imagine further that we have a particle with charge $-q$ that is constrained to move along the x axis.

(a) Argue that for small displacements from the origin such that $x \ll d$, the net x-force on the negative particle has the form of Hooke's law

$$F_x = -k_s x \tag{E1.11}$$

where k_s is some constant.

(b) Find an expression for the frequency of small oscillations in terms of k, q, d, and the mass m of the negative charge. (*Hint:* See chapter N11.)

(c) Say that the particles are actually small Styrofoam balls. Reasonable values of q, d, and m in such a case might be $q = 5$ nC, $d = 10$ cm, and $m = 0.05$ g. Compute the frequency of small oscillations of the negative ball.

ANSWERS TO EXERCISES

E1X.1 Both U tapes were prepared in the same way, so they really *should* have like charges. It is *conceivable* that they might get charges of random signs from the preparation process, but this would mean that the U tapes would sometimes attract and sometimes repel. Since we consistently observe U tapes to behave in the same way, the model that they get the *same* charges from the process makes more sense. Assuming that this is true, it must be like charges that repel.

E1X.2 See which tape is repelled by a charged comb (which is negatively charged). You should find that the T tape is negative (for most brands of tape).

E1X.3 *Model:* Avogadro's number of nucleons (protons and/or neutrons) has a mass of about 1 g. Copper has 29 protons per 63 nucleons (see figure E1.2).

Solution: The number of protons in the penny is about

$$5\,\text{g}\left(\frac{6.02\times10^{23}\,\text{nucleons}}{1\,\text{g}}\right)\left(\frac{29\,\text{protons}}{63\,\text{nucleons}}\right) = 1.4\times10^{24}\ \text{protons} \tag{E1.12}$$

The total charge of these protons is

$$1.4\times10^{24}\,\text{protons}\left(\frac{1.60\times10^{-19}\,\text{C}}{1\,\text{proton}}\right) = 220{,}000\ \text{C} \tag{E1.13}$$

Evaluation: This is an *enormous* amount of charge, about 10^{13} times the charge that one can put on an object that size in practice.

E1X.4 A decay process involves the disintegration of a particle into fragments. Conservation of relativistic energy implies that the fragments must have smaller total rest mass than the electron. But if there is no lighter charged particle than the electron, then any decay process that is consistent with conservation of relativistic energy will violate conservation of charge.

E1X.5 *Model:* We can calculate the force using Coulomb's law (approximating the pennies by point objects). *Solution:* According to the results of exercise E1X.3, if we remove 1 electron out of every 1000, each penny will end up with a positive charge of about 220 C. If we separate two such pennies by 1.0 m, the magnitude of the force acting on either will be

$$F_e = k\frac{|q_1 q_2|}{r^2} = 8.99 \times 10^9 \frac{\text{N} \cdot \cancel{\text{m}^2}}{\cancel{\text{C}^2}} \frac{(220\ \cancel{\text{C}})^2}{(1.0\ \cancel{\text{m}})^2}$$

$$= 4.4 \times 10^{14}\ \text{N} \tag{E1.14}$$

Evaluation: My weight is about 550 N, so this would be the weight of roughly a trillion people! (Note that the current world population is only 6 billion.)

E1X.6 Consider the forces acting on the charges in object A. The positive charge in A is slightly closer to the negative charge in object B than to the positive charge in B, so the positive charge in A will be somewhat attracted to B. On the other hand, the negative charge in A will be somewhat repelled by B, because the negative charge in B is closer. However, the negative charge in A is a bit farther from B as a whole than the positive charge in A is, so the repulsion it feels will be slightly weaker than the attraction that the positive charge in A feels. Therefore, these objects will very weakly attract each other.

E2 Electric Fields

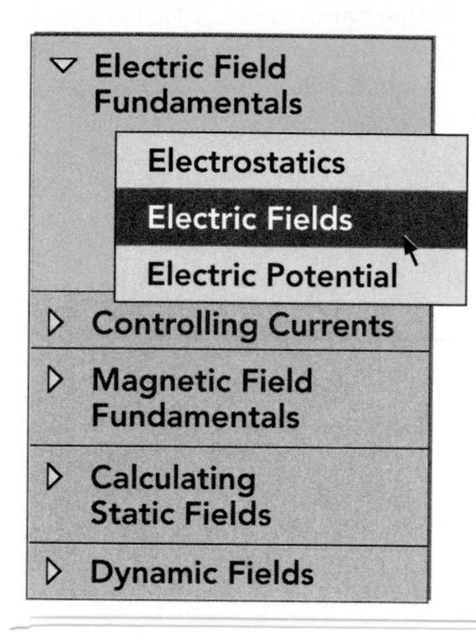

Chapter Overview

Introduction

Coulomb's law provides a straightforward model of how charged particles interact. However, this chapter introduces an alternative *field model* of electrostatic interactions that proves more useful in the long run.

Section E2.1: The Field Concept

Taken literally, Coulomb's law describes an **action-at-a-distance model** of electrostatic interactions where charged particles exert forces directly and instantaneously on one another across the distance separating them. A **field model** instead imagines that a charged particle creates a **field** in the space around it, and another particle responds to the field at its own location, not to the first particle directly:

$$\text{Charge} \leftrightarrow \text{field} \leftrightarrow \text{charge} \qquad (\text{instead of} \quad \text{charge} \leftrightarrow \text{charge}) \tag{E2.1}$$

A field (*unlike* a particle) exists not at a specific location but throughout space. Even so, it is a physical object that (*like* a particle) has energy, carries momentum, and obeys equations of motion.

We need a field model because instantaneous action at a distance violates the relativistic principle that no effect can travel faster than the speed of light c. A field can carry effects between particles at a speed $\leq c$.

Section E2.2: An Operational Definition of $\vec{E}$

We describe an electric field by attaching to every point in space an **electric field vector** $\vec{E}$ such that

$$\vec{E} \equiv \frac{\vec{F}_e}{q_{\text{test}}} \tag{E2.2}$$

Purpose: This equation defines the electric field vector $\vec{E}$ at a point in space and time.

Symbols: $\vec{F}_e$ is the electrostatic force experienced at that time by a small test particle with charge q_{test} held *at rest* at that point in space.

Limitation: The test charge q_{test} must be small enough that the forces that *it* exerts do not significantly push around the charges that are creating $\vec{E}$.

Notes: $\vec{E}$ points in the same direction as $\vec{F}_e$ if q_{test} is positive.

Performing this measurement at all points in space determines the entire field as a **vector function** $\vec{E}(x, y, z)$ of position.

This definition implies that once we know $\vec{E}(x, y, z)$, we can compute the electrostatic force that a particle feels at any point as follows:

$$\vec{F}_e = q\vec{E} \tag{E2.3}$$

Purpose: This equation describes the force $\vec{F}_e$ that an electric field exerts at a given instant of time on a particle with charge q.

Symbols: $\vec{E}$ is the electric field vector at the particle's location at the given instant.

Limitations: Again, q should be small enough that it doesn't disturb the charge distribution from the configuration for which we calculated $\vec{E}$.

Note: $\vec{F}_e$ points in the same direction as $\vec{E}$ when q is positive, and opposite to $\vec{E}$ when q is negative.

Section E2.3: The Field of a Point Charge

Coulomb's law and the definition of $\vec{E}$ imply that the electric field of a point particle is

$$\vec{E} = \frac{kq}{r_{PC}^2}\hat{r}_{PC} = \frac{1}{4\pi\varepsilon_0}\frac{q}{r_{PC}^2}\hat{r}_{PC} \quad \text{or equivalently,} \quad \vec{E} = \frac{kq}{r_{PC}^3}\vec{r}_{PC} \tag{E2.9}$$

Purpose: This equation describes the electric field vector $\vec{E}$ at an arbitrary point P created by a point charge q located at point C.

Symbols: $\vec{r}_{PC} \equiv \vec{r}_P - \vec{r}_C$ is the position of point P relative to C, $r_{PC} =$ mag$(\vec{r}_{PC})$, and $\hat{r}_{PC} \equiv \vec{r}_{PC}/r_{PC}$ is a directional (unit vector) standing for the direction at P that is "away from the charge q." $\varepsilon_0 \equiv 1/4\pi k$.

Limitations: The charge q must be a point particle at rest.

Note: The field vector $\vec{E}$ points away from q if $q > 0$ and *toward* q if $q < 0$.

Section E2.4: The Superposition Principle

Superposition implies that the net electric field $\vec{E}$ created at a given point P by a *set* of charged particles is simply

$$\vec{E} = \vec{E}_1 + \vec{E}_2 + \vec{E}_3 + \cdots \tag{E2.12}$$

Purpose: This equation describes how electric fields combine.

Symbols: $\vec{E}$ is the total electric field at point P; $\vec{E}_1$ is the electric field vector produced at P by particle q_1 alone; $\vec{E}_2$ is the same for q_2 alone; and so on.

Limitations: None are known!

We can therefore (in principle) calculate the field created by any object simply by summing the fields created by each charged particle in the object.

Section E2.5: The Field of a Dipole

An electric dipole, which is a pair of charges $+q$ and $-q$ separated by a small distance d, creates an electric field whose magnitude at a point a distance r from the dipole along a given line going through the dipole's center is proportional to kqd/r^3 when $r \gg d$. A dipole's field thus decreases with increasing r faster than that of a point charge.

One can use this result to argue that the *force* that a point charge exerts on a small polarizable object typically decreases with distance as $1/r^5$.

Section E2.6: Handling Charge Distributions

We can most easily calculate the electric field of a distribution of charge by dividing the distribution into tiny bits (each small enough to be accurately modeled as a point charge) and using the following five-step process (which examples in the section illustrate) to evaluate the sum of the fields over all the bits:

1. Find a variable that you can use to locate each bit in the distribution.
2. Express each bit's charge in terms of an infinitesimal change in that variable.
3. Express the distance between the bit and a point P in terms of the variable.
4. Use equation E2.9 to find the field vector $\vec{E}_b$ created at point P by the bit.
5. Convert the sum of $\vec{E}_b$ over all bits to an integral and evaluate.

E2.1 The Field Concept

As we saw in chapter E1, Coulomb's law provides a simple model for understanding how charged particles interact. Physicists in the mid-1800s took a crucial step toward a better understanding of the electromagnetic phenomena by developing a more sophisticated model that we will call a *field model* of the electromagnetic interaction. In this section, we will see why such a model is needed.

Coulomb's law expresses action at a distance

Coulomb's law presents a straightforward picture of how two point charges q_1 and q_2 interact: each exerts *directly* on the other a force parallel to the line connecting them. This model of the electrostatic interaction is called an **action-at-a-distance model,** since each charge is imagined to act *directly* on another charge across the distance between them. (Newton's law of universal gravitation implies the same kind of model for the gravitational interaction.)

Coulomb's law is accurate as long as the charged objects in question are essentially at rest, but problems arise if they are not. Taken literally, Coulomb's law implies that the force exerted by one charge on the other at a given instant of time depends on the distance between those charges *at that instant* and acts along the line connecting those charges *at that instant*. If this were really true, then we could make a device that would allow us to communicate instantaneously over large distances. For example, imagine that at point P, I suddenly wiggle a point charge. If you measure the electrostatic force acting on another charge at point Q, you will see the direction of that force wiggle a bit at exactly the same time I wiggle the charge. We could use this effect to send a message instantaneously between points P and Q, even over very large distances.

But this is inconsistent with the theory of relativity, which asserts that *no* signal can travel between two points faster than the speed of light. So if the theory of relativity is true, then Coulomb's law *cannot* be literally correct for wiggling particles.†

The field model presents an alternative to an action-at-a-distance model

The **field model** provides an alternative that avoids this problem. In the field model, we imagine that the space around a charged particle is filled with something called an *electric field*. Unlike a particle, a field does not occupy only a single point but rather exists at *all* points in space simultaneously. We in fact describe a field mathematically by assigning some kind of numerical quantity to every point in the space. Like a particle, though, a field (as we will see) is a real thing that has energy, carries momentum, and obeys equations of motion that describe how it evolves with time in response to its surroundings.

In a field model, the field mediates the interaction

In the field model, charges do not *directly* exert forces on each other, but rather each responds to the *field* that the other creates in its vicinity. The field thus *mediates* the interaction between the two charges. If we represent the action-at-a-distance model schematically by

$$\text{Charge} \leftrightarrow \text{charge} \qquad \text{action-at-a-distance model} \qquad (\text{E2.1}a)$$

(expressing the idea that in this model the charges interact *directly* with each other), then we can represent the field model schematically by

$$\text{Charge} \leftrightarrow \text{field} \leftrightarrow \text{charge} \qquad \text{field model} \qquad (\text{E2.1}b)$$

expressing the idea that each charge sets up a field to which the other responds.

---

†If Coulomb's law were literally true, then (according to chapter R8) it would be possible to find an inertial reference frame in which the force on the point charge at point Q changed before the charge at point P is wiggled, which is absurd!

How does this resolve the problem with regard to the instantaneous communication implied by Coulomb's law? In the field model, if one charged particle is wiggled, it does not *directly* affect a distant particle. Rather, the wiggling particle wiggles the values of the field in its immediate vicinity, and these wiggles in turn affect the field values at slightly more distant locations, and so on. The net effect is that ripples in the field move away from the wiggling particle at a finite speed as ripples on the surface of a pond do. Only when these ripples reach the distant charged particle will it feel a wiggling force. As long as these ripples move at a speed less than or equal to that of light, the field model will be consistent with relativity. Maxwell's theory of the electromagnetic field, which we will develop throughout this unit, provides a complete and relativistically consistent picture of how this all works.

This makes consistency with relativity possible

While this argument makes it clear why we need a field model, I also want to make it clear that the physics community did not *historically* arrive at the model this way. Rather it arose slowly as the community struggled (through an involved process of trial and error) to understand electromagnetic phenomena. Einstein invented the theory of relativity, decades after Maxwell's field theory was complete, to explain how Maxwell's theory was consistent with the principle of relativity (which states that the laws of physics should be the same in all inertial reference frames). Thus the historical process was actually the *reverse* of the argument I have presented! Even so, our present understanding of relativity helps us see why we need an electromagnetic field theory more clearly and immediately than following the historical path would.

This argument was not the historical path to the field model

By the way, the same problem arises with Newton's law of universal gravitation. General relativity solves the problem by providing a field theory for gravity analogous to Maxwell's field theory for the electromagnetic interaction. Both theories have surprising and exciting consequences unanticipated by the newtonian action-at-a-distance models.

General relativity is the analogous field theory for gravitation

E2.2 An Operational Definition of $\vec{E}$

What numerical quantity should we assign to each point in space to describe the **electric field** that is created by a charged particle according to this model? There is actually a very simple way to do this: *we define the field in terms of what it does*. If the basic action of a field is to exert a force on a charged particle, why not define the field in terms of the force that it exerts?

So, to evaluate the electric field at a given point, imagine taking a test particle with charge q_{test}, holding it *at rest* at the point in question, and measuring the electrostatic force $\vec{F}_e$ exerted on the charge. (Note that if we hold the charge at rest, the net force on the charge will be zero, so the magnitude of the electrostatic force will be equal to that of the opposing force we need to apply to the charge to keep it at rest.) We define the **electric field vector** $\vec{E}$ at that point and at that time to be

$$\vec{E} \equiv \frac{\vec{F}_e}{q_{\text{test}}} \tag{E2.2}$$

Purpose: This equation defines the electric field vector $\vec{E}$ at a point in space and time.

Symbols: $\vec{F}_e$ is the electrostatic force experienced at that time by a small test particle with charge q_{test} held *at rest* at that point in space.

The operational definition of the electric field vector at a point in space

Limitation: The test charge q_{test} must be small enough that the forces that *it* exerts do not significantly push around charges in the distribution whose field we are trying to measure.

Note: $\vec{E}$ points in the same direction as $\vec{F}_e$ if q_{test} is positive.

Exercise E2X.1

What are the SI units of the electric field vector?

Why divide the force by q_{test}? We find experimentally (and will shortly see theoretically) that the force a test charge experiences at a given location is *proportional* to q_{test}. Dividing by q_{test} thus produces a quantity $\vec{E}$ that depends *only* on one's *position* relative to the charges creating the field and not on the magnitude of the test charge we use.

Note we want q_{test} to be small enough that the electrostatic forces that *it* exerts do not significantly push around the set of charges creating $\vec{E}$, thus *changing* the charge distribution whose field we are trying to measure! (Technically, we should take the limit as $q_{\text{test}} \to 0$.) The reason that I am also insisting that the test charge be at rest will become clearer in chapter E8.

The concept of a *vector function*

This definition implies that the numerical quantity that we assign to every point in space to describe an electric field is in fact a *vector* $\vec{E}$. It is essential to understand, however, that a charged object's electric field is not *itself* a vector in the same way that its velocity is a vector. An object's electric field is rather an infinite *set* of vectors, with (potentially) a different vector for every point in space. One has to specify the field vectors at *all* points to describe the field fully. However, literally listing field vectors for *all* points in space is impossible. Whenever possible, physicists instead describe an object's total electric field at a given instant of time by using a **vector function** $\vec{E}(x, y, z)$. Just as an ordinary function $f(x)$ implicitly attaches a number to every point x on the number line by describing how to *calculate* that number in terms of x, so a vector function $\vec{E}(x, y, z)$ implicitly assigns a vector to every point in space by describing how to calculate $\vec{E}$ for every possible choice of that point's position coordinates x, y, and z. Such a function, when available, provides a compact and useful way to describe a field.

Given such a function $\vec{E}(x, y, z)$, equation E2.2 implies that we can determine the force exerted by the field on *any* charge q at a given location by using

How to use the field vector to compute the electric force on a given charge

$$\vec{F}_e = q\vec{E} \tag{E2.3}$$

Purpose: This equation states the force $\vec{F}_e$ that an electric field exerts on a charged particle.

Symbols: q is the charge on the particle, and $\vec{E}$ is the electric field at the particle's location at the instant of time in question.

Limitations: Again, q should be small enough that it doesn't disturb the charge distribution from the configuration when we calculated $\vec{E}$. (This limitation doesn't apply if we calculate $\vec{E}$ with q already in place.)

Note: $\vec{F}_e$ points in the same direction as $\vec{E}$ when q is positive, and opposite to $\vec{E}$ when q is negative.

A field having a magnitude of 1 N/C at a point thus exerts a force of 1 N on a 1-C charge placed at that point.

Field strength benchmarks

The magnitude of the electric field outdoors on a sunny day (due to various processes that separate atmospheric charges) is about 100 to 150 N/C. During a thunderstorm, this may exceed 10,000 N/C. Charges separated by moving water in a shower can create a 800 N/C field. If the field strength in air anywhere exceeds 3×10^6 N/C, air breaks down and becomes a conductor, and sparks fly.

Comparing $\vec{E}$ to the gravitational field vector $\vec{g}$

Note, by the way, that we define the gravitational field vector $\vec{g}$ at a point near a massive object in a way very similar to equations E2.2 and E2.3:

$$\vec{F}_g \equiv m\vec{g} \quad \Rightarrow \quad \vec{g}\,(\text{at a point } P) \equiv \frac{\vec{F}_g\,(\text{on } m_{\text{test}} \text{ at } P)}{m_{\text{test}}} \tag{E2.4}$$

Just as $\vec{E}$ is measured in newtons per coulomb, so $\vec{g}$ is measured in newtons per kilogram (which happens to be the same as meters per second squared). Thus $\vec{g}$ is to the gravitational field what $\vec{E}$ is to the electric field. This analogy may help make you more comfortable with what $\vec{E}$ represents.

E2.3 The Field of a Point Charge

Let's get practical. Consider a point charge q placed at a point C whose position is $\vec{r}_C = [x_C, y_C, z_C]$. What is the electric field produced by this charge at an arbitrary point P whose position is $\vec{r}_P = [x, y, z]$? Coulomb's law tells us that the electrostatic force on a test charge q_{test} at P is

$$F_e = \frac{k|q_{\text{test}}q|}{r_{PC}^2}, \quad \text{where} \quad \begin{array}{l} \vec{F}_e \text{ points } \textit{away} \text{ from charge } q \text{ if } q_{\text{test}}q > 0 \\ \vec{F}_e \text{ points } \textit{toward} \text{ charge } q \text{ if } q_{\text{test}}q < 0 \end{array} \tag{E2.5}$$

where $\vec{r}_{PC} \equiv \vec{r}_P - \vec{r}_C$ is the position of point P relative to the charge's position at point C, and r_{PC} is the distance between the two points: $r_{PC} \equiv \text{mag}(\vec{r}_P - \vec{r}_C)$ (see figure E2.1). Note that $q_{\text{test}}q > 0$ and $q_{\text{test}}q < 0$ imply that the charges are like and unlike respectively.

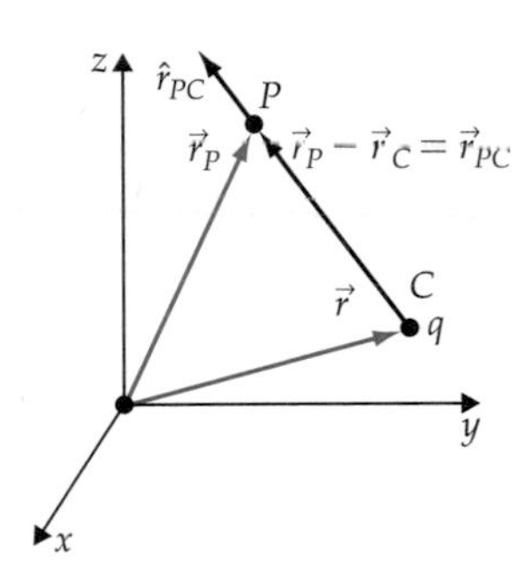

Figure E2.1
A diagram illustrating the definition of the vector $\vec{r}_P - \vec{r}_C$ and the unit vector $\hat{r}_{PC}$.

The indication of the direction in equation E2.5 is awkward, though. We can make this simpler if we note that the vector $\vec{r}_{PC} \equiv \vec{r}_P - \vec{r}_C$ points directly *away* from the charge q at point C. Therefore, the directional

Definition of the useful directional $\hat{r}_{PC}$

$$\hat{r}_{PC} \equiv \frac{\vec{r}_{PC}}{r_{PC}} = \frac{\vec{r}_P - \vec{r}_C}{\text{mag}(\vec{r}_P - \vec{r}_C)} \tag{E2.6}$$

is a vector of unit magnitude that points in the direction "away from C."

Exercise E2X.2

Take the magnitude of both sides of equation E2.6 and verify that $\hat{r}_{PC}$ has a magnitude of 1 (with no units).

Using this directional, we can express equation E2.5 in the compact form

$$\vec{F}_e = \frac{kq_{\text{test}}q}{r_{PC}^2}\hat{r}_{PC} \tag{E2.7}$$

Note that if $q_{\text{test}}q > 0$, this vector points in the direction $\hat{r}_{PC}$ = "away from C" and if $q_{\text{test}}q < 0$, it points in the direction $-\hat{r}_{PC}$ = "toward C."

Equation E2.2 then implies the electric field vector at point P due to q is

$$\vec{E}\text{ (at }P) \equiv \frac{\vec{F}_e}{q_{\text{test}}} = \frac{1}{q_{\text{test}}}\frac{kq_{\text{test}}q}{r_{PC}^2}\hat{r}_{PC} = \frac{kq}{r_{PC}^2}\hat{r}_{PC} \tag{E2.8}$$

Note how dividing by q_{test} has erased all references to q_{test}! Therefore, we see that

The formula for the electric field of a point charge

$$\vec{E} = \frac{kq}{r_{PC}^2}\hat{r}_{PC} = \frac{1}{4\pi\varepsilon_0}\frac{q}{r_{PC}^2}\hat{r}_{PC} \tag{E2.9a}$$

Purpose: This equation describes the electric field vector $\vec{E}$ at an arbitrary point P created by a point charge q located at point C.
Symbols: $\vec{r}_{PC} \equiv \vec{r}_P - \vec{r}_C$ is the position of point P relative to C; $r_{PC} = \text{mag}(\vec{r}_{PC})$; and $\hat{r}_{PC} \equiv \vec{r}_{PC}/r_{PC}$ is a directional (unit vector) standing for the direction at P that is "away from the charge q." $\varepsilon_0 \equiv 1/4\pi k$.
Limitations: The charge q must be a point particle at rest.
Note: $\vec{E}$ points away from q if $q > 0$ and *toward* q if $q < 0$.

About the constant ε_0

For historical reasons, physicists often write the Coulomb constant in terms of a constant ε_0 (the *permittivity* constant) where $\varepsilon_0 \equiv 1/4\pi k$. I do not like this, partly because it obscures useful connections between electrostatics and both gravity and magnetism, and partly because k has an easily remembered magnitude (9×10^9) in SI units. Therefore, I will use k in all formulas and calculations in this text. But to honor the convention, I will express the most important basic formulas (such as equation E2.9*a*) in terms of both k and ε_0. In any other formula, one only has to substitute $k = 1/4\pi\varepsilon_0$ to reexpress that formula in terms of ε_0.

Displaying explicitly how this vector function depends on x, y, and z

Equation E2.9*a* describes a vector *function* for the electric field in the sense that it tells us how to compute the field vector at an *arbitrary* point P. Figure E2.2 illustrates the total field created by positive and negative point

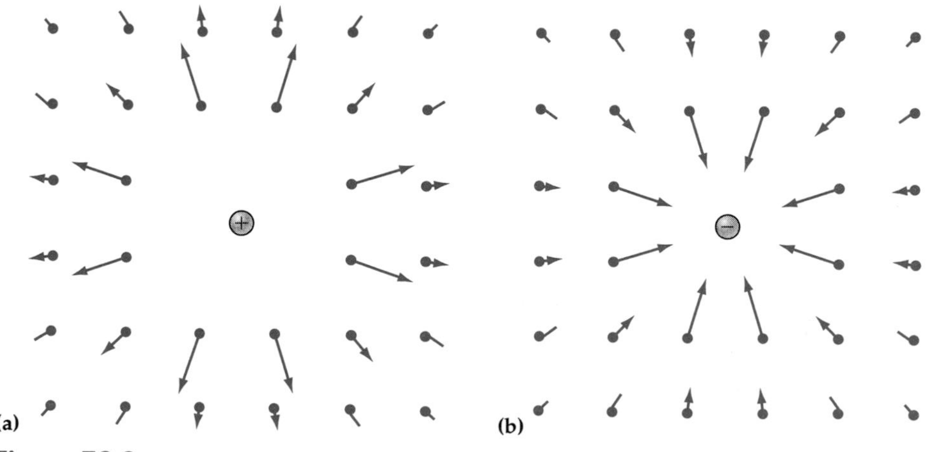

Figure E2.2
These drawings illustrate the electric fields of a positive point charge and a negative point charge. The arrow at a point represents the magnitude and direction of the electric field at that point (the arrowhead is omitted if the vector is too short.) Note that the vectors are longest near the charge; the magnitudes of the field vectors fall off as the inverse square of the distance from the charge.

charges by showing what the field vectors look like at a sampling of points surrounding the point charge. We call such a picture a **field diagram.**

If we want to display $\vec{E}(x, y, z)$ *explicitly* as a vector *function* of the point P's position coordinates x, y, and z, it helps to use the definition of $\hat{r}_{PC}$ to write equation E2.9*a* in the form

$$\vec{E} = \frac{kq}{r_{PC}^2}\hat{r}_{PC} = \frac{kq}{r_{PC}^2}\frac{\vec{r}_{PC}}{r_{PC}} = \frac{kq}{r_{PC}^3}\vec{r}_{PC} \tag{E2.9b}$$

Equation E2.9*a* has the advantage of clearly displaying that the magnitude of $\vec{E}$ at P depends on $1/r_{PC}^2$, but the form given by equation E2.9*b* is often more practical to use when we actually want to calculate $\vec{E}$. Since $\vec{r}_{PC} = \vec{r}_P - \vec{r}_C = [x - x_C, y - y_C, z - z_C]$,

$$\vec{E}(x, y, z) = \frac{kq}{r_{PC}^3}\vec{r}_{PC} = \frac{kq}{[(x - x_C)^2 + (y - y_C)^2 + (z - z_C)^2]^{3/2}}\begin{bmatrix} x - x_C \\ y - y_C \\ z - z_C \end{bmatrix} \tag{E2.10}$$

This makes it clear that equation E2.9 is indeed a vector *function* that implicitly defines the point charge's entire electric field.

E2.4 The Superposition Principle

According to the *superposition principle* discussed in chapter E1, the net electric *force* $\vec{F}_e$ exerted on a test charge q_{test} at an arbitrary point P by a set of other charges $q_1, q_2, q_3, \ldots$ is the *sum* of the forces that each individually exerts on q_{test}:

$$\begin{aligned} \vec{F}_e &= \vec{F}_1 + \vec{F}_2 + \vec{F}_3 + \cdots \\ &= \frac{kq_{\text{test}}q_1}{r_{P1}^2}\hat{r}_{P1} + \frac{kq_{\text{test}}q_2}{r_{P2}^2}\hat{r}_{P2} + \frac{kq_{\text{test}}q_3}{r_{P3}^2}\hat{r}_{P3} + \cdots \\ &= q_{\text{test}}\left(\frac{kq_1}{r_{P1}^2}\hat{r}_{P1} + \frac{kq_2}{r_{P2}^2}\hat{r}_{P2} + \frac{kq_3}{r_{P3}^2}\hat{r}_{P3} + \cdots\right) \end{aligned} \tag{E2.11}$$

Note that this $\vec{F}_e$ is always proportional to q_{test} (as I asserted in section E2.1). According to our definition of electric field, the net electric field vector $\vec{E}$ at an arbitrary point P is thus

$$\vec{E} \equiv \frac{\vec{F}_e}{q_{\text{test}}} = \frac{kq_1}{r_{P1}^2}\hat{r}_{P1} + \frac{kq_2}{r_{P2}^2}\hat{r}_{P2} + \cdots$$

But this is just the vector sum of the electric field vectors that charges $q_1, q_2, \ldots$ would individually produce at point P. Therefore

$$\vec{E} = \vec{E}_1 + \vec{E}_2 + \cdots \tag{E2.12}$$

The superposition principle for the fields of point charges

Purpose: This equation describes how electric fields combine.
Symbols: $\vec{E}$ is the total electric field at an arbitrary point P; $\vec{E}_1$ is the electric field vector produced at P by particle q_1 alone; $\vec{E}_2$ is the same for q_2 alone; and so on.
Limitations: None are known!

A charge does not exert a force on itself

Note that if we place a point charge q_0 at point P, the force it experiences is $\vec{F}_e = q_0\vec{E}$, where $\vec{E}$ is the total electric field created at P by charges *other* than q_0. No matter how a static point charge's field might contribute to the total electric field elsewhere, that charge's field never exerts a force on the charge itself.

Example E2.1

Problem Calculate the magnitude and directions of the field vectors at points A and B in figure E2.3, expressing your results in terms of k, q, and d.

Model Assume that the charges are point particles. We can then use equation E2.9*b* to compute the electric field created by each point particle at the point in question, and add the field vectors (as *vectors*) to find the total field vector at that point.

Solution In the given coordinate system, point A is a distance d in the $-x$ direction from q_1 and $2d$ in the same direction from q_2, so $\vec{r}_{A1} = [-d, 0, 0]$ and $\vec{r}_{A2} = [-2d, 0, 0]$. (We could have *calculated* $\vec{r}_{A1}$ using $\vec{r}_{A1} = \vec{r}_A - \vec{r}_1$, but this is actually more work in this case.) So

$$\vec{E}_A = \frac{kq_1}{r_{A1}^3}\vec{r}_{A1} + \frac{kq_2}{r_{A2}^3}\vec{r}_{A2} = \frac{kq}{d^3}\begin{bmatrix}-d\\0\\0\end{bmatrix} + \frac{k(-q)}{(2d)^3}\begin{bmatrix}-2d\\0\\0\end{bmatrix}$$

$$= \frac{kq}{d^2}\begin{bmatrix}-1\\0\\0\end{bmatrix} + \frac{kq}{4d^2}\begin{bmatrix}1\\0\\0\end{bmatrix} = \frac{kq}{d^2}\begin{bmatrix}-1+\frac{1}{4}\\0\\0\end{bmatrix} = \frac{3kq}{4d^2}\begin{bmatrix}-1\\0\\0\end{bmatrix} \qquad \text{(E2.13)}$$

So $\vec{E}_A$ points in the $-x$ direction and has a magnitude of $E_A = 3kq/4d^2$.

Similarly, we see that $\vec{r}_{B1} = [+\frac{1}{2}d, d, 0]$ and $\vec{r}_{B2} = [-\frac{1}{2}d, d, 0]$. Note that $r_{B1} = r_{B2} = [(\pm\frac{1}{2}d)^2 + d^2 + 0]^{1/2} = [(5/4)d^2]^{1/2} = (5/4)^{1/2}d$. Therefore,

$$\vec{E}_B = \frac{kq_1}{r_{B1}^3}\vec{r}_{B1} + \frac{kq_2}{r_{B2}^3}\vec{r}_{B2} = \frac{kq}{(5/4)^{3/2}d^3}\begin{bmatrix}+\frac{1}{2}d\\d\\0\end{bmatrix} + \frac{k(-q)}{(5/4)^{3/2}d^3}\begin{bmatrix}-\frac{1}{2}d\\d\\0\end{bmatrix}$$

$$= \frac{kq}{(5/4)^{3/2}d^3}\begin{bmatrix}\frac{1}{2}d+\frac{1}{2}d\\d-d\\0\end{bmatrix} = \left(\frac{4}{5}\right)^{3/2}\frac{kq}{d^2}\begin{bmatrix}1\\0\\0\end{bmatrix} = 0.72\frac{kq}{d^2}\begin{bmatrix}1\\0\\0\end{bmatrix} \qquad \text{(E2.14)}$$

So $\vec{E}_B = 0.72kq^2/d^2$ in the $+x$ direction.

Evaluation Both results have directions consistent with those obtained by *qualitatively sketching* the sum of the field vectors, as shown in figure E2.3.

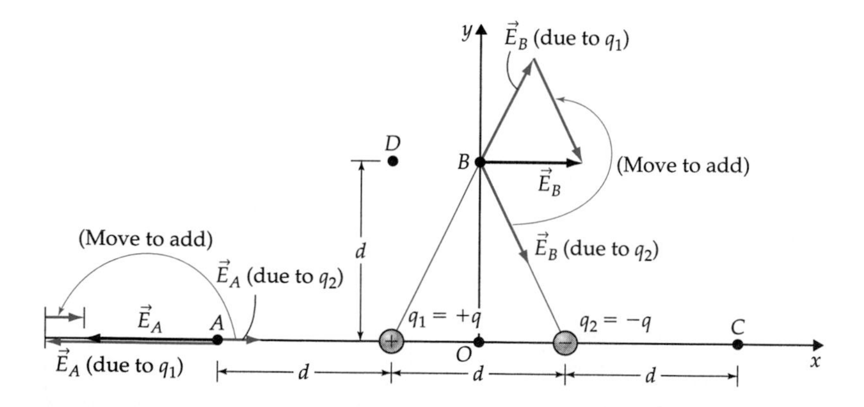

Figure E2.3
This drawing illustrates how we can apply the superposition principle to sketch the electric field vectors near a simple set of charges (an electric dipole in this case). To do the sketch, we have to estimate the relative magnitudes of the electric field vectors created at the point in question by each individual charge in the set of charges, then add them as vectors to get the total electric field.

Exercise E2X.3

Calculate the electric field vectors at points C and D on figure E2.3 and check your calculation by sketching arrows on the diagram.

Exercise E2X.4

Consider two point charges of arbitrary sign and (nonzero) magnitude. Explain why it is *impossible* for $\vec{E}$ to be zero at any point (not at infinity) that does not lie somewhere on the line connecting the two charges.

E2.5 The Field of a Dipole

What the field of an electric dipole looks like

We saw in chapter E1 that an *electric dipole* is a pair of particles with equal and opposite charges separated by a small distance d. The charges shown in figure E2.3 comprise a dipole, and in example E2.1 we evaluated this dipole's electric field at a few selected points. Figure E2.4 shows a more complete field diagram of a dipole's electric field. (The computer program that drew this figure calculated each field arrow essentially as we did in example E2.1.) Note how the electric field vectors point generally away from the positive charge and toward the negative charge; and at points on the plane midway between the charges, the electric field vectors point parallel to the line connecting the charges.

A dipole's field decreases with distance more rapidly than that of a point charge

Since a dipole *as a whole* is electrically neutral, it may be surprising that it has an electric field at all, but the slight *separation* of charge means that it does. Even so, the magnitude of a dipole's electric field does decrease with distance much more rapidly than that of a charged particle.

We can demonstrate this most easily at an arbitrary point P lying on the dipole's axis, as shown in figure E2.5. The net electric field $\vec{E}$ at P is the vector sum of the field vectors $\vec{E}_+$ and $\vec{E}_-$ created at point P by the dipole's positive

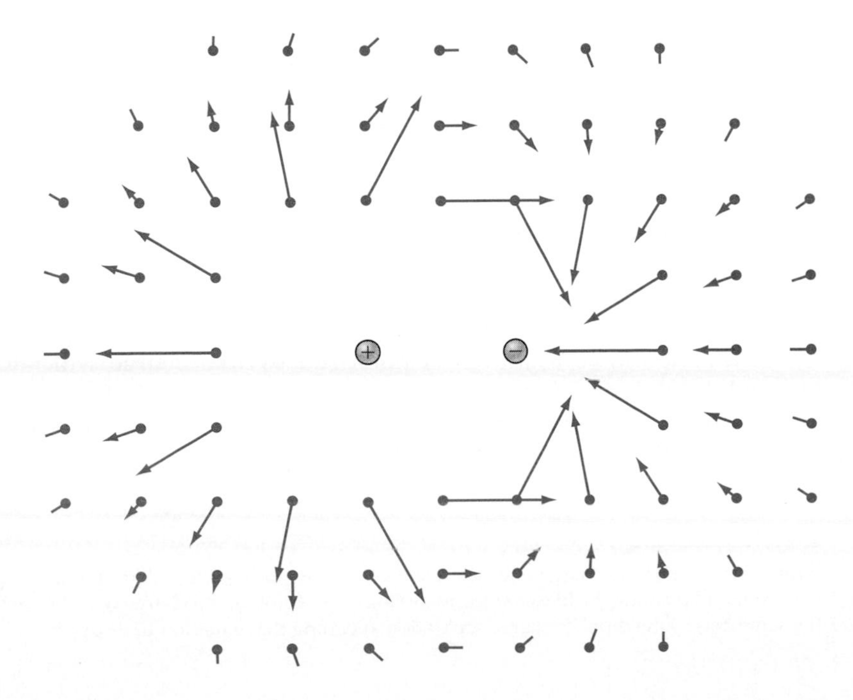

Figure E2.4
A field diagram of the electric field of a dipole. (As usual, the arrowhead on an electric field vector is not shown if the field vector is too short.)

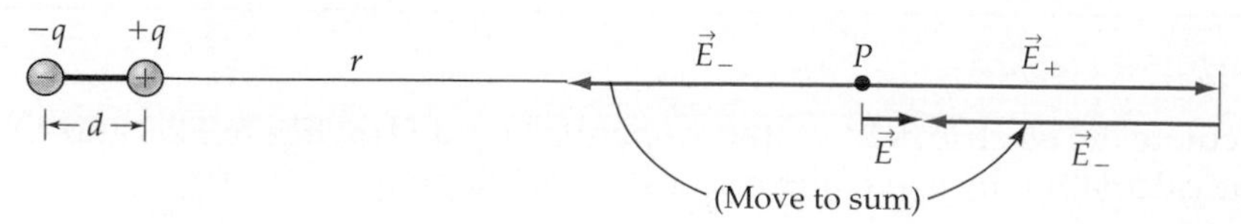

Figure E2.5
A diagram illustrating how we can calculate the electric field of a dipole at a point P along that dipole's axis.

and negative charges, respectively. In this particular case, $\vec{E}$ is nonzero because the field vector due to the more distant charge is not quite as long as that due to the nearer charge. Specifically, if we take the x direction to be the direction of point P relative to either charge and place the positive charge at the origin, then

$$\vec{E}(r,0,0) = \frac{kq}{r^2}\hat{x} + \frac{k(-q)}{(r+d)^2}\hat{x} = \frac{kq}{r^2}\left[1 - \frac{1}{(1+d/r)^2}\right]\begin{bmatrix}1\\0\\0\end{bmatrix} \quad \text{(E2.15)}$$

Now, in the limit that $r \gg d$ (that is, the limit that point P is very far from the dipole compared to the separation of the dipole's charges), $d/r \ll 1$, so we can use the binomial approximation to simplify the quantity in square brackets. Since $(1+x)^{-2} \approx 1 - 2x$ when $x \ll 1$, we see that

A dipole's electric field strength at points along the dipole's axis

$$\vec{E}(r,0,0) \approx \frac{kq}{r^2}\left[1 - \left(1 - \frac{2d}{r}\right)\right]\begin{bmatrix}1\\0\\0\end{bmatrix} = \frac{2kqd}{r^3}\begin{bmatrix}1\\0\\0\end{bmatrix} \quad (\text{when } r \gg d) \quad \text{(E2.16)}$$

Calculating $\vec{E}$ at an arbitrary point P is more complicated, but a more general calculation (see problem E2A.1) shows that along *any* line going through the dipole's center, the magnitude of $\vec{E}$ is proportional to kqd/r^3 (in the limit that $r \gg d$), just as it is along the x axis. Thus the magnitude of a dipole's electric field does indeed decrease *more rapidly* with increasing r than that of a point charge (as the latter only falls off as $1/r^2$). Since $\vec{F}_e = q\vec{E}$, this also means that the electric *force* that a dipole exerts on a point charge decreases as $1/r^3$.

The force exerted by a point charge on a small neutral object

As we discussed in chapter E1, the atoms in a typical substance become *polarized* in the presence of an external electric charge. Since the separating forces that the external charge's field exerts on an atom's nucleus and electrons are proportional to that external field, the separation d between the charges in an atom also turns out to be generally proportional to mag($\vec{E}_{\text{ext}}$) (see problem E2R.1). Since the magnitude of the electric field of a point charge decreases as $1/r^2$, equation E2.16† and $d \propto E_{\text{ext}} \propto 1/r^2$ imply that the magnitude of the electrostatic force exerted on an external point charge by an atomic dipole in an object polarized by that charge will fall off as $(1/r^2)(1/r^3) = 1/r^5$. If the polarized object is small enough that all its atoms are essentially the same distance r from the external charge and exert forces on that charge that all point in essentially the same direction, then the *net*

†Note that the external charge will polarize the atom by pulling the unlike charges in the atom toward it and pushing the like charges away from it, so the external charge will always lie along the same axis as the dipole's charge separation, meaning that equation E2.16 applies.

force that the polarized object exerts the external charge (and by Newton's third law, the force the charge exerts on the object) will also be proportional to $1/r^5$, as claimed in chapter E1. This is why one usually has to get a charged object quite close to a neutral object to observe much attraction.

E2.6 Handling Charge Distributions

A technique for computing the field of macroscopic charge distributions

Any macroscopic charged object involves a vast number of charged particles (usually electrons) distributed in some way over the object's surface or throughout its volume. We often cannot model such a **charge distribution** as a point charge, but *in principle* we can calculate the electric field vector it creates at any given point P in space by summing the field vectors that each of its actual charged particles creates at P. *In practice,* though, the huge number of charged particles in any macroscopic object makes doing this sum hopelessly impractical.

A more practical way to do this sum is to divide the charge distribution into a large number of tiny "bits," each of which contains a large number of charged particles but is small enough to model as a point charge. We can then use equation E2.9 to calculate the field vector $\vec{E}_b$ that each bit contributes at the point P and sum over all bits. One could easily use this approach to write a computer program to compute the electric field at a point near an arbitrary charge distribution.

We can learn even more in the special cases where we can convert the sum over the bits to a tractable *integral.* When it can be done, this process yields an essentially exact and *symbolic* result for the total electric field vector, which is generally much more informative than a numerical sum. Examples E2.2 and E2.3 illustrate the process.

Example E2.2

Problem Consider a very thin ring of radius r carrying a uniformly distributed positive total charge Q (meaning that all equal-length segments of the ring have the same charge). What is the electric field at a point P a given distance z from the center of the ring along the ring's central axis?

Translation Figure E2.6 defines an appropriate coordinate system and some useful symbols.

Model We can describe the position of each bit of the ring in terms of the angle ϕ that a line connecting the ring's center with that bit makes with the x axis. The infinitesimal length of each bit is then $dL = r\,d\phi$.

Solution If the charge Q on the ring is uniformly distributed, then the charge q_b of a bit of ring of infinitesimal length dL will be to the total charge Q as the bit's length dL is to ring's total circumference $2\pi r$. Therefore,

$$\frac{q_b}{Q} = \frac{dL}{2\pi r} = \frac{r\,d\phi}{2\pi r} = \frac{d\phi}{2\pi} \quad \Rightarrow \quad q_b = Q\frac{d\phi}{2\pi} \tag{E2.17}$$

Note that the distance between each bit on our ring and the point P has the fixed value $R \equiv (r^2 + z^2)^{1/2}$. According to equation E2.9 and the geometry of figure E2.6, the electric field vector $\vec{E}_b$ contributed by the bit located

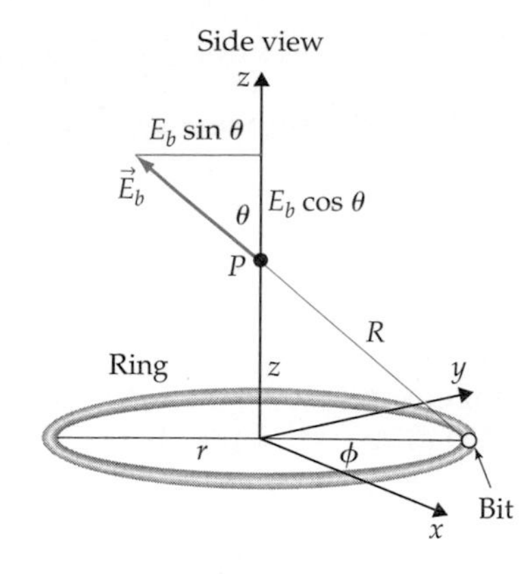

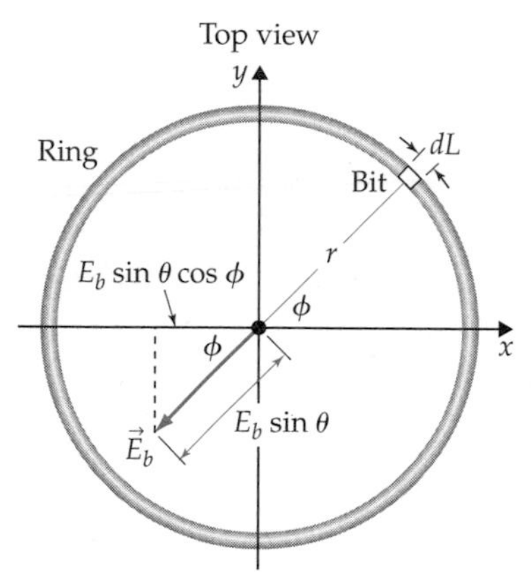

Figure E2.6
A side view and a top view of a uniformly charged circular ring.

at angle ϕ is

$$\vec{E}_b = E_b \begin{bmatrix} -\sin\theta\cos\phi \\ -\sin\theta\sin\phi \\ \cos\theta \end{bmatrix} \qquad \text{where } E_b = \frac{kq_b}{R^2} = \frac{kQ\,d\phi}{R^2\,2\pi} = \frac{kQ}{2\pi R^2}\,d\phi \tag{E2.18}$$

Note that $\sin\theta$ and $\cos\theta$ have the fixed values $\sin\theta = r/R$ and $\cos\theta = z/R$ in this case. Plugging these results into equation E2.18, summing $\vec{E}_b$ over all bits in the ring, and converting the sum to an integral over ϕ from 0 to 2π, we get

$$\vec{E} = \sum_{\text{all bits}} \vec{E}_b = \begin{bmatrix} \sum_{\text{all bits}} -\dfrac{kQr}{2\pi R^3}\cos\phi\,d\phi \\ \sum_{\text{all bits}} -\dfrac{kQr}{2\pi R^3}\sin\phi\,d\phi \\ \sum_{\text{all bits}} \dfrac{kQz}{2\pi R^3}\,d\phi \end{bmatrix} \to \begin{bmatrix} -\dfrac{kQr}{2\pi R^3}\displaystyle\int_0^{2\pi}\cos\phi\,d\phi \\ -\dfrac{kQr}{2\pi R^3}\displaystyle\int_0^{2\pi}\sin\phi\,d\phi \\ \dfrac{kQz}{2\pi R^3}\displaystyle\int_0^{2\pi}d\phi \end{bmatrix} \tag{E2.19}$$

These integrals are pretty easy:

$$\int_0^{2\pi}\cos\phi\,d\phi = \sin\phi|_0^{2\pi} = \sin 2\pi - \sin 0 = 0 - 0 = 0 \tag{E2.20a}$$

$$\int_0^{2\pi}\sin\phi\,d\phi = -\cos\phi|_0^{2\pi} = -\cos 2\pi + \cos 0 = -1 + 1 = 0 \tag{E2.20b}$$

$$\int_0^{2\pi}d\phi = \phi|_0^{2\pi} = 2\pi - 0 = 2\pi \tag{E2.20c}$$

So the electric field at point P a distance z from the center of the ring along its central axis is simply

$$\vec{E}(0,0,z) = \frac{kQz}{2\pi R^3}\begin{bmatrix} 0 \\ 0 \\ 2\pi \end{bmatrix} = \frac{kQz}{R^3}\begin{bmatrix} 0 \\ 0 \\ 1 \end{bmatrix} = \frac{kQz}{(r^2+z^2)^{3/2}}\begin{bmatrix} 0 \\ 0 \\ 1 \end{bmatrix} \tag{E2.21}$$

Evaluation It also makes sense that the horizontal components should add to zero: because of the ring's symmetry, for every bit one side of the ring that contributes a rightward component (say) to the sum, there is a bit on the opposite side that contributes an equal leftward component. You can easily check that the units of $\vec{E}$ in equation E2.21 are correct as well. (See problem E2S.7 for some other checks.)

Example E2.3

Problem Imagine a straight, very thin wire of length L that has a uniformly distributed charge Q. What is the electric field vector produced by this wire at an arbitrary point P a perpendicular distance r from the center of the wire?

Translation The first step is to set up a useful reference frame. Let us choose the x axis to coincide with the wire and set $x = 0$ to be the point on the axis closest to P. Let us also take advantage of our freedom to choose coordinates

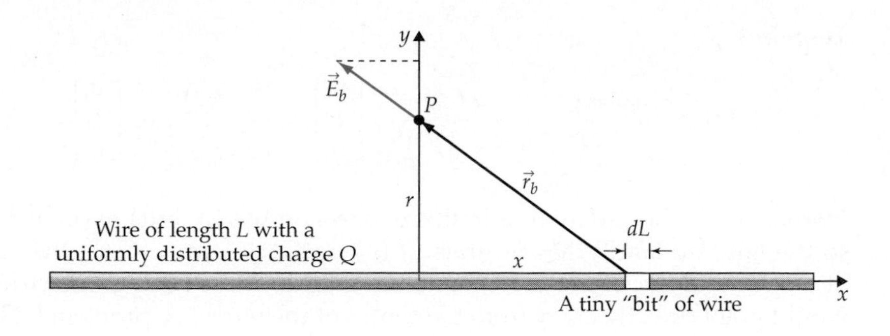

Figure E2.7
An appropriate coordinate system and useful symbols for evaluating the electric field vector produced at point P by a given tiny segment of a uniformly charged wire.

to choose our y axis so that it goes through the point P. Figure E2.7 illustrates the situation and defines some appropriate symbols.

Model We can express each bit's position in terms of its coordinate x on the x-axis and its size in terms of the increment dx in that variable.

Solution If charge is uniformly distributed on the wire, then the charge q_b of a bit is to the wire's charge Q what the bit's length dx is to the wire's length L. Therefore

$$\frac{q_b}{Q} = \frac{dx}{L} \quad \Rightarrow \quad q_b = \frac{Q}{L}\,dx \tag{E2.22}$$

The vector that specifies the position of point P relative to the bit of charge at position x is

$$\vec{r}_{Pb} = \vec{r}_P - \vec{r}_b = \begin{bmatrix} 0 \\ r \\ 0 \end{bmatrix} - \begin{bmatrix} x \\ 0 \\ 0 \end{bmatrix} = \begin{bmatrix} -x \\ r \\ 0 \end{bmatrix} \tag{E2.23}$$

Therefore, if we use equation E2.9*b* to compute the electric field created by each bit and sum over all bits, we get

$$\vec{E}(0, r, 0) = \sum_{\text{all bits}} \vec{E}_b = \sum_{\text{all bits}} \frac{kq_b}{r_{Pb}^3}\vec{r}_{Pb} = \sum_{\text{all bits}} \frac{k(Q/L)\,dx}{(r^2+x^2)^{3/2}} \begin{bmatrix} -x \\ r \\ 0 \end{bmatrix}$$

$$= \frac{kQ}{L} \begin{bmatrix} \displaystyle\sum_{\text{all bits}} \frac{-x\,dx}{(r^2+x^2)^{3/2}} \\ \displaystyle\sum_{\text{all bits}} \frac{r\,dx}{(r^2+x^2)^{3/2}} \\ 0 \end{bmatrix} \to \frac{kQ}{L} \begin{bmatrix} \displaystyle\int_{-\frac{1}{2}L}^{+\frac{1}{2}L} \frac{-x\,dx}{(r^2+x^2)^{3/2}} \\ \displaystyle r\int_{-\frac{1}{2}L}^{+\frac{1}{2}L} \frac{dx}{(r^2+x^2)^{3/2}} \\ 0 \end{bmatrix} \tag{E2.24}$$

These integrals are easy to look up in a table of integrals (see also the inside front cover). The results are

$$\int_{-\frac{1}{2}L}^{+\frac{1}{2}L} \frac{-x\,dx}{(r^2+x^2)^{3/2}} = \frac{1}{(r^2+x^2)^{1/2}}\Bigg|_{-\frac{1}{2}L}^{+\frac{1}{2}L} = \frac{1-1}{\left(r^2+\frac{1}{4}L^2\right)^{1/2}} = 0 \tag{E2.25a}$$

$$\int_{-\frac{1}{2}L}^{+\frac{1}{2}L} \frac{dx}{(r^2+x^2)^{3/2}} = \frac{x}{r^2(r^2+x^2)^{1/2}}\Bigg|_{-\frac{1}{2}L}^{+\frac{1}{2}L}$$

$$= \frac{\frac{1}{2}L - (-\frac{1}{2}L)}{r^2\left(r^2+\frac{1}{4}L^2\right)^{1/2}} = \frac{L}{r^2\left(r^2+\frac{1}{4}L^2\right)^{1/2}} \tag{E2.25b}$$

Therefore

$$\vec{E}(0, r, 0) = \frac{kQ}{L} \frac{rL}{r^2\left(r^2 + \frac{1}{4}L^2\right)^{1/2}} \begin{bmatrix} 0 \\ 1 \\ 0 \end{bmatrix} = \frac{kQ}{r\sqrt{r^2 + \frac{1}{4}L^2}} \begin{bmatrix} 0 \\ 1 \\ 0 \end{bmatrix} \tag{E2.26}$$

Evaluation The square root in this expression has SI units of $(\text{m}^2)^{1/2} = \text{m}$, so the entire quantity has SI units of $(\text{N} \cdot \text{m}^2\text{C}^{-2})\text{C}/\text{m}^2 = \text{N/C}$, the correct SI units for an electric field. It also makes intuitive sense that the electric field would point directly away from the center of the wire. See problem E2S.8 for some other checks.

Summary of the crucial steps in the technique

The point of these examples is not the *results* we get (the fields of uniformly charged rings and wires are rarely of practical interest) but rather the *method*. A summary of the crucial steps might look like this:

1. Choose a variable (θ in example E2.2, x in example E2.3) you can use to describe everything you need to know about the position and size of each bit.
2. Express the bit's charge q_b in terms of the variable and the infinitesimal change in that variable (see equations E2.17 and E2.22).
3. Express the position $\vec{r}_{Pb}$ of the point P relative to the bit in terms of that variable (see equation E2.23).
4. Use equation E2.9*a* or E2.9*b* to find the components of the electric field $\vec{E}_b$ created by each bit at point P (see equations E2.18 and E2.24).
5. Convert the sum over all bits to an integral over the variable, and evaluate to find the total electric field $\vec{E}$ (see equations E2.19 and E2.24).

Problems E2S.9 through E2S.12 will give you some practice in using this technique, which is valuable in a number of areas of physics.

TWO-MINUTE PROBLEMS

E2T.1 The electric field created by a negative point charge at any given position points in what direction?
A. Toward the charge.
B. Away from the charge.
C. The direction depends on how close you are.
D. The direction depends on the sign of k.

E2T.2 A negative point charge placed at point P experiences an electrostatic force in the $+x$ direction. The electric field $\vec{E}$ at P must point in the $-x$ direction, true (T) or false (F)?

E2T.3 Imagine that we place two particles a certain distance apart. The electric field at some point on the line *between* the particles is zero. What is the most general thing we can say about the (otherwise unknown) signs and magnitudes of the particles' charges? The charges have
A. The same sign and magnitude.
B. Opposite signs but the same magnitude.
C. The same sign, but may have different magnitudes.
D. Opposite signs, but may have different magnitudes.
E. Zero magnitude necessarily (both of them).
F. Other (specify).
T. We can say nothing about the signs or magnitudes.

E2T.4 Imagine that we place two particles a certain distance apart. The electric field is zero at some point that is *not* on the line connecting them. What is the most general thing we can say about the signs and magnitudes of the particles' charges? The charges have . . . (choose an answer from the list for problem E2T.3).

E2T.5 Imagine that we place two particles a certain distance apart. The electric field at some point on a line connecting the particles (but *not* between them) is zero. What is the most general thing we can say about the signs and magnitudes of the particles' charges? The charges have

A. The same sign and magnitude.
B. Opposite signs but the same magnitude.
C. The same sign, but *must* have different magnitudes.
D. Opposite signs, but *must* have different magnitudes.
E. Zero magnitude necessarily (both of them).
F. Other (specify).
T. We can say nothing about the signs or magnitudes.

E2T.6 The electric field vector created by a dipole at an arbitrary point near the dipole is parallel to that dipole's axis, T or F?

E2T.7 The charges in an electrostatically polarized atom separate along an axis parallel to the electric field $\vec{E}$ that polarizes it, T or F?

E2T.8 Imagine two dipoles oriented as shown below. These dipoles will

A. Attract each other.
B. Repel each other.
C. Exert zero force on each other.

E2T.9 Imagine two dipoles oriented as shown below. These dipoles will

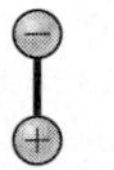

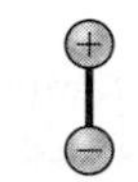

A. Attract each other.
B. Repel each other.
C. Exert zero force on each other.

E2T.10 Consider the finite piece of wire with a uniformly distributed *negative* charge shown below. What is the field direction at the point *P*? (Pick the vector that you think will be closest.) What about at point *S*?

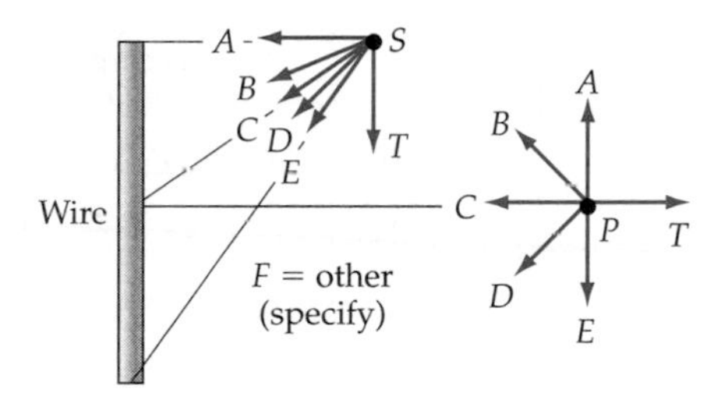

HOMEWORK PROBLEMS

Basic Skills

E2B.1 Imagine that when you place a point test charge with a magnitude of 1.0 nC at a certain point in space, you find that it experiences a net force of 5.0 μN. What is the magnitude of the electric field at this point?

E2B.2 Imagine that when you place a point test charge with a magnitude of 120 nC at a certain point in space, you find that it experiences a net force of 0.3 mN. What is the magnitude of the electric field at this point?

E2B.3 Imagine that at a certain time during a thunderstorm, the electric field near the earth's surface is measured to be 5200 N/C upward. If you place a Styrofoam ball with a charge of –10 nC in this field, what are the magnitude and direction of the electric force on the ball?

E2B.4 The maximum electric field that a Van de Graaff generator can produce at the surface of its upper sphere is about 3×10^6 N/C outward. If you place a Styrofoam ball with a charge of +11 nC on the surface of the upper sphere, what will be the magnitude and direction of the electric force on the ball?

E2B.5 If the magnitude of both charges in figure E2.8 is 15 nC and $d = 2.0$ cm, what is the magnitude of the electric field at point *A*?

E2B.6 What are the direction and magnitude of the electric field at the point marked *A* in figure E2.8 if $q = +11$ nC and $d = 1.0$ cm?

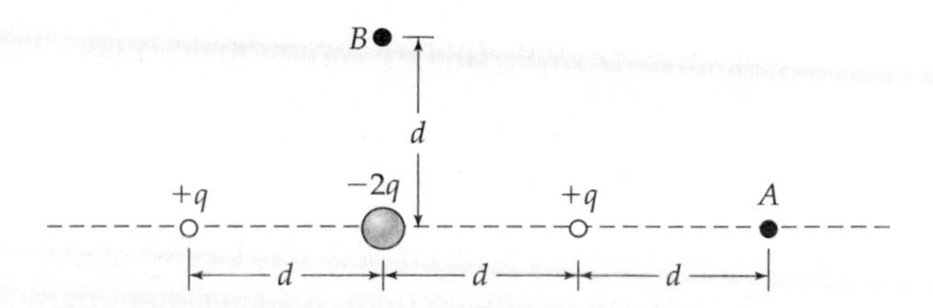

Figure E2.8
The charge arrangement discussed in problems E2B.6 and E2B.7.

E2B.7 What are the direction and magnitude of the electric field at the point marked B in figure E2.8 if $q = +11$ nC and $d = 1.0$ cm?

E2B.8 A stream of water can be noticeably deflected by a charged object. Imagine that a friend, upon seeing the demonstration, exclaims, "But this cannot be! The water molecules in the stream are certainly randomly oriented, so as many will be repelled by the object as attracted to it. Therefore, the net force exerted on the stream *must* be zero." Courteously correct the error in your friend's reasoning.

E2B.9 Calculate the components of the electric field at the point P shown in figure E2.9. Assume that the charge Q is uniformly distributed over the wire. (*Hint:* What would the limits of integration in equation E2.24 be in this case?)

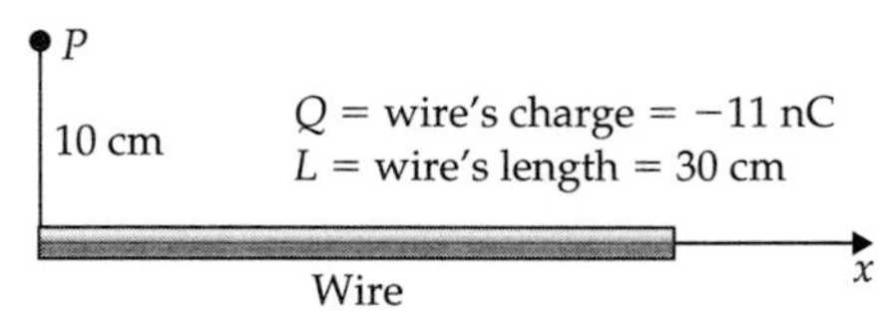

Figure E2.9
(See problem E2B.9.)

Synthetic

E2S.1 If the magnitude of both charges shown in figure E2.3 is 15 nC and $d = 2.0$ cm, what are the magnitude and direction of the electric field at the point D, if D is a distance d from the nearest charge? (*Hint:* Break down the electric field vectors into components.)

E2S.2 A square with sides d has charges q, $2q$, $3q$, and $4q$ arranged clockwise around the corners of the square. What are the magnitude and direction of the electric field at the center of the square? (*Hints:* Consider *pairs* of charges on opposite sides of the square. Choose a coordinate system that makes finding the field vectors' components easy.)

E2S.3 A point particle with charge $-3q$ (where $q = 12$ nC) is located at the origin of a coordinate system. Another point particle with charge $+q$ is located at $x = 2.0$ cm. Find the location of any point or points along the x axis where the electric field is zero.

E2S.4 A point particle with charge $-q$ is located at the origin of a coordinate system. Another point particle with charge $+q$ is located a distance d away along the x axis. Argue (in words, but possibly with the help of a diagram) that the electric field of two point charges along the x axis could only possibly be zero at some point along the x axis. Then prove that the electric field *cannot* be zero anywhere along the x axis either in this particular case.

E2S.5 Consider dipoles placed as shown in figure E2.10. Will they attract, repel, or exert zero force on each other? If they do exert a force on each other, how does the magnitude of this force vary with r in the limit that $r \gg d$?

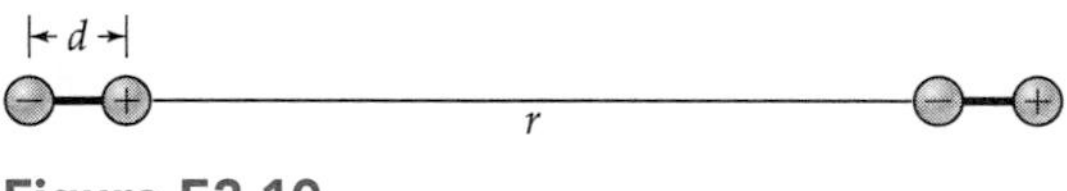

Figure E2.10
(See problem E2S.5.)

E2S.6 Consider dipoles placed as shown in figure E2.11. Will they attract, repel, or exert zero force each other? If they do exert a force on each other, how does the magnitude of this force vary with r in the limit that $r \gg d$?

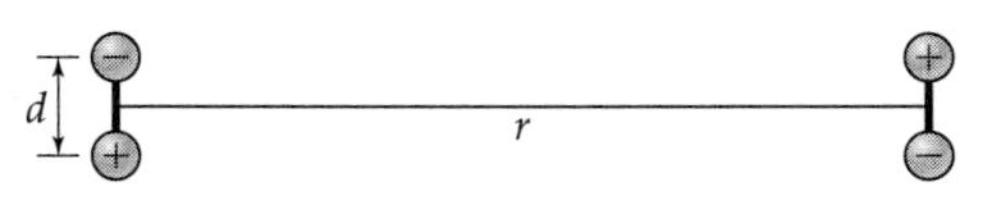

Figure E2.11
(See problem E2S.6.)

E2S.7 Consider equation E2.21, which specifies the electric field vector produced by a uniformly charged ring at any point along its central axis (the z axis).
(a) Verify that this equation has the correct units for an electric field vector.
(b) Intuitively, what do you think that the electric field ought to be at the exact center of a uniformly charged ring, and why? Check that equation E2.21 evaluated at this point yields the result predicted by your intuition.
(c) When z becomes extremely large, the ring's electric field ought to become approximately the same as that of a point charge, since the ring will become a tiny speck compared to z. Show that equation E2.21 is consistent with this expectation in the limit that $z \gg r$.

E2S.8 Consider equation E2.26, which specifies the electric field at points along the y axis produced by a uniformly charged thin wire lying along the x axis and centered on $x = 0$.
(a) When the distance r between the point and the wire's center becomes extremely large, the wire's electric field ought to become approximately the same as that of a point charge, since the wire will become a tiny speck compared to r.

Show that equation E2.26 is consistent with this expectation in the limit that $r \gg L$.

(b) To an observer at a point very close to the wire, the wire will stretch essentially to the horizon in both directions and will therefore look essentially as if it were infinite. Show that in the limit that $r \ll L$ equation E2.26 reduces to the field outside an infinite pipe described in table E3.1 in section E3.4. (Note that a sufficiently thin pipe will look like a wire.)

E2S.9 Consider the half-ring shown in figure E2.12. Assume that it carries a positive charge Q uniformly distributed along its length. Calculate the electric field vector at point P, using the five-step plan outlined at the end of section E2.6.

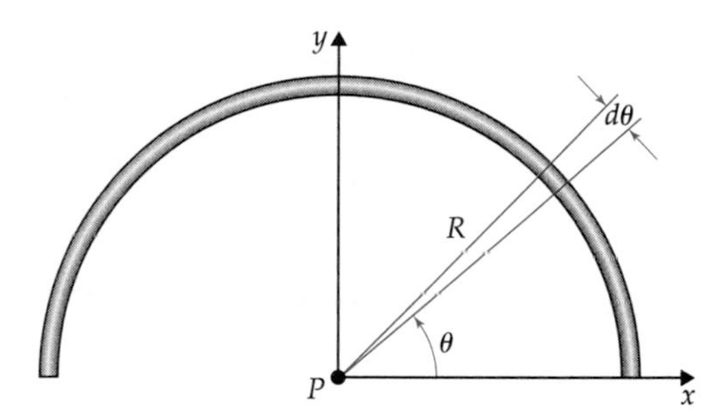

Figure E2.12
(See problem E2S.9.)

E2S.10 Calculate the electric field at a point on the x axis a distance r from one end of a uniformly charged wire of length L, using the five-step plan outlined at the end of section E2.6. (*Hint:* Think about using the variable $u \equiv r + x$. I also recommend defining $x = 0$ to be the end of the wire nearest to point P.)

E2S.11 Imagine that the charge on a straight wire of length L is not uniformly distributed, but the charge per unit length is bx, where b is some constant and $x = 0$ is the center of the wire. The charge on a bit is then $q_b = bx\,dL$. Use the five-step process described in section E2.6 to calculate the electric field of the wire at a point P a distance r from the center of the wire in this case. [*Hint:* Use x as the variable. My table of integrals tells me that if $r_b = (x^2 + r^2)^{1/2}$, then

$$\int \frac{x\,dx}{r_b^{3/2}} = -\frac{1}{r_b}$$
$$\int \frac{x^2\,dx}{r_b^{3/2}} = -\frac{1}{r_b} + \ln(x + r_b) \tag{E2.27}$$

You will probably find these integrals helpful.]

E2S.12 One can adapt the five-step method discussed at the end of section E2.6 for more complicated charge distributions if one relaxes the idea that a "bit" in the distribution has to be small enough to be like a point charge (so that we can use equation E2.9 to find its field). As long as we have *some* way to calculate the field of the bit, the bit can in fact be of any size or shape we want.

For example, imagine that we want to find the electric field at a point P along the central axis of a circular disk of radius R that has a uniformly distributed surface charge of Q. We know from equation E2.21 how to calculate the electric field of a ring centered on that axis, so let's divide the disk up into "bits" that are actually thin rings, as shown in figure E2.13. Calculate the total electric field due to the disk, using the previously mentioned five-step plan, except use equation E2.21 to calculate the electric field contributed by each bit (= ring) instead of equation E2.9. *Hints:* Using the variable $u \equiv r^2 + z^2$ as your integration variable makes the integral easier, although you can use r and look up the resulting integral in a table. You should find that

$$\vec{E} = 2\pi k\sigma \left[1 - \frac{1}{\sqrt{1 + (R/z)^2}}\right] \begin{bmatrix} 0 \\ 0 \\ 1 \end{bmatrix} \tag{E2.28}$$

where $\sigma \equiv Q/\pi R^2$ is the charge per unit area on the disk.

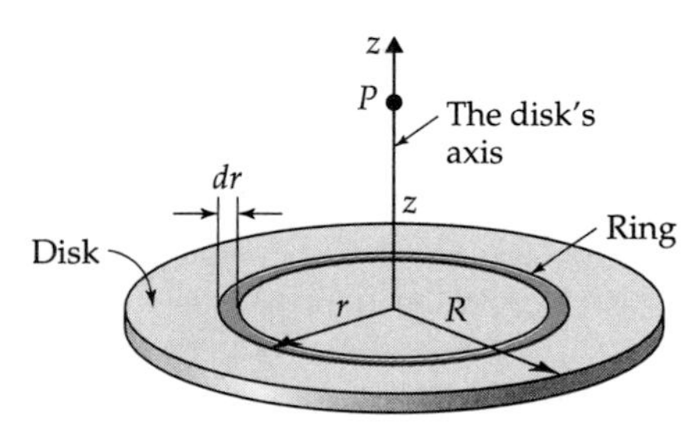

Figure E2.13
(See problem E2S.12.)

Rich-Context

E2R.1 We can very roughly estimate the strength of the polarization effect as follows. Consider an atom consisting of a negative electron cloud of approximate radius a bound to a positive, pointlike nucleus. The strength of the proton's electric field at the edge of the electron cloud is $E_0 \approx ke/a^2$, where e is the proton's charge. Now imagine that we apply external electric field that at the atom has some magnitude E_A. Let us *assume* that this field shifts the electron cloud by a distance d that is to the cloud radius a what the imposed field strength E_A

is to the proton's field strength E_0 at the outer edge of the cloud.

(a) Show that this means that

$$d \approx \frac{3V_{\text{atom}}}{4\pi k} E_A \qquad \text{(E2.29)}$$

where $V_{\text{atom}} = (4/3)\pi a^3$ is the atom's volume.

(b) Now, note from equation E2.16 that the electric field *created* by this polarized atom at a point a distance r from the atom along its polarization axis has a magnitude of $2ked/r^3$. Use this and equation E2.29 to show that the total electric field $\vec{E}_P$ created by the polarized atoms in a solid object of volume V at a point along a line parallel to their common polarization direction will have a magnitude of

$$E_P = \frac{3V}{2\pi r^3} E_A \qquad \text{(E2.30)}$$

assuming that the object is small enough that $\vec{E}_A$ is approximately constant in magnitude and direction throughout its volume. (*Hint:* The number of atoms in a solid object is $\approx V/V_{\text{atom}}$.)

Comment: The crude model discussed in this problem is better than it has any right to be: experiments show that the actual field is within a factor of 3 of that predicted by this model for a wide variety of solid substances. However, one has to be careful when using this result to, say, calculate the electric force between a point charge and a small polarizable object, because the applied electric field $\vec{E}_A$ that each atom in the object experiences is not just the electric field created by charges *outside* the object but that field superposed with the fields of nearby dipoles *inside* the object. This makes it trickier to calculate $\vec{E}_A$ in many circumstances.

E2R.2 As we will see in unit Q, light striking a metal plate can eject electrons from the plate's surface (this is called the *photoelectric effect*). Imagine that we place the plate in a vacuum and set up an electric field in the plate's vicinity that is uniform in magnitude and direction such that the electric field vector $\vec{E}$ at any point near the plate has a magnitude of 1000 N/C and points directly away from the plate, perpendicular to its surface. Electrons ejected by the photoelectric effect will have initial kinetic energies of no more than about 3 eV, where 1 eV $\equiv 1.6 \times 10^{-19}$ J. Roughly estimate the maximum distance that an ejected electron can get from the plate under these circumstances.

Advanced

E2A.1 Our task in this problem is to evaluate the electric field vector $\vec{E}$ of a dipole at an arbitrary point P far from the dipole compared to its charge separation d. Imagine that line drawn from the dipole's center to P makes an angle of θ with respect to a line drawn through the dipole charges in the direction from the negative charge to the positive charge. The situation is shown in figure E2.14. We assume that r is so large that $\phi \approx 0$, the lines from P to both charges are almost parallel, and the angles those lines make with the dipole's axis are both $\approx \theta$. We will describe the field vector $\vec{E}$ by stating its component E_r in the radial direction (that is, directly away from the dipole) and its component $E_\perp$ perpendicular to the radial direction to the left as we look toward P from the dipole. In this case, $\vec{E} \neq 0$ both because the contributions $\vec{E}_+$ and $\vec{E}_-$ do not point in the same direction and because they have different lengths.

(a) Argue that $E_\perp \approx kqd \sin\theta/r^3$, given our assumptions. (Be sure to check that this has the right sign for all θ.)

(b) Argue that $E_r \approx 2kqd \cos\theta/r^3$, given our assumptions. (Be sure to check that this has the right sign for all θ.)

(c) Find the magnitude of $\vec{E}$ in this limit, and argue that it is always proportional to kqd/r^3 for a given θ.

(d) Argue that these results are consistent with equation E2.16 and qualitatively consistent with figure E2.4.

(e) Argue that these results are enough to specify $\vec{E}$ at *any* point P (in three dimensions!) sufficiently far from the dipole.

E2A.2 Consider a dipole lying along the x axis with its positive charge at $x = +\frac{1}{2}d$ and its negative charge at $x = -\frac{1}{2}d$. Write down an exact expression for the dipole's field components $[E_x, E_y]$ in the xy plane as a function of position x, y. Then take the dot product of $\vec{E}$ with the radial unit vector $\hat{r} = [x/r, y/r]$ and the unit vector $\hat{e}_\perp = [-y/r, x/r]$ perpendicular to it to find the components E_r and $E_\perp$. Compare with the results of problem E2A.1 in the limit that $r \gg d$.

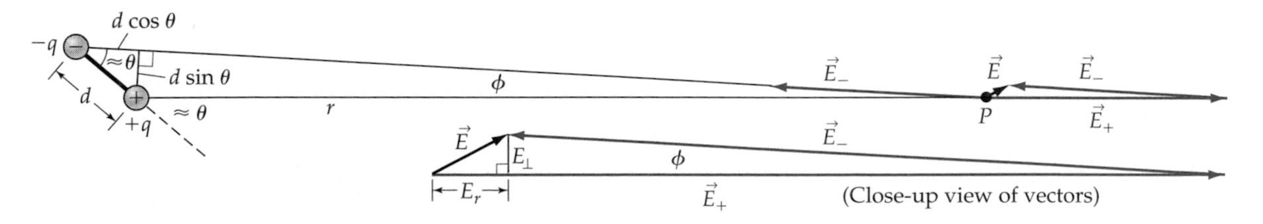

Figure E2.14
(See problem E2A.1.)

ANSWERS TO EXERCISES

E2X.1 For the units to be consistent on the right and left sides of equation E2.2, the electric field vector must have units of force over charge, or N/C in SI units.

E2X.2 We have

$$\text{mag}(\hat{r}_{PC}) = \text{mag}\left(\frac{\vec{r}_{PC}}{r_{PC}}\right) = \frac{\text{mag}(\vec{r}_{PC})}{r_{PC}} = \frac{r_{PC}}{r_{PC}} = 1 \quad \text{(E2.31)}$$

E2X.3 The vector $\vec{E}_C$ has the same magnitude and direction as $\vec{E}_A$. The vector $\vec{E}_D = [0.35kq/d^2, 0.64kq/d^2, 0]$, which is a vector about as long as $\vec{E}_A$ but oriented at an angle of about 30° to the right of vertical.

E2X.4 The vector $\vec{E}$ can be zero at a point P if and only if the two electric field vectors contributed by the two charges happen to cancel. If the two vectors are to cancel, they must lie on the same line. But since each of the two vectors lies on a line connecting P to the corresponding charge, it follows that the line connecting the first charge with P must be the same line as the line connecting the second charge with P. So the two charges and P must all lie on the same line.

E3 Electric Potential

▽ Electric Field Fundamentals
Electrostatics
Electric Fields
Electric Potential
▷ Controlling Currents
▷ Magnetic Field Fundamentals
▷ Calculating Static Fields
▷ Dynamic Fields

Chapter Overview

Introduction

We close the subdivision on static electric fields with a discussion of the *electric potential*, which provides an alternative way of describing the electric field created by a charge distribution and is essential background for the next subdivision.

Section E3.1: The Electric Potential

The **electric potential** ϕ at a point in space near a charge distribution is defined to be the *change in electrostatic potential energy per unit charge* that results when we move a charged test particle to that point from some reference point:

$$\phi(x, y, z) \equiv \frac{V_e(x, y, z)}{q_{\text{test}}} \tag{E3.2}$$

Purpose: This equation defines the electric potential ϕ at a point in space.
Symbols: $V_e(x, y, z)$ is the electrostatic potential energy, relative to some reference point, of a test particle with charge q_{test} placed at point $[x, y, z]$.
Limitations: The test particle charge q_{test} must be small enough that moving it does not significantly disturb the charge distribution creating the field.
Note: The SI unit of electric potential is the **volt:** $1\text{ V} \equiv 1\text{ J/C}$. Note also that $\phi(x, y, z)$ is a *scalar* function of position: it has no direction.

In combination with the potential energy formula for interacting charged particles we learned in unit C, this definition implies that the potential field of a charged particle is

$$\phi = \frac{kq}{r_{PC}} = \frac{1}{4\pi\varepsilon_0}\frac{q}{r_{PC}} \tag{E3.6}$$

Purpose: This equation describes the electric potential field created by a charged particle.
Symbols: ϕ is the potential at point P; q is the charge of a particle at point C; r_{PC} is the distance between P and C; k is the Coulomb constant; $\varepsilon_0 \equiv 1/4\pi k$.
Limitations: This equation strictly applies only to *particles at rest*. It also assumes that the reference point where $\phi = 0$ is infinity.

The superposition principle also applies to potentials:

$$\phi = \phi_1 + \phi_2 + \cdots + \phi_N \tag{E3.7}$$

Purpose: This equation describes the superposition principle for potentials.
Symbols: $\phi_1, \phi_2, \ldots$ are the potentials due to objects $1, 2, \ldots$ alone.
Limitations: This equation has no known limitations (this even applies to time-dependent fields)!

We can conveniently represent the potential field $\phi(x, y)$ on a two-dimensional slice of space around an object by using an **equipotential diagram** analogous to a topographical map of the earth's surface. **Equipotential curves** on such a diagram connect all points having the same potential.

Section E3.2: Calculating ϕ from $\vec{E}$

A simple energy argument implies that

$$d\phi = -\vec{E} \cdot d\vec{r} \tag{E3.11}$$

Purpose: This equation allows us to find the change $d\phi$ in the electric potential ϕ as we move a small displacement $d\vec{r}$ in a *static* electric field $\vec{E}$.

Limitations: The displacement $d\vec{r}$ must be so small that changes in $\vec{E}$ are negligible.

Therefore, if we know the electric field $\vec{E}(x, y, z)$ and the value of the potential ϕ at any point A, we can find ϕ at any other point B by dividing up a path between the points into small steps and summing $\phi = -\vec{E} \cdot d\vec{r}$ for all steps. Conservation of energy requires the result be independent of the path chosen.

Section E3.3: Calculating $\vec{E}$ from ϕ

Similarly, if one knows $\phi(x, y, z)$, one can calculate the electric field vector $\vec{E}$ at any point as follows:

$$\vec{E} = \left|\frac{d\phi}{dD}\right| \hat{D} \tag{E3.19}$$

Purpose: This equation specifies how to calculate the electric field vector $\vec{E}$ at an arbitrary point from the potential field $\phi(x, y, z)$.

Symbols: $\hat{D}$ is the "downhill" direction (the direction perpendicular to an equipotential and toward decreasing values of ϕ), and dD is the length of an infinitesimal step in that direction.

Limitation: This equation only applies to static fields

Section E3.4: Useful Theorems About Charge Distributions

Table E3.1 in this section summarizes the results of such calculations for various useful surface charge distributions. Make *sure* that you know the results in this table! In particular, you should know that the field *outside* a charge uniformly distributed on a spherical surface is the same as that of a particle having the same charge, and the field *inside* the surface is zero. Also, the field of an infinite plane of charge is nearly uniform.

Section E3.5: The Electric Field as Dispersed Energy

This section uses the result in table E3.1 for a spherical surface charge to argue that

$$\text{Electric field energy density} = \frac{E^2}{8\pi k} = \frac{\varepsilon_0 E^2}{2} \tag{E3.29}$$

Purpose: This equation expresses the energy density of an electric field in a region of space where the field vector has a magnitude E.

Symbols: k is the Coulomb constant, and $\varepsilon_0 \equiv 1/4\pi k$.

Limitations: There are no known limitations.

An electric field, like a particle, is therefore simply a form of energy, except dispersed rather than concentrated.

E3.1 The Electric Potential

In chapter E2, we defined the *electric field* $\vec{E}(x, y, z)$ at a point to be the electrostatic *force per unit charge* that a small test charge at rest experiences at that point:

$$\vec{E}(x, y, z) \equiv \frac{\vec{F}_e}{q_{\text{test}}} \quad \text{where } q_{\text{test}} \text{ is small and at rest at } [x, y, z] \tag{E3.1}$$

This is a straightforward way of describing the electrostatic field around a charge distribution because force is a fairly intuitive concept and the force experienced by a test charge is something easy to measure and imagine.

However, we can also describe the field created by a charge distribution at a point in space in terms of the *change in electrostatic potential energy per unit charge* required to move a small test charge from some reference point to that point. In fact we define the **electric potential** ϕ (usually called simply the **potential**) at a point as follows:

Definition of the electric potential $\phi(x, y, z)$

$$\phi(x, y, z) \equiv \frac{V_e(x, y, z)}{q_{\text{test}}} \tag{E3.2}$$

Purpose: This equation defines the electric potential ϕ at a point in space.

Symbols: $V_e(x, y, z)$ is the electrostatic potential energy, relative to some reference point (usually infinity), of a test particle with charge q_{test} placed at point $[x, y, z]$.

Limitations: The test particle charge q_{test} must be small enough that the charge distribution creating the field is not disturbed significantly as we move the test particle.

Note: The SI unit of the electric potential is the **volt:** $1 \text{ V} \equiv 1 \text{ J/C}$. Note also that $\phi(x, y, z)$ is a *scalar* function of position: it has no direction.

Knowing the electric potential ϕ at all points around a charged object is completely equivalent to knowing the electric field $\vec{E}$ around the object. Indeed, in sections E3.2 and E3.3, we will learn how to calculate $\phi(x, y, z)$ around a charged object given $\vec{E}(x, y, z)$, and vice versa.

Advantages of using the potential field approach

Although using ϕ to describe an electric field is somewhat more abstract than using $\vec{E}$ (because potential energy is less intuitive than force), it does have some offsetting advantages. First, ϕ is a *scalar* quantity instead of a vector quantity, which makes it quite a bit easier to calculate than $\vec{E}$ in most circumstances. Second, energy is an important concept that is useful in many situations, and describing the field in terms of ϕ makes it easy to calculate the energy implications of electric fields.

The similarity in terminology makes it easy to confuse electric *potential* ϕ with electrostatic *potential energy* V_e: keep in mind that potential ϕ is potential energy *per unit charge* and is measured in volts, not joules.

The analogy to gravitational potential

Note that we can define a **gravitational potential** ϕ_g for a gravitational field in a directly analogous way: the gravitational potential at a certain point in space is

$$\phi_g = \frac{V_g}{m} \tag{E3.3}$$

where V_g is the gravitational potential energy of a test object of mass m at that point relative to some reference point. At points near the surface of the earth where $V_g = mgh$, the gravitational potential (relative to whatever position we define to have $h = 0$) is simply

$$\phi_g = gh \tag{E3.4}$$

Therefore, electrostatic potential is to an electric field what gh is to the earth's gravitational field: $\Delta\phi$ describes the energy per unit charge, and $\Delta\phi_g$ describes the energy per unit mass that will be released if the object changes position in the field. Note that just as *electric potential* (which is measured in joules per coulomb) is not the same as *electrostatic potential energy* (which is measured in joules), so *gravitational potential* (which is measured in J/kg = m^2/s^2) is not the same as *gravitational potential energy*. Both potential *energies* depend on the characteristics of the test particle we use, but that information has been stripped out of the potentials so that they depend only on the fields.

The potential field of a point charge

Let's see how we can compute the potential field at a given point P near a point charge of magnitude q. As we saw in section C7.1, the potential energy associated with the interaction between a test charge q_{test} and a charge q at point C is (when we take V_e to be zero at infinity) simply

$$V_e = \frac{kqq_{\text{test}}}{r_{PC}} \tag{E3.5}$$

where r_{PC} is the distance between the point P and the charge q at C. According to equation E3.2, this means that the electric potential at point P due to q is

$$\phi = \frac{kq}{r_{PC}} = \frac{1}{4\pi\varepsilon_0}\frac{q}{r_{PC}} \tag{E3.6}$$

Purpose: This equation describes the electric potential field created by a charged particle.

Symbols: ϕ is the potential at point P; q is the charge of a particle at point C; r_{PC} is the distance between P and C; k is the Coulomb constant; and $\varepsilon_0 \equiv 1/4\pi k$.

Limitations: This equation strictly applies only to *particles at rest*. It also assumes that the reference point where $\phi = 0$ is infinity.

The potential field of a set of charged objects

Potentials also obey the superposition principle, so the total potential ϕ at a point due to a *set* of charged objects is simply

$$\phi = \phi_1 + \phi_2 + \cdots + \phi_N \tag{E3.7}$$

Purpose: It describes the superposition principle for potentials.

Symbols: $\phi_1, \phi_2, \ldots$ are the potentials due to objects 1, 2, ... alone.

Limitations: This equation has no known limitations (this even applies to time-dependent fields)!

For example, the total potential ϕ at point P due to N point charges $q_1, q_2, \ldots$ is

$$\phi = \frac{kq_1}{r_{P1}} + \frac{kq_2}{r_{P2}} + \cdots + \frac{kq_N}{r_{PN}} = \sum_{i=1}^{N}\frac{kq_i}{r_{Pi}} \tag{E3.8}$$

These quantities are all scalars, so we do not have to worry about directions or vector components! This is why ϕ is generally much easier to calculate than $\vec{E}$.

Exercise E3X.1

Consider two positive point charges with equal charges q that sit along the x axis at points $x = \pm a$. Find the total electric potential ϕ for these point charges as a function of y at all points along the y axis.

Using an equipotential diagram to display the field

Since the electric potential at any given point is simply a number, we can conveniently depict the potential field around a charge distribution by using a contour map, much as a topographical map describes the altitude of various points on the earth's surface. Figure E3.1 shows such a map for a dipole in the $z = 0$ plane. Each curve on the map connects points having the same electric potential. These curves are thus called **equipotential curves** or just **equipotentials,** and the diagram is called an **equipotential diagram.**

(Figure E3.1 ignores the fact that the potential can depend on z as well as x and y. Thus the equipotential curves shown in the diagram are really just two-dimensional slices through balloonlike **equipotential surfaces** that connect all points in *three*-dimensional space that have the same potential.)

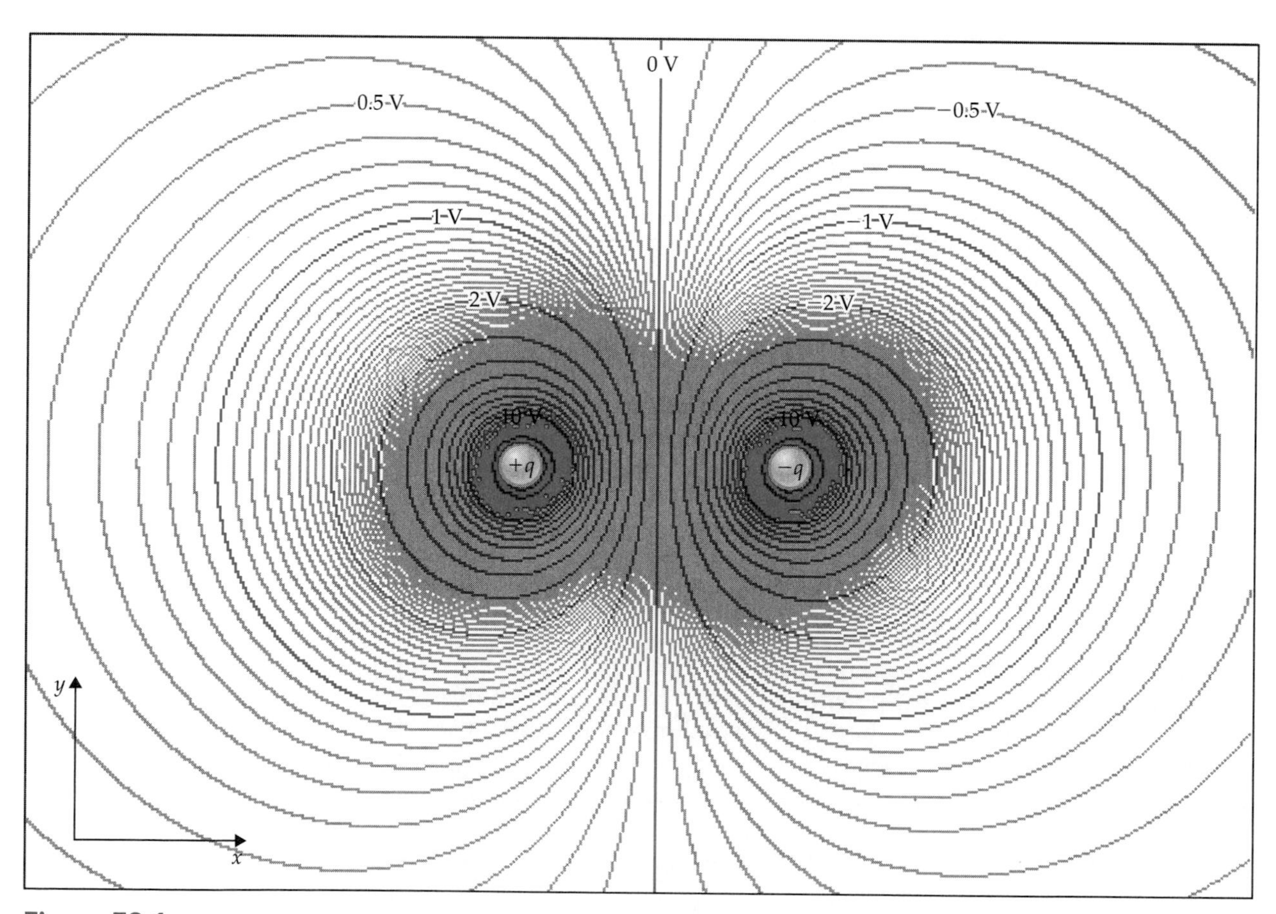

Figure E3.1
An equipotential diagram for a dipole. Each contour connects points that have the same electric potential.

An equipotential map for a gravitational field is essentially the same as a topographical map displaying height above some arbitrary reference altitude. Equipotential curves on a topographical map are simply the contour lines that connect points having a given elevation. The only thing that we technically need to do to convert a topographical map to a gravitational equipotential map is to relabel the contours to display the value of gh instead of just h.

E3.2 Calculating ϕ from $\vec{E}$

Knowing the potential field $\phi(x, y, z)$ created by a set of charges is equivalent to knowing the electric field $\vec{E}(x, y, z)$ around those charges: one can compute either representation of the field given the other. The purpose of this section is to show how to compute $\phi(x, y, z)$ from $\vec{E}(x, y, z)$.

Consider a point particle with charge q that experiences an electrostatic force $\vec{F}_e = q\vec{E}$ at a certain point in space. Now imagine we move this particle through an infinitesimal displacement $d\vec{r}$ that is so small that $\vec{E}$ is essentially constant during the displacement. Under these circumstances, according to chapter C8, the electrostatic interaction contributes a k-work

$$[dK] = \vec{F}_e \cdot d\vec{r} = q\vec{E} \cdot d\vec{r} \tag{E3.9}$$

to the particle during this displacement. (Note that this equation involves a *dot product* of vectors $\vec{E}$ and $d\vec{r}$. If you don't remember much about the dot product, now would be an excellent time to review section C8.2.) This energy transfer to the particle's kinetic energy must come at the expense of electrostatic potential energy, so the particle's electrostatic potential energy V_e must change by

$$dV_e = -[dK] = -q\vec{E} \cdot d\vec{r} \tag{E3.10}$$

during this displacement. According to the definition of the potential, though, this means that the change in the electric potential between the endpoints of the displacement is

$$d\phi = \frac{dV_e}{q} = \frac{-q\vec{E} \cdot d\vec{r}}{q} \quad \Rightarrow \quad d\phi = -\vec{E} \cdot d\vec{r} \tag{E3.11}$$

Purpose: This equation allows us to find the change $d\phi$ in the electric potential ϕ as we move a small displacement $d\vec{r}$ in a *static* electric field $\vec{E}$.

Limitations: The displacement $d\vec{r}$ must be so small that changes in $\vec{E}$ are negligible.

The change in potential along an infinitesimal displacement

Therefore, if we know the electric field $\vec{E}(x, y, z)$ and value of the potential ϕ at any specific point A, we can (in principle) find the potential at any other point B as follows:

A plan for calculating the potential at any point in an electric field

1. Define a convenient path through space connecting A to B.
2. Divide the path up into steps that are so small that $\vec{E} \approx$ constant for each step.
3. Compute $-\vec{E} \cdot d\vec{r}$ for each step along the path.
4. Add the results to find the total change in potential $\Delta\phi$.
5. Use the known value of the potential at A to find the value at B.

Doing this for all possible points B allows us to determine the entire potential field $\phi(x, y, z)$ from the electric field $\vec{E}(x, y, z)$. Example E3.1 illustrates the process.

Example E3.1

Problem Using equation E2.9 for the electric field $\vec{E}$ of a point particle, find the electric potential ϕ at an arbitrary point P near a particle with positive charge q.

Translation Let r_P be the distance between point P and the charged particle.

Model We know that the electric potential is zero at infinity, so we choose the path to be a radial line going from $r = r_P$ to $r = \infty$. We can divide this path into an infinite number of outward infinitesimal steps $d\vec{r}$. The displacement $d\vec{r}$ for any step of this path points directly away from the particle, and $\vec{E}$ also points in this direction, so the dot product becomes

$$\vec{E} \cdot d\vec{r} = [\text{mag}(\vec{E})][\text{mag}(d\vec{r})] \cos 0^\circ = +E\,dr \qquad \text{(E3.12)}$$

But equation E2.9 tells us that $E = \text{mag}(\vec{E}) = kq/r^2$ for a point particle. If we plug this into equation E3.12 and sum over all steps along the path, we get

$$\phi(\infty) - \phi(r_P) = \Delta\phi_{\text{path}} = \sum_{\text{all steps}} -\vec{E} \cdot d\vec{r} = \sum_{\text{all steps}} -\frac{kq}{r^2}\,dr \qquad \text{(E3.13)}$$

In the limit that the steps become infinitesimal, we can convert this sum to an integral:

$$\phi(\infty) - \phi(r_P) = \int_0^\infty -\frac{kq}{r^2}\,dr = +\frac{kq}{r}\bigg|_{r_P}^{\infty} = 0 - \frac{kq}{r_P} = -\frac{kq}{r_P} \qquad \text{(E3.14)}$$

Finally, since $\phi(\infty) = 0$ by definition, the potential at point P due to a point charge is

$$0 - \phi(r_P) = -\frac{kq}{r_P} \quad \Rightarrow \quad \phi(r_P) = \frac{kq}{r_P} \qquad \text{(E3.15)}$$

Since P was completely arbitrary, this essentially specifies the entire potential field around the charged particle.

Evaluation This is exactly the result given by equation E3.6, which is comforting.

Exercise E3X.2

Go through the argument again, assuming that q is negative, and prove that equation E3.15 still applies (even though the potential is now negative).

Exercise E3X.3

Consider a line that is perpendicular to the axis of a dipole and goes through that axis halfway between the dipole's charges (this is the vertical straight line in figure E3.1). Figure E2.4 implies that the electric field at all points along such a line is perpendicular to the line. Argue that the potential at all points along this line must be 0 V, as shown in figure E3.1.

The potential difference for a static field is path-independent

If this process is to yield a well-defined value for ϕ at a point B given $\phi(A)$, the potential difference $\Delta\phi$ that we compute between two points *cannot* depend on the path we choose to go from one point to the other. What right do we have to assume this? For a static electric field, it turns out that $\Delta\phi$ *is* path-independent, and the reason is related to the conservation of energy. Imagine that we go from A to B along one path and then back to A along a different path. Let us pretend, for the sake of argument, that the potential difference from A to B along the first path is smaller than the difference from A to B along the second path. If we transport the particle in a circle from A up the first path and then down the second, the particle will have a lower potential (and thus a lower potential energy) when it returns to A than it did initially. But this means it must have more kinetic energy than it did originally, and if we continue to transport it around in a circle, we can generate as much kinetic energy as we like! But where is this energy coming from? In a static electric field nothing *else* is changing, so there is nowhere that this energy can come from!

(The gravitational analogy would be a bike path in a park that goes in a closed loop away from and then back to a given location, and yet goes downhill all the way! You could generate unlimited amounts of kinetic energy by riding around and around such a loop without changing anything external, in violation of the law of conservation of energy.)

If, however, the potential difference is the *same* for both paths, then the particle will return to point A with exactly the same potential, and thus it neither gains nor loses energy in the cycle, preserving the conservation of energy. We see, therefore, that the law of conservation of energy demands that the potential difference $\Delta\phi$ between two points in a *static* field be *independent* of the path we use to compute it. (We will see in chapter E14, however, that the story is different in a time-dependent field.)

E3.3 Calculating $\vec{E}$ from ϕ

In section E3.2, we learned how to compute the potential field $\phi(x, y, z)$ of a charged object, given the electric field $\vec{E}(x, y, z)$ of that object. In this section, we will learn how to do the reverse.

Consider again the expression

$$d\phi = -\vec{E} \cdot d\vec{r} \tag{E3.16}$$

which follows directly from the definition of the potential. Imagine that we know how ϕ varies with position. How could we use this information and equation E3.16 to determine the magnitude and direction of $\vec{E}$ at an arbitrary point P?

Finding the direction of $\vec{E}$

Let's work first on the *direction* of $\vec{E}$. Since we know $\phi(x, y, z)$, at any point P we can find directions along which we can move a small displacement $d\vec{s}$ without changing the value of ϕ. (The gravitational analogy is that if we are walking on a mountainside, we can always find directions to walk where our elevation, and thus our gravitational potential, does not change.) By definition, such a displacement will be along whatever equipotential curve (or equipotential *surface* if we think in three dimensions) goes through the point in question. For such a "sideways" displacement

$$0 = -\vec{E} \cdot d\vec{s} \qquad \text{if } d\vec{s} \text{ is along an equipotential} \tag{E3.17}$$

But the dot product $0 = -\vec{E} \cdot d\vec{s} \equiv E\,ds \cos\theta$ of two nonzero vectors will be zero only if the two vectors are perpendicular ($\theta = 90°$). Therefore, *the electric field vector $\vec{E}$ at point P must point perpendicular to any equipotential curve or surface going through that point.*

Now imagine a different displacement $d\vec{D}$ that at point P is perpendicular to whatever equipotential curve or surface goes through P and points in the direction that ϕ decreases (the gravitational analogy would be a displacement in the steepest *downhill* direction). The field vector $\vec{E}$ must be either parallel to or opposite to such a displacement. Now equation E3.16 implies that $d\phi = -\vec{E} \cdot d\vec{D}$. Since $d\phi$ is negative by hypothesis, $\vec{E} \cdot d\vec{D}$ must be positive, which implies $\vec{E}$ must be *parallel* to $d\vec{D}$. Therefore, *$\vec{E}$ at point P points in whichever direction is perpendicular to the equipotential points toward decreasing values of ϕ*, that is, downhill.

Finding the magnitude of $\vec{E}$

Moreover, the definition of the dot product implies that

$$d\phi = -\vec{E} \cdot d\vec{D} = -E\,dD \cos 0° = -E\,dD \quad \Rightarrow \quad E = \left|\frac{d\phi}{dD}\right| \tag{E3.18}$$

We see that *the magnitude of the electric field vector $\vec{E}$ at point P is equal to the magnitude of the slope of ϕ.*

To summarize,

$$\vec{E} = \left|\frac{d\phi}{dD}\right| \hat{D} \tag{E3.19}$$

Purpose: This equation specifies how to calculate the electric field vector $\vec{E}$ at an arbitrary point from the potential field $\phi(x, y, z)$.
Symbols: $\hat{D}$ is the downhill direction (the direction perpendicular to an equipotential and toward decreasing values of ϕ), and dD is the length of an infinitesimal step in that direction.
Limitation: This equation applies only to static fields.

We can use this result to draw electric field vectors on an equipotential diagram of ϕ. At any given point on an equipotential curve, we should draw $\vec{E}$ perpendicular to the equipotential curve in the direction that ϕ is decreasing. The magnitude of $\vec{E}$ is proportional to how rapidly ϕ changes in that direction, and thus should be inversely proportional to the distance between equipotential curves in that direction. Figure E3.2 illustrates the construction of electric field vectors on an equipotential diagram.

Exercise E3X.4

Draw electric field arrows at various points on the equipotential diagram shown in figure E3.1, and compare your resulting picture of the electric field with that shown in figure E2.4.

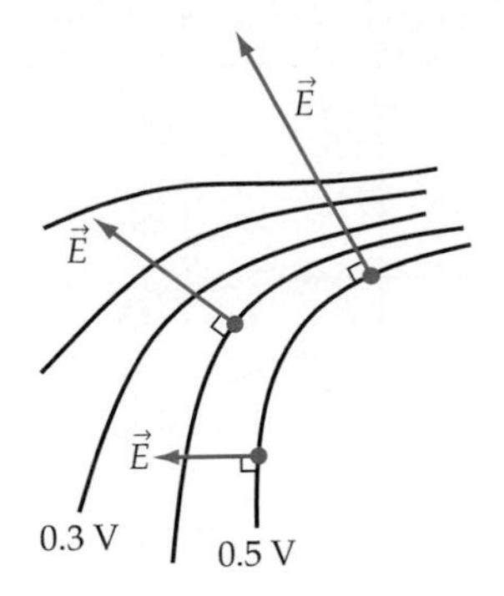

Figure E3.2
The electric field vector at a point on an equipotential diagram is perpendicular to the equipotential through that point and has a magnitude inversely proportional to the distance between curves near the point.

Example E3.2

Problem Consider a single particle with positive charge q located at point C. Use equation E3.6 for the potential of a charged particle to calculate the magnitude and direction of the particle's electric field at P.

Solution According to equation E3.6, the potential of a charged particle evaluated at point P is

$$\phi - \frac{kq}{r_{PC}} \tag{E3.20}$$

where r_{PC} is the distance between P and the particle's position C. This obviously decreases as r_{PC} increases (if q is positive), so the "downhill" direction at point P must point radially away from point C. Now the length of a small step $d\vec{D}$ in this direction will be the same as the change in the value of r_{PC}, so according to equation E3.19, the magnitude of the electric field at point P must be

$$E = \left|\frac{d\phi}{dr_{PC}}\right| = \left|\frac{d}{dr_{PC}}\left(\frac{kq}{r_{PC}}\right)\right| = \frac{kq}{r_{PC}^2} \tag{E3.21}$$

Therefore, using the directional $\hat{r}_{PC}$ to stand for the radially outward (= downhill) direction from C, we have

$$\vec{E} = \frac{kq}{r_{PC}^2}\hat{r}_{PC} \tag{E3.22}$$

Evaluation This is the same expression for the electric field of a point charge that we found in chapter E2.

It is often easier to calculate $\phi(x, y, z)$ than $\vec{E}(x, y, z)$ for a complicated charge distribution because we can add the *potential* fields contributed by bits of the distribution as scalars without having to worry about resolving these contributions into vector components and adding as vectors. Example E3.3 illustrates how much easier this can be (compare with example E2.2).

Example E3.3

Problem Consider a thin ring of radius R carrying a total positive charge Q. What is **(a)** the potential and **(b)** the electric field at a point P a distance z from the ring's center along the ring's central axis? (See figure E3.3.)

Figure E3.3
This diagram illustrates how we can calculate the electric potential at a point P along the central axis of a ring. Note that I have set up the coordinate system so that the ring's center is at the origin and its central axis coincides with the z axis. Note that each point on the ring is the same distance $r = (z^2 + R^2)^{1/2}$ from point P.

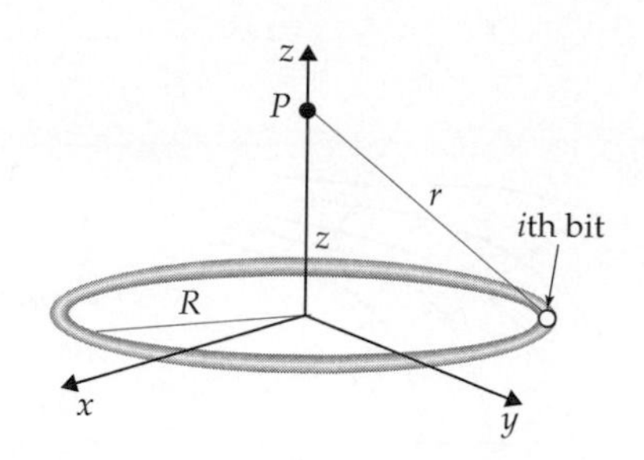

(a) *Model* The first step, as usual, is to divide the ring into a set of bits that are sufficiently small to model as point charges. If the ring is very thin compared to its radius, all these bits are essentially the same distance $r_P = (z^2 + R^2)^{1/2}$ from the point P.

Solution This means that the total potential at point P is very easy to calculate: we find

$$\phi(0,0,z) = \sum_{\text{all } i} \frac{kq_i}{r_P} = \frac{k}{r_P}\sum_{\text{all } i} q_i = \frac{kQ}{r_P} = \frac{kQ}{(z^2 + R^2)^{1/2}} \tag{E3.23}$$

(b) *Model* This only gives the potential along the z axis, so we cannot tell from this formula exactly how ϕ might vary as we move away from the z axis (and the calculation of ϕ for points off the axis is not trivial). Still, it makes intuitive sense that the downhill direction from any point P on the ring's central axis will point directly away from the ring along that axis. Moving in this direction increases the distance between P and *all* parts of the ring, thus reducing each bit's contribution to the total potential. Moving in any other direction will not increase the distance to some bits quite as rapidly. This suggests that $\vec{E}$ at point P should point along the z axis, away from the ring's center.

Alternatively, one can argue that because of the ring's symmetry, for every bit on the ring that contributes at P a horizontal component to $\vec{E}$ in one direction, there is a bit the same distance away on the opposite side of the ring that contributes a canceling horizontal component in the opposite direction. Therefore the horizontal components of the net electric field vector at P should add to zero, again implying that $\vec{E}$ will point along the z axis.

Solution The length of a small step along the axis will be simply $|dz|$. Equation E3.19 therefore implies that

$$\begin{aligned}\text{mag}(\vec{E}) &= \left|\frac{d\phi}{dz}\right| = kQ\left|\frac{d}{dz}(z^2 + R^2)^{-1/2}\right| \\ &= kQ\left|-\frac{1}{2}(z^2 + R^2)^{-3/2}(2z)\right| = \frac{kQ|z|}{(z^2 + R^2)^{3/2}}\end{aligned} \tag{E3.24}$$

Since the downhill direction is $+\hat{z}$ if z is positive (P is above the ring) and $-\hat{z}$ if z is negative (P is below the ring), the quantity $(z/|z|)\hat{z}$ is a directional of magnitude 1 that points in the correct downhill direction for all z. Therefore,

$$\vec{E} = \frac{kQ|z|}{(z^2 + R^2)^{3/2}}\frac{z}{|z|}\hat{z} = \frac{kQz}{(z^2 + R^2)^{3/2}}\hat{z} \tag{E3.25}$$

Evaluation This is the same as the result we found in example E2.2.

Exercise E3X.5

What if Q were negative? Go through the steps of the argument, assuming that Q is negative, and show that you *still* end up with equations E3.23 and E3.25.

E3.4 Useful Theorems About Charge Distributions

Calculating the potential field of a charge distribution is generally easier than calculating the electric field, but even simple shapes can still lead to difficult integrals. Rather than go through the gruesome calculations, let me simply describe some results for several useful kinds of charge distributions. These results will be useful in future chapters.

Practical charge distributions are often *surface* charge distributions

Rubbing an insulator will distribute charge on its surface, not its interior. In chapter E4, we will see that charge placed on a conductor also distributes itself on the conductor's surface. Therefore, many of the charge distributions we encounter in daily life are very thin surface charge distributions. Table E3.1 simply lists some useful results for infinitesimally thin uniform charge distributions placed on the surfaces of certain simple shapes. Note that in each case, the electric field vector $\vec{E}$ has the same direction as the specified directional if the surface charge is positive, but points in the opposite direction if the charge is negative. For example, $\vec{E}$ at a point outside a spherical surface charge distribution points directly *away* from the sphere's center (in the direction of $\hat{r}$) if Q is positive, but directly *toward* the sphere's center (in

Table E3.1 **Electric fields produced by selected surface charge distributions**

Shape	Fields *Outside* the Shape's Uniformly Charged Surface	Fields *Inside* That Charged Surface
Sphere	$\phi = \dfrac{kQ}{r}$ $\quad \vec{E} = \dfrac{kQ}{r^2}\hat{r}$ $Q \equiv$ total surface charge $r \equiv$ distance from sphere's center $\hat{r} \equiv$ directly away from sphere's center	$\phi = \dfrac{kQ}{R}$ $\quad \vec{E} = 0$ $Q \equiv$ total surface charge $R =$ radius of sphere
Infinite cylinder	$\phi - \phi_s = 2k\lambda \ln\left(\dfrac{r}{R}\right)$ $\quad \vec{E} = \dfrac{2k\lambda}{r}\hat{r}$ $\lambda \equiv$ charge per unit length on cylinder $r \equiv$ distance from cylinder's axis $R \equiv$ radius of cylinder's surface $\phi_s \equiv$ potential at surface $\hat{r} \equiv$ directly away from cylinder's central axis	$\phi - \phi_s = 0$ $\quad \vec{E} = 0$ $\phi_s \equiv$ potential at surface
Infinite flat surface	$\phi - \phi_s = 2\pi k\sigma r$ $\quad \vec{E} = 2\pi k\sigma\hat{r}$ $\sigma \equiv$ charge per unit area on surface $r \equiv$ distance from surface $\phi_s \equiv$ potential at surface $\hat{r} \equiv$ directly away from surface	(This shape has no "inside.")

the direction of $-\hat{r}$) if Q is negative. Note also that it is more convenient in the case of infinite surfaces to define the potential relative to the potential ϕ_s at the surface rather than relative to infinity.

We will prove these results in chapter E11

One can (with great effort) use the techniques of section E2.6 or of this chapter to verify that these results are correct. However, in chapter E10, we will learn a very powerful method for computing static electric fields that will make proving these results *much* easier. We will therefore wait until then to display formal proofs of these results. (If you can't wait, see problems E2S.12, E3S.6, E3A.1, and E3A.2 for some example calculations.)

The field of a spherical surface charge is like that of a point charge!

These idealized results provide useful models for more practical situations. For example, note that the result for the field outside a uniformly distributed spherical surface charge is *exactly* the same as if the sphere's total charge Q were concentrated at the sphere's center. This means that we can accurately model any uniformly charged spherical object as a point particle when we calculate its external field! This is very useful to know in many practical circumstances.

Application notes for the other distributions

While it is prohibitively expensive to construct infinite cylinders or flat planes, the fields of these idealized charge distributions provide reasonable approximations for the fields at points sufficiently close to long cables or flat plates, which are common shapes in practical applications. A point that is "sufficiently close" is one that is much closer to the object's nearest part than to the object's ends or edges.

When applying the infinite flat surface model to a finite plate, however, you should note that more complicated calculations show that the electric field vector $\vec{E}$ at a point near a *finite* plate has a component parallel to the plate (even if the point is *very* close to the plate) that increases steadily as you get closer to an edge of the plate (see figure E3.4a and b): only very near the exact center of the plate is this component essentially zero. The infinite flat surface model does not describe this component, but it does very accurately describe the component of $\vec{E}$ *perpendicular* to the plate.

The field near parallel plates with opposite charge

Actually, the most common application of the infinite flat surface model is to the situation where one places two oppositely charged parallel plates very close together compared to their size. In such a case, the components of the electric field that are *parallel* to the plates essentially cancel each other out, as illustrated in figure E3.4c. According to table E3.1, the perpendicular component of $\vec{E}$ near a truly infinite flat surface does not depend at all on the distance one is from the plate, and even for finite plates, this component depends only very weakly on distance. This means that in the region outside a pair of oppositely charged parallel plates, the plates' fields will essentially cancel entirely, while between the plates, the fields will reinforce each other to create an electric field with a nearly uniform magnitude $E = 4\pi k\sigma$ and uniform direction (away from the positive plate and toward the negative plate), as illustrated in figure E3.4c. This is the most practical way to create an essentially uniform electric field in a finite region of space, and for that reason, this result will be quite useful to us in future chapters.

Exercise E3X.6

Consider two concentric spherical shells, one with radius R and a uniformly distributed surface charge Q and the other with radius $2R$ and similarly distributed charge $-Q$. Describe the electric field at all points in space inside and outside these shells.

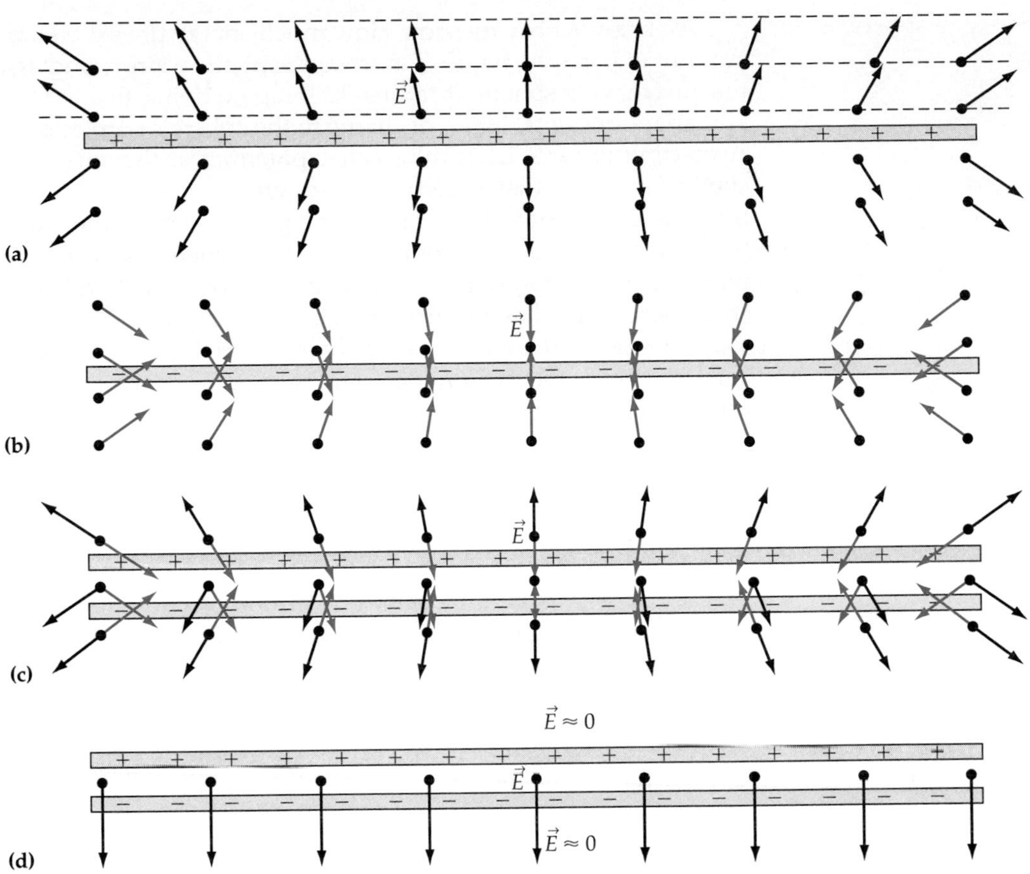

Figure E3.4
(a) Electric field vectors at points near a positively charged plate. The dashed lines help one see that the perpendicular components of these field vectors are nearly equal at all points. (b) The corresponding electric field vectors at points near a negatively charged plate. (c) When such plates are close together, the (colored) field vector contributed by the negative plate at a point *outside* the plates is nearly exactly equal and opposite to the (black) field vector contributed at that point by the positive plate. If the point is *between* the plates, the components *parallel* to the plates cancel, but the perpendicular components add. (d) The result is a strong and nearly uniform perpendicular electric field in the region between the plates and almost zero electric field outside.

Exercise E3X.7

Consider a thin insulating rod with a diameter d of 2 mm and a length L of 1.0 m. Imagine that this rod has a uniformly distributed total charge Q of 3.0 nC on its surface. What is the magnitude of the electric field at a point a distance $r = 0.5$ cm from the rod's center? (*Hint:* Justify the use of the infinite cylinder as a model.)

E3.5 The Electric Field as Dispersed Energy

In unit R, we saw that a particle's mass is a form of energy, meaning that a massive particle is just a compact form of energy. In this section, we will use the result for a spherical surface charge distribution to show that an electric field is a *dispersed* form of energy, and we will derive an expression for density of energy stored in an electric field.

The self energy of a spherical surface charge

We begin with a question. How much energy does it take to move a total amount of positive charge Q in from essentially infinity and assemble it onto the surface of a sphere of radius R? Imagine doing this in stages where we add an infinitesimal bit dq of charge to the sphere's surface during each step. According to table E3.1, the electric potential at the surface of a spherical shell of radius R that already contains charge q is $\phi = kq/R$. By definition, this ϕ is the electrostatic potential energy per unit charge that it would take to move a test charge in from infinity to the shell's surface. So the infinitesimal energy dU required to move in an infinitesimal charge dq to a spherical shell that already has a charge q is $dU = dV = \phi\, dq = (kq/R)\, dq$. To find the total energy required to assemble charge Q on this sphere, we simply need to sum the energy required in each step. In the limit that dq and dU are truly infinitesimal, this sum becomes the integral

$$U_{\text{self}} = \int dU = \int_0^Q \frac{kq\,dq}{R} = \frac{k}{R}\int_0^Q q\,dq = \frac{k}{R}\left[\frac{1}{2}q^2\right]_0^Q = \frac{kQ^2}{2R} \qquad \text{(E3.26)}$$

This is called the **self-energy** of the surface charge. This would also be the energy released if the surface charge were allowed to disperse itself to infinity.

The energy stored in the electric field

How is this related to the energy stored in the electric field? Imagine that we compress the sphere by an infinitesimal amount so that the change in its radius is $dR = -|dR|$. The self-energy of the surface charge will then increase by an infinitesimal amount

$$dU_{\text{self}} = \frac{dU_{\text{self}}}{dR}\,dR = \frac{kQ^2}{2}\left(-\frac{1}{R^2}\right)dR = +\frac{kQ^2}{2R^2}|dR| \qquad \text{(E3.27)}$$

since dR is negative. According to table E3.1, the electric field inside the shell is zero, and the electric field outside the original shell radius R is unchanged. The only thing that is different between the uncompressed case and the compressed case is that a small region of space of volume $dV = 4\pi R^2|dR|$ that originally contained no electric field (because it was *inside* the surface) now contains an electric field (because it is now *outside* the surface: see figure E3.5). The magnitude of the field in this region is very nearly $E = kQ/R^2$ throughout (because the region is very thin). Let us hypothesize that the energy that we must supply to compress the shell goes to creating the electric field in this region: after all, nothing else has changed! If this is so, then the density of energy contained by this newly created region of field must be

$$\frac{\text{Energy}}{\text{Volume}} = \frac{dU_{\text{self}}}{dV} = \frac{\frac{1}{2}(kQ^2/R^2)|dR|}{4\pi R^2|dR|} = \frac{1}{8\pi k}\left(\frac{kQ}{R^2}\right)^2 = \frac{E^2}{8\pi k} \qquad \text{(E3.28)}$$

where E is the electric field strength in that region.

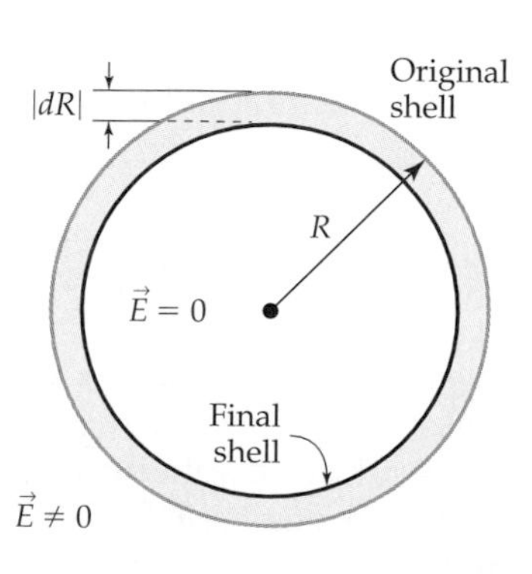

Figure E3.5
If we compress a shell of charge a distance $|dR|$, we create an electric field in the colored region where there was none before. This region has an area of $4\pi R^2$ (since that is the formula for the area of a sphere) and a thickness $|dR|$.

We have only shown that this formula makes sense in this particular context, but it in fact works in all known contexts:

$$\text{Electric field energy density} = \frac{E^2}{8\pi k} = \frac{\varepsilon_0 E^2}{2} \quad \text{(E3.29)}$$

The energy density of an electric field

Purpose: This equation expresses the energy density of an electric field in a region of space where the field vector has a magnitude E.
Symbols: k is the Coulomb constant, and $\varepsilon_0 \equiv 1/4\pi k$.
Limitations: There are no known limitations (this expression even works for nonstatic fields)!

We can interpret this result as meaning that an electric field in space, just like matter, is ultimately a form of energy. The only difference is that the energy in an electric field is dispersed smoothly over a volume instead of being concentrated into particles. But the fact that equation E3.29 usefully describes the energy contained in an electric field in a wide variety of circumstances makes it clear that an electric field is a real physical object, like matter, not just some kind of mathematical abstraction.

The electric field as a dispersed form of energy

Exercise E3X.8

If typical atmospheric electric fields are roughly 200 N/C on a cloudless day, roughly how much electric field energy is contained in 1 m^3 of atmosphere?

TWO-MINUTE PROBLEMS

E3T.1 The electric field is normally measured in newtons per coulomb. It could also be expressed in units of volts per meter, true or false (T or F)?

E3T.2 Consider two point particles, one with charge q and one with charge $-q$, separated by a distance d. The potential at a point halfway between the charges is
A. kq/d
B. $2kq/d$
C. $4kq/d$
D. 0
E. $-4kq/d$
F. Other

E3T.3 The electric potential at a point can have a negative value, T or F?

E3T.4 The equipotential curves in a certain region of an equipotential diagram for the xy plane are parallel to the y axis and are equally spaced, with the potential increasing in the $\pm x$ direction. An electric field vector in this region would point in which direction?
A. $+x$ direction
B. $-x$ direction
C. $+y$ direction
D. $-y$ direction
E. Other (specify)

E3T.5 Say that we drew the electric field vector $\vec{E}$ to be 1 cm long in the case mentioned in problem E3T.4. If in a different region of the same diagram the equipotentials are twice as widely spaced as they were in the first region, we should draw the electric field vector in this second region with a length of
A. 4 cm
B. 2 cm
C. 1 cm
D. 0.5 cm
E. 0.25 cm
F. Other (specify)

E3T.6 Consider two concentric spherical shells, one with radius R and one with radius $2R$. Both have the same charge Q. At a point just inside the outer shell, the magnitude of the electric field is
A. $2kQ/R$
B. kQ/R
C. $kQ/2R$
D. $2kQ/R^2$
E. kQ/R^2
F. $kQ/2R^2$
T. $kQ/4R^2$

E3T.7 Consider a spherical ball with a uniform surface charge density. In the absence of other charges, its surface is an equipotential surface, T or F?

HOMEWORK PROBLEMS

Basic Skills

E3B.1 Imagine that a particle with positive charge $+q$ is placed on the x axis at $x = +a$, and a particle with negative charge $-q$ is placed on the x axis at $x = -a$. Find the potential at all points along the y axis.

E3B.2 Imagine that a particle with positive charge $+q$ is placed on the x axis at $x = +a$, and a particle with negative charge $-2q$ is placed on the x axis at $x = -a$. Find the potential at all points along the y axis.

E3B.3 Use equation E3.11 and equation E2.16 for the electric field of a dipole at points along its axis to find the dipole's potential field points along that axis.

E3B.4 Use the Equipotential computer program (available from the *Six Ideas* website) to construct an equipotential diagram for two equally charged particles spaced some distance apart. On a printout of this diagram, draw at least 20 electric field vectors at various positions.

E3B.5 Imagine that in a certain region of an equipotential diagram, the equipotentials are vertical and spaced 2 mm apart, and the potential increases by 0.1 V per equipotential curve as one moves toward the left. What are the magnitude and direction of the electric field at that point?

E3B.6 Imagine that you have a ring 10 cm in diameter that has a uniformly distributed charge of 11 nC. What is the potential at the center of the ring (in volts)? What is the potential at a point on the ring's central axis that is 10 cm from the ring's center?

E3B.7 Imagine that you have a sphere 4 cm in diameter with a uniformly distributed charge of 33 nC. What is the potential (in volts) 5 cm from its center? What is E there?

E3B.8 Consider a square plate whose sides have length $L = 10$ cm with a total charge of $+5.0$ nC spread uniformly on its surface. What are the magnitude and direction of the electric field at a point 1.0 mm away from the plate's center? (*Hint:* Justify the use of the infinite flat sheet approximation.)

E3B.9 Consider two concentric metal spheres of radius R_1 and $R_2 > R_1$. Imagine that we put a uniformly distributed positive charge Q on the inner sphere and a uniformly distributed charge of $-Q$ on the outer sphere. Use the superposition principle and table E3.1 to find an expression for the electric field vector at a point a distance r away from the spheres' common center in each of the following three cases: $r < R_1$, $R_1 < r < R_2$, and $R_2 < r$. Explain your reasoning.

Synthetic

E3S.1 Consider a dipole consisting of two charged particles on the x axis, one with positive charge $+q$ located at $x = +\frac{1}{2}d$ and one with negative charge $-q$ located at $x = -\frac{1}{2}d$.
(a) Derive an exact expression for the three-dimensional potential field created by this dipole at a point P whose coordinates are $[x, y, z]$.
(b) Use the binomial approximation to show that for points whose distance r from the origin is very large compared to d, the electric potential is approximately given by

$$\phi(x, y, z) = \frac{kqd}{r^2}\frac{x}{r} = \frac{kqd}{r^2}\cos\theta \qquad \text{(E3.30)}$$

where $\cos\theta = x/r$ is the angle that the position vector of point P makes with the x axis. Note that the potential of a dipole falls off as $1/r^2$ compared to $1/r$ for a point charge. Note also that it is a lot easier to compute the electric field of a dipole using this formula than it is to calculate the field directly! [*Hints:* You should have an expression in the denominator of your answer for part (a) that involves $(x \pm \frac{1}{2}d)^2 + y^2 + z^2$. Write out the square involving d, and drop the d^2 term because it is very small compared to the other two terms. Then factor out r^2 to make the expression look like $r^2(1 \pm \text{something small})$. Now you can apply the binomial approximation.]

E3S.2 Consider the expression for the potential field of a dipole given by equation E3.30.

(a) Use this to find an expression for the electric field $\vec{E}$ of a dipole at a point P along the x axis a distance r from the dipole. (Your result should be consistent with equation E2.16.)

(b) At all points along the y axis, $\theta = 90°$, so the potential is zero. This means that the y axis is an equipotential line. Use this information and equation E3.30 to determine the electric field vector $\vec{E}$ (magnitude and direction!) at a point P along the y axis that is a distance r from the dipole.

E3S.3 Table E3.1 claims that the potential outside a uniform surface charge placed on an infinitely long cylinder of radius R is given by

$$\phi - \phi_s = 2k\lambda \ln \frac{r}{R} \tag{E3.31}$$

where λ is the charge per unit length placed on the cylinder's surface and ϕ_s is the potential at that surface.

(a) Prove that if equation E3.31 is correct, the electric field described in table E3.1 is also correct.

(b) Argue that the electric field described in table E3.1 is consistent with what one would get by taking the limit of equation E2.26 as $L \to \infty$.

E3S.4 Consider two circular parallel plates of radius R separated by a distance $d \ll R$. Assume that these plates have uniformly distributed surface charges of Q and $-Q$. As discussed in section E3.4, electric field vectors in the region between the plates will all point perpendicular to the plates from the positive plate to the negative plate and will have a magnitude of $4\pi k\sigma$.

(a) Use equation E3.11 and the information just given to evaluate the potential difference between the two plates in terms of Q, R, and d.

(b) Argue that you get the same results using the formula for the potential of an infinite plate given in table E3.1. (*Hint:* You will have to adjust the potential formula for one of the plates so that the reference point for both is the same plate, say, the negative plate.)

E3S.5 Use the Equipotential program (available on the *Six Ideas* website) to construct an equipotential diagram for two parallel square slabs with opposite charges. Set up the slabs so that the distance between them is about 20 percent of the width of the display. Submit a printout of the diagram with answers to the following questions.

(a) In section E3.4, I argued that the electric field of such plates is approximately uniform except near the edges of the plates (which are at the edges of the screen). Is this idea consistent with what you see on the diagram?

(b) In that section, I also argued that the electric field in the region outside the plates is very small. What is the approximate ratio between the electric field strength between the plates at the center and a point just outside the positive plate near its center? Does what you find support the assertions in section E3.4? Explain.

E3S.6 Consider a circular disk with radius R that has a uniformly distributed surface charge of Q.

(a) Calculate the electric potential ϕ at various points along the central axis of this disk. *Hints:* Figure E3.6 shows how you can break the disk up into thin rings of infinitesimal thickness dr. We know from the chapter how to compute the potential of each ring if we can determine the charge of each ring. The ratio of the charge q on the ring with a given radius r to the disk's total charge is the same as the ratio of the ring's surface area, which is $2\pi r\, dr$, to the area of the entire disk, which is πR^2. Express q in terms of Q, R, and dr; sum the potentials contributed by each disk; and convert the sum to an integral. According to my table of integrals,

$$\int \frac{x\,dx}{\sqrt{x^2 + a^2}} = \sqrt{x^2 + a^2} \tag{E3.32}$$

You may find this integral helpful.

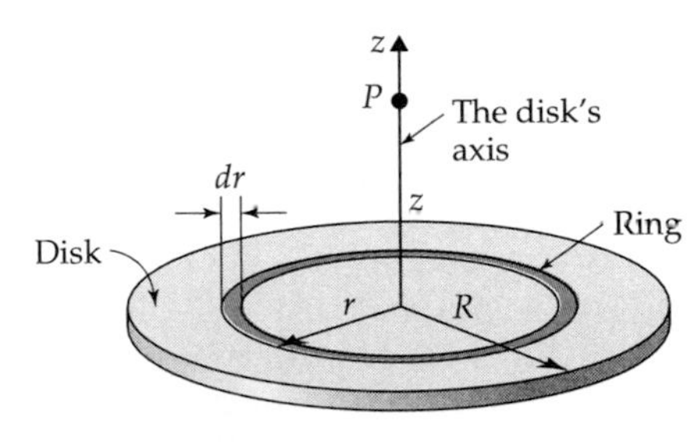

Figure E3.6
We can think of a charged disk as being a sum of charged rings, one of which is shown here.

(b) Use the result of part (a) to determine the electric field at points along the axis. *Hints:* Argue that the field vectors at such points should point along the axis. You should then be able to show that the magnitude of the electric field at such points is

$$E = 2\pi k\sigma \left[1 - \frac{1}{\sqrt{1 + (R/z)^2}} \right] \tag{E3.33}$$

where $\sigma \equiv Q/\pi R^2$ is the charge per unit area on the disk.

(c) Show that your results for both parts reduce to the appropriate formula in table E3.1 in the limit that $R \to \infty$.

E3S.7 If the magnitude of the electric field at a point in air exceeds about 3×10^6 N/C, air (which is normally an excellent insulator) will begin to conduct electricity. What happens is this. Such a strong electric field accelerates any free electrons that happen to be floating around (cosmic rays ensure that there are always a few) so violently that they crash into nearby air molecules with enough energy to ionize them, creating more free electrons, and so on. This creates an avalanche of ionization that frees a huge number of electrons for conduction.

(a) What is the maximum charge that you could put on the surface of a plastic sphere 1 cm in diameter without causing the air to conduct electricity (thus dissipating the charge)?

(b) Why do you think I say that the charges involved in typical static electricity experiments are on the order of magnitude of 10 nC?

E3S.8 Consider two concentric spherical shells, the inner one having radius R and a uniformly distributed positive charge Q and the outer one having radius $2R$ and a charge $-Q$.

(a) Find a symbolic expression for the energy stored in the electric field between the two shells.

(b) If $Q = 22$ nC and $R = 10$ cm, what is this energy in joules?

(*Hints:* See exercise E3X.6 for a description of the electric field involved here. One might think that one could find the energy stored in the field simply by using equation E3.29 for the energy density and multiplying by the volume occupied by the field. However, the density of energy is not uniform over this volume, since E is a function of the distance r that one is from the spheres' center. One can handle this by dividing up the empty space between the shells into spherical shell-like regions of thickness dr, where dr is small enough that the field strength E does not change much over the region. One then can use equation E3.29 to find the energy in this shell-like region and then integrate over all shells to get the total energy. Alternatively, one can approach this problem by using the definition of self-energy and imagining that we assemble the two charged shells from scratch.)

E3S.9 To an observer very close to a spherical surface charge distribution, the sphere's surface looks flat and stretches essentially to the horizon in all directions. It will therefore look to the observer as if it were an infinite flat charged sheet. Argue that if the observer's distance d from the spherical surface is very small compared to the sphere's radius R, the electric field strength predicted in table E3.1 for a uniformly distributed spherical surface charge is the same as that for an infinite sheet.

E3S.10 To an observer very close to a infinite cylindrical surface charge distribution, the cylinder's surface looks flat and stretches essentially to the horizon in all directions. It will therefore look to the observer as if it were an infinite flat charged sheet. Argue that if the observer's distance from the cylindrical surface is very small compared to the pipe's radius R, the electric field strength predicted in table E3.1 for a uniformly distributed charge on a cylindrical surface is the same as that for an infinite sheet.

Rich-Context

E3R.1 Electric power is shipped over large distances using high-tension wires that have been charged by the power company so that they have a potential relative to the ground of as much as 250 kV. For simplicity, let us assume that charges in the ground respond negligibly to the field of the wire (so that we don't have to worry about the field of such charges in our calculation) and that potential difference is static. (Neither of these assumptions is strictly true, but these assumptions are adequate for a crude estimate.)

(a) Very roughly how much surface charge is on 1 m of such a wire? (*Hint:* You will have to estimate both the diameter of such a wire and its distance above the ground. Consider figure E3.7 and your own memories in making such an estimate. Also explain what model of the wire you are using to arrive at your result.)

(b) Estimate the approximate magnitude of the electric field created by a single such wire in the vicinity of a person standing on the ground. How does this compare to typical atmospheric field strengths (see section E2.2)?

(c) If the field strength at the surface of the wire exceeds about 3×10^6 N/C, air will break down and conduct electricity away from the wire. Since this robs energy from the power company,

Figure E3.7
What is the electric field created by this power line? (See problem E3R.1.)

the company will design the wires so that this does not happen. Based on your estimates, how close is the wire to violating this limit? (You may want to readjust your estimate of the wire's diameter after doing this part.)

E2R.2 The result in table E3.1 for the electric field of a spherical surface charge distribution *also* applies (if we change $\vec{E}$ to $\vec{g}$, $k \to G$, and $Q \to M$) to the gravitational field created by a very thin spherical shell of mass M, because Newton's law of universal gravitation has the same mathematical form as Coulomb's law except for these substitutions. We can apply this to a simple newtonian model of the universe as follows. It is an observational fact that the universe is expanding: indeed a galaxy a distance r from us is observed to be moving away from us at a speed $v \approx Hr$, where the *Hubble constant* H has the same value for all galaxies. Model the universe as an essentially infinite spherical distribution of pointlike galaxies with average uniform density ρ centered on the earth.

(a) Divide the universe into a series of very thin nested shells centered on the earth. Argue that the gravitational force exerted by the whole universe on a given galaxy a finite distance r from the earth (see figure E3.8) is as if there were a point particle of mass $M = (4/3)\pi r^3 \rho$ located at the earth.

(b) The universe will thus expand forever if and only if each galaxy's speed exceeds that mass' escape speed at that particular galaxy's distance r from the earth (see unit C for a discussion of the escape speed). Show that this will be true for every galaxy if and only if ρ is less than a certain ρ_{crit} that depends only on G and H. (Surprisingly, a more sophisticated general relativistic model of the universe predicts the same result!)

(c) The current best experimental values of H and ρ are $2.3 \times 10^{-18}\ s^{-1}$ and $3 \times 10^{-30}\ g/cm^3$, respectively. If our model is correct, will the universe expand forever?

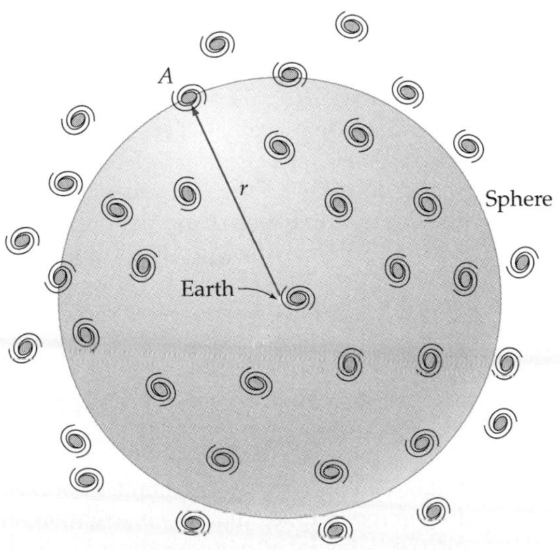

Figure E3.8
The gravitational field at the galaxy marked A is as if the mass of the galaxies within the sphere shown were concentrated at the sphere's center.

Advanced

E3A.1 Consider a very long straight wire of length L parallel to the x axis whose center is at $x = 0$. Find the potential ϕ as a function of distance r from the wire at points along the y axis. Show that $\vec{E}$ has the same r dependence as we found in chapter E2 in the limit that $L \gg r$. [*Hints:* Note that any particular point P on the y axis will have a y coordinate $y = +r$. It helps greatly here to recognize that each half of the wire contributes the same amount to ϕ (why)? This means that we can integrate from $x = 0$ to $x = \frac{1}{2}L$ and double the result to get the right answer.]

E3A.2 We can prove the sphere results on table E3.1 as follows. Pick an arbitrary point P outside the surface, and draw a line from P through the shell's center. We will model the surface charge as being comprised of a set of thin rings, each corresponding to a small increment of the angle θ from the line through P (see figure E3.9). The ring at a given θ has a radius $R_r = R \sin\theta$ and a width of $R\,d\theta$. We will assume that h and $R\,d\theta$ are both small enough that all points on a given ring are essentially the same distance u from point P. We will also assume that charge is uniformly spread over the sphere's surface.

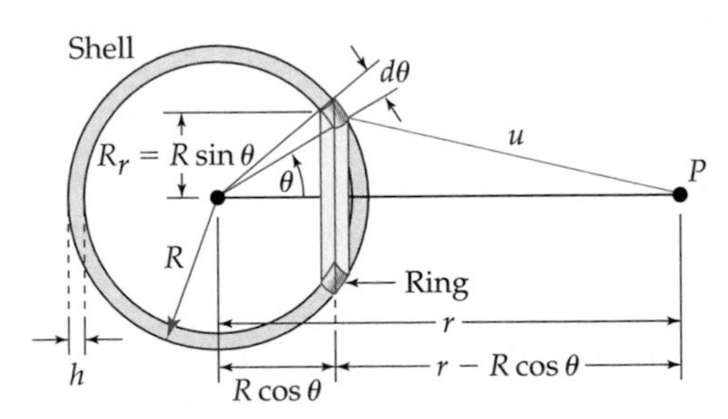

Figure E3.9
We can treat a shell as a sum of rings, one of which (along with its associated variables) is shown here.

(a) The last assumption implies that the infinitesimal charge q_r on any given ring is to the shell's total charge Q as the ring's surface area is to the shell's total surface area $4\pi R^2$. A given ring stakes out an area on the shell's surface equal to the ring's circumference times its width. Use these ideas to show that

$$q_r = \tfrac{1}{2} Q \sin\theta \, d\theta \tag{E3.34}$$

(b) Use the pythagorean theorem and the identity $\cos^2\theta + \sin^2\theta = 1$ to show that the distance u between all points on this ring and the point P is

$$u^2 = R^2 + r^2 - 2Rr\cos\theta \tag{E3.35}$$

(c) Now, at this point, I am going to use a nonobvious mathematical trick that ends up *greatly* simplifying the remainder of the proof. Note that u is basically a function of the ring's angle θ. If we take the θ derivative of both sides of equation E3.35, we get

$$2u\frac{du}{d\theta} = 0 + 0 + 2rR\sin\theta$$

$$\Rightarrow \quad u\,du = rR\sin\theta\,d\theta \tag{E3.36}$$

Combine this with equation E3.34 to show that

$$q_r = \frac{Q}{2Rr}u\,du \tag{E3.37}$$

(d) Show that the potential contributed by each ring is simply

$$\phi_r = \frac{kQ}{2Rr}du \tag{E3.38}$$

(e) Note that the values of u range from $r - R$ near the front of the shell to $r + R$ at the back. Show that if we sum over all rings and convert the sum to an integral with respect to u, we get

$$\phi = \frac{kQ}{r} \tag{E3.39}$$

as claimed in the table.

(f) Now consider a point P *inside* the surface. Carefully argue that every step of the derivation outlined above is exactly the same *except* that the limits of integration are now $R - r$ to $R + r$. Do the integral with these new limits to show that inside the sphere

$$\phi = \frac{kQ}{R} = \text{constant} \tag{E3.40}$$

as claimed in the table.

ANSWERS TO EXERCISES

E3X.1 Both charges are a distance $r = (a^2 + y^2)^{1/2}$ from a point at coordinate y on the y axis. The total potential at such a point is therefore

$$\phi = \frac{kq_1}{r_{P1}} + \frac{kq_2}{r_{P1}} = \frac{2kq}{r} = \frac{2kq}{\sqrt{a^2 + y^2}} \tag{E3.41}$$

Note that this potential is maximum at the origin.

E3X.2 If the charge is negative, its electric field points toward the charge and thus opposite to the displacement $d\vec{r}$. Therefore $\vec{E} \cdot d\vec{r} = [\text{mag}(\vec{E})][\text{mag}(d\vec{r})] \times \cos 180° = -E\,dr$ in this case. However, we also have $E \equiv \text{mag}(\vec{E}) = k|q|/r^2 = -kq/r^2$ when q is negative, because the magnitude of $\vec{E}$ must be positive. Therefore,

$$-\vec{E} \cdot d\vec{r} = -(-E\,dr) = +E\,dr = \left(-\frac{kq}{r^2}\right)dr \tag{E3.42}$$

Therefore,

$$\phi(\infty) - \phi(r_P) = \sum_{\text{all steps}} -\vec{E} \cdot d\vec{r}$$

$$= \sum_{\text{all steps}} -\frac{kq}{r^2}\,dr \tag{E3.43}$$

which is the same as equation E3.13. The rest of the steps follow as before.

E3X.3 Imagine dividing this straight line into a sequence of equal infinitesimal displacements $d\vec{s}$. Since figure E2.4 implies electric field $\vec{E}$ at every point along this line is perpendicular to the line, $\vec{E}$ will be perpendicular to any of these displacements, implying that the potential difference between the end points of any such displacement is $d\phi = -\vec{E} \cdot d\vec{s} = 0$. This means that ϕ must have the same value at all points along the line (this line is an equipotential curve in this situation). This line does go out to infinity, where $\phi = 0$ by definition. But if $\phi = 0$ anywhere along this particular line, it must be zero *everywhere* along that line, as stated.

E3X.4 The arrows you draw *should* look qualitatively like the arrows shown in figure E2.4.

E3X.5 Equation E3.23 is true independent of the sign of Q. It therefore implies that if Q is negative, $\phi(0, 0, z)$ is also negative for all z but becomes less negative (approaching zero) as z becomes very large. This means that the downhill direction is now *toward* the loop, that is, $-\hat{z}$ if z is positive and $+\hat{z}$ if z is negative; so $-(z/|z|)\hat{z}$ now indicates the correct direction. Also, when Q is negative, equation E3.24 becomes

$$\text{mag}(\vec{E}) = \frac{k|Qz|}{(z^2+R^2)^{3/2}} = \frac{-kQ|z|}{(z^2+R^2)^{3/2}} \qquad \text{(E3.44)}$$

Therefore, when Q is negative,

$$\vec{E} = \frac{-kQ|z|}{(z^2+R^2)^{3/2}}\left(-\frac{z}{|z|}\right)\hat{z} = \frac{kQz}{(z^2+R^2)^{3/2}}\hat{z} \qquad \text{(E3.45)}$$

which is the same as before.

E3X.6 The space inside the smaller sphere is inside both shells, so the electric field in this region is zero. The space outside the larger shell is outside both shells, so the field there is as if the total charge of the two shells were concentrated at their common center. Since the total charge is $Q + (-Q) = 0$, the field in the region outside the larger sphere is also zero. The only region that has a nonzero field is the region between the two shells. Since this region is inside the larger shell, that shell's contribution to the field in this region is zero. The field in this region is thus entirely due to the inner shell, which is the same as the field of a charge Q at the center. The field is thus radially outward with magnitude $E = kQ/r^2$ for points a distance r from the center such that $R < r < 2R$, and is zero elsewhere.

E3X.7 The distance $r = 0.5$ cm $= 0.005$ m is very small compared to the length of the rod, so the point at which we are asked to evaluate the field should be *sufficiently close* to the rod that we can use the infinite cylinder model. Assuming that the charge is uniformly distributed on its surface, the charge per unit length on this rod is $\lambda = Q/L$. The diameter of the rod is irrelevant except to indicate that the point we are interested in is *outside* the rod. The magnitude of the electric field at the point in question is therefore

$$\begin{aligned} E &\approx \frac{2k\lambda}{r} = \frac{2kQ}{Lr} \\ &= \frac{2(8.99\times10^9\ \text{N}\cdot\cancel{\text{m}^2}/\text{C}^{\cancel{2}})(3.0\times10^{-9}\ \cancel{\text{C}})}{(1.0\,\cancel{\text{m}})(0.005\,\cancel{\text{m}})} \\ &= 11{,}000\ \frac{\text{N}}{\text{C}} \end{aligned} \qquad \text{(E3.46)}$$

E3X.8 The energy contained in 1 m^3 is roughly the field energy density times the volume in question, or

$$\begin{aligned} &\frac{E^2}{8\pi k}(1\ \text{m}^3) \\ &= \frac{(200\,\cancel{\text{N}}/\cancel{\text{C}})^{\cancel{2}}}{8\pi(9\times10^9\ \cancel{\text{N}}\cdot\cancel{\text{m}^2}/\cancel{\text{C}^2})}(1\,\cancel{\text{m}^3})\left(\frac{1\,\text{J}}{1\,\cancel{\text{N}}\cdot\cancel{\text{m}}}\right) \\ &= 2\times10^{-7}\ \text{J} = 0.2\ \mu\text{J} \end{aligned} \qquad \text{(E3.47)}$$

E4 Conductors

▷ Electric Field Fundamentals
▽ Controlling Currents
Conductors
Driving Currents
Analyzing Circuits
▷ Magnetic Field Fundamentals
▷ Calculating Static Fields
▷ Dynamic Fields

Chapter Overview

Introduction

This chapter is the first of a three-chapter subdivision that explores how we can create and control electric currents. This subdivision is essential background for the subdivisions on magnetic fields and Maxwell's equations. This particular chapter explores how charges flow and distribute themselves in conductors.

Section E4.1: Introduction to Current

Any movement of charge comprises an **electric current.** We quantify a current by describing the *rate* at which charge flows through some usefully defined surface (such as the cross section of a wire at a given point):

$$I = \left|\frac{dQ}{dt}\right| \tag{E4.1}$$

Purpose: This equation defines the magnitude I of an electric current.
Symbols: dQ is the net amount of charge flowing through an imaginary or real surface during a tiny time interval dt.
Limitations: The interval dt must be short enough that $I \approx$ constant.
Notes: The SI unit of current is the **ampere,** where $1\text{ A} \equiv 1\text{ C/s}$.

So that we do not have to determine the sign of the actual **charge carriers** in a given conductor, we define the direction of **conventional current** flow to be the direction in which *positive* carriers would move to create the observed current.

Section E4.2: A Microscopic Model

In many metals, one electron per atom is free to roam as a charge carrier. According to the **Drude model,** these electrons move with large and random thermal velocities between collisions with atoms. An external electric field $\vec{E}$ gives each electron a velocity component opposite to $\vec{E}$ before the next collision rerandomizes its velocity. The *average* velocity that electrons gain between collisions is their **drift velocity** $\vec{v}_d$. On the average, collisions between electrons and lattice atoms transfer any energy that the electrons gain from the field into thermal energy in the lattice.

Section E4.3: Current Density

The **current density** $\vec{J}$ usefully describes current flow at a given point in a conductor:

$$\vec{J} \equiv nq\vec{v}_d = \rho\vec{v}_d \tag{E4.9}$$

Purpose: This equation defines the current density $\vec{J}$ at a given point in a conductor.
Symbols: q is the charge of a charge carrier, $\vec{v}_d$ is a carrier's drift velocity, n is the number of carriers per unit volume, and $\rho = nq$ is their charge density.

Limitations: This equation assumes a single type of charge carrier.

Notes: The current density $\vec{J}$ has units of current per area (A/m^2) and points in the direction of the conventional current: i.e., parallel to $\vec{v}_d$ if q is positive or opposite to $\vec{v}_d$ if q is negative.

If $\vec{J}$ is uniform across a conductor's cross section, then

$$\vec{I} \equiv \vec{J}A = nqA\vec{v}_d = \rho A\vec{v}_d \quad \text{(E4.10)}$$

Purpose: This equation links the vector current $\vec{I}$ flowing through a given cross-sectional area A of a conductor to the current density $\vec{J}$ and the drift velocity $\vec{v}_d$ of its charge carriers.

Symbols: q is the charge of a charge carrier, n is the number of carriers per unit volume, and $\rho \equiv nq$ is their charge density.

Limitations: This equation assumes that $\vec{J} \approx$ constant across the cross-sectional area and that there is only one type of charge carrier.

Note: The vector current $\vec{I}$ points in the direction of the conventional current.

We define a conductor's **conductivity** σ_c such that

$$\vec{J} = \sigma_c\vec{E} \quad \text{(E4.11)}$$

Purpose: This equation defines a conductor's **conductivity** σ_c.

Symbols: $\vec{E}$ is the electric field at a given point inside the conductor, and $\vec{J}$ is the current density at that point.

Limitations: This equation assumes that $\vec{J}$ is parallel to $\vec{E}$.

Notes: Typically, σ_c is independent of E for a given material, so $\vec{J} \propto \vec{E}$. Note that σ_c is *conductivity*, but σ is *charge per unit area*. Conductivity has SI units of $C^2 \cdot s \cdot m^{-3} \cdot kg^{-1}$.

Section E4.4: Static Charges on Conductors

If we put excess charges on a conductor, these charges, being free to move, rapidly rearrange themselves into a **static equilibrium** distribution where they can remain at rest. Since $\vec{v}_d \propto \vec{E}$, and $\vec{v}_d = 0$ if the charges are at rest, $\vec{E}$ must be zero *everywhere* inside the conductor in static equilibrium. This in turn implies that (1) the potential ϕ must be the same at all points inside the conductor, (2) $\vec{E}$ at points on its surface must be perpendicular to that surface, and (3) any excess charge ends up on that surface.

Section E4.5: Capacitance

A **capacitor** is a pair of initially neutral conductors called **plates.** If we move charge Q from one plate to the other, the potential difference $|\Delta\phi|$ between the plates is proportional to Q. The constant of proportionality, which in a vacuum depends only on the size, shape, and arrangement of the conductors, is

$$C \equiv \frac{Q}{|\Delta\phi|} \quad \text{(E4.15)}$$

Purpose: This defines the **capacitance** C of a two-conductor capacitor.

Symbols: Q is the charge we have moved from one initially neutral conductor to the other; $|\Delta\phi|$ is the resulting potential difference between them.

Limitations: The conductors must be isolated from other charges.

Note: The SI unit of capacitance is the **farad,** where $1\ F \equiv 1\ C/V$.

Table E4.2 in this section lists the capacitances of several conductor arrangements.

E4.1 Introduction to Current

In chapter E1, we saw that we can classify most substances as being either **conductors** or **insulators.** The electrons in an *insulator* are firmly attached to its molecules: even if we apply a strong electric field to an insulator, very few electrons move freely in response to the field. In a good *conductor,* though, some electrons are essentially detached from their atoms and are free to roam around inside the conductor in response to an applied electric field.

For example, imagine that we place a metal bar between two oppositely charged metal spheres (most metals are good conductors). Electrons from the negatively charged sphere, because they repel each other, will flow through the bar, filling in the deficit of electrons in the positively charged sphere to which they are attracted (see figure E4.1). For a brief time (a tiny fraction of a second in this case) an **electric current** will flow through the bar.

Any movement of charge is an electric current

Any net flow of charge is an electric current, whether such a flow of charge is brief or steady, whether it is driven *by* an electric field (as the electron flow in the bar in figure E4.1 is) or driven by other forces *against* the forces exerted by an electric field (as it is inside a battery), whether the moving charged particles in the current (**charge carriers**) are negatively charged (as are the free electrons in a metal) or positively charged (as in a proton beam produced by an accelerator) or both positively *and* negatively charged (as are the ions in a car battery or a plasma).

We can quantify an electric current by describing *the rate at which charge flows* through a real or imaginary surface oriented perpendicular to the flow (see figure E4.2). For example, in the situation illustrated in figure E4.1b, if the charge on the negative sphere becomes more positive by dQ (as a result of the depletion of negative charge) in an infinitesimal time interval dt, we say that the current (which is conventionally denoted by the symbol I) flowing through the left end face of the bar during that interval is

$$I = \left|\frac{dQ}{dt}\right| \tag{E4.1}$$

Purpose: This equation defines the magnitude I of an electric current.

Symbols: dQ is the net amount of charge flowing through an imaginary or real surface during a tiny time interval dt.

Limitations: The interval dt must be short enough that $I \approx$ constant during dt.

Notes: The SI unit of current is the **ampere,** where $1\text{ A} \equiv 1\text{ C/s}$.

Current flows of 1 to 50 A are common in household circuits; currents in electronic circuits are commonly expressed in *milliamperes* ($1\text{ mA} = 10^{-3}\text{ A}$) or *microamperes* ($1\ \mu\text{A} = 10^{-6}\text{ A}$).

The direction of a current is defined to be the direction in which *positive* charge flows

At the macroscopic level it is difficult to distinguish the flow of negative particles in one direction from the flow of positive particles in the opposite direction. For example, in figure E4.1b, the right sphere becomes less positive and the left sphere becomes more positive when current flows through the bar. If we did not know that it is actually electrons that are free to move in the bar, we could easily assume that positive charge was flowing to the left. The *direction* in which any current flows is conventionally defined to

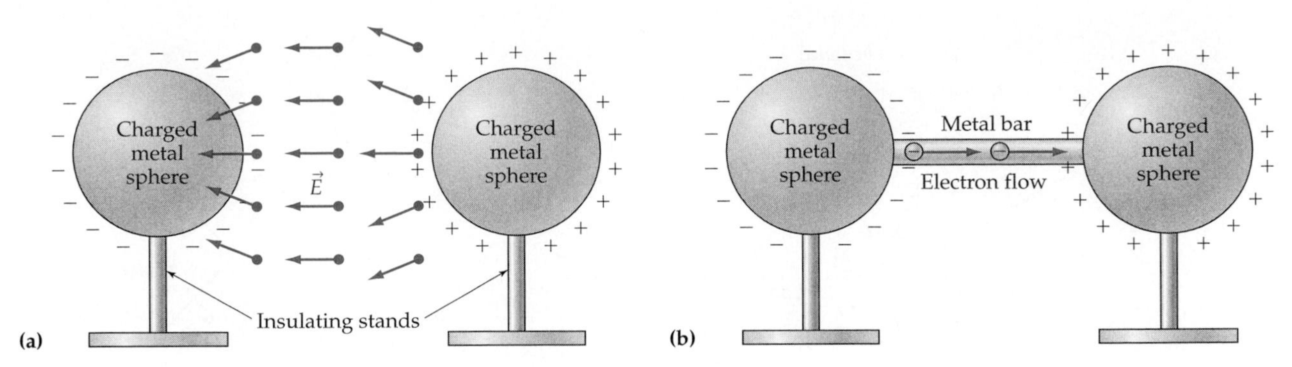

Figure E4.1
(a) A field diagram showing part of the electric field between two metal spheres with opposite charges. (b) If the spheres are connected by a metal bar, the field will drive electrons through the bar from the negative sphere to the positive sphere (the force is *opposite* to the field direction because electrons have negative charge).

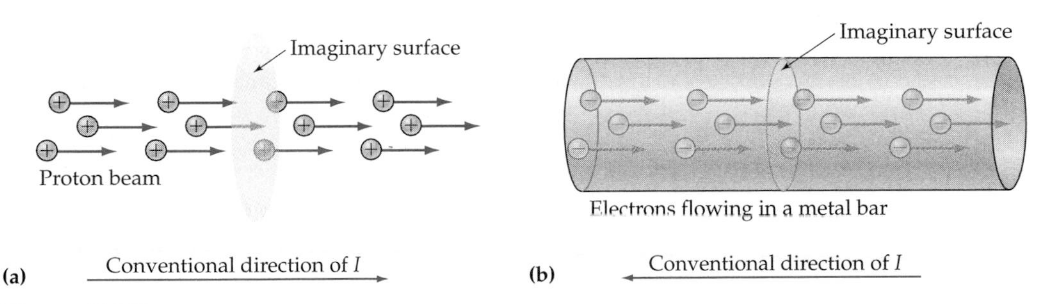

Figure E4.2
(a) A beam of protons produced by a particle accelerator. (b) Electrons flowing in a metal bar. The current in either case can be quantified by determining the magnitude of the charge $|dQ|$ that flows in time dt through an imaginary surface perpendicular to the flow that is large enough to enclose the entire flow: $I \equiv |dQ/dt|$.

be the direction in which *positive* charges would be flowing if they were carrying the charge. In a proton beam, the direction of the current represented by the beam is thus the same as that of the velocity of the (positively charged) protons. In the case of electrons flowing in a metal, however, the **conventional current** flows *opposite* to the velocity of the flowing (negatively charged) electrons. (The conventional definition of current direction was established long before anyone knew electrons existed!)

E4.2 A Microscopic Model

While I have emphasized that current can be used to describe *any* kind of flow of electric charge, in practice the kind of electric current that we will most commonly encounter in the laboratory (and in this course) is the flow of electrons through metals. In this section, we will consider this particular kind of electric current flow in greater depth by exploring a model of electron flow first proposed by P. K. Drude in 1900, shortly after the electron was discovered. This model is thus called the **Drude model** of electron conduction in metals.

The Drude model is a simple microscopic model of current flow in metals

In a typical metal, some electrons are detached from their parent atoms and are free to roam around the metal. The Drude model thus imagines a metal to consist of a fairly rigid lattice of positively charged atoms surrounded by

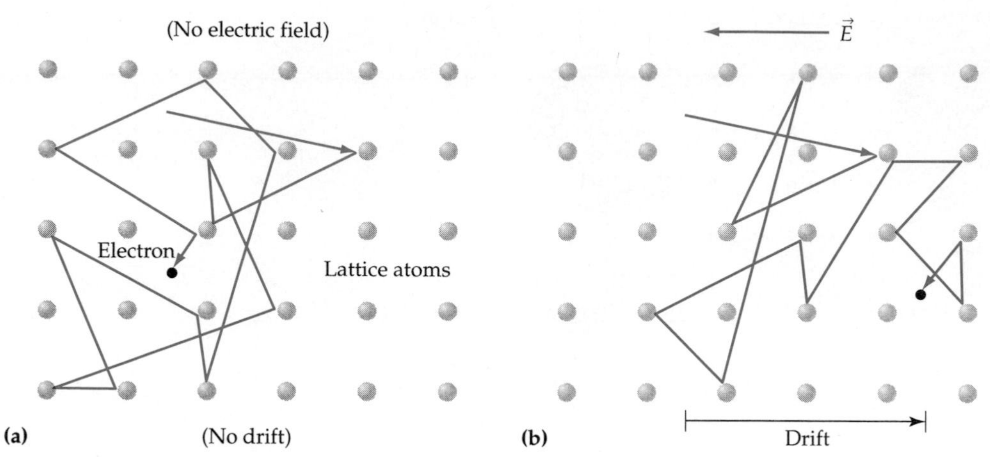

Figure E4.3
(a) When there is no applied external electric field, an electron in a metal bounces randomly from atom to atom in the lattice, getting nowhere on the average. (b) When an external electric field is applied to the metal, an electron will accelerate between collisions in the direction of the electric force on the electron. This causes the electron's average position to drift in that direction as time passes.

a "gas" of free electrons. As we saw in unit C, the particles of any substance with a nonzero absolute temperature are in constant random motion: the object's temperature reflects the average kinetic energy of this random motion. The atoms in a metal are thus constantly vibrating around their lattice positions, while the free electrons bounce randomly around the lattice at very high speeds (on the order of magnitude of 10^6 m/s!) as if they were balls in a pinball game. The Drude model *assumes* that electrons only interact with the lattice during short-duration collisions with individual atoms and otherwise move as free particles.

In the absence of an applied electric field, the motion of the bouncing electrons is completely random (see figure E4.3a). On the average, we will see as many electrons go through an imaginary surface in the metal one way as the other: the net flow across the surface (and thus the current in the metal) is thus zero. This would be like a *level* (and frictionless) pinball game where the balls carom off obstacles randomly and endlessly without any sense of direction.

An electric field causes bouncing electrons to drift slowly opposite to field

If we apply a uniform external electric field $\vec{E}$ to the metal, though, then in the time between collisions, the field causes an electron to accelerate at the rate $\vec{a} = \vec{F}_e/m = -e\vec{E}/m$, where $-e$ is the electron's charge and m is its mass. In the time between collisions, each electron will thus acquire a small velocity in a direction opposite to the direction of $\vec{E}$ that it wouldn't have had otherwise: this causes the electrons collectively to *drift* in that direction (see figure E4.3b). This is analogous to a *tilted* pinball game: while a given ball will still carom off obstacles in wild and unpredictable directions, it will *on the average* move downward toward the bottom edge of the playing area.

What is the magnitude of this drift velocity? Consider a single electron that at time $t = 0$ collides with an atom in the lattice and acquires an initial velocity $\vec{v}_0$. In the absence of an electric field, the electron would continue to have this velocity until the next collision. In the presence of the applied field, the electron's velocity when it reaches the next atom at time t is

$$\vec{v}(t) = \vec{v}_0 + \vec{a}t = \vec{v}_0 - \frac{e\vec{E}}{m}t \tag{E4.2}$$

assuming that the electric field $\vec{E}$ experienced by the electron is constant between collisions. When the electron then collides with the next atom, its velocity is again randomized (you can imagine that it "forgets" its original direction of motion), and the process starts over. The **drift velocity** $\vec{v}_d$ of this electron can be found by averaging its velocity over many collisions:

$$\vec{v}_d \equiv \vec{v}_{\text{avg}} = \vec{v}_{0,\text{avg}} - \frac{e\vec{E}}{m}t_{\text{avg}} = -\frac{e\vec{E}}{m}\tau \qquad \text{(E4.3)}$$

where $\tau \equiv t_{\text{avg}}$ is the average time between collisions.† Note that since we are assuming that the initial velocities $\vec{v}_0$ just after each collision are completely random, the *average* of $\vec{v}_0$ will be zero.

This result is very interesting. It tells us that the drift speed of an electron in a metal is proportional to the electron's charge-to-mass ratio e/m, the strength of the applied electric field $\vec{E}$, and the average time between collisions τ, which depends on the random thermal speed of the electrons, the average distance between lattice atoms, and other characteristics of the metal. For example, τ will be inversely dependent on how large a target an atom appears to be to a passing electron. In excellent conductors such as copper, the electrons travel a fairly long distance between collisions (see problem E4S.2), implying that lattice atoms must represent pretty small targets. The Drude model would thus suggest that the conduction electrons in copper have to get quite close to an atomic nucleus before they experience strong enough forces in any specific direction to cause a bounce.

Drift speed is proportional to the field strength and time between collisions

The most important thing to note is that this model predicts that the electron drift speed is *proportional* to the magnitude E of the applied field. If the electrons in a metal were *completely* free, they would experience uniform *acceleration* due to the applied field, and thus their velocities would increase with time, like freely falling objects. The fact that the electrons move at a *constant* (drift) velocity in the presence of a constant electric field means that they behave more as objects do that are falling at their terminal velocities through a viscous medium.

As electrons accelerate in response to the applied field, they also pick up some kinetic energy. On the average, though, these electrons lose any kinetic energy they have gained in collisions with lattice atoms (otherwise their average velocity would not be constant). This model therefore successfully explains why wires conducting large currents get hot.

Collisions transfer energy to the wire, making it hot

We can easily calculate the drift speed of electrons in a typical metal wire. Let the number of free electrons (also called **conduction electrons**) per cubic meter of the metal be n, and let their drift speed be v_d. In a short time interval Δt, these electrons will move a distance $v_d \Delta t$. This means that the volume of electrons that pass in time Δt through an imaginary surface of area A spanning the wire perpendicular to the flow will be $Av_d\Delta t$ (see figure E4.4). The magnitude of the current passing through that surface will thus be

Calculating the drift speed

$$I = \frac{|\text{charge}|}{\text{time}} = \left|\frac{\text{charge}}{\cancel{\text{electron}}}\right|\left(\frac{\cancel{\text{electrons}}}{\cancel{\text{volume}}}\right)\left(\frac{\cancel{\text{volume}}\text{ through }A}{\text{time}}\right)$$

$$= en\left(\frac{Av_d\cancel{\Delta t}}{\cancel{\Delta t}}\right) = neAv_d \qquad \text{(E4.4)}$$

†See E. M. Purcell, *Electricity and Magnetism*, 2d ed., McGraw-Hill, New York, 1985, pp. 136–137, for a careful argument about why $\vec{v}_d = -(e\vec{E}/m)\tau$ and not $-\frac{1}{2}(e\vec{E}/m)\tau$, as one might naively expect.

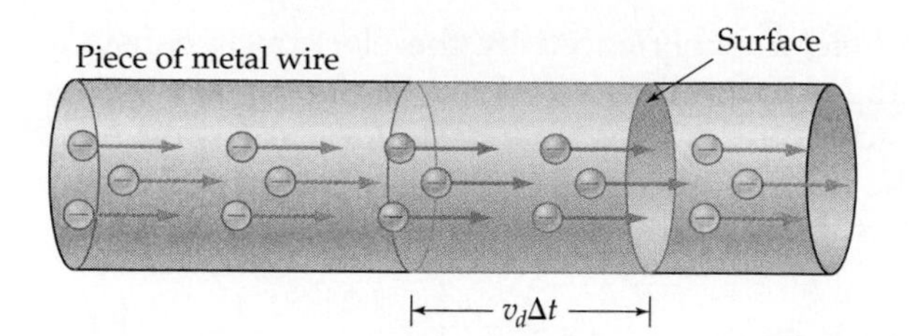

Figure E4.4
If we assume all the electrons in a conductor have the same drift speed, all the electrons in the shaded volume will go through the surface in time Δt.

In many metals including copper, silver, and gold, each atom contributes roughly *one* electron to the pool of free electrons. This means that the number density of free electrons n in such metals is the same as the number density of atoms in the metal. If we divide the metal's mass density (in kilograms per cubic meter) by its atomic weight (in kilograms per mole), we get the metal's molar density (in moles per cubic meter). Since 1 mol of any elemental metal contains Avogadro's number of atoms, multiplying the molar density by Avogadro's number gives the number density (atoms per cubic meter). For metals that contribute more free electrons per atom (such as aluminum, which contributes 3), we multiply this atomic number density by the number of free electrons per atom to get the number density of free electrons.

Example E4.1

Problem Consider a household copper wire (see figure E4.5) with a diameter of 1.0 mm carrying a current of $1.0 \text{ A} = 1.0 \text{ C/s}$ (typical values in such a case). What is the drift speed of the electrons in the wire?

Translation The cross-sectional area A of the wire is $\pi r^2 = \pi(0.0005 \text{ m})^2 = 7.9 \times 10^{-7} \text{ m}^2$.

Model Since roughly one electron is detached from every copper atom, the number density n of conduction electrons in copper will be the same as the number density of atoms.

Figure E4.5
How fast do electrons typically move through a wire like this?

Solution Copper's atomic weight is 63.5 g/mol and its density is 8900 kg/m^3, so the number density of atoms is

$$n = \left(\frac{8900\,\cancel{\text{kg}}}{\text{m}^3}\right)\left(\frac{1\,\cancel{\text{mol}}}{0.0635\,\cancel{\text{kg}}}\right)\left(\frac{6.02 \times 10^{23}\,(\text{atoms})}{1\,\cancel{\text{mol}}}\right) = \frac{8.4 \times 10^{28}}{\text{m}^3} \tag{E4.5}$$

So the electron drift velocity through this wire must be

$$v_d = \frac{I}{neA} = \frac{1.0\,\cancel{\text{C}}/\text{s}}{(8.4 \times 10^{28}\,\text{m}^{-3})(1.6 \times 10^{-19}\,\cancel{\text{C}})(7.9 \times 10^{-7}\,\text{m}^2)}$$
$$= 9.4 \times 10^{-5}\,\text{m/s} \approx 8.2\,\text{m/day (!)} \tag{E4.6}$$

Evaluation The units are right, but the speed is astonishingly slow!

The average thermal speed of electrons in a metal is very roughly 10^6 m/s. We see that the drift speed of the electrons in the wire is about a factor of 10^{10} smaller than the typical speed of the electrons due to their thermal motion! The fundamental reason why this drift speed is typically so small is that a wire contains a *huge* number of electrons: they don't have to move very fast to carry an adequate amount of charge past any given point in the wire.

Even though drift speed is very slow, circuits respond quickly to changes

This exercise's result implies if you threw a light switch 4 m away from the lightbulb, it would take about 12 *hours* for electrons going through the switch to reach the bulb! So how is it that the bulb goes on almost immediately? We will discuss the detailed mechanism in chapter E5, but for now you might think of an analogy to water in a pipe. Forcing water into one end of a pipe exerts pressure on the water already in the pipe, which pushes water out the other end immediately, even if it would take quite a while for a given water molecule to travel the pipe's length. The *mechanism* that drives electrons is more complicated, but the net effect is essentially the same.

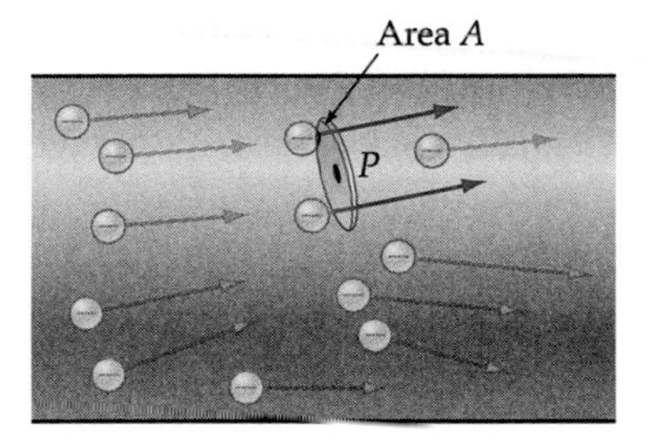

Figure E4.6
We can describe the flow rate at an arbitrary point P in an object by measuring how much current per unit area flows through a small surface with area A centered on point P, and then we take the limit as $A \to 0$.

Exercise E4X.1

An aluminum wire 1.5 mm in diameter carries a current of 5.2 A = 5.2 C/s. The density of aluminum is 2700 kg/m^3, and its atomic mass is 26.98 g/mol. What is the drift speed of the electrons in this wire?

E4.3 Current Density

We need a *local* measure of current flow

Note that in deriving equation E4.4, I assumed that the electron drift speed is constant everywhere in the wire. This turns out to be a pretty good assumption for thin metal wires, for reasons we will discuss in chapter E5. But just as the current in a river may not flow at the same rate at all places on the river, so the current flowing through an object may in principle flow at different rates at different places. Indeed the flow of charge *can* vary significantly from place to place in large and/or irregularly shaped conductors. In such circumstances, it helps us to be able to talk about the flow of charge at a given point in the object.

Imagine that we'd like to describe the current in the neighborhood of a point P in a metal object (see figure E4.6). Imagine a small surface with area A around P perpendicular to the direction of the current at that point. If we choose this surface to be small enough, all charges moving through it have essentially the same local drift velocity $\vec{v}_d(P)$ (read "the drift velocity at point P"). Equation E4.4 then implies that the current flowing through this small surface is

$$I = neAv_d(P) \tag{E4.7}$$

The approximation that all the charges have the same velocity $v_d(P)$ becomes more and more accurate as the area of the surface gets smaller, but the value of the localized current given in equation E4.7 unhelpfully goes to zero as $A \to 0$. The ratio I/A, though, approaches a well-defined, meaningful value in this limit:

$$\lim_{A \to 0} \frac{I}{A} = ne\, v_d(P) \tag{E4.8}$$

With this in mind, we define the **current density** $\vec{J}$ at a point P to be

The definition of *current density*

$$\vec{J} = nq\vec{v}_d = \rho\vec{v}_d \tag{E4.9}$$

Purpose: This equation defines the current density $\vec{J}$ at a given point in a conductor.
Symbols: q is the charge of a charge carrier, $\vec{v}_d$ is the carriers' drift velocity, n is the number density (number per unit volume) of carriers, and $\rho = nq$ is their charge density.
Limitations: This equation assumes a single type of charge carrier.
Notes: The current density $\vec{J}$ has units of current per area (A/m^2) and points in the direction of the conventional current: i.e., parallel to $\vec{v}_d$ if q is positive or opposite to $\vec{v}_d$ if q is negative.

The magnitude of $\vec{J}$ is the current flow *per unit area* in the limit that the area goes to zero around P (so $\vec{J}$ is an *area* density). If $\vec{J}$ is constant over and perpendicular to the conductor's *entire* cross-sectional area A at a point, we can define a vector $\vec{I}$ that represents the total current flowing past that point:

The relationship between current and current density

$$\vec{I} \equiv \vec{J}A = nqA\vec{v}_d = \rho A\vec{v}_d \tag{E4.10}$$

Purpose: This equation links the vector current $\vec{I}$ flowing through a given cross-sectional area A of a conductor to the current density $\vec{J}$ and the drift velocity $\vec{v}_d$ of its charge carriers.
Symbols: q is the charge of a charge carrier, n is the number density of carriers, and $\rho \equiv nq$ is their charge density.
Limitations: This equation assumes that $\vec{J}$ is perpendicular to the cross-sectional area and is approximately constant across that area, and that there is only one type of charge carrier.
Note: The vector current $\vec{I}$ points in the direction of conventional current flow.

Note that mag($\vec{I}$) here is the same as I in equation E4.4. The approximation that the current density (and thus the drift velocity) is constant over a conductor's entire cross section is good for thin wires but fails for large or irregular objects.

We can define a conductor's **conductivity** σ_c as follows:

$$\vec{J} = \sigma_c \vec{E} \tag{E4.11}$$

Definition of a conductor's *conductivity* σ_c

Purpose: This equation defines a conductor's **conductivity** σ_c.
Symbols: $\vec{E}$ is the electric field at a given point inside the conductor, and $\vec{J}$ is the current density at that point.
Limitations: This equation assumes that $\vec{J}$ is parallel to $\vec{E}$ (which is only rarely untrue).
Notes: Typically, σ_c is independent of E for a given material, so $\vec{J} \propto \vec{E}$. Note that σ_c is the symbol for *conductivity*, but σ is the symbol for *charge per unit area*. Conductivity has SI units of $C^2 \cdot s \cdot m^{-3} \cdot kg^{-1}$.

As the first note implies, a conductor's conductivity σ_c typically depends only on the conductor's characteristics. We can see that this is true for any conductor accurately described by Drude's model, as equations E4.3 and E4.9 imply that

$$\vec{J} = nq\left(\frac{q\vec{E}}{m}\right)\tau = \left(\frac{nq^2\tau}{m}\right)\vec{E} \quad \Rightarrow \quad \sigma_c = \frac{nq^2\tau}{m} \tag{E4.12}$$

How σ_c depends on a conductor's characteristics in the Drude model

where n, q, and m are the number density, charge, and mass, respectively, of the conductor's charge carriers and τ is the mean time between collisions. The conductivity of most conductors does vary modestly with temperature (because the thermal speed of the electrons increases with temperature, the mean time τ between collisions and thus the conductivity decrease), but σ_c does not generally vary significantly with the magnitude or direction of the applied field $\vec{E}$ as long as mag($\vec{E}$) is small. When this is true, the current density $\vec{J}$ is directly proportional to $\vec{E}$. Table E4.1 lists the conductivities of various substances at room temperature.

However, $\vec{J}$ is *not* always proportional to $\vec{E}$ (that is, σ_c is not always independent of E). For example, σ_c for air increases sharply as E goes above $\approx 3 \times 10^6$ N/C, because physical processes that ionize air molecules become possible at that point. Many other insulators have similar "breakdown" points. Many semiconductors have conductivities that vary fairly smoothly with E.

Circumstances where $\vec{J}$ is not very proportional to $\vec{E}$

Table E4.1 Conductivities of various substances (in $C^2 \cdot s \cdot m^{-3} \cdot kg^{-1}$) at room temperature

Substance	Conductivity	Note
Silver	6.3×10^7	Note that metals have conductivities basically on the order of magnitude of 10^7.
Copper	5.9×10^7	
Gold	4.1×10^7	
Aluminum	3.6×10^7	
Iron	1.0×10^7	
Lead	4.6×10^6	
Nichrome	6.7×10^5	"Nichrome" is a nickel-chromium alloy used in electric heating elements because of its low conductivity.
Carbon	4.9×10^4	
Seawater	4×10^{-2}	The conductivity of water depends significantly on the number and types of ions dissolved in it.
Pure water	4.0×10^{-8}	
Glass	$\approx 10^{-12}$	Note the stark difference in σ between metals and these typical insulators. This is mostly so because there are very few free charge carriers in insulators.
Rubber	$\approx 10^{-13}$	

Using a conductor's empirical conductivity and equation E4.12, we can estimate the mean time between collisions τ predicted by the Drude model. For copper, we get about 2.5×10^{-14} s (see problem E4S.2). This may not seem long, but a typical electron in copper has a thermal speed of roughly 10^6 m/s, so it will travel an average of ≈ 25 nm between collisions. This is a *long* way in a lattice where copper atoms are ≈ 0.2 nm apart! Copper atoms must present pretty small targets to passing electrons (assuming the Drude model is correct).

Example E4.2

Problem Estimate the electric field strength inside a silver wire 0.05 mm in diameter that is conducting a total current of 5.2 A.

Translation Let $I = 5.2$ A be the current in the wire and $r = 0.025$ mm be its radius.

Model Assume that the current density is the same at all points inside the wire. If we take the magnitude of both sides of equations E4.10 and E4.12 and combine them, we get

$$I = JA = (\sigma_c E)A = \sigma_c E \pi r^2 \tag{E4.13}$$

where in the last step, I have used the fact that the wire's cross-sectional area is $A = \pi r^2$.

Solution Solving for the electric field strength E, we get

$$\begin{aligned} E &= \frac{I}{\pi r^2 \sigma_c} = \frac{5.2\,\cancel{\text{C}}/\text{s}}{\pi(0.025 \times 10^{-3}\,\text{m})^2(6.3 \times 10^7\,\text{C}^{\cancel{2}} \cdot \text{s} \cdot \text{m}^{-3} \cdot \text{kg}^{-1})} \\ &= 42\,\frac{\cancel{\text{kg}} \cdot \cancel{\text{m}}}{\cancel{\text{s}^2} \cdot \text{C}}\left(\frac{1\,\text{N}}{1\,\cancel{\text{kg}} \cdot \cancel{\text{m}}/\cancel{\text{s}^2}}\right) = 42\,\text{N/C} \end{aligned} \tag{E4.14}$$

Evaluation Note that the units are right. We see that a pretty small field is able to drive a pretty big current in silver.

Exercise E4X.2

Use equation E4.11 to verify that conductivity has SI units of $\text{C}^2 \cdot \text{s} \cdot \text{m}^{-3} \cdot \text{kg}^{-1}$.

Exercise E4X.3

Estimate mag($\vec{E}$) inside a copper wire 1.0 mm in diameter conducting a steady current of 1.0 A (if $\vec{J} \approx$ constant throughout the wire).

E4.4 Static Charges on Conductors

When we place charges on or inside an insulating object, they stay put. When we place charges on or inside a conducting object, however, they are free to distribute themselves on the object as they please. We find empirically that eventually (typically within a fraction of a microsecond), any excess charge placed on a conducting object rapidly arranges itself into a **static equilibrium** distribution where all charges can remain at rest. In this section, we will explore what such distributions look like in various cases.

In sections E4.2 and E4.3, we have seen that charges will move in a conductor whenever the electric field $\vec{E}$ in that conductor is nonzero. When the charges have found a *static* equilibrium distribution, they are at rest by definition. Therefore, the most basic characteristic of the equilibrium distribution on any conductor is as follows:

The basic characteristic of equilibrium charge distributions on conductors

In static equilibrium, the charges on a conducting object will have arranged themselves so that $\vec{E} = 0$ *everywhere* inside that object.

This statement has some interesting consequences: in static equilibrium,

1. The potential ϕ is the same at all points inside a conductor. (*Proof:* According to equation E3.16, $d\phi = -\vec{E} \cdot d\vec{r} = 0$ for a displacement $d\vec{r}$ between any two points inside the conductor if $\vec{E} = 0$.)
2. The electric field at points just barely outside a conductor's surface must point perpendicular to that surface. (*Proof:* The conductor's surface is an equipotential surface. We saw in chapter E3 that $\vec{E}$ is always perpendicular to an equipotential surface.)
3. Any excess charge on a conductor ends up on its surface. (We will prove this in chapter E10, where we will see that a clump of excess charge in a conductor's interior necessarily creates a nonzero $\vec{E}$ in its vicinity.)

The equilibrium charge distribution on various conducting shapes

For example, imagine putting some excess charge Q on a conducting sphere. All our conditions for static equilibrium are met if this excess charge ends up *uniformly distributed on the sphere's surface* (see figure E4.7a). Such a distribution obviously satisfies the third statement, and table E3.1 tells us that $\vec{E} = 0$ and $\phi = kQ/R =$ constant inside the sphere's surface. Moreover, since the *external* field of such a distribution is as if the excess charge Q were concentrated at the sphere's center, $\vec{E}$ outside the surface will point directly toward or away from that center and so will be perpendicular to the sphere's surface, consistent with the second statement.

Finding the equilibrium charge distribution on nonspherical conductors is much harder. Qualitatively, though, excess charges on a conductor will distribute themselves to be as far from each other as possible. On a nonspherical conductor, this means that the surface charge density in equilibrium tends to be highest in parts of the conductor farthest from the rest of the conductor. For example, figure E4.7b shows the exact solution for the equilibrium charge distribution on a thin circular conducting disk (σ is the charge per unit area on the disk's surface). Note how the charge density is highest at the disk's edge.

Polarization of conductors by external fields

Fields from external charges can affect the equilibrium charge distribution. For example, imagine bringing a positive point charge close to a neutral metal sphere (see figure E4.8a). The positive charge attracts electrons to the side of the sphere closest to it, leaving the opposite side of the sphere positively charged. Bringing the charge closer makes the crowding more extreme (figure E4.8b). This is like the process of electrostatic polarization discussed in chapter E1 except more extreme, since the charges responding to the field are not trapped in

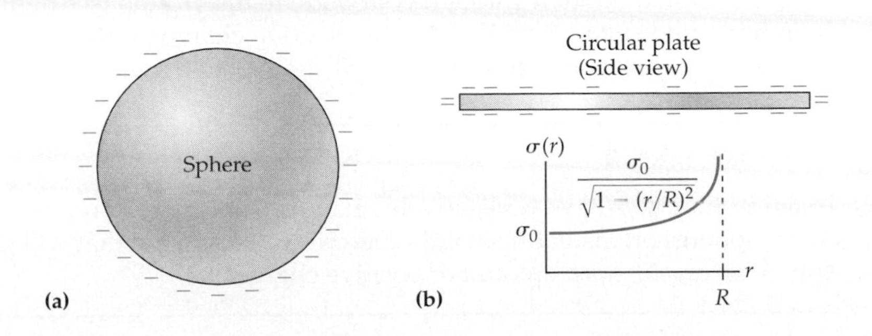

Figure E4.7
(a) In static equilibrium, excess charge placed on a sphere will end up uniformly distributed on its surface. (b) The equilibrium charge distribution on a flat, circular conducting plate (which is shown in side view here) has a surface density σ that increases toward the disk's edge.

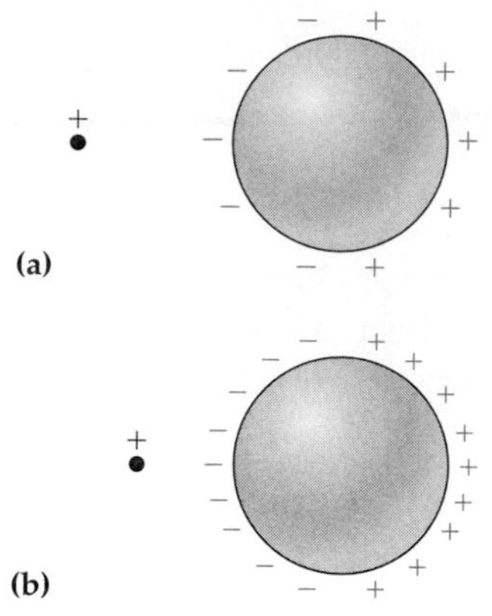

Figure E4.8
(a) Bringing a positive point charge near a conducting sphere creates a surface charge distribution that shields the conductor's interior from the field.
(b) Bringing the charge closer increases the separation of charge on the conductor.

atoms or molecules but in fact can move all the way across the sphere. The interaction between a charge and a neutral conductor is therefore typically much stronger than that between a charge and a neutral insulator.

(Note, by the way, that the presence of the external point charge changes the sphere's surface charge distribution and thus the sphere's external electric field. This is the kind of thing I was worrying about in chapter E2 when I said that we should measure an object's electric field by using a test charge "small enough that its presence does not disturb the charge distribution" creating the field.)

As before, though, the equilibrium distribution of charge on the sphere's surface produce an electric field that *completely cancels* the point charge's field inside the conductor so that $\vec{E} = 0$ there. The surface charge distribution's countervailing electric field thus has the effect of completely shielding the conductor's interior from external electric fields!

A conducting enclosure shields its interior from electric fields

This works even if the conductor is *hollow*. Imagine that we magically remove most of the interior of the sphere in figure E4.8a, making it a hollow shell. Since the charge distribution that ensures that $\vec{E} = 0$ inside the sphere is on the sphere's surface, removing the sphere's neutral core does not change this distribution, so $\vec{E}$ is *still* zero in the newly created cavity. If you were to sit in this cavity, you would be completely shielded from the effects of outside electrostatic fields by the canceling field created by the surface charge.

The same argument applies to nonspherical hollow conductors. Indeed, it turns out that regions where the net electric field is significantly nonzero only penetrate a negligible distance through any holes in the hollow conductor as long as they are small compared to the size of the enclosed cavity. Even a cage of fairly coarse wire mesh works almost as well as a solid hollow conductor.

Even time-dependent external electric fields are effectively screened out as long as they vary slowly compared to the time it takes the charges to redistribute themselves on the conductor (which is typically a fraction of a microsecond). You may have noticed that when your car crosses a metal-framed bridge, your radio may cut out, particularly if you are tuned to an AM station. Charges on the bridge's metal frame are able to move quickly enough to cancel out the comparatively slowly varying electromagnetic field that carries the radio signal.

Example E4.3

Problem Imagine you have a metal sphere on an insulating stand and a negatively charged rubber rod. Describe at least one way that you could use the rod to give the sphere a net *positive* charge. (*Hint:* You can take advantage of the fact that your body is a conductor.)

Solution If you bring the rod close to one side of the sphere while you touch the other side with your hand, the negatively charged rod will repel electrons from the sphere through your body to the ground (see figure E4.9a). Removing your hand then isolates the positive charge on the sphere: it will remain even after you withdraw the rod (figure E4.9b and c).

Exercise E4X.4

Imagine that you have a negatively charged rubber rod and *two* identical metal spheres on insulating stands. Describe a procedure that will give each sphere an *exactly equal* amount of positive charge.

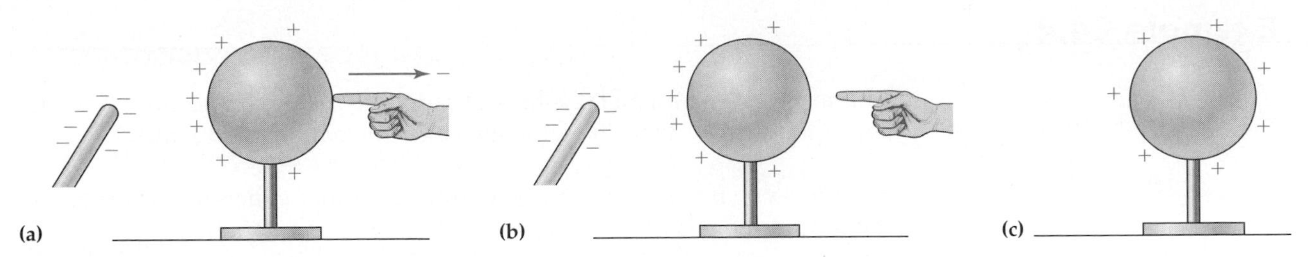

Figure E4.9
How to use a charged rod to give a conducting sphere an opposite charge.

E4.5 Capacitance

Definition of a capacitor

Consider two initially neutral conductors A and B that are isolated from other charged objects. Imagine that we move a certain amount of charge Q from A to B, and we allow everything to come to static equilibrium. Since each object has a certain fixed potential throughout its interior, the potential difference $\Delta\phi \equiv \phi_B - \phi_A$ between the two objects is precisely defined. We call such a pair of objects a **capacitor.** We call the two conductors that comprise the capacitor the **plates** of the capacitor (though they often do not have the shape of plates).

We define a capacitor's **capacitance** C to be

$$C \equiv \left| \frac{Q}{\Delta\phi} \right| \qquad \text{(E4.15)}$$

Purpose: This equation defines the capacitance C of a two-conductor capacitor.
Symbols: Q is the charge we have moved from one initially neutral conductor to the other; $|\Delta\phi|$ is the resulting potential difference between them.
Limitations: This equation assumes that the conductors are isolated from other charges.
Note: The SI unit of capacitance is the **farad,** where $1\text{ F} \equiv 1\text{ C/V}$.

A capacitor with a larger capacitance thus holds more charge on its plates for a given potential difference than one with a smaller capacitance. This is the root idea behind the terms *capacitor* and *capacitance.*

Exercise E4X.5

What is the farad in terms of more basic units of coulomb, kilogram, meter, and second?

Capacitance C depends only on the size, shape and separation of the conductors

A capacitor's capacitance C turns out to depend on only the size, shape, and arrangement of the plates, not on the magnitude of the potential difference $|\Delta\phi|$ between them. Generally C is hard to calculate for a given arrangement of conductors, but example E4.4 illustrates one case where it can be done pretty easily. Table E4.2 lists capacitances for some common conductor arrangements.

Table E4.2
Capacitances for various conductor arrangements

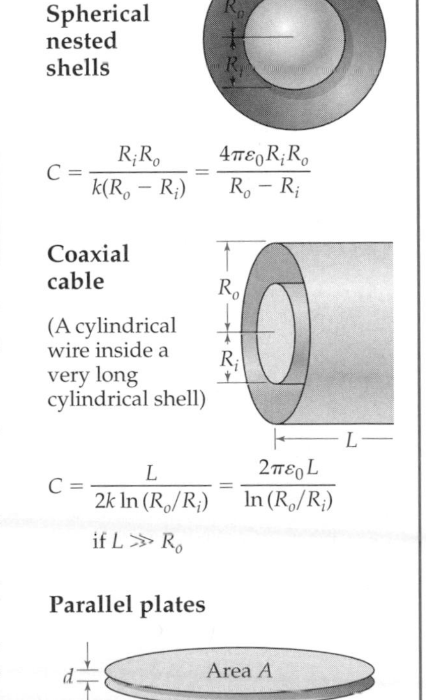

Example E4.4

Problem Imagine a metal sphere of radius R_i inside a concentric thin metal shell of radius R_o. Find the capacitance of this pair of conductors.

Translation If we pump charge Q from the inner sphere to the outer shell, the inner sphere is left with charge $-Q$.

Model According to the shell theorem, the potential relative to infinity due to the outer shell alone is a constant kQ/R_o everywhere inside that shell. The potential due to the inner sphere alone is $-kQ/r$ at all points outside $r = R_i$ (r is the distance from the sphere's center). According to the superposition principle, the total potential at points where r is between R_i and R_o is $\phi(r) = kQ(1/R_o - 1/r)$.

Solution This means that the potential difference between the capacitor's "plates" here is

$$\Delta\phi = \phi(R_o) - \phi(R_i) = kQ\left(\frac{1}{R_o} - \frac{1}{R_o}\right) - kQ\left(\frac{1}{R_o} - \frac{1}{R_i}\right)$$

$$= kQ\left(\frac{1}{R_i} - \frac{1}{R_o}\right) = \frac{kQ(R_o - R_i)}{R_i R_o} \tag{E4.16}$$

The capacitance is therefore

$$C = \left|\frac{Q}{\Delta\phi}\right| = \left|\frac{\not{Q}}{k\not{Q}}\frac{R_i R_o}{R_o - R_i}\right| = \frac{R_i R_o}{k(R_o - R_i)} \tag{E4.17}$$

Evaluation Note that the capacitance does indeed depend on the size and separation of the plates but not on the potential difference between them.

Exercise E4X.6

What is R_o if $R_o - R_i = 1.0$ mm and $C = 1.0$ F?

An isolated conductor as a capacitor

One can also talk about the capacitance of a *single* isolated conductor if one assumes that the other conductor is essentially a spherical shell at infinity. For example, the capacitance of an isolated sphere of radius R is found by setting $R_i = R$ and taking the limit of equation E4.17 as $R_o \to \infty$: the result is $C = R/k$.

The parallel-plate capacitor

Another important capacitor configuration is the **parallel-plate capacitor,** which consists of two flat conducting plates separated by a small distance. If two oppositely charged conducting plates have area A and are separated by distance d such that $d \ll A^{1/2}$, then it turns out (see problem E3S.5) that in static equilibrium the charges are *uniformly* distributed on each plate (instead of being concentrated at the periphery as for a single isolated plate, shown in figure E4.7b). Because of this, we can use the results for uniformly charged parallel plates (see section E3.4) to calculate the potential difference between the plates for a given amount of charge, and so calculate the capacitance. The result is

$$C = \frac{A}{4\pi kd} = \frac{\varepsilon_0 A}{d} \tag{E4.18}$$

For the derivation of this result, see problem E4S.9.

TWO-MINUTE PROBLEMS

E4T.1 Consider two wires, both 2.0 cm long. Wire A has a diameter of 1.0 mm and B has a diameter of 2.0 mm, but they are otherwise identical. Both have the same electric field acting in them. How do the *currents* in these wires compare?
A. $I_A = 4I_B$
B. $I_A = 2I_B$
C. $I_A = I_B$
D. $I_B = 2I_A$
E. $I_B = 4I_A$
F. Other (specify)

E4T.2 Consider two wires. Wire A is 10 cm long, and wire B is 5 cm long. Both wires are otherwise identical, and both have the same electric field acting in them. How do the *currents* in these wires compare?
A. $I_A = 4I_B$
B. $I_A = 2I_B$
C. $I_A = I_B$
D. $I_B = 2I_A$
E. $I_B = 4I_A$
F. Other (specify)

E4T.3 Consider two wires. Imagine that wire A has a diameter of 0.5 mm and wire B has a diameter of 1.0 mm, but the wires are otherwise identical. Both wires carry the same current of 1.0 A. How do the magnitudes of the *electric fields* in each wire compare?
A. $E_A = 4E_B$
B. $E_A = 2E_B$
C. $E_A = E_B$
D. $E_B = 2E_A$
E. $E_B = 4E_A$
F. Other (specify)

E4T.4 If you increase the temperature of a metal, what happens to its conductivity, assuming the Drude model is accurate? Its conductivity should
A. Increase
B. Decrease
C. Remain the same
D. It is impossible to tell

E4T.5 Imagine that we have two isolated metal spheres. The one on the left has a negative charge, while the one on the right has a positive charge. If these spheres are connected with a straight *metal* wire, in what direction do we conventionally consider the current to flow?
A. Left to right.
B. Right to left.
C. Because there is no complete circuit, no current flows.
D. Because there is no electric field, no current flows.

E4T.6 What if each copper atom in a wire were to contribute *two* conduction electrons instead of just one? If this were to happen, the conductivity σ of copper would
A. Quadruple
B. Double
C. Not change
D. Decrease by factor of 4
E. Decrease by factor of 2
F. Exhibit other change (specify)

E4T.7 Due to electrostatic polarization, a sufficiently charged rubber balloon will attract a neutral horizontal insulating plate strongly enough that it sticks firmly to the plate's bottom. If this neutral horizontal plate is a conductor instead of an insulator, the charged balloon will
A. Still stick
B. Stick only if the conductor is connected to the earth
C. Stick only if the conductor is insulated from the earth
D. Not stick at all

E4T.8 Imagine that you have two oppositely charged metal spheres on insulating supports. If you slowly bring these spheres closer and closer together, the surface charge density σ on each sphere will
A. Remain uniform over each sphere's entire surface
B. Be largest at the point nearest the other sphere
C. Be smallest at the point nearest the other sphere
D. Eventually no longer shield the sphere's interior from the other sphere's field

HOMEWORK PROBLEMS

Basic Skills

E4B.1 A lightning flash during a thunderstorm (see figure E4.10) can consist of one to tens of strokes, each with a duration between 0.01 s and 0.1 s. Peak currents in a stroke can be as high as 20 kA. If a certain lightning flash has 10 separate strokes, each lasting 0.05 s and carrying an average current of 8 kA, about how much charge gets transferred from the earth to the sky (or vice versa)?

Figure E4.10
How much charge is transferred? (See problem E4B.1.)

E4B.2 Figure E4.11 shows a spark leaping from one Van de Graaff generator to another.
(a) If the spark transfers a charge of 200 μC from one sphere to the other and lasts about 0.1 ms, what is the average current that the spark carries?
(b) If the left sphere was initially positive relative to the other, which way did electrons flow in the spark, left to right or right to left? Which way did conventional current flow? Explain.

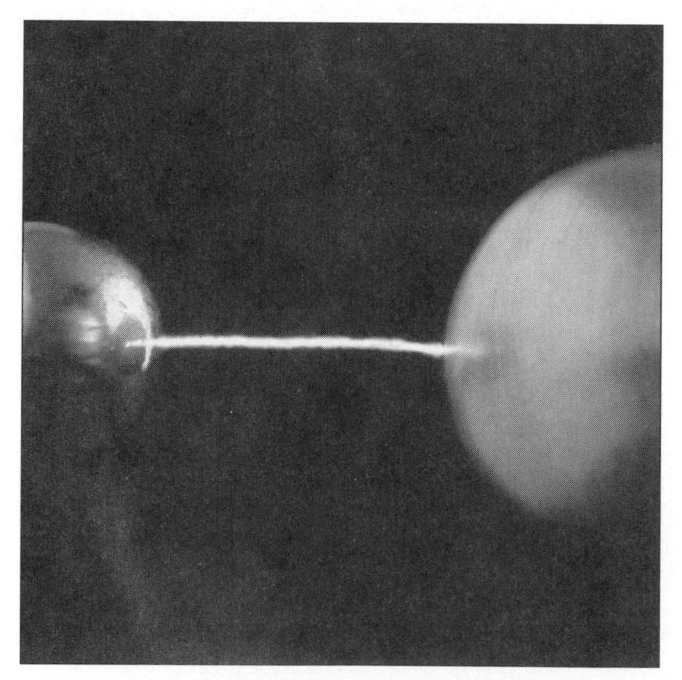

Figure E4.11
How much current flows in this spark? (See problem E4B.2.)

E4B.3 What is the drift speed of electrons in a 22-cm length of silver wire that carries a current of 5.0 A and has a cross-sectional area of 0.2×10^{-6} m^2? (Silver has an atomic mass of 107.9 g/mol, a density of 10,500 kg/m^3, and a conductivity of about 6.3×10^7 C$^2 \cdot$ s $\cdot$ kg$^{-1} \cdot$ m^{-3}.)

E4B.4 What is the drift speed of electrons in a 1.0-m length of copper wire that carries a current of 50 mA and has a cross-sectional area of 0.4×10^{-6} m^2? About how long will it take an electron to travel the wire's length?

E4B.5 A 10.0-cm length of copper wire has a uniform cross-sectional area of 1.0 mm^2. If the drift speed of electrons in this wire is 2.0×10^{-4} m/s and the electrons all move directly along the wire, what is the magnitude of the current density in the wire? What is the total current carried by the wire?

E4B.6 A 2.0-m length of silver wire has a uniform cross-sectional area of 0.6 mm^2. If the drift speed of electrons in this wire is 5.0×10^{-5} m/s and the electrons all move in the same direction along the wire, what is the magnitude of the current density in the wire? What is the total current carried by the wire? (Silver has an atomic mass of 107.9 g/mol, a density of 10,500 kg / m^3, and a conductivity at room temperature of about 6.3×10^7 C$^2 \cdot$ s $\cdot$ kg$^{-1} \cdot$ m^{-3}.)

E4B.7 A parallel-plate capacitor holds 0.10 μC of charge on its positive plate (and -0.10 μC on its negative plate) when the potential difference between the plates is 10 V. If the plates are separated by 1.0 mm, what is their area?

Synthetic

E4S.1 Imagine that in a certain proton accelerator, roughly 10^{11} protons at a time go around a ring having a radius of 1.0 km. The speed of the protons is essentially the same as the speed of light. What is the current flowing in the ring, in amperes?

E4S.2 Compute the average time between collisions τ for conduction electrons in copper, using the information available in example E4.1. Show that an electron traveling at a typical thermal speed of 10^6 m/s will cover about 25 nm of distance on the average between collisions.

E4S.3 Compute the average time between collisions τ for silver, assuming that each atom contributes one conduction electron per atom. (Pure silver metal has an atomic mass of 107.9 g/mol, a density of 10,500 kg/m^3, and a room-temperature conductivity of about 6.3×10^7 C$^2 \cdot$ s $\cdot$ kg$^{-1} \cdot$ m^{-3}.)

E4S.4 A silver wire that has a length of 28 cm and a diameter of 0.20 mm is carrying a current of 230 mA. The electrons in the wire are moving to the left. What are the magnitude and direction of the electric field in that wire?

E4S.5 A nichrome wire that has a length of 2.4 m and a diameter of 0.40 mm is carrying a current of 2.5 A. The electrons in the wire are moving to the right. What are the magnitude and direction of the electric field in that wire?

E4S.6 Imagine that you have two identical neutral metal spheres on insulating stands and a charged rubber rod. Using diagrams, carefully describe a step-by-step process that uses the rod to give the two spheres *exactly* opposite charges (all without touching either sphere with the rod). Take care to construct your procedure so that it ensures that the charges on each sphere will be *exactly* equal in magnitude, not just approximately equal.

E4S.7 Consider two initially neutral metal spheres on insulating stands. Imagine that we give one a fairly strong positive charge and then bring it slowly closer and closer to the other sphere. Experimentally, we find that at a certain separation distance, a spark jumps from one sphere to the other, carrying some of the excess charge from the charged sphere to the other. Draw a set of diagrams showing the charge distribution on each sphere as they approach each other, and explain why the spark jumps only when they are a certain distance apart. (*Hint:* The electric field just outside a conductor is strongest where the surface charge density is highest.)

E4S.8 Imagine a small, initially neutral metal ball hanging as a pendulum bob between two oppositely charged metal spheres on insulating stands. It happens that if the spheres have a large enough charge, the ball will begin to swing rapidly back and forth, hitting first one sphere and then the other. After a certain time (typically a few seconds) the ball stops. Carefully explain (with the help of appropriate diagrams) both what drives the ball's swinging initially and why the ball eventually stops.

E4S.9 According to section E3.4, the electric field $\vec{E}$ between two parallel plates with opposite, uniformly distributed charges is (as long as the plates are very large compared to their separation d) nearly uniform and has magnitude $E = 4\pi k\sigma$, where σ is the charge per unit area on the plates. Use this result to verify equation E4.18.

E4S.10 According to table E3.1, the electric field outside a uniformly charged, infinite cylindrical conductor is the same as if the cylinder's charge were concentrated in a thin wire along the cylinder's axis. Moreover, the potential inside a uniformly charged infinite cylindrical pipe, like that inside a spherical shell, is a constant. These statements are approximately true for finite cylinders and pipes as long as their length is much larger than their radius. Use these facts to calculate the capacitance of the coaxial cable arrangement shown in table E4.2.

E4S.11 Check that the capacitance of a nested-shell capacitor becomes the same as that of a parallel-plate capacitor of the same area in the limit that the distance between the inner and outer conductors becomes much smaller than the radius of the outer shell.

E4S.12 Imagine that we place a slab of polarizable insulating material between the two plates of a parallel-plate capacitor. If we charge the plates, the uniform electric field between the plates will polarize the material, turning its atoms into dipoles.

(a) Figure E4.12 depicts the positive charges in the material as a block of uniform gray, and the negative charges as a block of uniform color. Figure E4.12a shows the situation before the plates are charged: the positive and negative charges are mixed together, yielding a uniform neutral tinted gray inside the slab. The electric field of the charged plates, however, shifts the negative charge slightly upward and the positive charge slightly downward. As figure E4.12b shows, the interior of the slab remains a neutral tinted gray (shifting the charges a bit past each other does not change the fact that on the macroscopic scale, the slab's interior remains completely and uniformly neutral), but what *is* different is that there is now some uncanceled positive charge (uncolored gray) on the slab's bottom surface and uncanceled negative surface charge (pure color) on the slab's top surface. Use the superposition principle to argue that the net effect of these induced surface charges is to *reduce* the strength of the electric field in the slab by a certain amount.

(b) The strength of this effect depends on the degree to which the insulating material is polar-

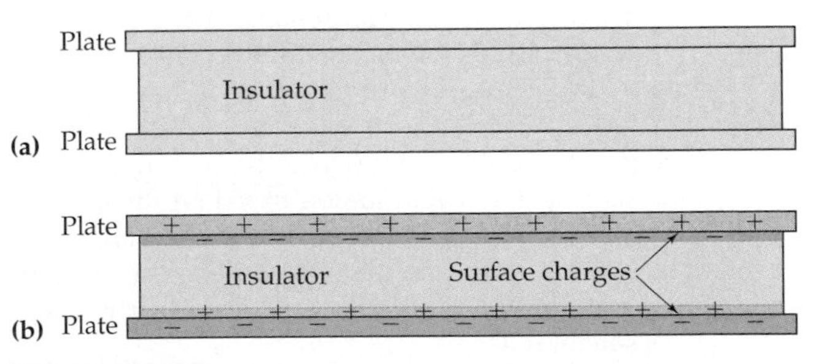

Figure E4.12
This drawing shows a slab of insulating material placed between two parallel plates. (a) When the plates are uncharged, the set of all atomic positive charges (gray) and the set of atomic negative charges (color) are uniformly intermixed, yielding no net charge (tinted gray) in the slab's interior. (b) When the plates are charged, the plates' electric field will shift atomic positive charges slightly downward and atomic negative charges slightly upward. The effect of this shift is to leave the interior neutral (tinted gray) but to create what amounts to surface charges (pure gray and pure color) on the slab's surfaces.

ized by the original field between the plates. Let's define the **permittivity** ε of an insulating substance so that

$$\vec{E} = \frac{\varepsilon_0}{\varepsilon}\vec{E}_{\text{vac}} \tag{E4.19}$$

where ε_0 is the permittivity of the vacuum, $\vec{E}$ is the electric field inside the insulator, and $\vec{E}_{\text{vac}}$ is the electric field that would exist at the same point if the entire insulator were replaced by a vacuum. It turns out that this formula applies very generally, not just when we have an insulating slab between flat plates. The ratio $\varepsilon/\varepsilon_0$ is sometimes called the insulator's **dielectric constant:** note that since $E < E_{\text{vac}}$, this ratio is always greater than 1. Use equation E4.19 to carefully argue that when the plates of a capacitor are separated by an insulating material with permittivity ε instead of a vacuum, the capacitor's capacitance becomes

$$C = \frac{\varepsilon}{\varepsilon_0}C_{\text{vac}} \tag{E4.20}$$

Since $\varepsilon/\varepsilon_0$ can be quite a bit greater than 1 for some materials, this presents a way of dramatically increasing the capacitance represented by a given configuration of plates.

Rich-Context

E4R.1 It is not clear that the Drude model should describe the flow of ions through a liquid with any accuracy, but it is interesting to see whether the average time between collisions τ that we calculate while assuming that the Drude model does apply to such systems is physically reasonable. Pure water has about 6×10^{19} H^+ ions (or more technically H_3O^+ ions) and the same number of OH^- ions in 1 m^3. What is τ? If the ions have an average thermal speed of about 500 m/s at room temperature, about how far would an ion travel on the average between collisions? How does this compare to the size of a water molecule (which is on the order of magnitude of 0.1 nm)? Is your answer physically reasonable?

E4R.2 This problem was adapted from Purcell, *Electricity and Magnetism,* 2nd ed. McGraw-Hill, New York, 1985.

(a) Estimate the capacitance of the human body relative to infinity, and attach an appropriate uncertainty to your estimate. Please explain your reasoning by answering the following questions: Can the capacitance be as large as that of a conducting sphere whose diameter equals the person's height? Can it be as small as a sphere that would fit inside the person's torso? Would it be better to model the person as being a long cylinder (see table E3.1 for discussion of the field of a cylinder)?

(b) We will see in chapter E5 that a capacitor whose plates are charged so that their potential difference is $\Delta\phi$ stores an electric energy of $\frac{1}{2}C|\Delta\phi|^2$. When you shuffle your feet on a rug on a dry day, you can build up a potential difference of several thousand volts. About how much electric energy is released, therefore, in the spark that jumps when you touch something?

E4R.3 Imagine two circular flat plates of radius R that are separated by a small distance s. Imagine that we connect the plates to a power supply that puts opposite charges on the plates of whatever magnitude Q is necessary to set up a fixed potential difference $|\Delta\phi|$ between them.

(a) Argue that the magnitude of the total force that each plate exerts on the other is given by

$$F = \frac{2\pi k Q^2}{A} \tag{E4.21}$$

where A is the area of each plate. What assumptions do you have to make?

(b) Show that you can eliminate the unknown charge Q from this expression to get

$$F = \frac{1}{8k}\left(\frac{R\Delta\phi}{s}\right)^2 \tag{E4.22}$$

(c) If you have an actual setup of this experiment available, calculate this force for the measured values of R, $\Delta\phi$, and s for this experiment (otherwise, use $R = 10$ cm, $\Delta\phi = 1000$ V, and $s = 2.0$ mm). Make sure that the units work out correctly.

(d) If you have an actual setup of this experiment available, carefully measure the actual force between the plates, and specify its uncertainty.

Advanced

E4A.1 (a) Using Coulomb's law and the superposition principle, argue that if the total electric field created by any charge distribution is zero at any point P, it will still be zero at P if the distribution's total charge is multiplied by some factor without changing the distribution of charge.

(b) Use the result of part (a) to argue that if a certain charge distribution on a pair of conductors yields zero electric field inside both conductors, then multiplying the charge in the distribution by some factor without changing the distribution will also yield zero electric field inside both conductors.

(c) There is a mathematical theorem asserting that if we move a given amount of charge Q from one conductor to another, there is a *unique* static charge distribution that will yield zero electric field inside both conductors. Argue, then, that if we pump *more* charge from one conductor to the other, the charge *distribution* on the conductors will *not* change (the charge density at each point on each conductor's surface is simply multiplied by a common factor).

(d) Argue, therefore, that the potential difference $\Delta\phi$ between two conductors in static equilibrium in a vacuum must always be directly proportional to Q.

(e) Finally, argue from this that the capacitance C of any arbitrary pair of conductors in a vacuum must be independent of $\Delta\phi$.

ANSWERS TO EXERCISES

E4X.1 Calculating n as in equation E4.5 (except noting that aluminum contributes *three* conduction electrons per atom) yields a value of $n = 1.8 \times 10^{29}\ \text{m}^{-3}$. The cross-sectional area of a wire that is 1.5 mm in *diameter* is $\pi(0.00075\ \text{m})^2 = 1.8 \times 10^{-6}\ \text{m}^2$. Calculating $v_d = I/neA$ as in equation E4.6, we get about 1.0×10^{-4} m/s.

E4X.2 Since $\vec{J}$ has SI units of amperes per square meter and $\vec{E}$ has SI units of newtons per coulomb, if equation E4.11 is to have self-consistent units, the SI units of conductivity σ_c must be $\text{A}\cdot\text{m}^2/(\text{N/C})$, or

$$\frac{\cancel{\text{A}}/\text{m}^2}{\cancel{\text{N}}/\cancel{\text{C}}}\left(\frac{1\ \cancel{\text{N}}}{1\ \text{kg}\cdot\text{m/s}^{\cancel{2}}}\right)\left(\frac{1\ \cancel{\text{C}}/\cancel{\text{s}}}{1\ \cancel{\text{A}}}\right) = \frac{\text{C}^2\cdot\text{s}}{\text{kg}\cdot\text{m}^3} \tag{E4.23}$$

E4X.3 According to equations E4.10 and E4.11,

$$\begin{aligned} E &= \frac{J}{\sigma_c} = \frac{I}{\sigma_c A} \\ &= \frac{1.0\ \cancel{\text{C}}/\text{s}}{(5.9\times10^7\ \text{C}^2\cdot\text{s}\cdot\text{kg}^{-1}\cdot\text{m}^{-3})\pi(0.0005\ \text{m})^2} \\ &= 0.022\,\frac{\cancel{\text{kg}}\cdot\cancel{\text{m}}}{\text{C}\cdot\cancel{\text{s}^2}}\left(\frac{1\ \text{N}}{1\ \cancel{\text{kg}}\cdot\cancel{\text{m}}/\cancel{\text{s}^2}}\right) \\ &= 0.022\ \text{N/C} \end{aligned} \tag{E4.24}$$

E4X.4 Charge one sphere, using the technique described in example E4.3. Then move the rod very far away, and touch the uncharged sphere to the charged sphere. In an effort to further disperse itself, part of the charge on the charged sphere will move to the other sphere, and since the spheres are identical, each sphere will end up with the same charge in static equilibrium. If we now separate the two spheres, we will end up with two equally charged spheres.

E4X.5 We have

$$\text{F} = \frac{\text{C}}{\cancel{\text{V}}}\left(\frac{1\ \cancel{\text{V}}}{1\ \text{J/C}}\right)\left(\frac{1\ \cancel{\text{J}}}{1\ \text{kg}\cdot\text{m}^2/\text{s}^2}\right) = \frac{\text{C}^2\cdot\text{s}^2}{\text{kg}\cdot\text{m}^2} \tag{E4.25}$$

E4X.6 ***Translation and Model:*** Assume for the sake of argument that the distance $d \equiv R_o - R_i = 1.0$ mm is much smaller than the radii R_i and R_o. According to equation E4.17, we have

$$C = \frac{R_o R_i}{k(R_o - R_i)} = \frac{R_o(R_o - d)}{kd} \approx \frac{R_o^2}{kd} \tag{E4.26}$$

Since we know C, k, and d, we can solve for R_o.

Solution: Doing this yields

$$\begin{aligned} R_o &= \sqrt{Ckd} \\ &= \sqrt{(1\ \text{F})(8.99\times10^9\ \text{N}\cdot\text{m}^2/\text{C}^2)(1.0\times10^{-3}\ \text{m})} \\ &= 3000\sqrt{\frac{\cancel{\text{N}}\cdot\text{m}^3\cdot\cancel{\text{F}}}{\cancel{\text{C}^2}}\left(\frac{1\ \cancel{\text{C}}/\cancel{\text{V}}}{1\ \cancel{\text{F}}}\right)\left(\frac{1\ \cancel{\text{V}}}{1\ \cancel{\text{J}}/\cancel{\text{C}}}\right)\left(\frac{1\ \cancel{\text{J}}}{1\ \cancel{\text{N}}\cdot\text{m}}\right)} \\ &= 3000\ \text{m} \end{aligned} \tag{E4.27}$$

Evaluation: This has the right units, but the capacitor is huge, nearly 4 mi in diameter! (The farad is a large unit!) Note that our original assumption that the plate separation is very small is amply justified here. One could solve for R_o exactly by rewriting equation E4.17 in the form of a quadratic equation

$$Ckd = R_o^2 - R_o d \quad\Rightarrow\quad 0 = R_o^2 - R_o d - Ckd \tag{E4.28}$$

and solving for R_o, but the result is virtually identical.

E5 Driving Currents

▷ Electric Field Fundamentals
▽ Controlling Currents
Conductors
Driving Currents
Analyzing Circuits
▷ Magnetic Field Fundamentals
▷ Calculating Static Fields
▷ Dynamic Fields

Chapter Overview

Introduction

In chapter E4, we saw that a nonzero electric field in a conductor will cause current to flow through it, at least for a short time. In this chapter, we will see how we can set up electric fields that drive *steady* currents through a conductor.

Section E5.1: A Mechanical Model of a Battery

If we connect a capacitor's charged plates with a wire, the plate's external electric field will drive a current through the wire, but only long enough to drain the capacitor's charge. To keep the current flowing steadily, we need to continually replenish the charge on the plates. A **battery** is like a capacitor with an internal "conveyor belt" that transports positive charge from the negative plate to the positive plate (*against* the plates' field) as rapidly as needed to maintain the charge on each plate.

Section E5.2: Surface Charges Direct Currents

Even if we replenish the charge on the plates, the nonuniform electric field created by those plates alone cannot drive a *steady* current through a wire connected to them. However, just after the wire is connected, initially nonuniform current flows nearly instantly push excess charges into a *dynamic* equilibrium distribution on the wire's surface that creates an electric field inside the wire that always points directly along the wire and (if the wire is uniform) has a uniform magnitude. A feedback mechanism keeps these surface charges in just the right configuration to maintain this field.

Section E5.3: The Emf of a Battery

The magnitude of this field is determined by the battery's **emf** $\mathscr{E}$ (pronounced "ee-yemehf"), which is the *energy per unit charge* that the process serving as the battery's "conveyor belt" contributes as it transports charge between the battery's plates (its **electrodes**) against the electric field created by those plates. This emf is a *fixed characteristic* of most batteries and power supplies. Conservation of energy requires that

$$|\Delta\phi| = \mathscr{E} - \mathscr{E}^{th} \tag{E5.4}$$

Purpose: This equation links the potential difference $|\Delta\phi|$ between a battery's electrodes to the characteristic emf $\mathscr{E}$ of the process driving charge through the battery.

Symbols: $\mathscr{E}^{th}$ is the energy per unit charge lost to thermal energy.

Limitations: This assumes no energy loss other than to thermal energy.

Note: The characteristic emf $\mathscr{E}$ is fixed for most batteries, but $\mathscr{E}^{th}$ is usually proportional to the current flowing through the battery. In the **ideal battery approximation,** we assume that $\mathscr{E}^{th} = 0$, independent of the current.

Note that while emf and potential have the same SI units (volts), they are conceptually distinct. Potential ϕ refers specifically to electrostatic potential energy per unit charge, while emf $\mathscr{E}$ refers to *other* kinds of energy per unit charge.

Section E5.4: Resistance

The fact that the electric field in a uniform wire is uniform and points directly along the wire implies that the potential difference between a wire's ends is

$$|\Delta\phi_{\text{wire}}| = EL \qquad \text{(E5.7}a\text{)}$$

Purpose and Symbols: This equation links the electric field strength E in a thin wire to its length L and the potential difference $|\Delta\phi|$ between its ends.
Limitations: This equation assumes that the wire is uniform and thin, and that a steady-state current flow has been established.

We define any conductor's *resistance* to be

$$R \equiv \frac{V}{I} \qquad \text{(E5.9)}$$

Purpose: This equation defines a conducting object's resistance R.
Symbols: I is the magnitude of current the object conducts when the potential difference between its ends is $V \equiv |\Delta\phi|$.
Limitations: The object must have well-defined "ends."
Note: The SI unit of resistance is the **ohm** (Ω), where $1\ \Omega \equiv 1\ \text{V/A}$.

(See the section for some crucial notes about using the symbol V for the magnitude of a potential difference.) **Ohm's law** asserts that R is independent of V and I for many conductors. We say that conductors obeying Ohm's law are **ohmic.**

Section E5.5: The Power Dissipated in a Conductor

Collisions between a conductor's atoms and a current's moving charge carriers convert the carriers' electrostatic potential energy to thermal energy at the following rate:

$$P = IV = I^2R = \frac{V^2}{R} \qquad \text{(E5.12)}$$

Purpose: This equation expresses the power P associated with the conversion of electrostatic potential energy to thermal energy in a conductor.
Symbols: V is the potential difference between the conductor's ends, I is the magnitude of current it carries, and R is its resistance.
Limitations: The conductor must have well-defined ends.

Section E5.6: Discharging a Capacitor

A capacitor can drive current much as a battery does, except that the potential difference V_C between a capacitor's plates decreases with time as they discharge. The definitions of capacitance, resistance, and current together imply that in such a case

$$V_C(t) = V_0 e^{-t/RC} \qquad \text{(E5.22)}$$

Purpose: This equation specifies how the potential difference V_C between a capacitor's plates varies with time t as it discharges through a wire of resistance R.
Symbols: C is capacitor's capacitance, and V_0 is the initial value of V_C.
Limitations: This equation assumes that the connecting wire is ohmic.

E5.1 A Mechanical Model of a Battery

Imagine that we give the plates of a parallel-plate capacitor charges that are equal in magnitude and opposite in sign. If the plates have a finite size, the nearest plate's field *slightly* dominates over the farther plate's field at a point outside the plates. The plates will therefore create a very weak external electric field that looks qualitatively as shown in figure E5.1.

A parallel-plate capacitor will briefly drive a current

Now imagine that we suddenly connect the plates with a wire. The dark colored arrows in figure E5.1 show the electric field vectors produced by the plates evaluated at various points along such a wire at the instant of connection. You can see that there is always a nonzero component of the electric field in the direction of the wire in the clockwise direction, so the field will begin to drive a current through the wire from the positive plate to the negative plate. (Note that we are talking about *conventional current* here: the field drives *electrons* counterclockwise from the negative plate to the positive plate.)

Exercise E5X.1

Qualitatively, *why* is the electric field to the right of the positive plate so much weaker than the field between the plates? Why does it point in the opposite direction?

Exercise E5X.2

The lengths of the field vectors at points A and B of figure E5.1 indicate that the electric field is stronger at point A than at point B. Why?

When we connect the plates with the wire, current will only flow for a short time, because it won't take very long for all the charge on one plate to flow to the other. The only way to keep charge flowing steadily is to maintain the charges on the plates by physically transporting electrons from the positive plate to the negative plate *against* the electrostatic force that the field

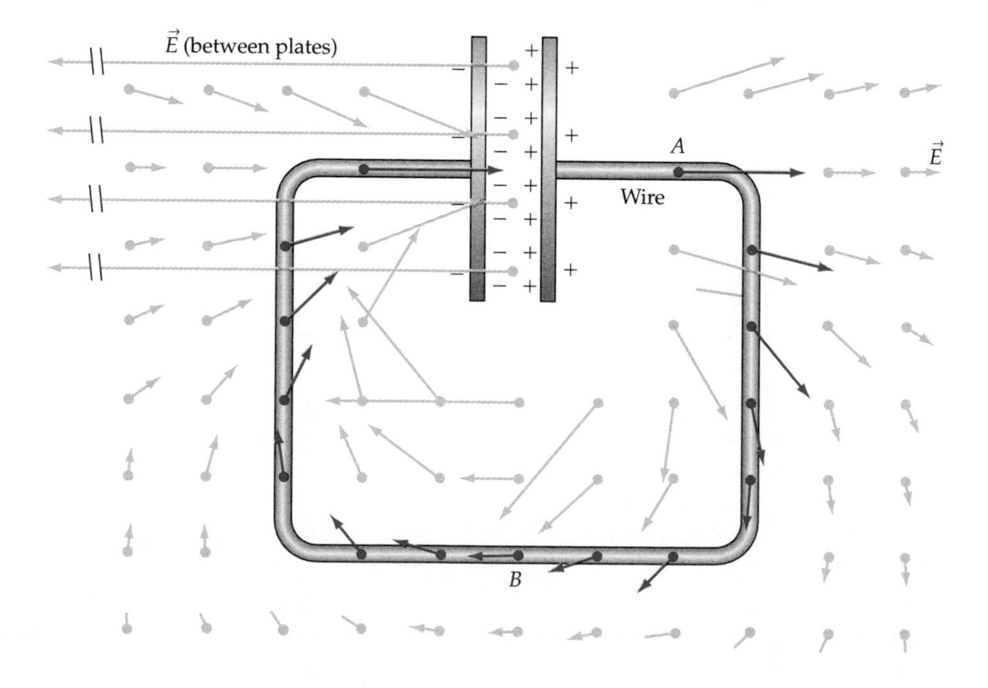

Figure E5.1
The electric field of a parallel-plate capacitor at the instant that its plates are connected by the wire shown (before any significant amount of charge flows in the wire). The dark colored arrows indicate electric field vectors at selected points inside the wire, while the lighter arrows indicate electric field vectors at selected points outside the wire.

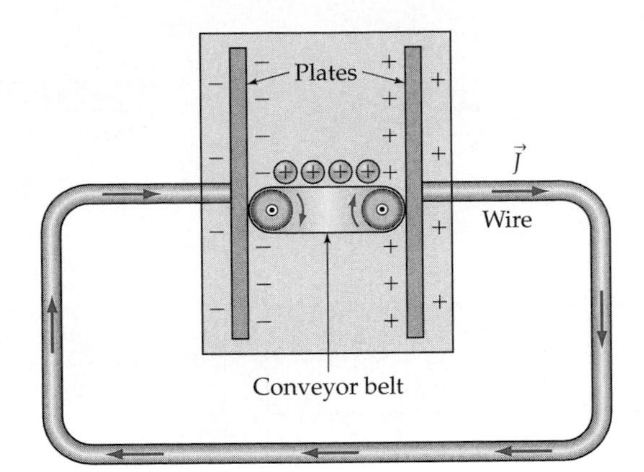

Figure E5.2
In this idealized *mechanical battery*, a motor-driven conveyor belt transports charges from the negative to the positive plate against the leftward force exerted by the plates' electric field. This maintains a current density $\vec{J}$ in the wire, whose magnitude and direction are indicated by the arrows. (This battery metaphor is adapted from R. W. Chabay and B. A. Sherwood, *Electric and Magnetic Interactions*, Wiley, New York, 1995.)

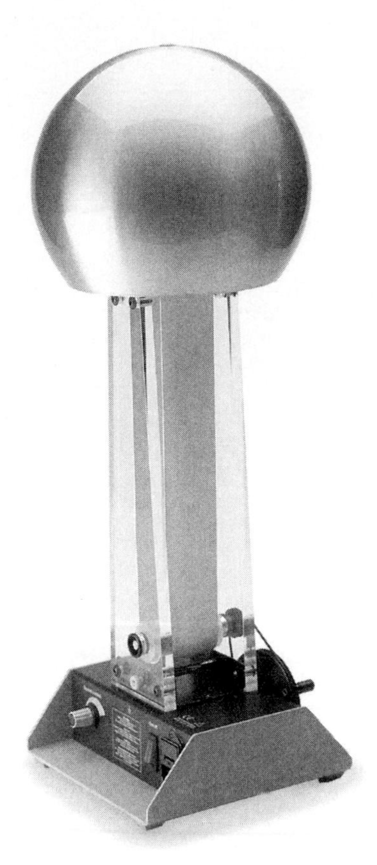

Figure E5.3
A Van de Graaff generator whose clear central column shows the rubber belt that carries charge to the upper sphere.

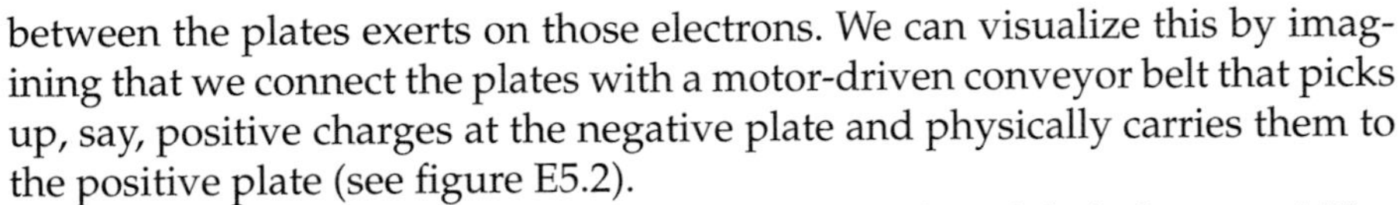

between the plates exerts on those electrons. We can visualize this by imagining that we connect the plates with a motor-driven conveyor belt that picks up, say, positive charges at the negative plate and physically carries them to the positive plate (see figure E5.2).

The mechanical model for a battery

The device shown in figure E5.2 is an idealized model of a **battery.** A Van de Graaff generator (see figure E5.3) uses an actual rubber conveyor belt to transport charge between its base and a metal sphere at the top, and so is an example of such a mechanical battery. The batteries we normally encounter in daily life are *chemical* batteries, which use chemical reactions as the "motor" that transports ions of various signs between the plates. The thing to understand at this point is that *a battery is a device that maintains a potential difference between two plates by transporting charge from one plate to the other against the electric forces acting on that charge.* Connecting the two ends of a battery with a wire creates the simplest possible example of an **electric circuit** (a closed path around which electrons are driven by an electric field).

E5.2 Surface Charges Direct Currents

How can the nonuniform battery plate field drive a uniform current in a wire?

Figure E5.2 shows the battery-driven conventional current flowing everywhere with the same speed and always directly along the wire. But according to $\vec{J} = \sigma_c \vec{E}$, the current flow should be proportional to and parallel to $\vec{E}$, and figure E5.1 shows that the electric field created by the battery plates does *not* necessarily point along the wire and has a nonuniform magnitude.

Even so, figure E5.2 *must be right!* A steadily flowing current in a circuit *must* flow directly along the wire at all points with the same drift speed everywhere. If this were *not* true, electrons would pile up at the edges of wires and at places in the wire where the drift speed is slower, a process that clearly cannot be sustained forever. So the electric field in the wire *cannot* be as shown in figure E5.1 once a steady current flow has been established.

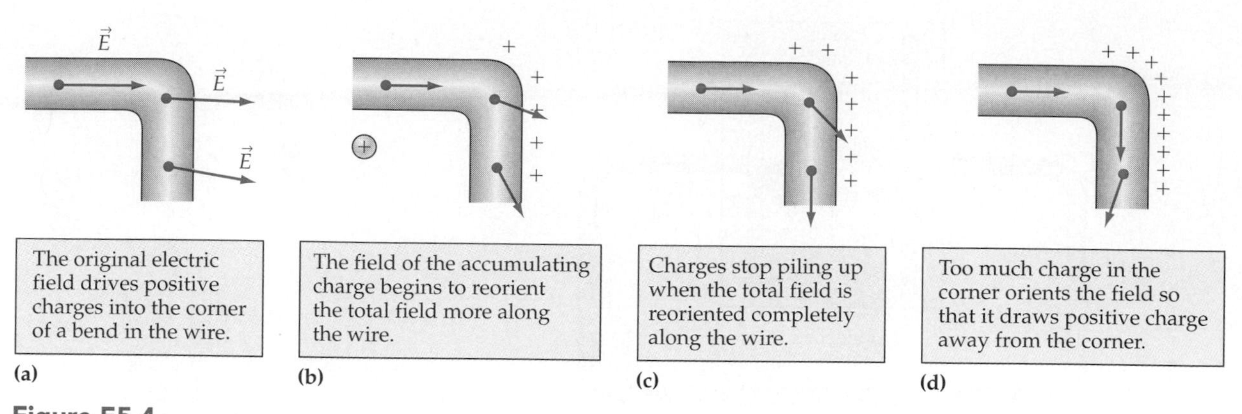

Figure E5.4
How accumulating surface charge reorients the original electric field on a wire to point directly along the wire.

What actually happens is this. At the instant the wire is connected, conventional current *starts* to flow parallel to $\vec{E}$, as shown in figure E5.1. But the initial nonuniform flow driven by this field causes electrons ultimately to pile up at places along the wire's surface in just the arrangement required to establish eventually a *net* electric field $\vec{E}$ that everywhere inside the wire has a uniform magnitude and points directly along the wire. This *dynamic* equilibrium situation is directly analogous to the way that charges on a conductor redistribute themselves in *static* equilibrium situations to make $\vec{E} = 0$ inside the conductor.

How surface charges on a wire control current flow

The complex charge concentrations along the wire required to create this simple electric field are maintained by a straightforward feedback mechanism. For example, consider the upper right-hand corner of the wire, as shown in figure E5.4. The field created by the *plates* at this corner is directed to the right: this causes positive charge to flow initially to the right, as shown in figure E5.4a (actually, *electrons* are driven by the field to the *left*, but the effect is the same). When these positive charges run into the corner, they cannot go any farther and so pile up there. The field produced by this pile of charge adds to the field produced by the plates to produce a net field more in the direction of the wire (figure E5.4b). This process continues until the concentration of charges at the corner creates a net field that directs all incoming charge carriers cleanly around the bend in the wire. When this happens, charge will no longer be added to the pile in the corner, and no more change takes place (figure E5.4c). If by chance a bit of extra charge is added to the corner by some fluctuation in the current flow, the field created by this extra charge will push positive charge away from the corner until balance is again established (figure E5.4d). In the long run, then, exactly the right charge concentration will be maintained on the corner's surface to direct the current flow around the bend.

This feedback process works for the straight parts of the wire as well. Notice that in figure E5.1, the electric field strength in parts of the wire distant from the plates is weaker than it is near the plates. This means that charge initially flows rapidly near the plates, piling up positive charge on wire's surface to the right of the plates and negative charge on the wire's surface to the left. This ultimately creates a charge distribution on the wire's surface that looks qualitatively as shown in figure E5.5. At every point in the circuit, a positive charge carrier will see less positive charge (or more negative charge) in the forward direction than in the reverse direction, and thus is drawn forward. The same kind of feedback mechanism described in figure E5.4 ensures that the electric field at every point in the wire points directly along the wire and (assuming the wire is uniform) has a uniform magnitude.

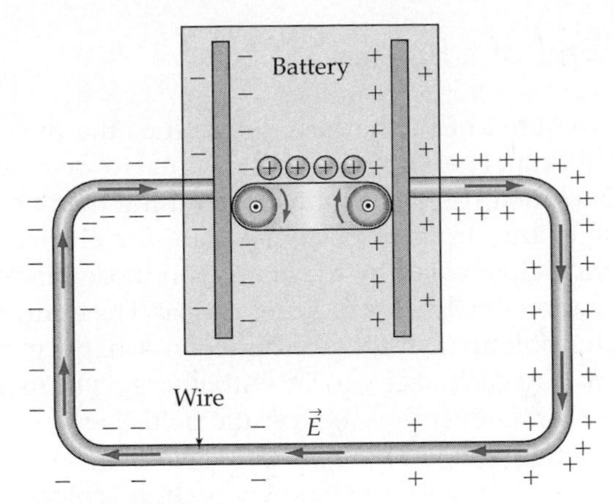

Figure E5.5
The initially uneven flow of electrons in a wire creates charge concentrations on the wire's surface that redirect the current until current flows uniformly along the wire. Excess charges on a corner's surface direct the current around a bend in the wire; a surface charge distribution that gradually becomes more negative drives the current forward in any straight segments. (The arrows show the direction of $\vec{E}$ and thus the direction of conventional current flow.)

The surface charge distributions established in this dynamic equilibrium process turn out to be *much* more important than the field of the actual battery plates in actually determining the flow of current in the wire. This is primarily so because they are much closer to the charged particles in the current than the plates are.

This rearrangement of surface charge occurs very rapidly (typically within a few tens of nanoseconds) because information about changes in an electric field flows from place to place along the wire at nearly the speed of light (≈ 1 ft/ns), and electrons do not have to move very far to set up the correct surface distributions. Therefore, it will seem to us that a steady flow of charge is established essentially instantly after the wire is connected to a battery.

In summary, the key idea of this section is that once a steady state is attained, concentrations of surface charge ensure that *the electric field everywhere in a uniform section of a current-carrying wire has the same magnitude and points directly along the wire*. We will use this idea extensively in what follows.

The result is that $\vec{E}$ everywhere inside a uniform wire points along the wire and has a constant magnitude

---

Exercise E5X.3

Imagine that as a (conventional) current begins to flow in a straight segment of wire, positive charges flow faster on the left part of the wire than on the right, as shown here. What happens to the surface charge on the wire as a result? How does this help even out the current flow?

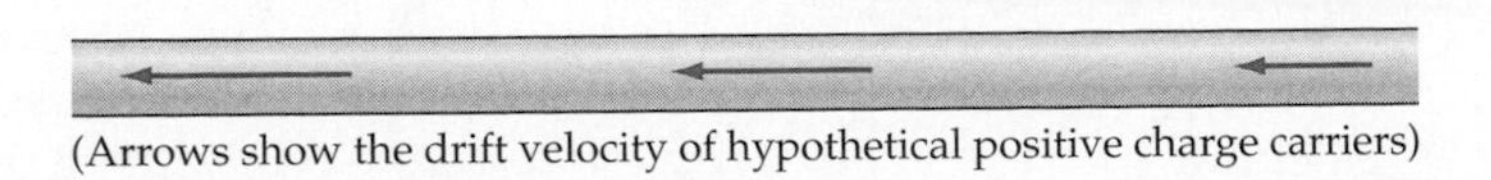

(Arrows show the drift velocity of hypothetical positive charge carriers)

E5.3 The Emf of a Battery

Why *emf* usefully characterizes a battery's strength

Plausibly, the "strength" of the battery determines the overall *magnitude* of the electric field ultimately established in the wire. How can we quantify this?

As figure E5.5 illustrates, the fundamental thing that a battery does is to transport electric charge from one battery plate (or **electrode**) to the other against the electric field created by the charge on those plates. Like raising a weight in a gravitational field, this requires *energy*. The definition of potential implies that if the potential difference between a battery's electrodes at a given time is $|\Delta\phi|$, then a charge carrier with charge q transported from one electrode to the other against the electrostatic field they create gains an electrostatic potential energy of $|q\,\Delta\phi|$.

This energy has to come from somewhere. In a typical chemical battery (say, a household alkaline battery) a chemical reaction occurs at the positive electrode that absorbs electrons and another at the negative electrode that releases electrons. Together, these reactions release a characteristic total amount of energy ΔU^{ch} for every electron absorbed and released. In a well-designed battery, this energy is almost completely converted to electrostatic potential energy, so

$$|\Delta U^{\text{ch}}| \approx e|\Delta\phi| \quad \Rightarrow \quad |\Delta\phi| \approx \frac{|\Delta U^{\text{ch}}|}{e} \equiv \mathscr{E} \tag{E5.1}$$

where e is the magnitude of the charge on the electron and $\mathscr{E}$ is the total *energy per unit charge* or **emf** transferred to each electron by whatever drives charge through the battery (chemical reactions in this case). The term *emf* is short for **electromotive force.** This is a misnomer, though, since emf is not a *force* but rather an *energy delivered per unit charge*. So I would like you to imagine that emf is simply a new word (which for historical reasons we pronounce "eeyemehf") that expresses a battery's *strength* (one might say "oomph") in terms of the characteristic energy per unit charge delivered by a battery's chemical reactions.

Note that equation E5.1 implies that the SI unit of emf is the same as that for potential difference (that is, the *volt*), but the two *concepts* are somewhat different. A potential difference specifically describes a change in *electrostatic potential energy per unit charge,* while emf describes energy per unit charge contributed by something *other* than an electric field, such as a chemical reaction.

The emf's of batteries in series add

Our model implies that if we put two batteries in **series** (so that the positive electrode of one battery touches the negative electrode of the other, as shown in figure E5.6), we get a battery whose emf is the *sum* of the original batteries' emf's. This is so because a charge carrier with charge q moving through the first battery receives an energy $q\mathscr{E}_1$ from that battery and then

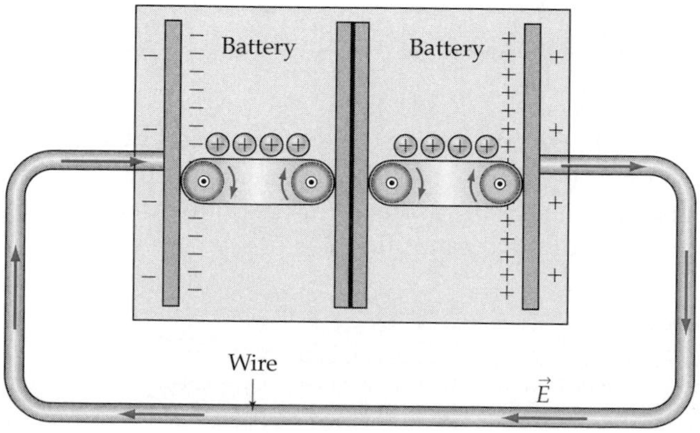

Figure E5.6
Two batteries in series give a charge carrier a total emf that is the sum of the two batteries' emf's. As we will see in section E5.4, this increases the electric field in the wire, driving more current through it.

receives an additional energy $q\mathcal{E}_2$ as it goes through the second battery. The total energy per unit charge that the carrier receives as it moves from first battery's negative electrode to the second battery's positive electrode is therefore $\mathcal{E}_{tot} = \mathcal{E}_1 + \mathcal{E}_2$. This is analogous to someone lifting a box from the ground to a height of 2.0 m and then handing it to someone else on a ladder who lifts it an additional 1.5 m: the total height through which the people lift the box is the sum of the heights contributed by each.

A practical example: a lead-acid battery

A practical example of a battery is a lead-acid car battery. A car battery consists of six lead-acid *cells* wired in series. (A **cell** is the simplest possible battery configuration, consisting of a single positive electrode and a single negative electrode immersed in the appropriate chemicals.) A lead-acid *cell* consists of a positive electrode of lead dioxide (PbO_2) and a negative electrode of pure lead immersed in an **electrolyte** of sulfuric acid (H_2SO_4) dissolved in water. Sulfuric acid in water disassociates into hydronium ions (H_3O^+) and bisulfate ions (HSO_4^-). Both electrodes chemically react with the electrolyte as follows:

$$PbO_2 + HSO_4^- + 3H_3O^+ + 2e^- \rightarrow PbSO_4 + 4H_2O \quad \text{(E5.2}a\text{)}$$

$$Pb + HSO_4^- + H_2O \rightarrow PbSO_4 + H_3O^+ + 2e^- \quad \text{(E5.2}b\text{)}$$

The first reaction grabs two electrons from the positive electrode, causing it to become more positive. The second produces two electrons, which pile up on the negative electrode. The net effect of the combined reactions is thus to transport two electrons from the positive electrode to the negative electrode. Note, however, that charge is *actually* transported by hydronium and bisulfate ions diffusing through the electrolyte: this is what serves as the conveyor belt that carries charge against the electric field created by the electrodes. Figure E5.7 illustrates the process.

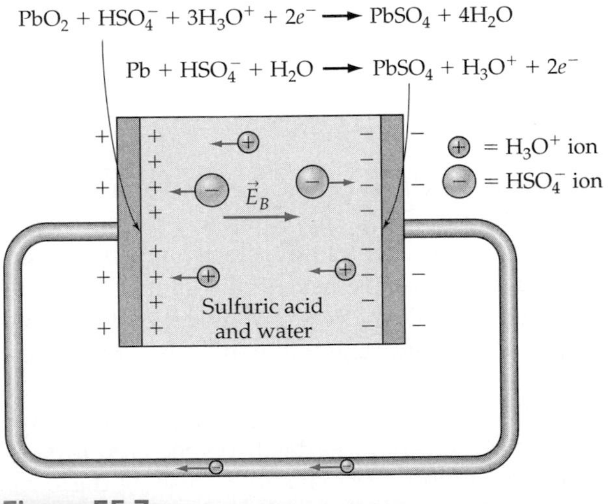

Figure E5.7
This figure illustrates what happens inside a lead-acid cell when it is conducting a current. Note that for the reaction to take place, hydronium ions have to diffuse to the left and bisulfate ions to the right against the electrostatic forces acting on them due to electrical charges on the electrodes. (I have drawn the electric field as if it filled the whole space between the electrodes, but the electric field is technically only very strong in a thin film of fluid surrounding each electrode: see problem E5S.7.)

The combined chemical reactions happen to release an energy of about 6.7×10^{-19} J while effectively transporting two electrons' worth of charge between the electrodes, so this cell's characteristic energy released per unit charge (emf) is

$$\mathscr{E} = \frac{6.7 \times 10^{-19}\ \cancel{\text{J}}}{2(1.6 \times 10^{-19}\ \cancel{\text{C}})}\left(\frac{1\ \text{V}}{1\ \cancel{\text{J}}/\cancel{\text{C}}}\right) = 2.1\ \text{V} \tag{E5.3}$$

This reaction goes very quickly (piling up charges on the electrodes) until enough charge has piled up that the energy cost per unit charge of transporting any more ions (which is the potential difference $|\Delta\phi|$ between the electrodes, by definition) would exceed the energy per unit charge $\mathscr{E}$ supplied by the reaction. If a flowing current depletes the charges on the electrodes, the reaction goes fast enough to essentially instantly resupply charge to electrodes so that the potential difference remains nearly $|\Delta\phi| = \mathscr{E}$. Therefore, we see that whether current is flowing or not, the reactions in a lead-acid cell strive to maintain a fixed potential difference of 2.1 V between the cell's electrodes. The potential difference across the terminals of a six-cell car *battery* is thus $6(2.1\ \text{V}) = 12.6\ \text{V}$.

How the potential difference between a battery's electrodes is linked to its emf

Under realistic conditions, not *all* the energy released by the chemical reaction gets converted to electrostatic potential energy in this way: some is converted to thermal energy as ions moving through the battery collide with stationary molecules. Conservation of energy implies that equation E5.1 should really be written

$$|\Delta\phi| = \mathscr{E} - \mathscr{E}^{\text{th}} \tag{E5.4}$$

Purpose: This equation links the potential difference $|\Delta\phi|$ between a battery's electrodes to the characteristic emf $\mathscr{E}$ of the process driving charge through the battery.

Symbols: $\mathscr{E}^{\text{th}}$ is the energy per unit charge lost to thermal energy.

Limitations: This equation assumes no energy loss other than to thermal energy.

Thus, while a battery's characteristic chemical emf $\mathscr{E}$ is *independent* of the current flowing through it, $\mathscr{E}^{\text{th}}$ is approximately *proportional* to that current (the more rapidly ions move through the electrolyte, the more energy is lost in collisions). This means that the potential difference $|\Delta\phi|$ between a battery's electrodes is equal to the battery's chemical emf only when *no* current flows through the battery, and $|\Delta\phi|$ will generally "droop" a bit as the current increases.

One can minimize this problem by supplying the battery with plenty of electrolyte (having more ions reduces the drift speed required to carry a given current, which reduces thermal losses) and increasing the surface area of the electrodes (so that the ions can more easily get to the electrodes). Battery designers generally work hard to keep $\mathscr{E}^{\text{th}}$ as small as economically possible. In many situations, $\mathscr{E}^{\text{th}}$ is small enough that we can use the **ideal battery approximation,** where we assume that $\mathscr{E}^{\text{th}} = 0$ independent of the current. A good car battery can supply tens of amperes without a significant decrease in the potential between its electrodes.

The *capacity* of a battery

A battery can provide only so much charge to a circuit. For example, when all the lead and/or lead dioxide and/or sulfuric acid in a lead-acid cell has been used up, the reaction cannot continue, and the battery is "dead." The total number of electrons that the battery can send through a circuit before dying thus depends on how many molecules of reactant are available and thus on how large the battery is.

One of the nice things about a lead-acid battery is that it can be *recharged* simply by forcing a current through the battery in the reverse direction (with a higher-voltage battery or power source). This causes the chemical reactions described in equations E5.2 to go in reverse, until all the lead sulfate is converted back to lead dioxide and lead. Not all batteries can be recharged in this way: in many cases, the reaction does not work well in reverse. For example, normal alkaline cells (with a characteristic voltage of 1.55 V) do not recharge well. Nickel-cadmium cells (with a characteristic voltage of 1.2 V) can be recharged many times, but fail eventually because the reverse reaction does not place its reaction products exactly where they started.

Other sources of electrical energy generally behave as batteries do

Other sources of electrical energy often behave as chemical batteries do. The power company that supplies electricity to your home uses nonchemical physical processes to transfer energy to electrons, but these processes also transfer a characteristic amount of energy to each electron. The two vertical terminals in a household electrical outlet behave (for our present purposes) as if they were the electrodes of a battery with a characteristic emf of 120 V. Electronic power supplies also generally behave as if they were batteries with a certain fixed emf.

E5.4 Resistance

So we see that a battery will strive to maintain a fixed potential difference of $|\Delta\phi| = \mathscr{E}$ between those terminals. How is this related to the electric field it establishes in a wire connected between its terminals?

According to equation E3.11, the potential difference $d\phi$ between two points in the wire separated by a small displacement $d\vec{r}$ is given by $d\phi = -\vec{E} \cdot d\vec{r}$. To find the potential difference between the wire's two ends, we can divide the wire (no matter how it bends or loops) into a set of small displacements $d\vec{r}_i$, each of which points along the wire in the direction that conventional current flows (see figure E5.8). Since we discovered in section E5.2 that $\vec{E}$ *also* points parallel to the wire in this direction, the angle between $\vec{E}$ and $d\vec{r}_i$ is zero for each such displacement. The change in the potential due to the ith displacement is thus

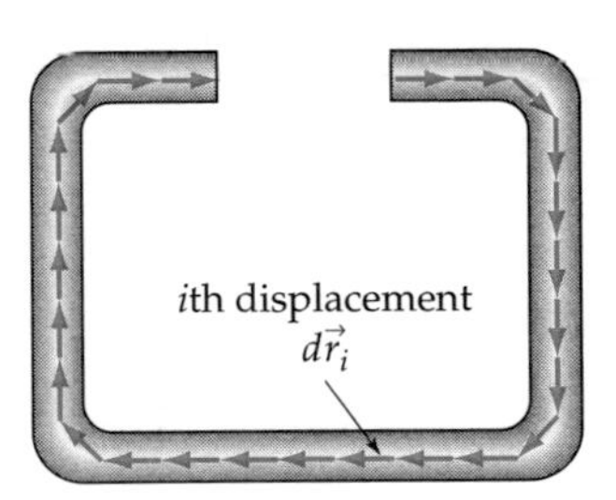

Figure E5.8
Dividing the wire into a set of small displacements.

$$d\phi_i = -\vec{E} \cdot d\vec{r}_i = -E\,dr_i \cos 0° = -E\,dr_i \tag{E5.5}$$

Note that $E \equiv \text{mag}(\vec{E})$ is the same at all points in the wire. The total potential difference between the wire's ends is thus

$$\Delta\phi = \sum_{\text{all } i} d\phi_i = \sum_{\text{all } i} (-E\,dr_i) = -E \sum_{\text{all } i} dr_i \tag{E5.6}$$

How the electric field strength in a wire is connected to its length

But the sum over all the displacement lengths dr_i is simply the wire's length L. So for all thin wires (no matter how they bend), we have

$$\Delta\phi_{\text{wire}} = -EL \quad \Rightarrow \quad |\Delta\phi_{\text{wire}}| = EL \tag{E5.7a}$$

Purpose and symbols: This equation links the electric field strength E in a thin wire to its length L and the potential difference $|\Delta\phi|$ between its negative and positive ends.

Limitations: This equation assumes that the wire is uniform and thin, and that a steady-state current flow has been established.

The potential difference here is negative because current flows in the direction that *decreases* its potential energy (just as a river flows downhill).

Exercise E5X.4

Imagine that we connect a wire that is 0.5 m long and has a diameter of 0.2 mm to a 1.5-V alkaline battery. What is the magnitude of the electric field in the wire? Does it depend on the wire's diameter?

The conventional notation for a potential difference

In the context of electric circuits, the potential difference between the two ends of a conductor is such a commonly used quantity that the notation $|\Delta\phi|$ becomes cumbersome. Conventionally the symbol V (with a subscript specifying the object) is used instead.

Problems with this notation

This convention is problematic for a number of reasons. (1) The symbol V is also commonly used for volume and potential energy (and a handwritten V might also be mistaken as a symbol for speed or the unit of volts!). Since potential is conceptually linked to potential energy, the risk that the notation will lead people to confuse the two is especially high. (2) The absence of a Δ in the notation may lead one to forget that V refers to a potential *difference*. (3) To make matters worse, there is no established convention specifying whether V is a signed quantity or a strictly nonnegative quantity.

My additional conventions for the notation for potential difference

However, were I to invent a notation to avoid these problems, you would be unprepared to read most of the available literature about electric circuits. Therefore, having duly warned you of the dangers, I *will* use the conventional notation in what follows. I will, however, try to avoid as many problems as possible by adding some minor conventions of my own. Specifically, in the remaining chapters in this unit (but *only* in this unit),

1. $V_{\text{object}} \equiv |\Delta\phi_{\text{object}}|$ = the *absolute value* of the potential *difference* between the specified object's two ends. The potential difference across an object is never negative in this text.
2. If a and b are two points in space or in an electric circuit, $V_{ab} \equiv |\phi_a - \phi_b|$.
3. I will use $q\phi$ instead of V_e for the electrostatic potential *energy* of a particle with charge q from now on. Therefore V will *never* refer to potential energy in the remainder of this unit.

How the current in a wire depends the potential difference between its ends

When expressed in terms of these conventions, equation E5.7*a* becomes

$$V_{\text{wire}} = EL \tag{E5.7b}$$

This only gives the *magnitude* of the potential change; remember also that the potential *decreases* in the direction in which conventional current flows in the wire.

Now, we saw in chapter E4 that $J = \sigma_c E$ and (assuming the current density $\vec{J}$ is constant across the wire's cross section) $I = JA$, where I is the magnitude of the current flowing in the wire and A is its cross-sectional area. If you combine these equations with equation E5.7*b*, you can easily show that

$$V_{\text{wire}} = I\left(\frac{L}{\sigma_c A}\right) \tag{E5.8}$$

Exercise E5X.5

Verify equation E5.8.

Note that if the wire's conductivity σ_c is constant, the current I flowing in the wire is *proportional* to the potential difference V_{wire} that the battery establishes between its ends.

We define the electrical **resistance** R of any conducting object to be

The definition of a conductor's *resistance*

$$R \equiv \frac{V}{I} \tag{E5.9}$$

Purpose: This equation defines a conducting object's resistance R.

Symbols: I is the magnitude of current that the object conducts when the potential difference between its ends is V.

Limitations: This equation assumes that the object has well-defined ends.

Note: The SI unit of resistance is the **ohm,** where $1\ \Omega \equiv 1$ V/A.

If we combine equations E5.8 and E5.9, we see that

The resistance of a thin, uniform wire

$$R_{\text{wire}} = \frac{L}{\sigma_c A} \approx \text{constant} \quad \text{for uniform wire with fixed } \sigma_c \tag{E5.10}$$

A wire of fixed length, cross-sectional area, and conductivity thus has a constant resistance independent of the current that it carries. Many other kinds of objects also have approximately constant resistance. We call the statement that

$$R \equiv \frac{V}{I} = \text{constant} \tag{E5.11}$$

Ohm's law

for a given object **Ohm's law,** and we describe an object whose resistance is indeed independent of the current it carries as being **ohmic.**

Exercise E5X.6

Express the ohm in terms of the more basic SI units of kilogram, meter, second, and coulomb, and show that the units of conductivity are $(\Omega \cdot \text{m})^{-1}$.

Exercise E5X.7

What is the resistance of a piece of copper wire 1.0 m long and 0.2 mm in diameter? If we connect it to the ends of a 1.5-V battery, what current flows through the wire?

E5.5 The Power Dissipated in a Conductor

The power dissipated when a current flows through a conductor

Assume that the potential difference between a conducting object's two ends is V and that its charge carriers have charge q. The definition of potential difference then implies that a charge carrier's electrostatic potential energy *decreases* by $|q|V$ as it flows through the conductor. We noted in chapter E4 that any energy a charge carrier picks up from the electrostatic field that drives it through a wire is eventually converted to thermal energy in the lattice; so this lost electrostatic potential energy shows up as increased thermal energy in the conductor.

Imagine now that the object is carrying current $I = |dQ/dt|$. The *rate* at which carriers in the conductor convert electrostatic potential energy to thermal energy (that is, the *power* P associated with the energy conversion) is

$$P = \frac{|dQ|V}{dt} = IV \tag{E5.12a}$$

Purpose: This equation expresses the power P associated with the conversion of electrostatic potential energy to thermal energy in a conductor carrying current of magnitude I.

Symbols: V is the potential difference between the conductor's ends.

Limitations: This equation assumes that the conductor has well-defined ends.

You can use equation E5.9 to express either V or I in terms of the other variable and the conducting object's resistance R: the results are

$$P = I^2R = \frac{V^2}{R} \tag{E5.12b}$$

Exercise E5X.8

Verify both equalities in equation E5.12*b*.

The power in these equations comes out in watts

Note that the SI unit for current is the ampere, while the SI unit of potential difference is the volt, so the SI units of $P = VI$ are

$$\cancel{\text{V}} \cdot \cancel{\text{A}} \left(\frac{1\ \text{J}/\cancel{\text{C}}}{1\ \cancel{\text{V}}}\right)\left(\frac{1\ \cancel{\text{C}}/\text{s}}{1\ \cancel{\text{A}}}\right) = \frac{\cancel{\text{J}}}{\cancel{\text{s}}}\left(\frac{1\ \text{W}}{1\ \cancel{\text{J}}/\cancel{\text{s}}}\right) = \text{W} \tag{E5.13}$$

So P comes out in *watts,* the standard SI unit for power. A 100-W lightbulb thus converts electrical energy to thermal energy at a rate of 100 J/s, a 2000-W toaster converts energy at a rate of 2000 J/s, and so on.

The *kilowatt-hour*

Power companies measure electrical energy in **kilowatt-hours,** where

$$1\ \text{kW} \cdot \text{h} = \left(\frac{1000\ \text{J}}{\cancel{\text{s}}}\right)(1\ \cancel{\text{h}})\left(\frac{3600\ \cancel{\text{s}}}{\cancel{\text{h}}}\right) = 3.6 \times 10^6\ \text{J} \tag{E5.14}$$

One kilowatt-hour of electrical energy costs very roughly \$0.15 (in southern California in 2002), or about 4.2¢ per megajoule.

We can use equations E5.12 to calculate interesting and useful quantities for simple household circuits. For example, when a 100-W lightbulb is connected to such a standard 120-V household outlet, equation E5.12*a* implies that it conducts a current of roughly

$$I_{\text{bulb}} = \frac{P_{\text{bulb}}}{V_{\text{bulb}}} = \frac{100\ \cancel{\text{W}}}{120\ \cancel{\text{V}}}\left(\frac{1\ \cancel{\text{V}} \cdot \text{A}}{1\ \cancel{\text{W}}}\right) = 0.83\ \text{A} \tag{E5.15}$$

and its effective resistance is

$$R_{\text{bulb}} = \frac{V_{\text{bulb}}}{I_{\text{bulb}}} = \frac{120\ \cancel{\text{V}}}{0.83\ \cancel{\text{A}}}\left(\frac{1\ \Omega}{1\ \cancel{\text{V}}/\cancel{\text{A}}}\right) = 140\ \Omega \tag{E5.16}$$

Exercise E5X.9

An electric hand-drier in a public bathroom (see figure E5.9) advertises that its heating element draws 20 A of current. How much power does this drier use? How much does it cost to run it for a 3-min cycle?

Figure E5.9
An electric hand-drier of the type commonly seen in public bathrooms.

E5.6 Discharging a Capacitor

Discharging a capacitor through a wire

Let's now return to the situation we considered in section E5.1 where we connect the ends of a wire to the plates of a charged capacitor (see figure E5.1). As we said, a current will only flow through the wire for a short time. But the time required for a typical capacitor to discharge through a wire (microseconds to seconds) is so long compared to the time required for surface charges to adjust on the wire (tens of nanoseconds) that at each instant of time during the discharge, the circuit behaves as if it were a wire connected to a battery, except that the potential difference between the plates of the capacitor "battery" slowly varies with time. According to the definition of capacitance, this potential difference is

$$V_R = V_C = \frac{Q}{C} \tag{E5.17}$$

where V_R is the potential difference between the ends of the wire (which we will assume has resistance R), V_C is the same between the plates of the capacitor (which we will assume has capacitance C), and Q is the charge on the capacitor's positive plate. Note that V_C will therefore decrease as charge flows from one plate to the other. According to the definition of resistance, we have

$$R = \frac{V_R}{I} \quad \Rightarrow \quad V_R = IR \tag{E5.18}$$

where I is the current flowing through the wire. Since this current comes entirely at the expense of the charge on the capacitor's plates, and the current is simply the rate at which charge moves past a certain point (say, the junction between the wire and the capacitor's positive plate), the current at any instant will be

$$I = -\frac{dQ}{dt} \tag{E5.19}$$

The minus sign is necessary because we need I to be positive in equation E5.18, but Q is decreasing with time in this situation.

In this situation $V_R = V_C$, Q, and I are all functions of time t. If you combine equations E5.17 through E5.19 in such a way as to eliminate V_R and Q, you can show fairly easily that

$$\frac{dV_C}{dt} = -\frac{1}{RC}V_C \tag{E5.20}$$

Exercise E5X.10

Verify that equation E5.20 is correct.

The potential difference between a discharging capacitor's plates decreases exponentially with time

This is a common type of differential equation in nature (and in social sciences) where the change in a quantity is proportional to that quantity's value. In this case, the equation tells us that $V_C = V_C(t)$ is a function of time such that when we take its time derivative, we get back the *same* function multiplied by a (negative) constant (assuming that the wire's resistance R is constant). The only mathematical function whose derivative is equal to itself is the exponential function. By taking the time derivative of

$$V_C(t) = Ae^{-t/RC} \tag{E5.21}$$

(where A is some unknown constant), you can check that this function $V_C(t)$ satisfies equation E5.20. (Note that $e = 2.7183$ here, not the magnitude of the charge on the electron.)

Exercise E5X.11

Verify that the function given in equation E5.21 satisfies equation E5.20. (Problem E5S.10 discusses another approach for solving equation E5.20.)

Now, at time $t = 0$, equation E5.21 implies that $V_C(0) = Ae^{-0/RC} = A$. If we define V_0 to be the initial potential difference between the plates, then $A = V_0$. Our final equation for the voltage across the capacitor as it discharges is thus

$$V_C(t) = V_0 e^{-t/RC} \tag{E5.22}$$

Purpose: This equation specifies how the potential difference V_C between a capacitor's plates varies with time t as it discharges through a wire of resistance R.

Symbols: C is capacitor's capacitance and V_0 is the initial value of V_C.

Limitations: This equation assumes that the connecting wire is ohmic.

We see that the potential difference between the capacitor's plates decreases exponentially with time from its initial value of V_0, decreasing by a factor of $e = 2.7183$ during each time interval of duration RC (see figure E5.10). Note that as either R or C increases, the time required for the capacitor to discharge increases. Note also that the capacitor never quite gets fully discharged: the potential difference between its plates asymptotically approaches zero. However, after a time equal to 4 or 5 times RC has passed, the potential difference is zero for all practical purposes.

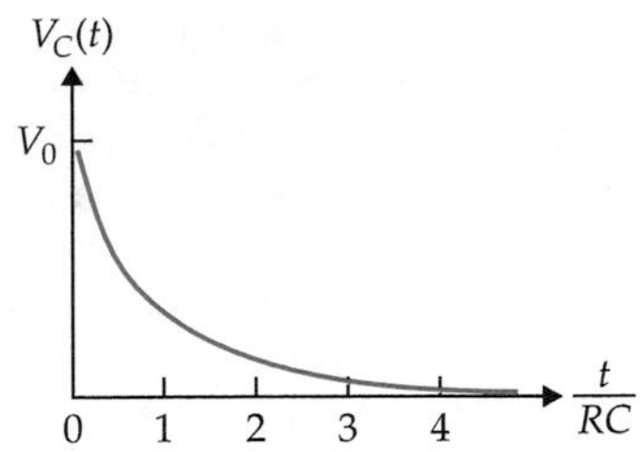

Figure E5.10
How the voltage across a capacitor decreases with time as it discharges through an ohmic wire.

Exercise E5X.12

Check that RC has units of time. (It had *better* have these units: the exponent of e must be unitless.)

One can show that the total thermal energy that has been dissipated in the wire when the capacitor has completely discharged is the same as the energy initially stored in the capacitor's electric field (see problems E5S.11 and E5S.12). This is another piece of evidence supporting the idea that an electric field is a form of energy.

Since the potential difference between a discharging capacitor's plates depends on time in such a predictable and controllable manner, one can essentially use a discharging capacitor as a clock. This is a common role for capacitors in electronic circuits. Capacitors are also used in electronic circuits to temporarily store charge and thus even out fluctuations in current flows. Figure E5.11 shows a selection of capacitors that might be used in electronic circuits.

Figure E5.11
A selection of capacitors of the type commonly used in electronic circuits to generate or regulate time-dependent current flows.

TWO-MINUTE PROBLEMS

E5T.1 The magnitude of the steady-state electric field in a given wire connected to a battery is proportional to the voltage difference between the battery's plates, true or false (T or F)?

E5T.2 A battery always drives the same amount of current through a wire connecting its terminals (no matter what the characteristics of that wire might be), T or F?

E5T.3 We have seen that we can construct a battery with a high potential difference between its terminals by connecting lower-potential batteries in series (see figure E5.6). We can also construct a battery by connecting the cells in *parallel*, that is, connecting all the cells' positive terminals together and all the negative terminals together. What might be the advantage of doing this?

A. Again, the resulting battery will have a higher potential difference between its terminals than each individual cell.
B. The resulting battery will last longer than an individual cell in the same circuit would.
C. The resulting battery would be more ideal ($\mathscr{E}^{th}$ would be closer to zero for a given current) than a single cell.
D. Both A and B.
E. Both B and C.
F. A, B, and C are all true.
T. There is no advantage to connecting batteries in this way.

E5T.4 If the electrodes of a 1.5-V alkaline D cell are connected to a certain 1.5-m wire, the wire conducts 2.0 A. What is the magnitude of the electric field in the wire?

A. 0.75 N/C
B. 1.0 N/C
C. 3.0 N/C
D. 4.5 N/C
E. 30 N/C
F. Some other field (specify)

E5T.5 If the length of the wire described in problem E5T.4 is doubled (while its other characteristics remain the same), the voltage difference between the battery's electrodes changes by a factor of x, where x is (select from the list below). By what factor does the value of the electric field change? The value of the current the wire conducts?

A. 4
B. 2
C. $\frac{1}{2}$
D. $\frac{1}{4}$
E. There is no change.
F. Some other factor (specify)

E5T.6 If we connect the two ends of a given wire to 1.5-V alkaline D cell, the magnitude of the steady-state electric field in the wire is 3.0 N/C. If we connect the two ends of the same wire to two D cells in series, the magnitude of the electric field in the wire will be

A. 3.0 N/C
B. 1.5 N/C
C. 6.0 N/C
D. Other (specify)

E5T.7 Wire 1 is 25 cm long and has a diameter of 0.6 nm, and wire 2 has double the length and diameter. Both are copper. How do their resistances compare?

A. $R_1 \approx 4R_2$
B. $R_1 \approx 2R_2$
C. $R_1 \approx R_2$
D. $R_2 \approx 2R_1$
E. $R_2 \approx 4R_1$
F. Some other relationship (specify)

E5T.8 Imagine we connect the ends of a copper wire with a given length and diameter to the electrodes of one alkaline 1.5-V D cell, and we connect the ends of an iron wire with the same length and diameter to another D cell. The electric field is higher in the copper wire, T or F? The current flowing is higher in the copper wire, T or F?

E5T.9 A light-emitting diode (LED) has the property that the potential difference between its ends is approximately 1.5 V when it conducts currents ranging from 1 mA to 50 mA. This device is ohmic, T or F?

Figure E5.12
A light-emitting diode (see problem E5T.9).

E5T.10 When it is conducting 10 mA of current, the LED described in problem E5T.9 has a resistance of

A. 0.015 Ω
B. 0.15 Ω
C. 150 Ω
D. 1500 Ω
E. We cannot calculate R because the LED is not ohmic.

E5T.11 Imagine that you have two plates of area A separated by a small distance d that are connected to a battery with fixed emf $\mathscr{E}$. If you were to reduce

the distance between the plates by a factor of 2, the magnitude Q of the charge on each plate would
A. Quadruple
B. Double
C. Remain the same
D. Be halved
E. Be quartered
F. Change in some other way (specify)

HOMEWORK PROBLEMS

Basic Skills

E5B.1 The terminals of a 12.6-V automobile battery are connected by a piece of wire 63 m long. The wire conducts a current of 25 A under these circumstances. What is the magnitude of the electric field in the wire?

E5B.2 The terminals of a 1.5-V alkaline D cell are connected by a piece of very thin wire 5 cm long. The wire only conducts a current of 12 mA under these circumstances. What is the magnitude of the electric field in the wire?

E5B.3 The electric field inside a wire connected between the terminals of a 9-V battery is 36 N/C. How long is the wire? What else would we need to know to determine the current in the wire?

E5B.4 Imagine that we reverse the right-hand battery in figure E5.6, so that its negative terminal is in contact with the negative terminal of the left-hand battery. Would a current still flow in the wire? Why or why not?

E5B.5 Imagine that if we connect a 5.0-m length of an unknown type of wire with a diameter of 0.75 mm to a 1.5-V battery, we find that it conducts 3.0 A. What is the resistance of this wire? What is the electric field in the wire?

E5B.6 Power cords for electric lamps and the like are often made of #18 copper wire, which has a diameter of about 1.01 mm. What is the resistance of 50 ft of such wire?

E5B.7 Imagine that you need to create a circuit element having a resistance of 12 Ω. You have a spool of nichrome wire that has a diameter of 0.82 mm. How long would a piece of such wire have to be to have a resistance of 12 Ω?

E5B.8 Imagine that a certain lightbulb has a tungsten filament with a length of 8.0 mm and a cross-sectional area of 1.2×10^{-7} m^2. This bulb conducts 0.24 A of current when connected to a 2.0-V battery. What is the magnitude of the electric field in the filament?

E5B.9 The terminals of a 1.5-V alkaline D cell are connected by a piece of wire that is 25 cm long, which conducts a current of 30 A under the circumstances. How much chemical energy is converted to thermal energy each second?

E5B.10 The terminals of a 12.6-V automobile battery are connected by a piece of wire whose resistance is 6.3 Ω. At what rate does this system convert chemical energy in the battery to thermal energy in the wire?

E5B.11 One of the many expenses associated with a home swimming pool is the electrical power required to run the pool filter. The motor that pumps pool water through the filter typically uses a power of 1.5 hp and has to run about 8 h/day during the swimming season. About how much would this add one's electric bill in southern California?

Synthetic

E5S.1 A wire of length L is connected between the terminals of a certain battery and is found to conduct a current I. If this wire is replaced by wire which is either one-half the length or double the cross-sectional area (but which is otherwise identical), the current will go up by a factor of 2. In each case, explain *why* in terms of the basic physics principles introduced *before* equation E5.7.

E5S.2 Imagine that we measure the magnitude of the potential difference between the electrodes of a certain alkaline cell to be 1.55 V when its electrodes are unconnected and 1.49 V when they are connected by a wire carrying a current of 1.0 A. (a) *Why* does the potential difference decrease when the cell conducts the current? (b) How much thermal energy is being produced in the battery every second in the second case?

E5S.3 Imagine that you have a wire made of an unknown metal. The wire is 1.0 m long and has a diameter of 1.0 mm, and it conducts 1.0 A when connected to the terminals of a 2.0-V battery. What would you guess the metal to be? (*Hints:* Can you determine the electric field magnitude in this wire? The current density? How might knowing these quantities help?)

E5S.4 Imagine that we connect a long wire to the negative terminal of a battery, but we do not connect the other end to the positive terminal. Does a current flow in the wire immediately after it is connected to the battery? If so, does the current continue to flow indefinitely, or does it stop after a short time? If the current flows indefinitely, where does the charge go? If it stops, why does it stop? Explain your reasoning carefully, using pictures if possible.

E5S.5 Consider a battery connected to a wire bent in the shape shown in figure E5.13. Just before the wire is connected, the electric field of the plates will tend to push positive charges at points *A*, *B*, and *C* toward the *left*. Explain in detail how connecting the wire to the battery produces surface charges that redirect the current at point *B* so that it flows "uphill" against the field produced by the battery plates. Sketch these charges on a copy of figure E5.13.

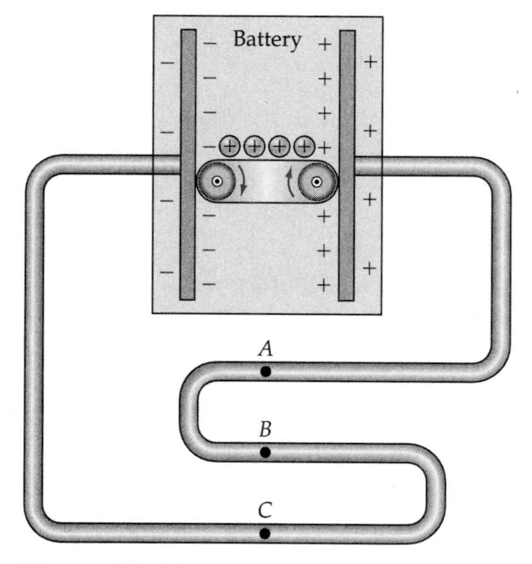

Figure E5.13
How can we create an electric field that drives the current in opposite directions at points *A*, *B*, and *C*? (See problem E5S.5.)

E5S.6 A chemical battery's *capacity* expresses the total amount of charge that it sends through a circuit, which is proportional to the total number of molecules of chemical reactants it contains. Thus a large battery generally has greater capacity than a small battery of the same type. A battery's capacity is generally expressed in ampere-hours (Ah), which is numerically equal to the number of hours that the battery could supply 1 A of current.

(a) Express 1 Ah in coulombs, as a number of electrons, and in terms of moles of electrons.
(b) Imagine that you are designing a battery-powered toaster that uses six alkaline D cells in series. While the toaster is on, it must produce at least 900 W of thermal energy. D cells have a capacity of about 5 Ah. Roughly how long will your set of batteries run your toaster?

E5S.7 Consider a lead-acid cell of the type discussed in section E5.3.

(a) Explain why there must be essentially *no* electric field inside the conducting sulfuric acid solution except in microscopically thin regions around the electrodes (this is pretty much true even if the battery is conducting a significant current). You might draw a picture showing how concentrations of various molecules vary near each electrode.
(b) What are the charge carriers in the sulfuric acid solution, and which way do they move? (These carriers travel by diffusion.)
(c) Battery capacity is measured in ampere-hours, which specifies how long the battery can supply a current of 1 A (see problem E5S.6). Show that 1 Ah is equivalent to 0.0373 mol of electrons.
(d) If we connect the cell's electrodes with a wire, electrons will flow through the wire only if there are lead and lead dioxide and bisulfate ions left in the cell to react. Lead, sulfur, oxygen, and hydrogen have atomic weights of 207, 32, 16, and 1 g/mol, respectively. Estimate the minimum mass of a cell having a capacity of 100 Ah, the approximate capacity of a car battery.
(e) A typical car battery consists of six lead-acid cells wired in series. What must its mass be to have 100 Ah of capacity? Explain your reasoning.

E5S.8 A nuclear battery might be constructed as follows. Imagine that an atom of a certain kind of radioactive metal emits an electron when it decays. Each electron carries away a kinetic energy of 1.2 MeV, where 1 MeV $= 10^6$ eV, and 1 eV is equivalent to the amount of electrostatic potential energy one electron gains when traveling across a potential difference of 1 V. (Energies per reaction in the mega-electronvolt range are typical of nuclear processes.)

(a) Argue from its definition that 1 eV $= 1.60 \times 10^{-19}$ J.
(b) Imagine that we make a plate of this metal, and we place it near a plate of an ordinary non-radioactive metal. Which plate becomes positively charged? Which plate becomes negatively charged? Why will the potential difference between the plates reach a maximum, and what will be the maximum potential difference be?
(c) Describe some reasons (besides the danger of dealing with radioactive substances) why such a battery might be impractical. (*Hint:* Air breaks down and conducts electricity when the electric field strength exceeds about 3 MN/C = 3 MV/m. Also typical transistors and integrated circuits are destroyed by potential differences exceeding a few hundred volts.)

E5S.9 Imagine that you have a lightbulb that has a resistance of about 10 Ω and that can tolerate a maximum voltage of 3 V. Imagine that you want to connect this to a charged capacitor large enough to keep the bulb glowing reasonably brightly for more than 10 s. Roughly what should the capacitor's capacitance be?

E5S.10 If you are uncomfortable with the way that we guessed the solution for equation E5.20, perhaps you will be happier with the following method. Divide both sides of equation E5.20 by $-RC V_C$ and multiply both sides by dt. You should get

$$\frac{-dt}{RC} = \frac{dV_C}{V_C} \tag{E5.23}$$

If you integrate this expression with appropriate limits on both sides, you should get an expression that with a little more manipulation can be shown to be equal to equation E5.21. Show how this can be done.

E5S.11 Consider a nested-shell capacitor whose inner and outer conductors are *very* close together compared to their radii.

(a) Argue that the energy stored in the electric field of such a capacitor is $U_{\text{field}} = \frac{1}{2}CV_0^2$, if V_0 is the potential difference between the capacitor's plates. (*Hint:* Argue that between the plates, the field strength is $E \approx V_0/d$, where $d \equiv R_o - R_i$ is the distance between the conductors. Use the equation for the capacitance of a nested-shell capacitor and the fact that $R_o \approx R_i$ to eliminate d in the equation for E in favor of C. Roughly how much volume does this field occupy?)

(b) Imagine we discharge this capacitor by connecting its plates with a wire of resistance R. The rate at which thermal energy is produced in the wire at any given instant of time is

$$\frac{dU^{\text{th}}}{dt} = P = \frac{V_R^2}{R} = \frac{V_C^2}{R} \tag{E5.24}$$

where V_C is the time-dependent potential difference across the capacitor's plates. If we integrate both sides with respect to t, we find that the total energy converted to thermal energy in the wire as the capacitor discharges is

$$U^{\text{th}} = \frac{1}{R}\int_0^{\infty} V_C^2\, dt = \frac{1}{R}\int_0^{\infty} (V_0 e^{-t/RC})^2\, dt \tag{E5.25}$$

Do this integral, and show that $U^{\text{th}} = \frac{1}{2}CV_0^2$, which is consistent with the energy initially stored in the field according to part (a).

E5S.12 Consider a parallel-plate capacitor whose plates have an area A and are separated by a distance d such that $d \ll A^{1/2}$.

(a) Argue that the energy stored in the electric field of such a capacitor is $U_{\text{field}} = \frac{1}{2}CV_0^2$, if V_0 is the potential difference between the capacitor's plates. (*Hint:* Argue that between the plates, the field strength is $E \approx V_0/d$, where d is the distance between the plates. Use the equation for the capacitance of a parallel-plate capacitor given in table E4.2 to eliminate d in the equation for E in favor of C. Roughly how much volume does this field occupy?)

(b) Follow the path outlined in part (b) of problem E5S.11 to show that the thermal energy produced when the capacitor discharges is $U^{\text{th}} = \frac{1}{2}CV_0^2$, which is consistent with the energy initially stored in the field according to part (a).

E5S.13 Imagine that you would like to design a parallel-plate capacitor that could store 1 J of electrical energy in the electric field between its plates when connected to a 1.5-V battery. Imagine that you can arrange things so that the plates can be separated by a distance d as small as 0.5 mm. How big do the plates have to be? Is this realistic? (*Hint:* Argue that the electric field strength between the plates is $E \approx V/d$, where V is the potential difference between the plates.)

E5S.14 Nerve cells in the human body have long extensions called axons that connect to other cells. An axon typically maintains a resting potential difference of about 80 mV across its membrane (with the inside negative) by actively excluding positive sodium ions and retaining negatively charged protein ions. This builds up opposite concentrations of charge on either side of the membrane, making the membrane like a capacitor whose plates are the conducting fluid inside and outside the membrane.

We can model how a signal moves down an axon by imagining the axon to be divided into short segments. When the nerve cell "fires," it makes the membrane on the axon segment closest to the cell porous to sodium ions. This allows a current to flow across the membrane, effectively discharging this segment's capacitor (the technical term is *depolarizing the membrane*). The discharge instigates chemical changes that make the next segment's membrane porous, causing it to discharge even as the first segment recovers by actively pumping potassium ions out of its interior (thus making the interior negative again). The second segment's depolarization makes the next segment's membrane porous, and so on. The depolarization thus moves segment by segment down the axon.

The speed with which a signal moves down the axon depends on how rapidly a segment's capacitor can be discharged. The membranes of some axons are covered with a fatty insulator called *myelin* that effectively makes the membrane thicker (thus increasing the distance between the plates of each segment's capacitor). Will this increase or decrease the speed at which signals move down the axon? Present an argument for your conclusion. Also support your conclusion by a footnote or URL to some document that contrasts the speeds of signals traveling along myelinated and unmyelinated axons. (This problem is based on information in Cameron et al., *Physics of the Body*, 2d ed., Medical Physics Publishing, Madison, Wis., 1999.)

Rich-Context

E5R.1 An inventor claims to have created a chemical cell that involves nickel and iron plates immersed in a secret solution. The cell has a characteristic voltage of 1.3 V. The cell has a mass of 100 g, and the inventor claims a capacity of 250 Ah. A cell with this mass and capacity would sell like hotcakes. But is the inventor's claim physically impossible? Defend your response carefully (there is some serious money at stake here!). (*Hint:* Problem E5S.6 discusses the meaning of the ampere-hour as a unit of battery capacity.)

E5R.2 Imagine that you are trying to design a high school physics lesson where students test hypotheses about the quantity that is a fixed characteristic for a given battery. Assume that according to the literature you have read, the three most common hypotheses about batteries that high school students believe are the following: (1) A battery produces a characteristic constant *current* I. (2) A battery produces a characteristic *power* P. (3) A battery maintains a characteristic potential difference V between its electrodes. The students know that the brightness of a lightbulb conducting electricity is related to the power it dissipates, and that the power dissipated by an object is $P = VI$, where V is the potential difference across the object and I is the current it conducts. They know that current is the number of electrons flowing per unit time through the object. Finally, they know that when you connect two cells in series, the resulting battery has twice the potential difference across its free terminals than that across either cell. You give them two 1.5-V alkaline D cells, two 3-V lightbulbs, and the means to connect them in arbitrary ways.

(a) Describe an experiment or pair of experiments that allows a person knowing only these things to distinguish between these three hypotheses by simply observing qualitative bulb brightnesses. That is, design the experiment(s) so that the three hypotheses lead to easily distinguished and mutually incompatible results.

(b) If your instructor provides you with such equipment, run the experiment(s) for yourself, and check that the results are consistent with the hypothesis you believe.

Advanced

E5A.1 Capacitors are often constructed by constructing a four-layer sandwich consisting of a thin layer of plastic, a thin layer of metal foil, a second thin layer of plastic, and a second thin layer of metal foil. When lying flat, such a sandwich is essentially a parallel-plate capacitor, but has an awkward shape for practical use.

(a) Argue from first principles that if you fan-fold such a sandwich as shown in figure E5.14a, the capacitance of the result is essentially the same as the original unfolded sandwich.

(b) Argue from first principles that if you roll the sandwich into a cylinder, as shown in figure E5.14b, the capacitance of the result should

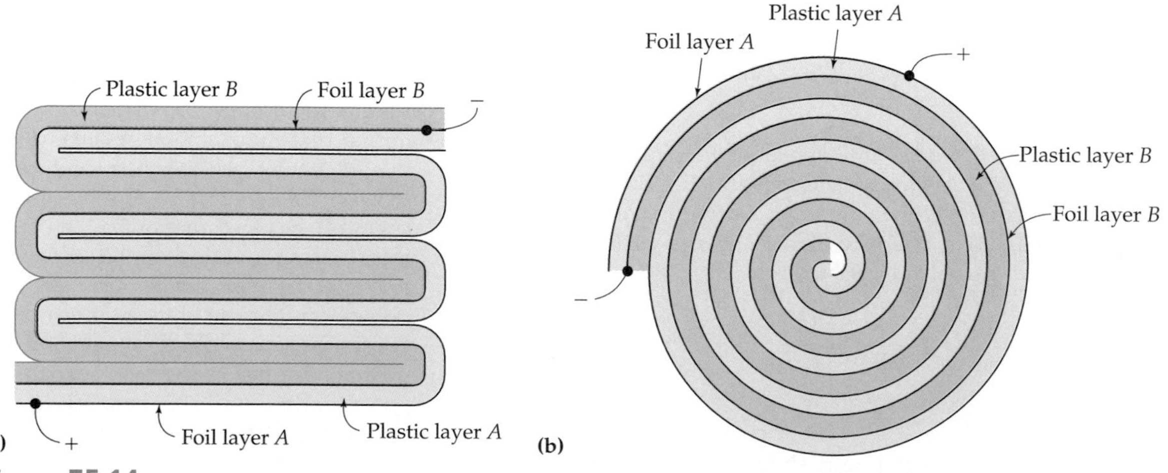

Figure E5.14

(a) Fan folding a sandwich consisting of a pair of foil layers alternating with a pair of plastic layers. (b) Rolling such a sandwich into a cylinder.

still be roughly the same as the original sandwich.

(c) Imagine that your sandwich is 12 cm wide and consists of plastic layers 20 μm thick and foil layers 5 μm thick. The dielectric constant of the plastic is about 3.0 (see problem E4S.12). What would be the approximate diameter of a rolled cylinder with a capacitance of 0.01 F?

ANSWERS TO EXERCISES

E5X.1 The electric field of the positive plate points away from that plate while the electric field of the negative plate points toward that plate. Between the plates, both electric fields point toward the left and so reinforce each other. At a point to the right of both plates, the fields almost cancel, but the field of the nearer plate is slightly stronger, so the total field is small but points to the right.

E5X.2 Point B is much farther from the charges on the plates than point A is.

E5X.3 The positive charge in the faster-moving current to the left will not be fully replaced by positive charge coming from the right. This will leave a deficit of positive charge on the left side of the wire, creating a negative surface charge on the wire there. This surface charge creates an electric field that will speed up positive charges coming in from the right and slow down the charge going away to the left, thus evening out the current flow.

E5X.4 According to equation E5.7,

$$E = \frac{|\Delta\phi|}{L} = \frac{1.5\,\cancel{\text{V}}}{0.5\,\cancel{\text{m}}}\left(\frac{1\,\cancel{\text{J}}/\cancel{\text{C}}}{1\,\cancel{\text{V}}}\right)\left(\frac{1\,\text{N}\cdot\cancel{\text{m}}}{1\,\cancel{\text{J}}}\right) = 3.0\text{ N/C} \tag{E5.26}$$

The diameter of the wire is irrelevant.

E5X.5 Since $J = \sigma_c E$, $E = J/\sigma_c$. Also $I = JA$ implies that $J = I/A$. Plugging these into equation E5.7b, we get

$$V_{\text{wire}} = EL = \frac{J}{\sigma_c}L = \frac{I/A}{\sigma_c}L = I\frac{L}{\sigma_c A} \tag{E5.27}$$

E5X.6 We have $1\ \Omega = 1$ V/A

$$= 1\frac{\cancel{\text{V}}}{\cancel{\text{A}}}\left(\frac{1\,\cancel{\text{J}}/\text{C}}{1\,\cancel{\text{V}}}\right)\left(\frac{1\text{ kg}\cdot\text{m}^2/\text{s}^2}{1\,\cancel{\text{J}}}\right)\left(\frac{1\,\cancel{\text{A}}}{1\text{ C}/\cancel{\text{s}}}\right)$$

$$= 1\frac{\text{kg}\cdot\text{m}^2}{\text{C}^2\cdot\text{s}} \tag{E5.28}$$

The units of conductivity are $\text{C}^2\cdot\text{s}\cdot\text{m}^{-3}\cdot\text{kg}^{-1}$, which we can see are the same as $\Omega^{-1}\cdot\text{m}^{-1}$.

E5X.7 According to equation E5.10,

$$R = L/\sigma_c A$$

$$= \frac{1.0\,\cancel{\text{m}}}{(5.9\times10^7\ \Omega^{-1}\cdot\cancel{\text{m}^{-1}})\pi(1.0\times10^{-4}\,\cancel{\text{m}})^2}$$

$$= 0.54\ \Omega \tag{E5.29}$$

If we connect this wire to a 1.5-V battery,

$$I = \frac{V_{\text{wire}}}{R_{\text{wire}}} = \frac{1.5\,\cancel{\text{V}}}{0.54\,\cancel{\Omega}}\left(\frac{1\,\cancel{\Omega}}{1\,\cancel{\text{V}}/\text{A}}\right) = 2.8\text{ A} \tag{E5.30}$$

E5X.8 Equation E5.9 implies that $I = V/R$ and that $V = IR$, so equation E5.12a implies that

$$P = IV = I(IR) = I^2R \tag{E5.31a}$$

and

$$P = IV = \frac{V}{R}V = \frac{V^2}{R} \tag{E5.31b}$$

E5X.9 The power is $P = VI = (120\text{ J}/\cancel{\text{C}})(20\,\cancel{\text{C}}/\text{s}) = 2400\text{ J/s} = 2400$ W. If the drier runs for 3 min or 180 s, it uses up about 430,000 J of energy which at 15¢ per 3.6×10^6 J is about 1.8¢.

E5X.10 Substituting $Q = CV_C$ into equation E5.19, we get

$$I = -\frac{dQ}{dt} = -\frac{d(CV_C)}{dt} = -C\frac{dV_C}{dt} \tag{E5.32}$$

Substituting this into equation E5.18, we get

$$V_C(t) = \left(-C\frac{dV_C}{dt}\right)R = -RC\frac{dV_C}{dt} \tag{E5.33}$$

Dividing both sides by $-RC$ gives equation E5.20.

E5X.11 Taking the derivative of equation E5.21, we get

$$\frac{dV_C}{dt} = \frac{d}{dt}(Ae^{-t/RC}) = Ae^{-t/RC}\left(\frac{-1}{RC}\right) \tag{E5.34}$$

by the chain rule. But $Ae^{-t/RC} = V_C$, so this is the same as equation E5.20.

E5X.12 The units of RC are

$$\Omega\cdot\text{F} = \frac{\cancel{\text{V}}}{\cancel{\text{A}}}\cdot\frac{\cancel{\text{C}}}{\cancel{\text{V}}}\left(\frac{1\,\cancel{\text{A}}}{1\,\cancel{\text{C}}/\text{s}}\right) = \text{s}$$

E6 Analyzing Circuits

▷ Electric Field Fundamentals
▽ Controlling Currents
Conductors
Driving Currents
Analyzing Circuits
▷ Magnetic Field Fundamentals
▷ Calculating Static Fields
▷ Dynamic Fields

Chapter Overview

Introduction

In chapter E5, we studied basic circuits involving a battery or capacitor driving current through a single wire. In this chapter (the last in the subdivision on controlling currents), we will draw on the basic concepts developed there to explore how we can analyze the behavior of more complicated electric circuits involving multiple elements.

Section E6.1: Two Wires in Series

If we connect a sequence of *two* wires to a battery, the steady-state surface charge distribution on the wires ensures that *the same current flows through each*, even if they have different properties. If the wires have different resistances, the potential differences across the wires are not the same (since $V = IR$), but their sum must be equal that across the battery ($V_{bat} = V_1 + V_2$).

A charge carrier moving through a circuit is analogous to a roller-coaster car moving along a track whose height is analogous to the potential at the corresponding point in the circuit. A battery is analogous to the motorized part of the track that lifts the car up to the highest point, while the other parts of the circuit correspond to unpowered portions of the track.

Since potential (like potential energy) is only defined up to a constant, we can define the potential ϕ at a point in a circuit only if we arbitrarily choose a reference point where we define ϕ to be zero. In *circuits*, we conventionally choose this point to be the battery's negative electrode.

Section E6.2: Circuit Elements in Series

A set of **circuit elements** is connected in **series** if every charge carrier that flows through any *one* element eventually flows through *all* elements before returning to the battery. For such a set,

$$I = I_1 = I_2 = \cdots \quad \text{(E6.4}a\text{)}$$

$$V_{set} = V_1 + V_2 + \cdots \quad \text{(E6.4}b\text{)}$$

$$R_{set} = R_1 + R_2 + \cdots \quad \text{(E6.4}c\text{)}$$

Purpose: These equations characterize a set of circuit elements in series.

Symbols: $I_1, I_2, \ldots$ and $V_1, V_2, \ldots$ and $R_1, R_2, \ldots$ are the magnitudes of the currents flowing through, the potential differences across, and the resistances of circuit elements $1, 2, \ldots$, respectively. I is the common current flowing through the elements, V_{tot} is the potential difference between the two ends of the entire set, and R_{set} is the total resistance of the set.

Limitations: These equations assume that the circuit has settled into a steady state.

Section E6.3: Circuit Diagrams

A **circuit diagram** is a visual tool for describing a circuit. Circuit elements are represented in such a diagram by conventional symbols such as those illustrated in figure E6.4. We often use the generic **resistor** symbol to represent any reasonably ohmic device. We use straight black lines (that we assume to have zero resistance) to indicate connections between elements. Do *not* assume that such lines connect at a point where they cross unless a connection is specifically indicated by a black dot.

Section E6.4: Circuit Elements in Parallel

We say that a set of circuit elements is connected in **parallel** if (1) a given charge carrier flowing through the set flows through only *one* of its elements and (2) the elements' ends are connected so that the potential difference across each element is *necessarily* the same as that across the entire set. For such a set,

$$I_{set} = I_1 + I_2 + \cdots \quad \text{(E6.11}a\text{)}$$

$$V_{set} = V_1 = V_2 = \cdots \quad \text{(E6.11}b\text{)}$$

$$\Rightarrow \quad \frac{1}{R_{set}} = \frac{1}{R_1} + \frac{1}{R_2} + \cdots \quad \text{(E6.11}c\text{)}$$

Purpose: These equations characterize a set of circuit elements in parallel.

Symbols: $I_1, I_2, \ldots$ and $V_1, V_2, \ldots$ and $R_1, R_2, \ldots$ are the magnitudes of the currents flowing through, the potential differences across, and the resistances of circuit elements 1, 2, ..., respectively. I_{set} is the current flowing through the set as a whole, V_{set} is the potential difference between the set's ends, and R_{set} is the set's total resistance.

Limitations: These equations assume that the circuit has settled into a steady state.

Section E6.5: Analyzing Complex Circuits

We can analyze almost any complex circuit by grouping its elements into series and parallel sets, each of which we replace by a *single* equivalent resistor whose value we calculate by using equation E6.4*c* or E6.11*c*. By repeating this process, we can eventually reduce *all* elements to a single resistor. If the elements are at least approximately ohmic, then we can calculate the current in that equivalent resistor by using Ohm's law, and then work backward to calculate currents and potential differences for all the elements.

Section E6.6: Realistic Batteries

We can realistically model a battery by treating it as an ideal battery in series with an ohmic internal resistance R_{int}. This internal resistance causes the potential difference across the battery's electrodes to "droop" a bit as current flow through the battery increases.

Section E6.7: Electrical Safety Issues

A prolonged electric current above 50 mA flowing through a person's body can kill by overriding nerve signals to the heart. High voltages are dangerous only because they can drive large currents. The relatively high resistance of dry skin *usually* limits currents driven by 120-V household sources to sublethal levels, but 120 V can easily drive a lethal current through wet skin. A Van de Graaff generator can produce potential differences exceeding 100,000 V, but is not dangerous because it can only produce a very brief current before becoming discharged. Note that a person's body does not need to complete a circuit for current to flow: current can flow from a high-potential part of a circuit through a body to the ground.

E6.1 Two Wires in Series

In chapter E5, we studied the case of a single uniform wire connected to a battery. Consider now the circuit shown in figure E6.1a, which consists a battery whose electrodes are linked by a sequence of two uniform wires with the same length L and conductivity σ_c but different cross-sectional areas A_1 and A_2. Figure E6.1b shows a similar circuit except that the wires have the same length L and cross-sectional area A but different conductivities σ_1 and σ_2.

Current carried by wires in series must be *equal*

The feedback effects discussed in chapter E5 will ensure that in both cases *the currents carried by the two wires will be equal.* Why? Imagine that the wire on the left initially conducts less current than the one on the right. This would cause positive charge to pile up at the junction of the wires, because the right wire supplies positive charge to the junction faster than the other can carry it away. The excess of positive charge on the junction will slow down incoming positive charges from the right wire and speed up departing charges in the other, reducing the difference between the currents. The accumulation of surface charge at the junction will only stop when the two wires' currents are equal.

But this means that the electric field strength E in the left wire must be greater than that in the right wire in both cases (as illustrated in the diagrams). Why? Since the current I carried by any reasonably thin wire of cross-sectional area A is $I = JA$ (where J is the magnitude of the current density) and $J = \sigma_c E$ (where σ_c is the wire's conductivity), we have

$$I = A\sigma_c E \tag{E6.1}$$

for all the wires. Therefore, for I to be the same in both wires in figure E6.1a, we must have $E \propto 1/A$, since both wires have the same conductivity. (Qualitatively, E is bigger in the thin wire because the thin wire has a smaller volume than the thick wire and so has fewer charge carriers. The few charge carriers it has thus must move faster in order to carry the same total current.) Similarly, for I to be the same in both wires in figure E6.1b, we must have $E \propto 1/\sigma_c$, since both wires have cross-sectional area A. (Qualitatively, E must be bigger to push carriers through the left wire at the same speed as carriers flow through the more conductive right wire.)

The potential differences across the wires must be *unequal*

If the two wires have roughly the same length L, the fact that E is different in the two wires implies that the potential difference $V_1 = E_1 L$ between the first wire's ends is *not* the same as that for the second wire $V_2 = E_2 L$ (we can see, in fact, that $V_2 > V_1$ in both cases above).

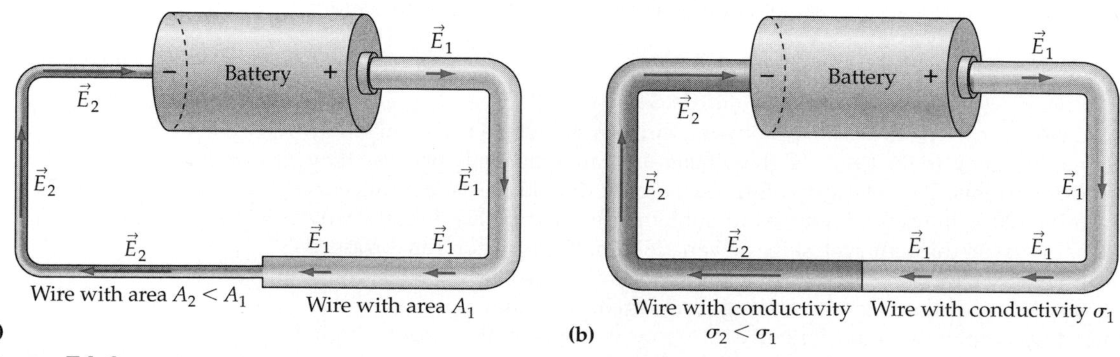

Figure E6.1
Circuits that involve a wire connected in series with an otherwise identical wire having (a) a smaller diameter, or (b) a smaller conductivity.

The potential differences across the wires add

Now, once the surface charges on the wires have settled down to their stable steady-state distribution, the electrostatic potential at any point P in space near or inside the wires will have a certain well-defined value ϕ relative to that at some reference point. Imagine that point P is the negative terminal of the battery. As the battery transports a given charge carrier to the positive plate, the carrier's electrostatic potential *increases* by V_{bat} (the potential difference across the battery). As it moves through the first wire, its potential decreases by V_1, and as it moves through the second wire back to point P, its potential decreases by the further amount V_2. Since the carrier cannot have a different electrostatic potential at P after completing the circuit than it had before, it follows that

$$V_{bat} - V_1 - V_2 = 0 \quad \Rightarrow \quad V_1 + V_2 = V_{bat} \tag{E6.2}$$

That is, the potential that a carrier gains in the battery must be equal to the potential that it loses as it goes through the wires.

The roller-coaster analogy

I find it helpful to visualize this by using a gravitational analogy. In chapter E3, we saw that the electrostatic potential ϕ at any point P is analogous to the gravitational potential ϕ_g, which near the earth is $\phi_g = gh$, where h is height above an arbitrary position where ϕ_g is defined to be zero. This means that we can visualize a circuit using a roller-coaster analogy, as shown in figure E6.2. A charge carrier is analogous to a roller-coaster car. A battery is analogous to the motorized part of the track that lifts the car to the track's highest point, and the rest of the circuit is analogous to the unpowered part of the track. Just as the charge carrier gains electrostatic potential energy from the battery and loses an amount equal to $q|\Delta\phi| = qV$ to thermal energy in the wires, the car gains gravitational potential energy from the motor and loses an amount equal to $m(gh)$ to thermal energy as it coasts down the rest of the track. Just as the roller-coaster car returns to the same height after completing one complete trip around the track, so a charge carrier returns to the same potential after going once around the circuit. Since $V/L = E$ for a wire, the electric field magnitude E in a wire corresponds to the *slope* of the corresponding section of roller coaster track. (Note that the section of track corresponding to each wire has a *constant* slope, which is atypical for a real roller coaster.)

The potential (or "voltage") at a point in a circuit

In many situations it is useful to talk about the potential, or **voltage,** at a given point in a circuit. Like the electrostatic potential energy on which it is based, the potential ϕ at any point in space near a charge distribution is only defined up to an overall additive constant, which we can choose by defining an arbitrary point to have zero potential. Once we have defined the potential to be zero at a specified point, the potentials of all other points are uniquely defined. In an electric circuit, it is conventional to define the negative electrode of the battery or power supply to have $\phi = 0$.

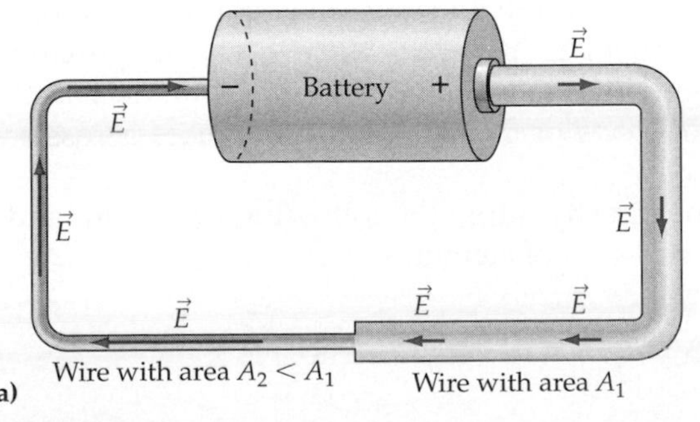

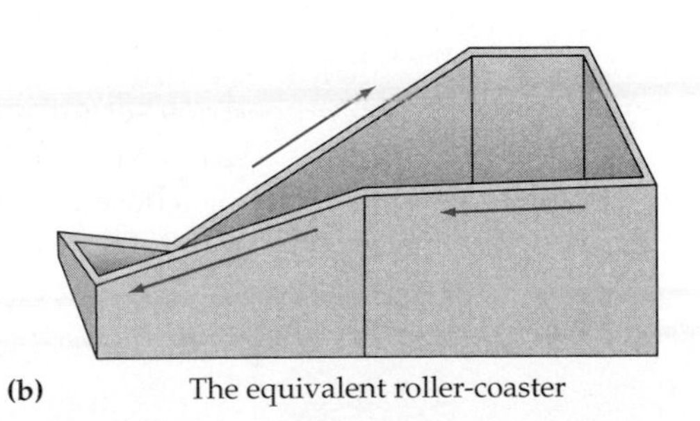

Figure E6.2
The roller-coaster analogy for an electric circuit.

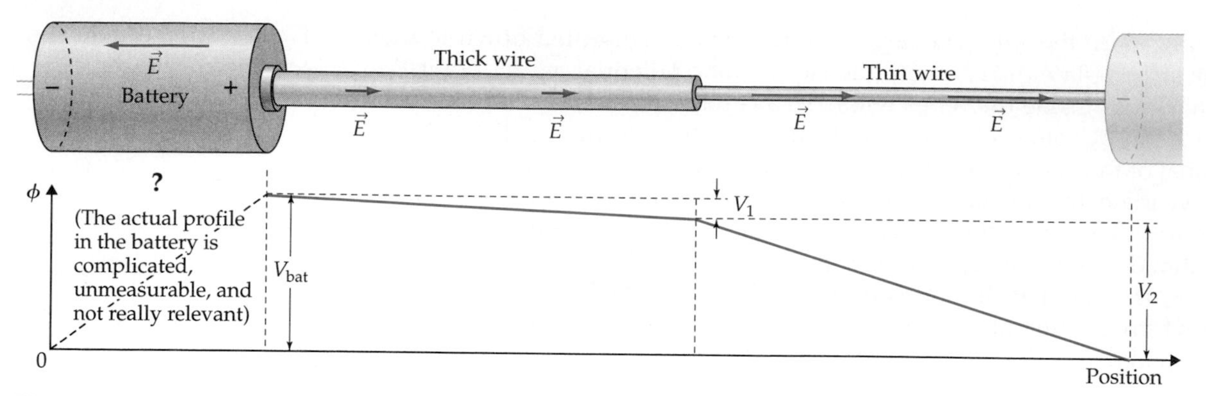

Figure E6.3
A graph of voltage versus position for the circuit shown in figure E6.1a.

Many books use the symbol V for the potential at a point in a *circuit* as well as for the potential difference across an object, though ϕ is conventional for the same concept in noncircuit contexts. We have enough trouble with multiple meanings of the V symbol already, so I am going to stick with the symbol ϕ for the potential at a point: just remember that some texts use V for this concept.

Figure E6.3 shows a graph of potential ϕ versus position for the circuit shown in figure E6.1a. We can make the the graph easier to draw by mentally unwrapping the circuit into a straight line as shown. Note that the slope of ϕ is $V/L = E =$ constant for each wire.

E6.2 Circuit Elements in Series

General definition of a series circuit

The two-wire circuits shown in figure E6.1 are simple examples of circuits in **series.** We say that the members of a set of wires and/or other devices (**circuit elements**) excluding batteries are connected "in series" if every electron that flows through any *one* element flows through *all* elements in the set in sequence before returning to the battery. In such a case (as we saw in the two-wire case), surface charges on the conducting elements in the set will adjust themselves so that each element in the set carries the same current: $I = I_1 = I_2 = \cdots$.

This in turn implies that the potential difference between the ends of a given element (which we more often call "the potential difference *across*" or "the voltage drop *across*" that element) is generally *different* from that across another element. Indeed, the definition of resistance $R \equiv V/I \Rightarrow V = IR$ implies that if the resistance of the ith circuit element is R_i, the potential difference across that element is $V_i = IR_i$. Since each charge carrier must go through each element in sequence, the potential difference across the entire series set must be the *sum* of the potential differences across each element: $V_{set} = V_1 + V_2 + \cdots$.

These equations directly imply the following important and useful relation: total resistance of a set of elements in series is

$$R_{set} \equiv \frac{V_{set}}{I_{set}} = \frac{V_1 + V_2 + \cdots}{I} = \frac{V_1}{I} + \frac{V_2}{I} + \cdots$$
$$= \frac{V_1}{I_1} + \frac{V_2}{I_2} + \cdots = R_1 + R_2 + \cdots \tag{E6.3}$$

Therefore, these equations characterize a set of circuit elements in series:

Equations characterizing a set of elements in series

$$I = I_1 = I_2 = \cdots \tag{E6.4a}$$

$$V_{set} = V_1 + V_2 + \cdots \tag{E6.4b}$$

$$R_{set} = R_1 + R_2 + \cdots \tag{E6.4c}$$

Purpose: These equations characterize a set of circuit elements in series.

Symbols: $I_1, I_2, \ldots$ and $V_1, V_2, \ldots$ and $R_1, R_2, \ldots$ are the magnitudes of the currents flowing through, the potential differences across, and the resistances of circuit elements $1, 2, \ldots$, respectively. I is the current flowing through all the elements in common, V_{tot} is the potential difference between the two ends of the entire set, and R_{set} is the total resistance of the entire set.

Limitations: These equations assume that the circuit has settled into a steady state.

If you look back at the derivation of these equations, you will see that they are based on *very* basic principles. The current must be the same in each element, or charges would pile up somewhere in the circuit. The potential differences across the elements must add by the definition of potential difference. The last equation follows directly from the definition of resistance. We have *not* assumed that the circuit elements are ohmic. These results are *very* general.

We can use these equations to determine completely the current flowing in a series circuit and the potential difference across each element.

Example E6.1

Problem Suppose that we connect a 50-Ω wire in series with a 10-Ω wire and a 12-V battery. What is the current flowing in this circuit? What is the potential difference across each element?

Translation The series set of nonbattery elements here is the two wires. Let $R_1 = 50\ \Omega$, $R_2 = 10\ \Omega$, and $V_{set} = V_{bat}$.

Model We will assume that the battery is ideal, so $V_{bat} = \mathscr{E}_{bat} = 12$ V independent of I.

Solution Equation E6.4*c* implies that the set's total resistance is $R_{set} = R_1 + R_2 = 60\ \Omega$. The current flowing through the set is therefore

$$I = \frac{V_{set}}{R_{set}} = \frac{12\ \text{V}}{60\ \Omega}\left(\frac{1\ \Omega}{1\ \text{V/A}}\right) = 0.20\ \text{A} \tag{E6.5}$$

This means that the potential differences across each wire are

$$V_1 = I_1 R_1 = IR_1 = (0.20\ \text{A})(50\ \Omega)\left(\frac{1\ \text{V/A}}{1\ \Omega}\right) = 10\ \text{V} \tag{E6.6a}$$

$$V_2 = IR_2 = (0.20\ \text{A})(10\ \Omega)\left(\frac{1\ \text{V/A}}{1\ \Omega}\right) = 2.0\ \text{V} \tag{E6.6b}$$

Evaluation The units are right. Note that $V_1 + V_2 = 12\ \text{V} = V_{bat}$, as expected.

Exercise E6X.1

Say that the battery in figure E6.1a has an emf of 3.0 V, and the thin and thick wires have resistances of 10 Ω and 2 Ω, respectively. What are the potential differences across these wires?

E6.3 Circuit Diagrams

Circuit diagrams

A **circuit diagram** is a tool that helps us describe and visualize an electric circuit. A real circuit may be a complicated three-dimensional nest of circuit elements, and it can be difficult to see which elements are connected and how. A circuit diagram represents the essential characteristics of the circuit in a two-dimensional diagram that clearly shows how its elements are connected.

Circuit elements are represented in a circuit diagram by conventional symbols that are both suggestive of the element in question and easy to draw. Figure E6.4 illustrates the conventional symbols for a *cell*, a *battery*, a low-resistance connecting wire, a *lightbulb*, and a *resistor*. (Conventional symbols also exist for many other kinds of electronic circuit elements that we will not consider in this course.)

Technically, a **resistor** is a circuit element manufactured to have a certain specified constant resistance. Such a resistor usually looks like a small cylinder with two connecting wires that emerge along the cylinder's axis; its resistance in ohms is often encoded as a sequence of three to five colored bands (see figure E6.5). Resistors are used in electric circuits primarily to limit currents and/or control the value of the potential at a certain point in the circuit.

However, we often use the resistor symbol in a circuit diagram to represent any reasonably ohmic device (such as a wire, lightbulb, electric motor, heating element, or the like) whose resistance is much larger than that of the connecting wires. As long as the device is reasonably ohmic, all the potential differences and currents in a circuit would be the same if we replaced it by an equivalent resistor.

The flexibility one has in drawing connecting lines

Connecting wires in a circuit typically have negligible resistances compared to other circuit elements. On a circuit diagram, we use black lines to show the connections between elements. These connecting lines are assumed to have *zero* resistance and thus zero potential difference between

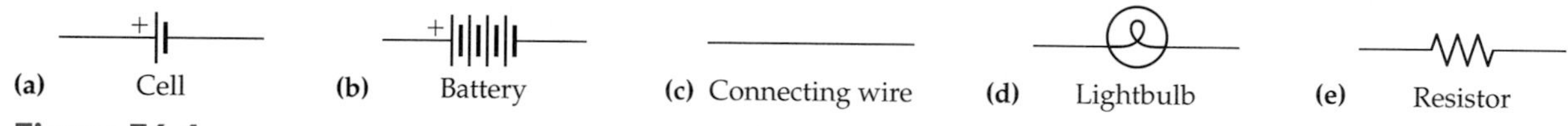

Figure E6.4
Conventional circuit diagram symbols for various circuit elements. (Note that in the cell symbol, the negative side of the cell is indicated by a shorter fat line.)

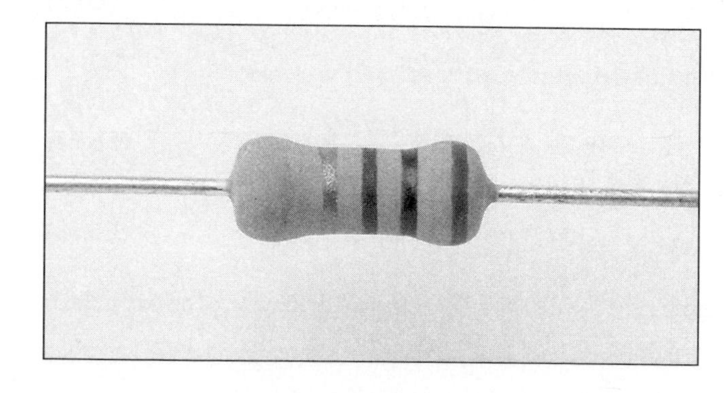

Figure E6.5
A manufactured resistor. The colored bands indicate the resistance (and the uncertainty range for that resistance).

their ends ($V = IR = I \cdot 0 = 0$). The entire purpose of these lines is to display clearly how circuit elements are connected: the shape or length of the lines on a circuit diagram does *not* generally correspond to the shape or length of actual connecting wires in the real circuit. Indeed, if it improves the layout in a circuit diagram, we can either add lines that don't exist in the real circuit or choose not to show wires that do exist as long as the electrical connections between circuit elements are accurately represented.

If a wire's resistance is nonnegligible, we indicate its resistance by using a resistor symbol. For example, we can represent the circuit in figure E6.1a (the one involving a thick wire, thin wire, and battery in series) by using a circuit diagram like figure E6.6. Note that lines have been added in the diagram where no connecting wires exist in the real circuit: this makes it easier to draw than if we had to connect the resistor symbols directly to the battery symbol. Note also that connection lines are almost always drawn as vertical or horizontal straight lines that change direction in right angles (this also makes the diagram easier to draw).

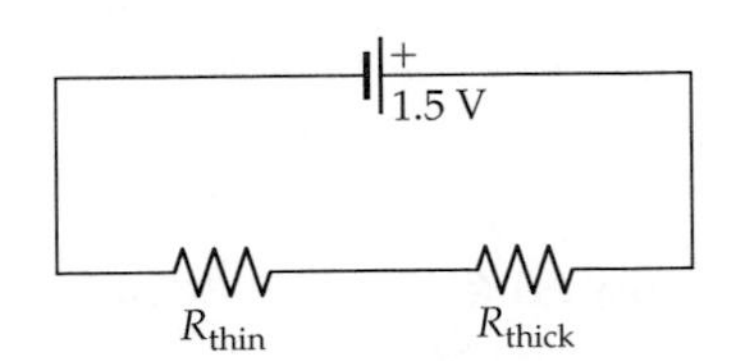

Figure E6.6
A circuit diagram for the circuit in figure E6.1a.

Example E6.2

Problem Draw a circuit diagram of the circuit illustrated in figure E6.7a.

Solution The circuit diagram is shown in figure E6.7b. Note that I have straightened out and uncrossed the connecting wires, and reoriented the devices to make the circuit's structure clearer. I have ignored the lightbulb sockets: they merely facilitate the connections indicated on the diagram. The black dots on the diagram indicate connections: for example, one socket terminal for bulb C is connected to both one socket terminal for bulb B and the battery's positive electrode. It is irrelevant that the wire from the battery *actually* goes to the B socket first and a connection is made from there to bulb C: if the connecting wires really have zero resistance, it doesn't matter exactly *how* the two bulbs are connected to the battery, it just matters that they are.

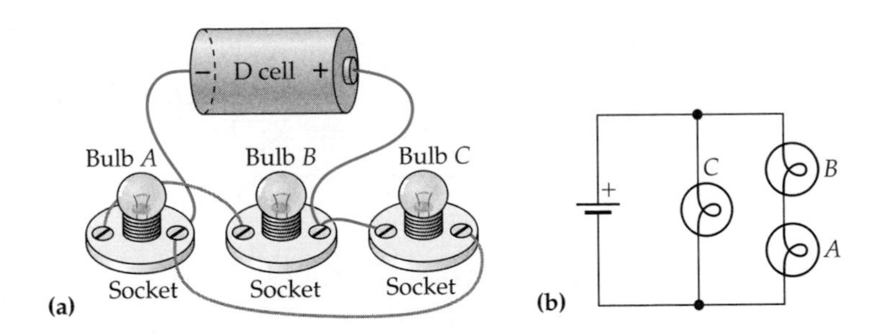

Figure E6.7

There is no one correct way to draw a circuit diagram. I could have reoriented the diagram so that the cell and lightbulbs were horizontal. I could have drawn the part of the circuit containing bulb C to be outside of the part containing bulbs A and B. Any diagram that correctly represents how the elements are *connected* is equally valid: the *connections* are the important thing.

Exercise E6X.2

Draw a circuit diagram for the circuit in figure E6.8 on the next page.

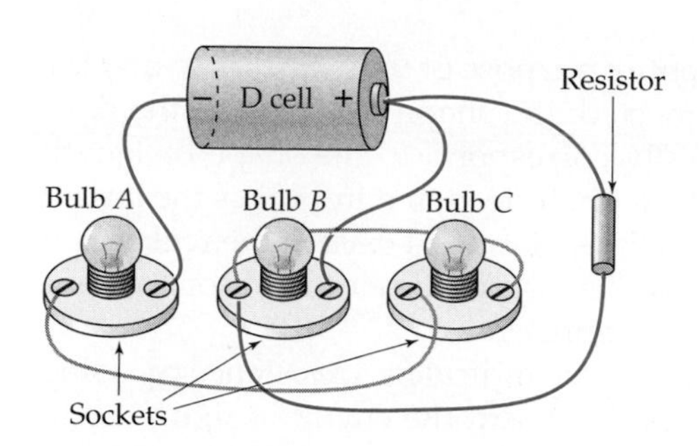

Figure E6.8
See exercise E6X.2.

In diagrams for particularly complicated circuits, it may be impossible to indicate the connections between circuit elements without having lines cross each other. If lines do cross on a diagram, they are usually assumed *not* to be connected (see figure E6.9a) unless a black dot at the intersection specifically indicates a connection. A *bridge* (see figure E6.9b) is sometimes used to more clearly communicate that crossing lines are not connected. It is better yet to *avoid* drawing crossed unconnected lines, if possible (see figure E6.9c).

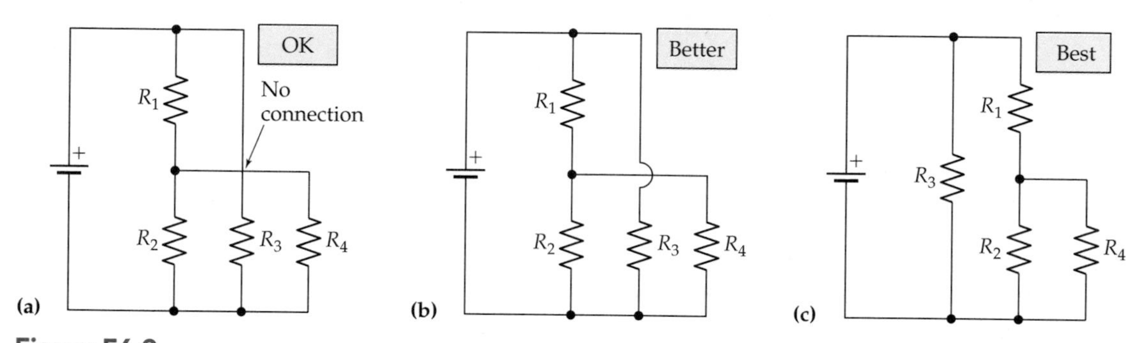

Figure E6.9
Three equivalent circuit diagrams illustrating conventions about crossing lines.

Exercise E6X.3

In figure E6.9, we exchanged the vertical columns involving R_3 and R_1 without affecting the diagram's validity. Can we exchange R_2 and R_4 without changing its validity? How about R_1 and R_2? Explain.

E6.4 Circuit Elements in Parallel

Two resistors in parallel

In a circuit similar to the one shown in figure E6.10, a given electron does *not* go through each of the resistors in sequence. Rather, an electron flowing out of the battery may go through one resistor *or* the other on the way back to the battery, but not both. This is clearly not a *series* circuit of the type described earlier: we say that the resistors in this kind of circuit are connected in **parallel.**

The potential differences across parallel elements are the same

The potential difference V_{ba} between points a and b on the diagram will be essentially the same as the potential difference across each resistor (assuming the potential differences across the connecting wires are negligible):

$$V_{ba} = V_1 = V_2 \qquad (= V_{\text{bat}} \text{ in this case}) \qquad \text{(E6.7)}$$

But the currents are not!

Although the potential difference driving current through each resistor is the same, if $R_1 \neq R_2$ the *currents* driven through each resistor are *not:* the

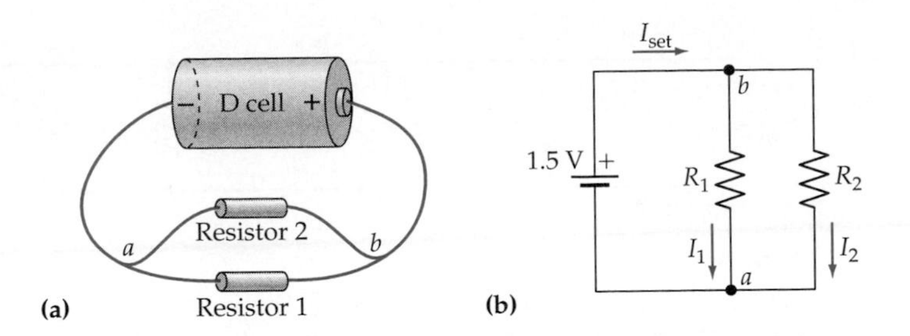

Figure E6.10
(a) A circuit involving two resistors in parallel. (b) A circuit diagram of the same circuit.

definition of resistance implies that $V = IR$ for any object, which in this case means that

$$I_1 R_1 = V_1 = V_2 = I_2 R_2 \quad \Rightarrow \quad \text{if } R_1 \neq R_2, \quad I_1 \neq I_2 \tag{E6.8}$$

So when the current comes to point b, it divides into two currents, one that goes through R_1 and one that goes through R_2 (these currents rejoin at point a). Equation E6.8 implies that *the current flowing in each parallel path is inversely proportional to that path's resistance* (once a steady state has been reached): *more* current flows through the path that has the *smaller* resistance. This is the most important thing to understand about circuits involving elements in parallel.

Surface charges direct the correct amount of current along each path

How do the flowing electrons figure out how to choose among these paths in the correct proportions? The explanation is similar to the explanations offered in chapter E5: any deviation from the "correct" current flow sets up surface charges on the circuit wires that oppose the incorrect flow, pushing the current flows back toward the balance indicated by equation E6.8.

For example, assume that $R_1 > R_2$. Imagine that when the circuit is first connected, electrons moving from the battery's negative electrode split randomly but equally at point a, so that equal numbers initially move into resistors R_1 and R_2. But electrons move more slowly through R_2, so they will pile up at the entrance of R_2 as cars pile up at a freeway entrance ramp when the freeway is jammed. The pileup of electrons at the entrance of R_2 repels electrons that are still coming toward a, directing more and more of them toward R_1. Eventually (within a few nanoseconds) the balance described by $I_1 R_1 = V_1 = V_2 = I_2 R_2$ is attained where most of the electrons approaching a are pushed by these surface charges through R_1 but a few continue to flow through R_2.

Equivalent resistance of a parallel pair of resistors

What is the *total* current flowing in this circuit? Once a steady-state current flow has been established, we won't have charge piling up anywhere, so the rate at which charge flows *into* any junction must be equal to the rate at which it flows out. Since "the rate at which charge flows" is simply the current, in figure E6.10, this means that the total current flowing through the battery must be equal to the sum of the current flowing through each resistor

$$I_{tot} = I_1 + I_2 \tag{E6.9}$$

We can determine the total effective resistance of the set of parallel resistors in figure E6.10 as follows. If I_{set} is the total current flowing through the set and V_{set} is the potential difference between its ends, its resistance is defined to be

$$R_{set} \equiv \frac{V_{set}}{I_{set}} \quad \Rightarrow \quad \frac{1}{R_{set}} = \frac{I_{set}}{V_{set}} = \frac{I_{tot}}{V_{ba}} \tag{E6.10a}$$

Using equations E6.7 and E6.9, you can easily show that

$$\frac{1}{R_{set}} = \frac{I_1}{V_1} + \frac{I_2}{V_2} = \frac{1}{R_1} + \frac{1}{R_2} \tag{E6.10b}$$

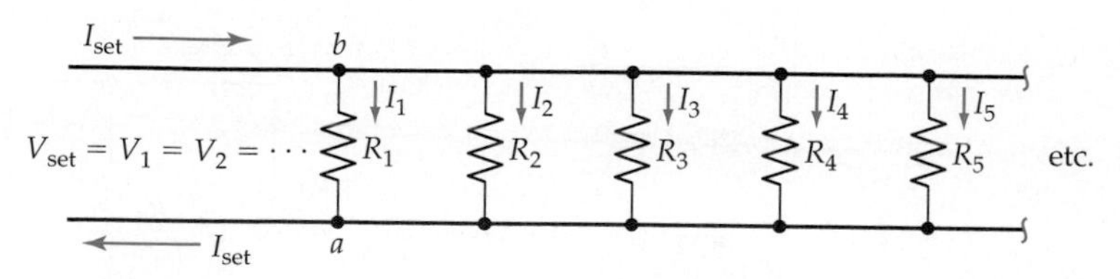

Figure E6.11
Circuit elements connected in parallel. The way that the ends of each element are connected ensures that the voltage difference across each element is *necessarily* the same, and the current flowing through one element will *not* flow through any other element.

Exercise E6X.4

Verify equation E6.10*b*.

Equations characterizing a set of elements in parallel

In general, we say that the members of a set of circuit elements are connected *in parallel* if (1) a given electron flowing through the set flows through only *one* element in the set and (2) the elements' ends are connected so that the potential difference across each element is *necessarily* the same as the potential difference across the entire set (see figure E6.11). Therefore, the fundamental equations that characterize a set of parallel circuit elements are

$$I_{set} = I_1 + I_2 + \cdots \qquad \text{(E6.11}a\text{)}$$

$$V_{set} = V_1 = V_2 = \cdots \qquad \text{(E6.11}b\text{)}$$

$$\Rightarrow \qquad \frac{1}{R_{set}} = \frac{1}{R_1} + \frac{1}{R_2} + \cdots \qquad \text{(E6.11}c\text{)}$$

Purpose: These equations characterize a set of circuit elements in parallel.

Symbols: $I_1, I_2, \ldots$ and $V_1, V_2, \ldots$ and $R_1, R_2, \ldots$ are the currents flowing through, the potential differences across, and the resistances of circuit elements 1, 2, . . . , respectively. I_{set} is the current flowing through the set as a whole, V_{set} is the potential difference between the set's ends, and R_{set} is the total resistance of the entire set.

Limitations: It assumes that the circuit has settled into a steady state.

Exercise E6X.5

Two nichrome wires with resistances of 3 Ω and 1 Ω are connected in parallel to a 1.5-V alkaline D cell. What is the combined resistance of the wires? What total current flows through the battery?

E6.5 Analyzing Complex Circuits

Equations E6.4 and E6.11 provide powerful tools for analyzing more-complex circuits. The elements in virtually any kind of circuit you are likely to encounter can be grouped into sets of parallel or series elements. Each of these sets can then be replaced by an equivalent *single* resistor whose value can be calculated by using equation E6.4*c* or E6.11*c*. By repeatedly reducing sets to single resistors, we eventually reduce *all* elements to a single resistor. If the elements are at least approximately ohmic, then we can predict the current in the resulting simple circuit by using Ohm's law, and then work backward to determine currents and voltages in any part of a circuit. Figures E6.12 and E6.13 illustrate the process. However, some circuits cannot be analyzed this way. In such cases, one can fall back to a more difficult but general approach based on the laws of conservation of charge and energy. See problem E6A.2 for more details.

Example E6.3

Problem Assume that the lightbulbs in figure E6.12a are reasonably ohmic and all have resistances of 10 Ω. Determine the current flowing through every lightbulb and the potential difference across every lightbulb in the circuit.

Translation Figure E6.12 defines some useful symbols.

Model We are assuming that the bulbs are ohmic and the battery is ideal.

Solution The first step is to find the equivalent resistance R_{all} of all the bulbs. According to equation E6.11*c*, the equivalent resistance R_{BD} of the parallel set consisting of bulbs B and D is

$$\frac{1}{R_{BD}} = \frac{1}{R_B} + \frac{1}{R_D} = \frac{1}{10\ \Omega} + \frac{1}{10\ \Omega} = \frac{2}{10\ \Omega} = \frac{1}{5\ \Omega} \qquad \text{(E6.12)}$$

So $R_{BD} = 5\ \Omega$. Now this equivalent resistor is in series with bulbs A and C. Equation E6.4*c* implies that the equivalent resistance of the set of all bulbs is

$$R_{\text{all}} = R_{BD} + R_A + R_C = 5\ \Omega + 10\ \Omega + 10\ \Omega = 25\ \Omega \qquad \text{(E6.13)}$$

Now we can begin the reverse analysis illustrated in figure E6.13. The voltage across the single resistor representing all the bulbs is the same as that across the battery. Since V_{bat} for a D cell is 1.5 V, the total current I flowing through this equivalent resistor is

$$I = \frac{V_{\text{bat}}}{R_{\text{all}}} = \frac{1.5\ \cancel{\text{V}}}{25\ \cancel{\Omega}}\left(\frac{1\ \cancel{\Omega}}{1\ \cancel{\text{V}}/\text{A}}\right) = 0.06\ \text{A} = 60\ \text{mA} \qquad \text{(E6.14)}$$

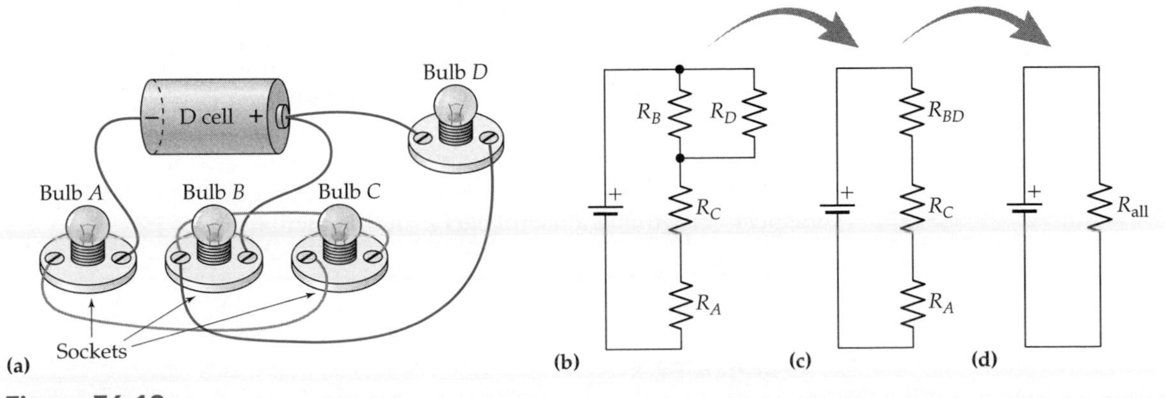

Figure E6.12
(a) A complex circuit. (b) A schematic diagram for this entire circuit (treating the bulbs as resistors). (c) Bulbs B and D are in parallel, and so they can be reduced to the single equivalent resistance R_{BD} (whose value can be found by using equation E6.11c). (d) R_{BD} is in series with bulbs A and C, so the entire set can be replaced by a single equivalent resistor R_{all}, whose value can be found by using equation E6.4c.

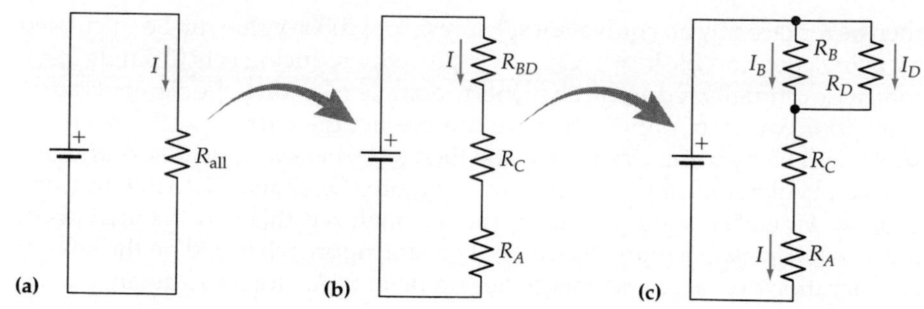

Figure E6.13
Reverse analysis of the circuit in figure E6.12. (a) $I = V_{bat}/R_{all}$ gives us the total current I flowing in the circuit. (b) We can then use $V = IR$ to find the potential differences across resistors A and C and the set BD. Note that $V_{BD} = V_B = V_D$ by equation E6.11*b*. (c) We can use $V_B = I_B R_B$ and $V_D = I_D R_D$ to find I_B and I_D. When we have done this, we have found the currents flowing through and potential differences across every element in the circuit!

This same current flows through all elements in a series set (see equation E6.4*a*), so $I_A = I_C = I_{BD} = 60$ mA. We can now apply $V = IR$ individually to each of these three series elements to determine the potential difference across each:

$$V_A = I_A R_A = (0.06\ \cancel{A})(10\ \cancel{\Omega})\left(\frac{1\ \text{V}/\cancel{A}}{1\ \cancel{\Omega}}\right) = 0.6\ \text{V} \tag{E6.15a}$$

$$V_C = I_C R_C = (0.06\ \cancel{A})(10\ \cancel{\Omega})\left(\frac{1\ \text{V}/\cancel{A}}{1\ \cancel{\Omega}}\right) = 0.6\ \text{V} \tag{E6.15b}$$

$$V_{BD} = I_{BD} R_{BD} = (0.06\ \cancel{A})(5\ \cancel{\Omega})\left(\frac{1\ \text{V}/\cancel{A}}{1\ \cancel{\Omega}}\right) = 0.3\ \text{V} \tag{E6.15c}$$

Equation E6.11*b* implies that the potential difference across the parallel pair of bulbs B and D is the same as the potential difference across each individual bulb: $V_B = V_D = V_{BD} = 0.3$ V. This allows us to compute the current through each:

$$I_B = \frac{V_B}{R_B} = \frac{0.3\ \cancel{V}}{10\ \cancel{\Omega}}\left(\frac{1\ \cancel{\Omega}}{1\ \cancel{V}/\text{A}}\right) = 0.03\ \text{A} = 30\ \text{mA} \tag{E6.16a}$$

$$I_D = \frac{V_D}{R_D} = \frac{0.3\ \cancel{V}}{10\ \cancel{\Omega}}\left(\frac{1\ \cancel{\Omega}}{1\ \cancel{V}/\text{A}}\right) = 0.03\ \text{A} = 30\ \text{mA} \tag{E6.16b}$$

(Because the resistances of these bulbs happen to be the same, the currents through these bulbs also happen to be the same.)

Evaluation All the units came out right. Note also that $V_A + V_{BD} + V_C = 1.5\ \text{V} = V_{bat}$ and $I_B + I_D = I$ as required by equations E6.4*b* and E6.11*a*, respectively (this is a useful check on the calculations).

Exercise E6X.6

Imagine that in the circuit shown in figure E6.12a, we remove bulb D from the circuit. What is the total current through the circuit now? What happens to the potential difference across bulb B?

E6.6 Realistic Batteries

Equation E5.4 implies that the potential difference between a realistic battery's electrodes is given by $V_{bat} = \mathscr{E} - \mathscr{E}^{th}$, where $\mathscr{E}$ is the battery's fixed emf and $\mathscr{E}^{th}$ is the energy per unit charge lost to thermal energy when current flows through the battery. In the ideal battery approximation, we assume that $\mathscr{E}^{th} = 0$, but a somewhat better model takes advantage of the empirical fact that $\mathscr{E}^{th} \propto I$, where I is the current flowing through the battery. Thus $R_{int} \equiv \mathscr{E}^{th}/I \approx$ constant. Note that since $\mathscr{E}^{th}$ has the same units as potential difference, R_{int} has the same units as a resistance. We can in fact rewrite equation E5.4 to read

$$V_{bat} = \mathscr{E} - IR_{int} \quad \Rightarrow \quad V_{ibat} \equiv \mathscr{E} = V_{bat} + IR_{int} = V_{rest} + IR_{int} \tag{E6.17}$$

where V_{ibat} would be the potential difference across the battery if it were ideal and V_{rest} is the potential difference across the rest of the circuit. If we compare this with equation E6.4*b*, we see that we can model a real battery as an ideal battery in series with an ohmic internal resistance R_{int}, as shown figure E6.14. This model is useful when the ideal battery approximation is too crude.

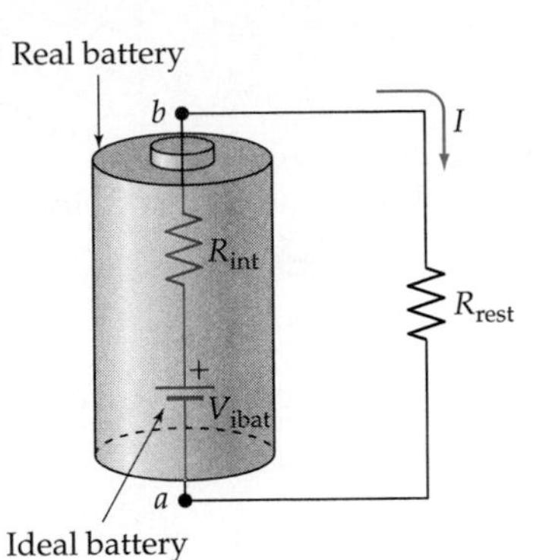

Figure E6.14
We can model a real battery as being an ideal battery in series with an approximately ohmic resistor R_{int}. The potential difference between *a* and *b* is $V_{bat} = V_{ibat} - IR_{int}$. This is the potential difference across the rest of the circuit (which is represented here by its equivalent resistor R_{rest}).

Exercise E6X.7

Imagine that the potential difference V_{bat} across an alkaline D cell is 1.55 V when it conducts no current and 1.47 V when it conducts a current of 4.0 A. What is its effective internal resistance R_{int}?

E6.7 Electrical Safety Issues

Now that we know how to analyze electric circuits in some detail, we have the background needed to understand electrical safety. Everyone knows that sometimes the electricity provided by a household electrical outlet can kill. Yet many of you have probably had the experience of being shocked by household electrical devices and lived to tell the tale. What makes accidental contact with household electricity lethal at times and only unpleasant at other times?

Current is what is lethal

It turns out that *current* is what kills. A current flowing through one's body (particularly through one's torso) disrupts electrical signals in one's nerves. Small currents (<5 mA) may feel uncomfortable but do no lasting harm. Currents as large 50 mA to 100 mA, on the other hand, override nerve signals controlling muscles (including the heart) and cause them to freeze up, making it impossible to breathe, pump blood, or even let go of whatever is driving the current. If this lasts for more than a few seconds, death can result.

The resistance of the tissue *inside* a human body is on the order of a few hundred ohms (salt water conducts electricity fairly well). If the resistance of the internal tissue in a person's body is, say, 200 Ω and a lethal current is 50 mA = 0.05 A, how large a potential difference is needed drive a lethal current? According to the definition of resistance, $V = IR = (0.050\text{ A})(200\ \Omega) = 10\text{ A}\cdot\Omega = 10\text{ V}$ (since $1\ \Omega = 1\text{ V/A}$). This means that even a 12-V automobile battery could push a lethal current through a person (if directly connected to a person's internal tissues), and the 120-V potential difference between the plugs in a household outlet is more than sufficient to kill.

Dry skin's high resistance makes household voltages painful but nonlethal

However, everyone is encased in a layer of skin that when *dry* has a much higher resistance (on the order of tens of thousands of ohms). If a person's dry skin resistance is, say, 20,000 Ω (the value depends on the amount of perspiration and other factors), the potential difference needed to drive a lethal current is more like $V = IR = (0.050\text{ A})(20{,}000\ \Omega) = 1000$ V. This is why one can sometimes survive a shock from a 120-V outlet.

But wet skin resistance is low enough to make household voltages lethal

Wet skin, however, has a resistance of only a few thousand ohms, so the lethal potential difference would be $V = IR = (0.050\text{ A})(2000\ \Omega) = 100$ V. Under these conditions, household potential differences of 120 V are very dangerous. This is why using electrical appliances while in the bath or in the rain or even while sweating profusely can be *extremely* dangerous.

A person can conduct a current to the *ground*

Moreover, you do not need to put two wet fingers into an outlet to conduct a dangerous current. The potential in household circuits is measured *relative* to the ground, that is, the surface of the earth. So if you touch a *single* bare wire at a point in a circuit where its potential difference with respect to the ground is sufficiently large, your body can and will conduct a current from the wire into the ground on which you stand (note, for example, that if electrons are crowded into your body, they can disperse themselves by flowing into the ground).

High voltages are only dangerous because of their potential for driving large currents. For example, touching a 1000-V power line while you are in contact with the ground will almost certainly kill you (even if your skin is dry) because a utility line can supply plenty of current. On the other hand, a Van de Graaff generator produces potential differences in excess of 100,000 V, but is relatively safe because it cannot supply a sizable current for any length of time (only brief sparks). You can build up a potential of more than 10,000 V on your body simply by shuffling your feet on a carpet on a dry day. The shock produced when you touch something is not dangerous because the current flow is so brief.

Normal batteries are safe

While high voltages are usually (but not always) dangerous, normal batteries (1.5 V to 12 V) can always be handled safely, because they simply cannot drive a dangerous current even through someone who is sopping wet.

Exercise E6X.8

Why are birds perched on a high-voltage power line not fried instantly? (*Hint:* What is the likely potential difference between a bird's feet?)

TWO-MINUTE PROBLEMS

E6T.1 Two identical lightbulbs are connected to a battery as shown in the circuit diagram here. The dark black lines represent connecting wires with negligible resistance, and the arrows indicate the direction of conventional current flow. Which bulb will conduct the greater current?

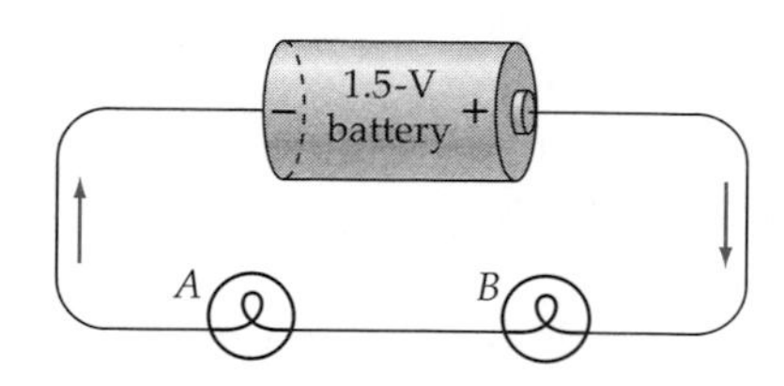

A. Bulb *A*.
B. Bulb *B*.
C. Each conducts the same *I*.
D. Impossible to tell (could be either).

E6T.2 Bulbs *A* and *B* have resistances of 40 Ω and 10 Ω, respectively, in the diagram above. What is the potential difference across bulb *A*?
A. 1.5 V
B. 1.2 V
C. 0.75 V
D. 0.5 V
E. 0.3 V
F. 0.15 V

E6T.3 Imagine in the circuit described in problem E6T.2 that we define ϕ to be zero in between the two bulbs. What is the potential at the battery's negative electrode? (Specify the magnitude of this voltage according to the list above, then specify the sign by $T = +, F = -$.)

E6T.4 Consider the circuit shown in figure E6.15. Which of the circuit diagrams below that figure correctly represents the circuit shown?

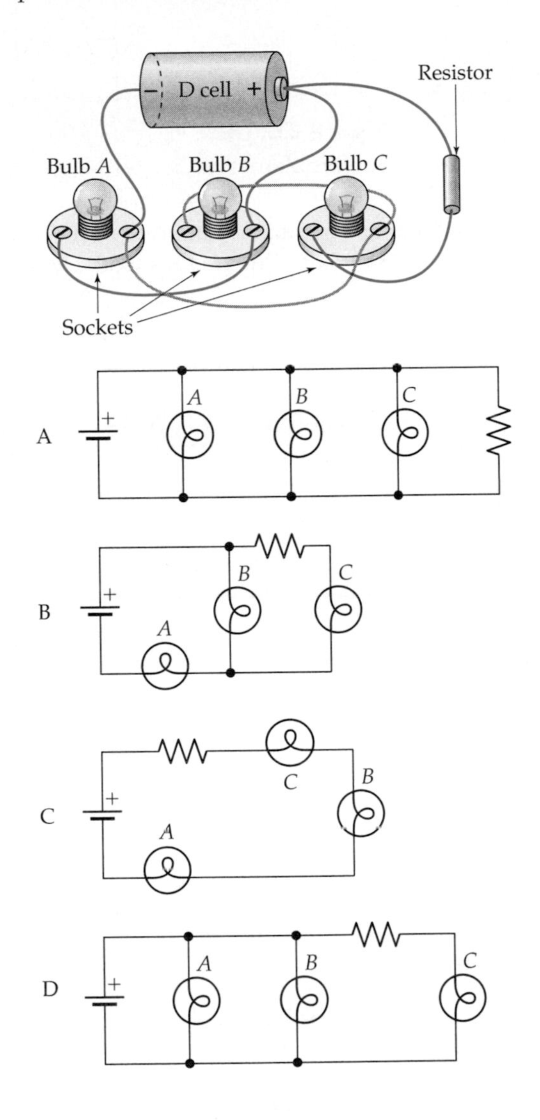

Figure E6.15
See problems E6T.4 through E6T.10.

E6T.5 Consider the circuit represented by answer B in figure E6.15. If you were to unscrew bulb *A* from its socket (effectively removing it from the circuit and leaving a gap in its place), bulb *B* would
A. Get brighter
B. Get dimmer
C. Remain the same
D. Go out

E6T.6 Consider the circuit represented by answer B in figure E6.15. If you were to unscrew bulb *B* from its socket (effectively removing it from the circuit and leaving a gap in its place), bulb *A* would
A. Get brighter
B. Get dimmer
C. Remain the same
D. Go out

E6T.7 Consider the circuit represented by answer D in figure E6.15. If you were to unscrew bulb *B* from its socket (effectively removing it from the circuit and leaving a gap in its place), bulb *A* would
A. Get brighter
B. Get dimmer
C. Remain the same
D. Go out

E6T.8 Consider the circuit diagram given as answer B in figure E6.15. If each element in this circuit has a resistance of 12 Ω, what is the resistance of the entire circuit?
A. 48 Ω
B. 32 Ω
C. 24 Ω
D. 20 Ω
E. 12 Ω
F. 4 Ω

E6T.9 In the circuit diagram shown in answer D in figure E6.15, the three bulbs are in parallel, true (T) or false (F)?

E6T.10 In the circuit diagram shown in answer B in figure E6.15, bulbs *A* and *B* are in series, T or F?

E6T.11 Imagine that two wires with the same length and conductivity are connected in *parallel* across a battery. Wire *A* has one-half the diameter of wire *B*. How do the strengths of the electric fields in each wire compare?
A. $E_A > E_B$
B. $E_A = E_B$
C. $E_A < E_B$

HOMEWORK PROBLEMS

Basic Skills

E6B.1 Imagine that we connect a 10-Ω lightbulb and a 20-Ω light bulb in series with a 1.5-V battery (using low-resistance wires that we can essentially ignore). What is the current flowing in this circuit? What is the voltage across the terminals of the 20-Ω lightbulb?

E6B.2 Imagine that we connect a 1.5-V battery in series with two copper wires. Wire A has a resistance of 0.85 Ω, and wire B has a resistance of 2.15 Ω. What current flows in this circuit? What is the voltage difference across wire A?

E6B.3 You and a partner are doing an experiment on a circuit involving three objects in series with a 1.5-V battery. Your partner has measured the potential differences across each object and claims they are 0.45 V, 0.92 V, and 0.23 V, respectively. Explain why these cannot possibly be right.

E6B.4 Draw a circuit diagram for the circuit shown in figure E6.16.

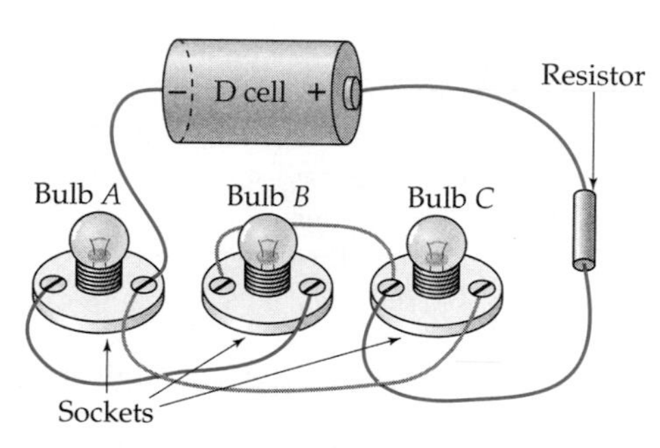

Figure E6.16
See problems E6B.4 and E6S.7.

E6B.5 Draw a circuit diagram for the circuit shown in figure E6.17.

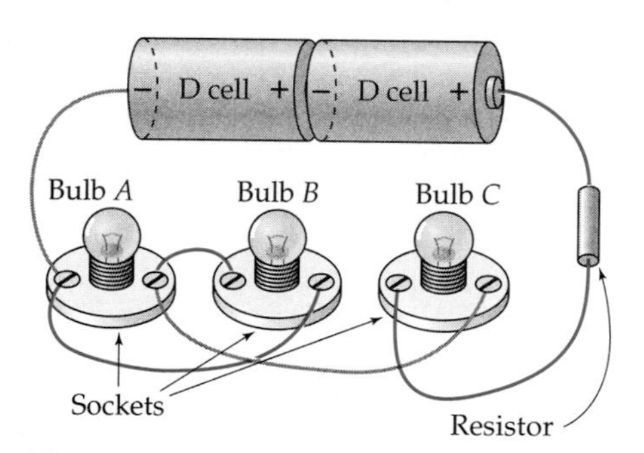

Figure E6.17
See problems E6B.5 and E6S.8.

E6B.6 Imagine that we connect a 12-Ω lightbulb and an 8-Ω lightbulb in parallel and connect the set to a 12-V battery. What is the total current flowing through the battery?

E6B.7 Imagine that we connect a 10-Ω lightbulb to a 6-V battery. What is the current flowing through the battery? Now imagine that we put a 50-Ω resistor in parallel with the lightbulb. By what factor does the current through the battery increase?

E6B.8 If all the lightbulbs in figure E6.18 have a resistance of 10 Ω, what is the circuit's total resistance?

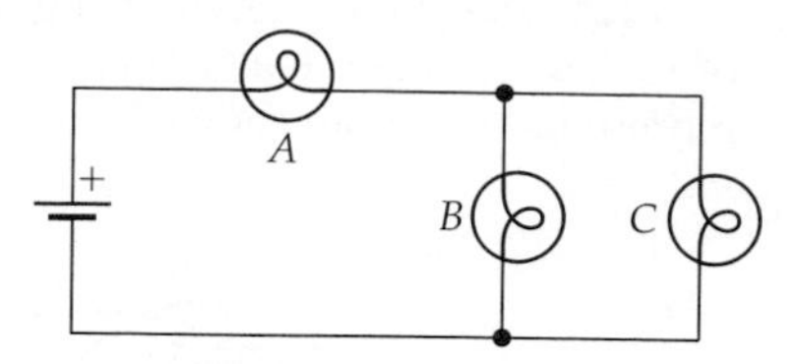

Figure E6.18
See problems E6B.8, E6B.11, and E6S.5.

E6B.9 If all the lightbulbs in figure E6.19 have a resistance of 12 Ω, what is the circuit's total resistance?

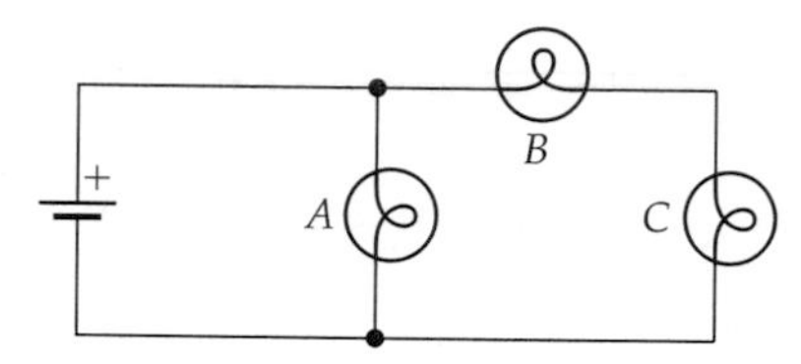

Figure E6.19
See problems E6B.9, E6B.12, and E6S.6.

E6B.10 Three lightbulbs are connected in parallel, and the set is connected to a 6-V battery. If one bulb is removed, what happens to the brightness of the other bulbs? Explain your response.

E6B.11 In figure E6.18, are bulbs A and B in series? Explain your response.

E6B.12 In figure E6.19, are bulbs A and B in parallel? Explain your response.

Synthetic

E6S.1 A 100-W household lightbulb is plugged into a wall outlet using a 100-ft extension cord made with #18 copper wire (which has a diameter of 1.01 mm). If the voltage difference at the outlet is 120 V, what is the voltage difference across the lightbulb?

E6S.2 Imagine a circuit consisting of three lightbulbs with resistances of 10 Ω, 20 Ω, and 30 Ω connected in series with a 1.5-V battery, with the 10-Ω bulb closest to the battery's positive terminal. These circuit elements are all connected with low-resistance wires. Say that we define the terminal of the 20-Ω bulb closest to the positive end of the battery to be where $\phi = 0$. Draw a *quantitatively* accurate graph of potential versus position for this circuit.

E6S.3 Imagine that you connect two circuit elements with resistances R_1 and R_2 in series with a battery having a potential difference V_{bat} across its electrodes. Imagine that we define the battery's negative electrode to have potential $\phi = 0$, which means that the battery's positive electrode has a potential $\phi = V_{bat}$. Assume that R_2 is attached to the battery's negative end. Find a formula for the potential at a point in the circuit between the two resistive circuit elements in terms of V_{bat}, R_1, and R_2. You may assume that all circuit elements are connected with wires with essentially zero resistance. Be sure to show your work. (This kind of circuit is called a *voltage divider*, and it can be used to establish at the point between the two resistors a potential of any desired value between 0 and V_{bat}.)

E6S.4 Imagine an electric lawn mower has a motor that can produce 2.0 hp of mechanical power when connected directly to a normal electrical outlet. When using this motor outside, however, one is forced to connect the mower to an outlet using a 100-ft power cord consisting of two strands of #14 copper wire (each strand has a diameter of 1.63 mm). How much mechanical power can one expect out of the mower under these conditions?

Figure E6.20
What is the power loss if we use a long extension cord for this electric mower? (See problem E6S.4.)

E6S.5 Assume that all the bulbs in the circuit shown in figure E6.18 have a resistance of 10 Ω, and the cell has a voltage of 1.5 V. (a) Find the potential difference across and current conducted by each bulb. (b) If bulb *C* were to be removed (leaving a nonconductive gap in its place), what would happen to the brightness of the other bulbs? Explain your reasoning carefully.

E6S.6 Assume that all the bulbs in the circuit shown in figure E6.19 have a resistance of 10 Ω, and the cell has a voltage of 1.5 V. (a) Find the potential difference across and the current conducted by each bulb. (b) If bulb *A* were to be removed (leaving a nonconductive gap in its place), what would happen to the brightness of the other bulbs? Explain your reasoning carefully.

E6S.7 Assume that all the elements in the circuit shown in figure E6.16 have a resistance of 12 Ω. Find the potential difference across and current conducted by each.

E6S.8 Assume that all the elements in the circuit shown in figure E6.17 have a resistance of 12 Ω. Find the potential difference across and the current conducted by each.

E6S.9 When you turn on your room air conditioner in the summer, you note that the lights in your room dim slightly. But in room wired the normal way, the lights and the air conditioner will be in parallel, and so they should have the same 120-V household voltage difference between their terminals no matter what is turned on, shouldn't they? Explain this phenomenon. Might the degree to which your lights dim depend on where in the room the air conditioner is plugged in? Explain. (*Hints:* This has nothing to do with the power company, which you can assume is dutifully supplying exactly 120 V to the house or dorm. A typical lightbulb draws from 60 to 100 W of power, while a room air conditioner will draw closer to 1000 W of power.)

E6S.10 As you draw more and more current from a realistic battery, the voltage difference between its electrodes droops slightly because of its effective internal resistance R_{int} (see section E6.6). Imagine that you have a battery with unknown emf ≈ 6 V, a 1-Ω resistor, a 2-Ω resistor, and a voltmeter of unknown resistance R_m that can measure the potential difference between the battery's electrodes. How can you use these items to measure the battery's R_{int}?

(a) We can attach the resistors to the battery in a number of different configurations to yield different total effective resistances. How many different effective resistances can we obtain? What are the values of these configurations?

(b) Imagine that when attaching a resistor configuration that has resistance R_1 to the battery's electrodes, we find the potential difference across the battery's terminals to be V_1. Show that

$$V_{ibat} = V_1\left[\frac{R_{int}(R_m + R_1)}{R_m R_1} + 1\right] \tag{E6.18}$$

(c) For a typical voltmeter, $R_m \gg R_1$. If this is true, show that equation E6.18 reduces to

$$V_{ibat} \approx V_1\left(\frac{R_{int}}{R_1} + 1\right) \tag{E6.19}$$

Also, find the minimum value of R_m/R_1 that ensures that this approximation will be valid to within 0.01 percent.

(d) Show that if we attach a second resistor configuration with resistance R_2 to the battery and find that the potential difference between the battery's electrodes is V_2, then we can eliminate the unknown V_{ibat} and solve for R_{int}, getting

$$R_{int} = \frac{R_1 R_2 (V_2 - V_1)}{R_2 V_1 - R_1 V_2} \tag{E6.20}$$

(e) Collect your own data (or assume that $V_1 = 5.48$ V when $R_1 = 1\ \Omega$ and $V_2 = 5.25$ V when $R_2 = 3\ \Omega$) to calculate the battery's internal resistance R_{int}.

E6S.11 Imagine that you have a set of batteries with the same ideal battery potential V_{ibat} and the same internal resistance R_{int}.

(a) Imagine that you connect N such batteries in parallel and connect the set to a circuit that has total effective resistance R (as shown in figure E6.21a). What is the effective total internal resistance of the set of batteries?

(b) Find the total current that flows through the resistor R in this case.

(c) Now imagine connecting the batteries in series and then to the circuit, as shown in figure E6.21b. What is the total effective internal resistance of the series set of batteries?

(d) Calculate the total current flowing through the resistor R in this case.

(e) Argue that if $R \gg R_{int}$ (as is normally the case), more current will flow through the circuit if the batteries are connected in series than in parallel; but if $R = R_{int}$, the same current flows whether we connect the batteries in series or parallel (!).

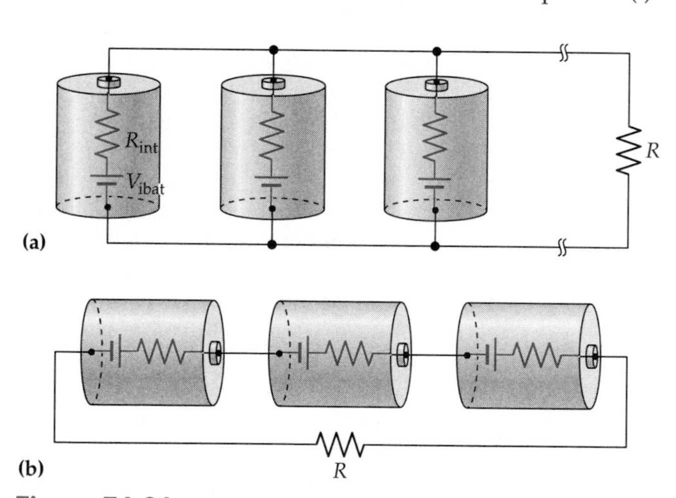

Figure E6.21
(a) A set of N realistic but identical batteries connected in parallel to a load resistor R. (b) The same set connected in series to the load. (See problem E6S.11.)

Rich-Context

E6R.1 Imagine that your uncle has an electric winch that he uses to haul his sailboat out of the water up into a trailer. The winch is connected to the car's 12-V battery and draws 240 W of power. One day, you and your uncle are using the sailboat at a place where you can't get the car close to the landing. So you go to the convenience store and buy four standard thickness (1.01-mm-diameter) 25-foot, two-wire extension cords and connect them, giving you essentially a 100-ft extension cord. You wheel the trailer down to the landing and park the car on the road about 80 ft away. You then use jumper cables to connect the two prongs on one end of the cord to the car battery, and you connect the two wires at the other end of the extension cord to the winch. But when you turn on the winch, nothing happens. You carefully verify that one end of the battery is indeed connected to one end of the winch through the extension cord, and the other end of the battery is connected to the other terminal of the winch through the other wire in the two-wire extension cord. As you are doing this, you notice that the extension cord is getting *hot*. What is going on? Why doesn't the winch work? Why is the cord getting hot? Explain with the help of a calculation.

Advanced

E6A.1 Imagine that you are designing a spotlight that is to be connected to your car battery using copper wire with a *total* length of L and a diameter of d. The light produced by an electric lamp is proportional to the power it dissipates. Derive a formula for the resistance of the lightbulb that will shine the brightest in terms of L, d, and σ_c for copper, and the voltage difference V_{bat} between the car battery's terminals.

E6A.2 Not *all* circuits can be analyzed by combining the elements into series and parallel sets: the drawing below shows one that *cannot*. However, one can completely analyze any arrangement of resistors by using *Kirchhoff's laws:*

1. **The loop rule:** The sum of (signed) potential differences around any closed loop must be zero.
2. **The node rule:** The current flowing into any junction must equal the current flowing out.

The first law is required if the potential at a point is to be well defined. The second expresses conservation of charge in a steady-state situation.

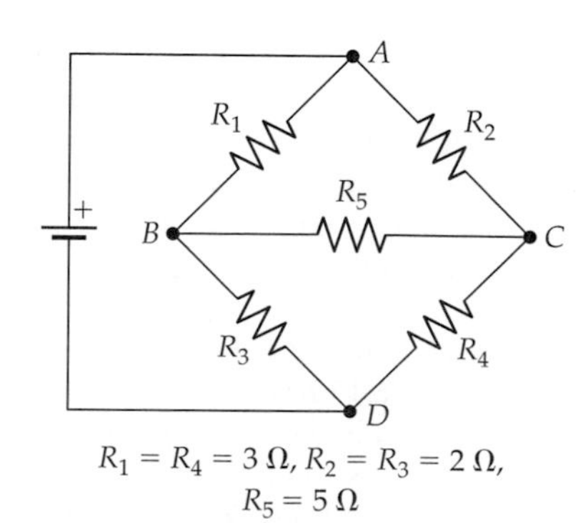

Figure E6.22
What is the resistance of this set of resistors? (See problem E6A.2.)

Use these laws to find the equivalent resistance of the set of resistors shown in figure E6.22. (*Hints:* Write an equation expressing the node rule for each of points A, B, C, and D. For example, if I is the total current flowing in the circuit, then the node rule at point A implies that $I = I_1 + I_2$. Then write an equation expressing the loop rule for each loop in the circuit. For example, the loop rule for triangle ABC says that $\Delta\phi_{BA} + \Delta\phi_{CB} + \Delta\phi_{AC} = 0$, where $\Delta\phi_{BA}$ is the signed potential difference $\Delta\phi_{BA} = \phi_B - \phi_A$ and so on. Note that $|\Delta\phi_{BA}| = V_1$ and so on. Finally, write down all the equations expressing $V = IR$ for the various elements. Can you solve this mass of equations for I? Don't put in numbers until you absolutely have to, or you will almost certainly get lost. Also, be very careful with signs.)

ANSWERS TO EXERCISES

E6X.1 Because resistances in series add, $R_{tot} = 12\ \Omega$. Remember that $1\ \Omega = 1$ V/A. The total current in the circuit is thus $V_{bat}/R = 3.0\text{ V}/12\ \Omega = (3.0\ \cancel{V})/(12\ \cancel{V}/\text{A}) = 0.25$ A. Since this current flows through each wire individually, the potential difference across the thick wire is $V_{thick} = IR_{thick} = (0.25\ \cancel{A}) \times (2\text{ V}/\cancel{A}) = 0.5$ V. Similarly, $V_{thin} = 2.5$ V.

E6X.2 One possible circuit diagram looks as shown in figure E6.23. It is possible to reorient the elements in different ways, but the basic connections in your diagram should be the same as in this diagram.

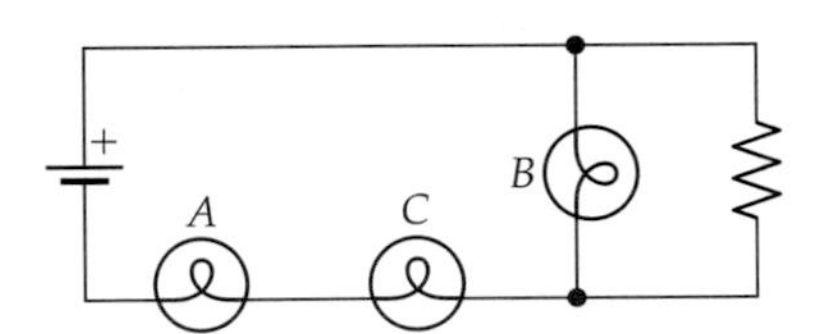

Figure E6.23
See the answer to exercise E6X.2.

E6X.3 It does not matter whether R_2 or R_4 is on the right. It does matter whether it is R_2 or R_1 that is connected to the battery's positive electrode, though, so exchanging these resistors would make the diagram invalid.

E6X.4 Since $I_{set} = I_1 + I_2$ and $V_{set} = V_1 = V_2$, we have

$$\frac{1}{R_{set}} = \frac{I_{set}}{V_{set}} = \frac{I_1 + I_2}{V_{set}} = \frac{I_1}{V_{set}} + \frac{I_2}{V_{set}}$$
$$= \frac{I_1}{V_1} + \frac{I_2}{V_2} = \frac{1}{R_1} + \frac{1}{R_2} \qquad \text{(E6.21)}$$

E6X.5 The combined resistance is

$$R_{set} = \left(\frac{1}{R_1} + \frac{1}{R_2}\right)^{-1} = \left(\frac{1}{3\ \Omega} + \frac{1}{1\ \Omega}\right)^{-1}$$
$$= \left(\frac{4}{3\ \Omega}\right)^{-1} = \frac{3\ \Omega}{4} \qquad \text{(E6.22)}$$

The current through the battery is thus

$$I_{bat} = I_{set} = \frac{V_{set}}{R_{set}} = \frac{1.5\ \cancel{V}}{3\ \cancel{\Omega}/4}\left(\frac{1\ \cancel{\Omega}}{1\ \cancel{V}/\text{A}}\right) = 2.0\text{ A} \qquad \text{(E6.23)}$$

E6X.6 *Model:* Assume that all the bulbs are ohmic and have a resistance of $10\ \Omega$ (as in example E6.3). *Solution:* Then if we remove bulb D, the remaining bulbs are in series and so have a total resistance of

$$R_{set} = R_A + R_B + R_C = 30\ \Omega \qquad \text{(E6.24)}$$

The total current is thus $I = V_{set}/R_{set} = 0.05$ A. The voltage across B was originally 0.3 V (see equation E6.15c), but now has increased to

$$V_B = I_B R_B = (0.05\ \cancel{A})(10\ \cancel{\Omega})\left(\frac{1\text{ V}/\cancel{A}}{1\ \cancel{\Omega}}\right) = 0.5\text{ V} \qquad \text{(E6.25)}$$

E6X.7 When $I = 0$, $V_{bat} = V_{ibat} + 0 = V_{ibat}$. So V_{ibat} must be 1.55 V. When $I = 4.0$ A, $IR_{int} = V_{ibat} - V_{bat} = 0.08$ V, so $R_{int} = (V_{ibat} - V_{bat})/I = 0.08\text{ V}/4.0\text{ A} = 0.02\ \Omega$.

E6X.8 The bird is not in contact with the ground or anything except the wire. Any current that flows through the bird would have to flow in one foot and out the other. But both feet are on the same wire, so they will have virtually no voltage difference between them (assuming the wire is a good conductor). So no significant current will flow through the bird.

E7

Magnetic Fields

▷ Electric Field Fundamentals
▷ Controlling Currents
▽ Magnetic Field Fundamentals
Magnetic Fields
Currents and Magnets
▷ Calculating Static Fields
▷ Dynamic Fields

Chapter Overview

Introduction

This chapter opens a new subdivision on static magnetic fields by discussing the behavior of magnets, the definition of the magnetic field vectors $\vec{B}$ and $\vec{\mathbb{B}}$, and how magnetic fields affect moving charges.

Section E7.1: The Phenomenon of Magnetism

The following properties of magnets have been known since at least the late 1500s:

1. Every **magnet** has exactly two **magnetic poles,** a **north** pole and a **south** pole.
2. Like magnetic poles repel but unlike poles attract.
3. A magnet's poles cannot be isolated.
4. Magnets strongly attract certain (**ferromagnetic**) substances.
5. A freely suspended magnet (a **magnetic compass**) will align itself so that its north pole points toward the earth's geographic north pole.

Section E7.2: The Definition of the Magnetic Field

We can describe a magnet's **magnetic field** by assigning to every point in space a magnetic field vector $\vec{B}$ whose *direction* is the "northward" direction indicated by a compass needle placed at that point and whose *magnitude* reflects how vigorously the compass responds to the field. We could have defined $\vec{E}$ in a similar way, using a suspended electric dipole instead of a compass.

Indeed, there are many similarities between magnets and electric dipoles: (1) Magnets repel and attract each other just as dipoles attract and repel each other. (2) A magnet's magnetic field is very similar to a dipole's electric field at points far from each. (3) A magnet responds to an external magnetic field as a dipole responds to an external electric field. This *analogy* is useful, but magnets are not the *same* as electric dipoles. Unlike dipoles, for example, magnetic poles cannot be separated, and magnets have no effect on stationary charges.

Section E7.3: Magnetic Forces on Moving Charges

However, magnetic fields *do* exert forces on *moving* charges. We can experimentally examine these forces by using a **cathode-ray tube (CRT).** An **electron gun** at the rear of a CRT creates an electron beam that travels down the tube and hits a phosphor-coated screen at the front. Since the screen glows where the beam hits, we can monitor how the beam deflects in response to an external magnetic field.

Experiments show that the following equation accurately describes magnetic force on a moving charged particle:

$$\vec{F}_m = q\vec{v} \times \vec{B} = q\left(\frac{\vec{v}}{c} \times \vec{\mathbb{B}}\right) \tag{E7.6}$$

Purpose: This equation specifies the magnetic force $\vec{F}_m$ acting on a charged particle with charge q moving with velocity $\vec{v}$ at a point where the magnetic field vector is $\vec{B}$ or $\vec{\mathbb{B}}$.

Symbols: c is the speed of light.

Limitations: This equation only works for charged *particles*.

Notes: This equation works in relativistic contexts as long as we define $\vec{F}_m$ to be the rate at which the interaction delivers relativistic momentum. The SI unit for $\vec{B}$ is the *tesla* (T), where 1 T = 1 (N/C)(m/s)$^{-1}$; the SI units for $\vec{\mathbb{B}}$ are merely newtons per coulomb, and $\vec{\mathbb{B}} = c\vec{B}$.

This equation defines the magnitudes of the magnetic field vectors $\vec{B}$ and $\vec{\mathbb{B}}$ by linking them quantitatively to the previously defined quantities of force, charge, and velocity. Magnetic fields are usually described by using $\vec{B}$, but we will find the alternative magnetic field vector $\vec{\mathbb{B}}$ very useful because it has the same units as $\vec{E}$.

Section E7.4: A Free Particle in a Uniform Magnetic Field

Equation E7.6 also implies that an otherwise free particle moving in a uniform magnetic field follows a helical trajectory around and along the magnetic field direction. The following equations describe this trajectory.

$$R = \frac{pc\sin\theta}{|q|\mathbb{B}} = \frac{p\sin\theta}{|q|B} \qquad \text{(E7.11}a\text{)}$$

$$T = \frac{2\pi mc}{|q|\mathbb{B}\sqrt{1-v^2/c^2}} = \frac{2\pi m}{|q|B\sqrt{1-v^2/c^2}} \qquad \text{(E7.11}b\text{)}$$

$$p_{\parallel} = \text{constant} \qquad \text{(E7.11}c\text{)}$$

Purpose: These equations describe features of the helical motion of a particle with charge q and (possibly relativistic) momentum $\vec{p}$ moving in a reasonably uniform magnetic field described by $\vec{\mathbb{B}}$ or $\vec{B}$.

Symbols: R is the radius of the circular part of the helical motion, T is the time required to go once around the helix, θ is the angle between $\vec{p}$ and $\vec{\mathbb{B}}$, c is the speed of light, $p_{\parallel}$ is the component of the particle's momentum parallel to $\vec{\mathbb{B}}$, and $B \equiv \mathbb{B}/c$ is the magnetic field strength expressed in teslas.

Limitations: The magnetic field must be nearly uniform over the region spanned by one cycle of the helix.

Note: The period T is essentially independent of v for $v \ll c$.

If $\theta = 90°$, the particle moves in a circular path in a plane perpendicular to $\vec{\mathbb{B}}$.

Section E7.5: The Magnetic Force on a Wire

A magnetic field exerts on a current-carrying wire a force given by

$$\vec{F}_{m,\text{seg}} = \frac{L}{c}\vec{I} \times \vec{\mathbb{B}} = L\vec{I} \times \vec{B} \qquad \text{(E7.14)}$$

Purpose: This equation describes the magnetic force $\vec{F}_{m,\text{seg}}$ exerted on a straight segment of wire of length L in a magnetic field described by $\vec{\mathbb{B}}$ or $\vec{B}$.

Symbols: $\vec{I}$ is the direction of the steady conventional current in the wire.

Limitations: The wire must be straight and thin, and L must be small enough that $\vec{\mathbb{B}}$ is nearly uniform over the segment.

This implies that a current-carrying loop will seek to orient itself in a magnetic field just as a magnet perpendicular to the loop's plane would. (The **loop-to-magnet rule** says that if you curl your right fingers in the direction the loop's current flows, your thumb indicates the south-to-north direction of the equivalent magnet.) Such a loop will continually rotate in a magnetic field if we switch the direction of its current every half rotation: this is how an electric motor works.

E7.1 The Phenomenon of Magnetism

The phenomenon of magnetism

Ancient texts show that magnetism was known and described in Greece as early as 800 B.C. *Magnetite,* a certain oxide of iron found in many parts of the world (but especially in the ancient Greek province of Magnesia), can exhibit substantial magnetic effects even in its natural state. In western Europe before the Renaissance, people considered magnets to be occult objects because of the uncanny nature of the forces they exerted. However, magnetic compasses were one of several new navigational tools that helped European sailors freely roam the globe in the 1500s and 1600s, and daily use by sailors eventually undercut superstition about magnetism. The British physicist William Gilbert was one of the first people to investigate magnetism scientifically: because of the comprehensive and careful work described in his book *De Magnete* (1600), magnetism was better understood than electricity for more than a century.

Empirically observed properties of magnets

Gilbert described these important empirical properties of magnets:

1. Every **magnet** has exactly two **magnetic poles,** a **north** pole and a **south** pole (we will discuss the reason for these names shortly).
2. North poles repel north poles, and south poles repel south poles; but unlike poles attract.
3. A magnet's poles cannot be isolated: breaking a magnet in half merely creates two magnets (each with its own north and south poles).
4. Either pole of a magnet can strongly attract certain (**ferromagnetic**) objects that are not themselves magnets, but most substances are only *very* weakly affected by magnets.
5. A freely suspended magnet (a **magnetic compass**) will align itself so that its north pole points toward the earth's geographic north pole (hence the names for a magnet's poles).

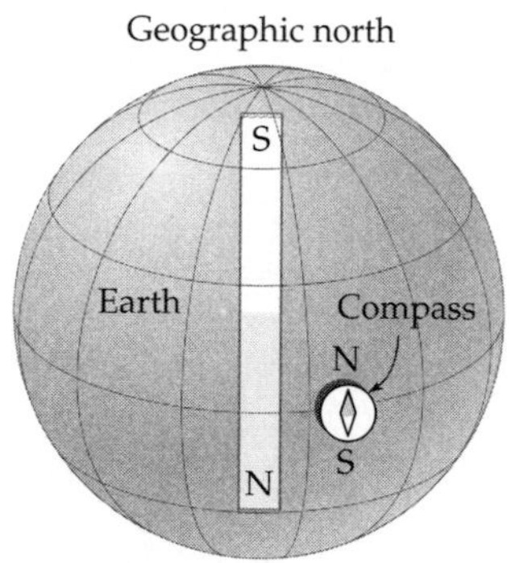

Figure E7.1
The conventional rule for naming magnetic poles says that when a magnet is used as a compass, the pole that ends up pointing toward the earth's geographic north pole is the magnet's north pole by definition. This means that the earth's geographic north pole is a magnetic *south* pole, since it attracts the compass magnet's north pole.

Gilbert seems to have been the first to recognize that the last item implies that the earth *itself* is a magnet. Since the north pole of a compass magnet is attracted to a magnetic *south* pole, the earth's geographic north pole must be a magnetic south pole, as illustrated in figure E7.1.†

E7.2 The Definition of the Magnetic Field

Just as we can describe the electrostatic interaction between two charged objects by using a field theory that assigns an electric field vector $\vec{E}$ to every point in space, so we can describe the magnetic interaction between magnetic poles by using a field theory that assigns a magnetic field vector $\vec{B}$ to every point in space. Indeed, we *need* a field theory for magnetic interactions for the same reason we need a field theory for electrostatic interactions: only a field theory can be consistent with special relativity.

Definition of the magnetic field vector at a point

In the case of magnetism, we can define the *direction* of $\vec{B}$ at a given point to be the direction a compass needle at that point would indicate is "northward," and we qualitatively define the *magnitude* of $\vec{B}$ in terms of how vigorously that compass quivers in response to the field when disturbed, as illustrated in figure E7.2.

†Actually, the earth's magnetic south pole is located just north of Hudson's Bay in Canada, about 1300 km from the earth's geographic north pole. The location, strength, and even polarity of the earth's magnetic poles are known to vary over geological time, and the details about exactly how the earth maintains its magnetism remain unclear.

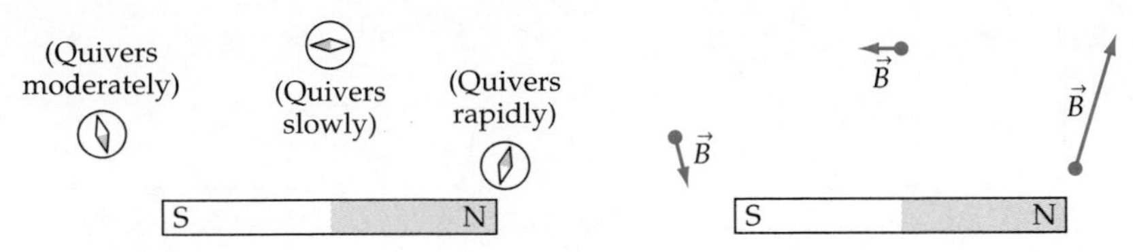

Figure E7.2
The magnetic field at a point in space is defined to point northward as indicated by a compass at that point and has a magnitude proportional to how strongly the compass is affected by the field at that point.

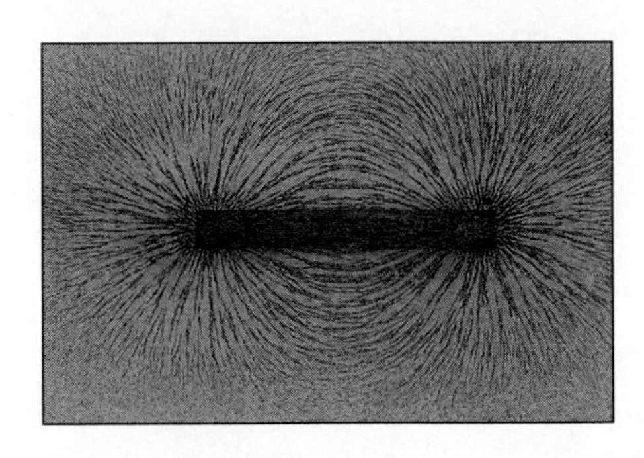

Figure E7.3
When iron filings are sprinkled on a piece of glass, they display the magnetic field of a magnet placed behind the glass by aligning themselves (as compasses do) with the field direction at every point.

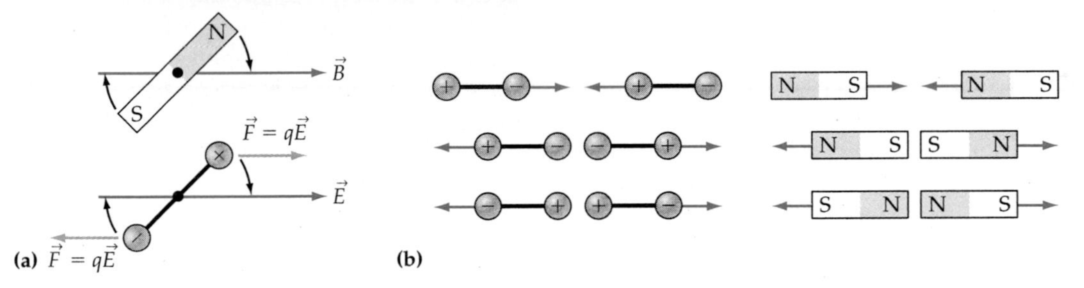

Figure E7.4
(a) Just as a magnet seeks to align itself with the magnetic field direction in its vicinity (by definition), an electric dipole will seek to align itself with the electric field in its vicinity. (b) Electric dipoles attract and repel each other just as magnets do.

A common way to display the magnetic field of a magnet is to place a magnet behind a sheet of paper and then sprinkle iron filings on the paper. The filings at a given point will align themselves with the direction of the magnetic field at that point, as shown in figure E7.3. This is so because the magnet's field actually magnetizes each iron filing, making it a weak magnet that aligns itself with the magnetic field just as a compass would.

In chapter E2, we defined the electric field vector $\vec{E}$ at a point in terms of the electrostatic force the field exerts on a motionless point charge; but we *could* have defined $\vec{E}$ in a way very similar to the way we have just defined $\vec{B}$. Imagine that we suspend an electric dipole in an electric field. Figure E7.4a illustrates that the electrostatic forces acting on the charges at the dipole's ends will cause the dipole to align itself with the direction of $\vec{E}$ just as a compass aligns itself with the direction of $\vec{B}$. We see, therefore, that the definitions of $\vec{E}$ and $\vec{B}$ are really quite similar.

Exercise E7X.1

Why would it be awkward to define $\vec{B}$ in terms of the magnetic force exerted on an isolated magnetic pole?

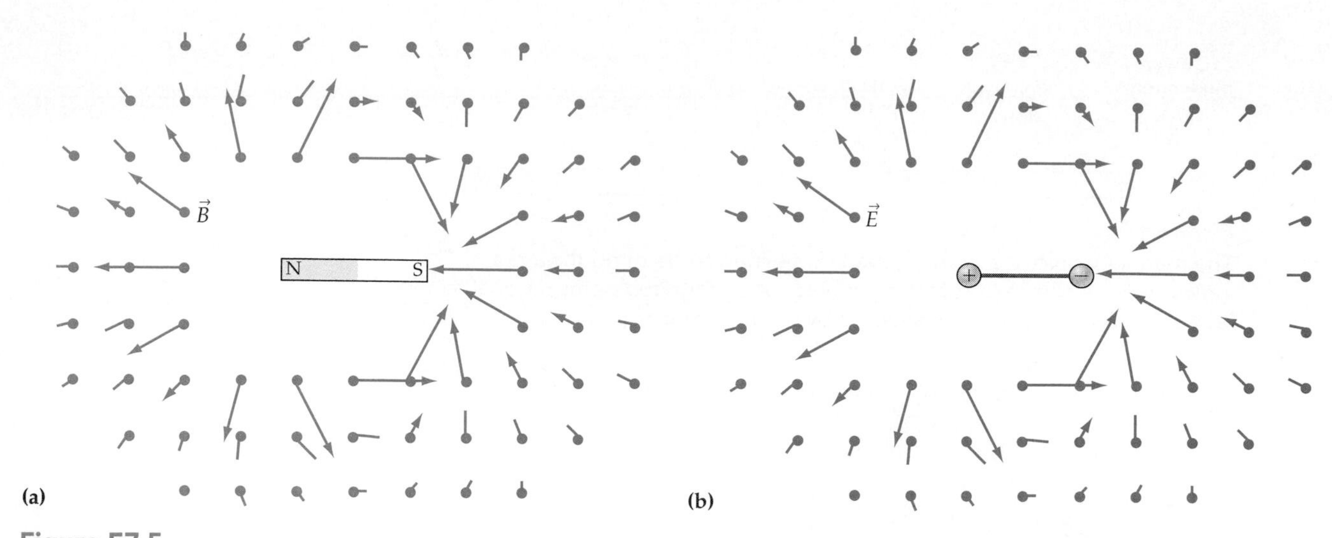

Figure E7.5
(a) The external magnetic field of a bar magnet. (b) The external electric field of a dipole.

Magnets are *analogous* to electric dipoles

Magnets and an electric dipole are similar in other ways. The like poles of two magnets repel each other and the unlike poles attract, just as the like ends of two dipoles repel and the unlike ends attract (see figure E7.4b). Indeed, the forces that two magnets exert on each other depend on their relative separation and orientation exactly as the forces between two similarly shaped electric dipoles would. Moreover, when we use a compass to plot out the magnetic field of a bar or horseshoe magnet, we find that (as long as the magnet's pole faces are small compared to the distance separating them) its *external* magnetic field is almost identical to the electric field of an electric dipole, as shown in figure E7.5. I recommend memorizing the fact that magnetic field vectors near to (but outside) a magnet point *away* from its north pole and *toward* its south pole, as shown in figure E7.5 (just $\vec{E}$ points away from a positive charge and toward a negative charge).

But magnets are not the *same* as electric dipoles

We will find the dipole analogy for magnets very useful in what follows. However, it is crucial to understand that this is just an *analogy:* magnets are not the *same* as electric dipoles, nor are magnetic poles the same as electric charges. For example, magnetic poles do not exert forces on electric charges at rest, while the ends of an electric dipole do. The kinds of neutral substances that respond strongly to magnets are not correlated with those that respond strongly to electric dipoles. Moreover, while we can easily find or make isolated objects that have purely positive or negative charge, it does *not* seem to be possible to find or make an object having an isolated magnetic pole (a **magnetic monopole**). Finally, the magnetic field *inside* a magnet is very different from the electric field in the region between the two charges of an electric dipole. Magnets and electric dipoles may be *analogous,* but they are not the same.

E7.3 Magnetic Forces on Moving Charges

While a magnetic field does not exert any force on a stationary electric charge, it turns out that it *can* exert a force on a *moving* charge. We will see in this section that this provides a tool for more precisely defining the magnitude of $\vec{B}$.

Description of a cathode-ray tube

One can create a beam of moving electrons by using an **electron gun** (see figure E7.6), which consists of two metal plates (one is called a **cathode** and

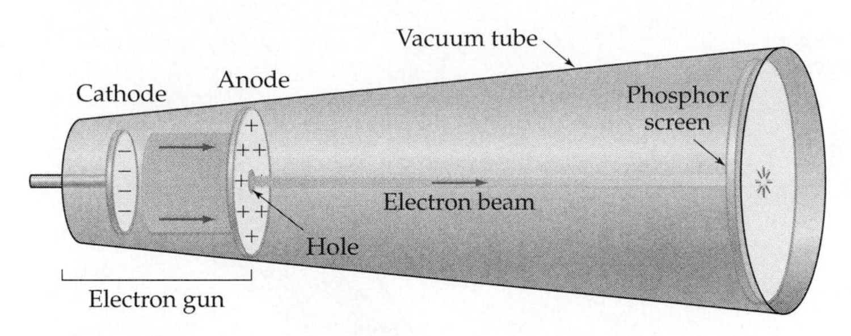

Figure E7.6
A schematic diagram of a cathode-ray tube. Electrons boil off a hot, negatively charged cathode and accelerate toward a positively charged anode. Some go through a hole in the anode, creating an electron beam. A phosphor-coated screen glows when the electron beam hits it.

the other an **anode**) enclosed in a vacuum tube. A power supply connected between the anode and cathode gives these plates a positive and negative charge, respectively. An electric heater makes the cathode sufficiently hot that some electrons have thermal energies large enough to allow them to escape the metal. These free electrons accelerate away from the negatively charged cathode toward the positive anode. Most hit the anode and are absorbed, but a few fly through a hole in the anode, creating a beam of electrons. A phosphor-coated screen placed at the other end of the vacuum tube displays a glowing spot where the electron beam hits it. The resulting device (called a **cathode-ray tube,** or **CRT**) is used in TVs, computer monitors, oscilloscopes, radar screens, and many other devices to convert electronic signals to visible images.

Observations about the magnetic forces on moving charges

We can use a simple CRT to investigate what a magnetic field does to moving charges. Figure E7.7 (on the next page) illustrates what happens when such a CRT is placed in magnetic fields with various orientations. The first four cases display the empirical result that when $\vec{v}$ and $\vec{B}$ are not collinear, *the magnetic force on the electron beam is perpendicular to both* $\vec{v}$ *and* $\vec{B}$. We can in fact use the following **right-hand rule** to predict the direction of deflection: If you point your right index finger in the direction of $\vec{v}$ and your second finger in the direction of $\vec{B}$, your right thumb will point in the direction *opposite* to the magnetic force that deflects the negatively charged electrons. If we do analogous experiments with beam of *positive* ions, we find experimentally that the magnetic force on the beam points in the *same* direction as your right thumb.

A possible formula for the magnetic force

Now, where have we seen a right-hand rule like this before? In chapter C13 we learned that the *cross product* $\vec{u} \times \vec{w}$ of two vectors $\vec{u}$ and $\vec{w}$ is a vector that is perpendicular to both $\vec{u}$ and $\vec{w}$ in the sense indicated by *exactly the same right-hand rule:* if you point your right index finger in the direction of $\vec{u}$ and second finger in the direction of $\vec{w}$, your thumb indicates the direction of the cross product. (If you feel rusty about the definition of the cross product, now would be an excellent time to review section C13.2.) Therefore, the observations we have discussed so far suggest that a possible mathematical expression for the magnetic force $\vec{F}_m$ exerted on a particle with charge q moving with velocity $\vec{v}$ at a point where the magnetic field vector is $\vec{B}$ might be

$$\vec{F}_m \propto q(\vec{v} \times \vec{B}) \tag{E7.1}$$

Note that since q is negative for electrons, the force would be opposite to the direction indicated by your right thumb if you pointed your right index

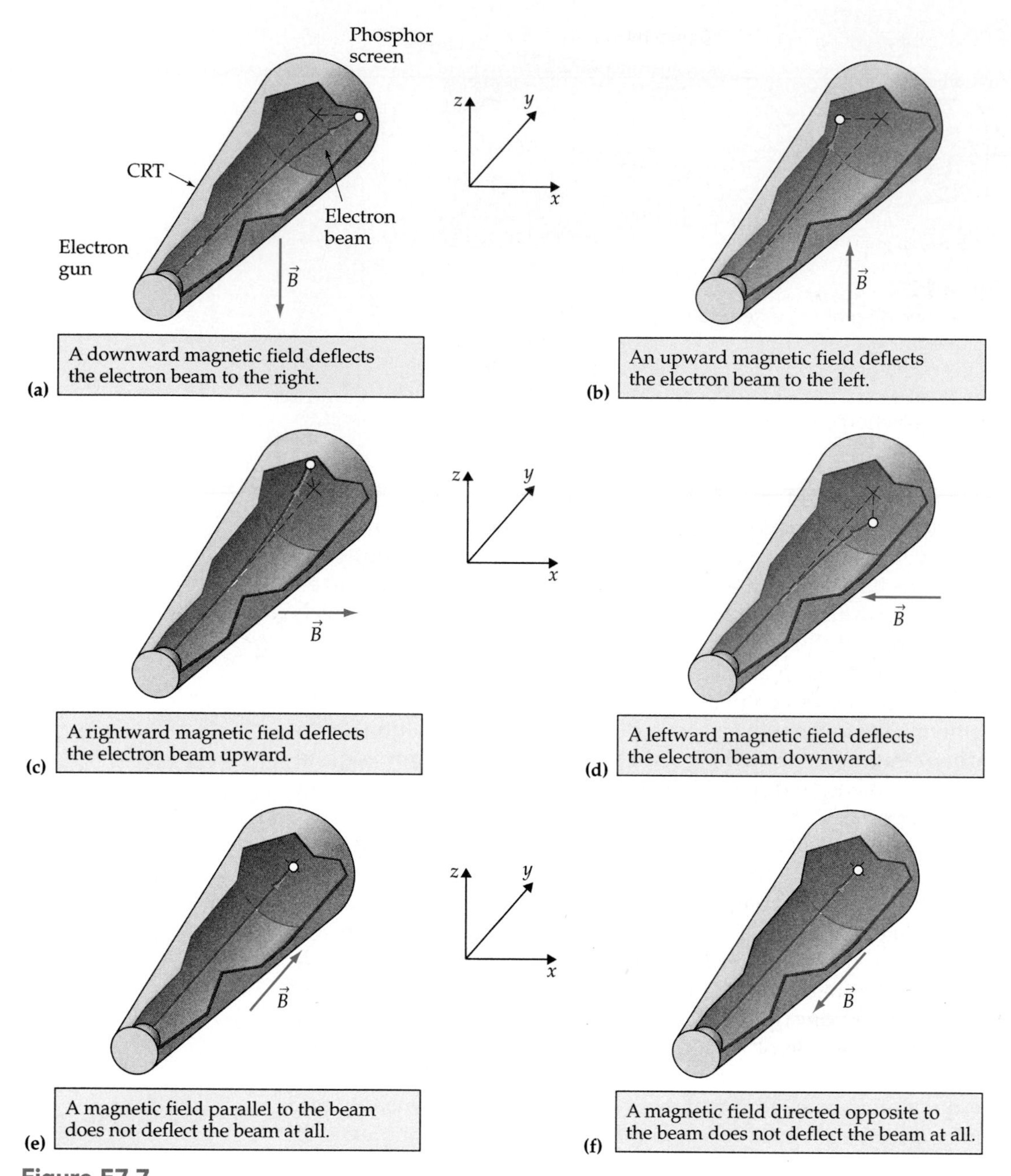

Figure E7.7

How magnetic fields affect an electron beam. In each drawing, the electron gun is closer to you than the phosphor screen is, and the electron beam moves in the $+y$ direction.

finger in the direction of $\vec{v}$ and your right second finger in the direction of $\vec{B}$, consistent with what we observe.

This expression would explain the results illustrated in figure E7.7e and E7.7f as well, because the definition of the magnitude of the cross product implies that $\text{mag}(\vec{v} \times \vec{B}) \equiv vB\sin\theta$, where θ is the angle between the vectors $\vec{v}$ and $\vec{B}$, so

$$\text{mag}(\vec{F}_m) \propto qvB\sin\theta \tag{E7.2}$$

In the cases shown in figure E7.7e and E7.7f, the θ is either 0° or 180°, so $\sin\theta = 0$ and the magnetic force should be zero, as is observed. Equation E7.2

also predicts (and experiments verify) that the magnetic force is directly proportional to q, v, and the magnetic field strength (as qualitatively indicated by the vigor of compass response).

Defining the magnitude of $\vec{B}$

So equation E7.1 appears to have solid experimental support: all that remains is to determine the constant of proportionality. We know how to measure the *force* acting on a particle and the particle's *charge* and *velocity*, but our compass-based definition of $\vec{B}$ is too fuzzy to define numerical value for $\vec{B}$. The historical solution to this problem was to *define* the numerical magnitude of $\vec{B}$ so that the constant of proportionality in equation E7.1 is 1, so that

$$\vec{F}_m = q\vec{v} \times \vec{B} \tag{E7.3}$$

Equation E7.3 therefore qualitatively expresses empirical observations about how moving charges respond to magnetic fields, but quantitatively amounts to a definition of the magnitude of $\vec{B}$. The SI units of $\vec{B}$ so defined must be $\text{N} \cdot \text{s}/(\text{C} \cdot \text{m})$ for the force to come out in newtons: therefore, the **tesla** (T), where $1\ \text{T} \equiv 1\ \text{N} \cdot \text{s}/(\text{C} \cdot \text{m})$, is the standard SI unit for magnetic field strength.

An alternative definition

This is, however, not the only possible choice for the constant of proportionality. If we choose that constant to be $1/c$ (where c is the speed of light), then the equation becomes

$$\vec{F}_m = q\left(\frac{\vec{v}}{c} \times \vec{\mathbb{B}}\right) \tag{E7.4}$$

where $\vec{\mathbb{B}}$ is an alternative version of the magnetic field vector. This definition has a crucial advantage: because $\vec{v}/c$ is unitless, $\vec{\mathbb{B}}$ has SI units of newtons per coulomb, which are the *same* as those for the electric field vector $\vec{E}$! We will find later in this unit that because of this symmetry, the equations of electromagnetism that link electric and magnetic fields are simpler and more beautiful when expressed in terms of $\vec{\mathbb{B}}$ rather than $\vec{B}$. Therefore, even though $\vec{B}$ is the conventional notation found in most texts, we will use $\vec{\mathbb{B}}$ (which we will call B-bar) instead of $\vec{B}$ to describe the magnetic field throughout this unit. (You can easily draw $\vec{\mathbb{B}}$ by hand by drawing two vertical bars on the left side of the symbol instead of one.) The simplicity and clarity gained come at only a small cost, because we can easily change any formula involving $\vec{\mathbb{B}}$ (in newtons per coulomb) to one involving $\vec{B}$ (in teslas) by substituting

$$\vec{\mathbb{B}} = c\vec{B} \tag{E7.5}$$

anywhere that $\vec{\mathbb{B}}$ occurs.

Table E7.1 lists some magnetic field strength benchmarks in units of newtons per coulomb, teslas, and another commonly used unit called the **gauss** (where $1\ \text{gauss} \equiv 10^{-4}\ \text{T} = 30{,}000\ \text{N/C}$). The magnetic field near a typical refrigerator magnet has a magnitude of roughly $200\ \text{gauss} = 0.02\ \text{T} = 6 \times 10^6\ \text{N/C} = 6\ \text{MN/C}$. The gauss is used partly because it is a nice "laboratory-sized" unit. (The meganewton per coulomb would also be a convenient unit if it were not so hard to say.†)

So to summarize, the following **magnetic force law** both describes the empirically observed character of the magnetic forces on a moving charged particle and operationally defines the magnitudes of the magnetic field vectors $\vec{B}$ and $\vec{\mathbb{B}}$:

†Perhaps we can informally call 1 MN/C a "meganick," which is not so hard.

Table E7.1 Some magnetic field strength benchmarks

Situation	$\mathbb{B}$	B (in gauss)	B (in teslas)
A magnetically shielded room	$\sim 10\ \mu$N/C	$\sim$0.3 ngauss	3×10^{-14} T
In interstellar space in our galaxy	$\sim$0.03 N/C	$\sim 1\ \mu$gauss	$\sim$0.1 nT
In a magnetic field with a strength of 1 N/C	$\equiv$ 1 N/C	33.3 μgauss	3.33 nT
At the surface of the earth	15 kN/C	0.5 gauss	50 μT
Near an interstate power line (above earth's field)	$\sim$15 kN/C	$\sim$0.5 gauss	$\sim$50 μT
In a magnetic field with a strength of 1 gauss	30 kN/C	$\equiv$ 1 gauss	$\equiv 10^{-4}$ T
At the surface of the sun	$\sim$3 MN/C	$\sim$100 gauss	$\sim$0.01 T
Near a refrigerator magnet	$\sim$6 MN/C	$\sim$200 gauss	$\sim$0.02 T
Within a sunspot	$\sim$90 MN/C	$\sim$3000 gauss	$\sim$0.3 T
In a magnetic field with a strength of 1 T	300 MN/C	$\equiv$ 10 kgauss	$\equiv$ 1 T
Inside a moving-coil loudspeaker	$\sim$450 MN/C	$\sim$15 kgauss	1.5 T
Strongest superconducting electromagnets	$\sim$6000 MN/C	$\sim$0.2 Mgauss	$\sim$20 T
Inside a large laboratory electromagnet	$\sim$10 GN/C	$\sim$350 kgauss	$\sim$35 T
Near the surface of a neutron star	$\sim 3 \times 10^{16}$ N/C	$\sim 10^{12}$ gauss	$\sim 10^{8}$ T

The magnetic force law

$$\vec{F}_m = q\vec{v} \times \vec{B} = q\left(\frac{\vec{v}}{c} \times \vec{\mathbb{B}}\right) \tag{E7.6}$$

Purpose: This equation specifies the magnetic force $\vec{F}_m$ acting on a charged particle with charge q moving with velocity $\vec{v}$ at a point where the magnetic field vector is $\vec{B}$ or $\vec{\mathbb{B}}$.

Symbols: c is the speed of light.

Limitations: This equation only works for charged *particles*.

Notes: The SI unit for $\vec{B}$ is the *tesla*, where 1 T = 1 (N/C)(m/s)$^{-1}$; the SI units for $\vec{\mathbb{B}}$ are merely newtons per coulomb, and $\vec{\mathbb{B}} = c\vec{B}$. This equation works in relativistic contexts as long as we define $\vec{F}_m$ to be the rate at which the interaction delivers relativistic momentum $\vec{p} = m\vec{v}[1 - v^2/c^2]^{-1/2}$.

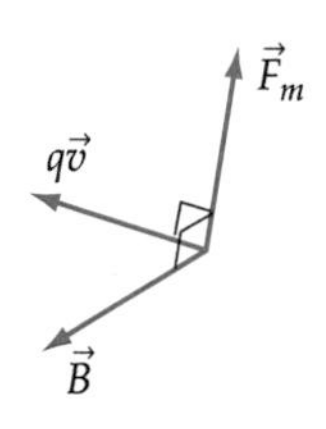

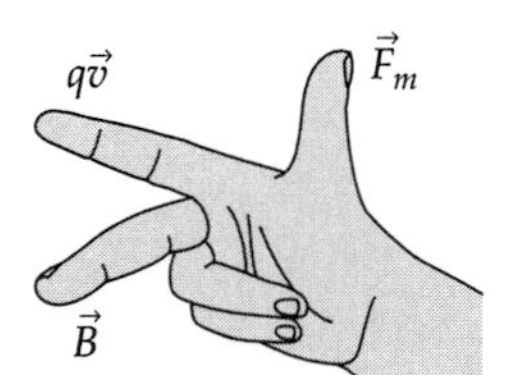

Figure E7.8
The **right-hand rule** for determining the direction of the magnetic force on a moving charged particle.

Figure E7.8 summarizes the right hand rule for determining the direction of the magnetic force $\vec{F}_m$. As illustrated, I generally find it easiest to think of $q\vec{v}$ as being a single vector that is parallel to $\vec{v}$ if q is positive and opposite to $\vec{v}$ when q is negative. If I point my right index finger in the direction of $q\vec{v}$ and my second finger in the direction of $\vec{B}$, then my right thumb always correctly indicates the direction of $\vec{F}_m$.

Exercise E7X.2

Given equation E7.6, prove that $\vec{\mathbb{B}} = c\vec{B}$.

Exercise E7X.3

Given equation E7.1, determine the initial direction of the force acting on the charged particles shown below as they enter regions of space where the magnetic field vectors have the directions shown. The colored vectors in these drawings indicate the direction of the particle's velocity. *Note*: The symbol × is conventionally used to indicate a magnetic field that is directed *into* the page perpendicular to the plane of the drawing (you are looking at the "feathers" of the magnetic field arrows). A dot is conventionally used to indicate a magnetic field that is directed out of the page (you are looking at the arrows' sharp ends in this case).

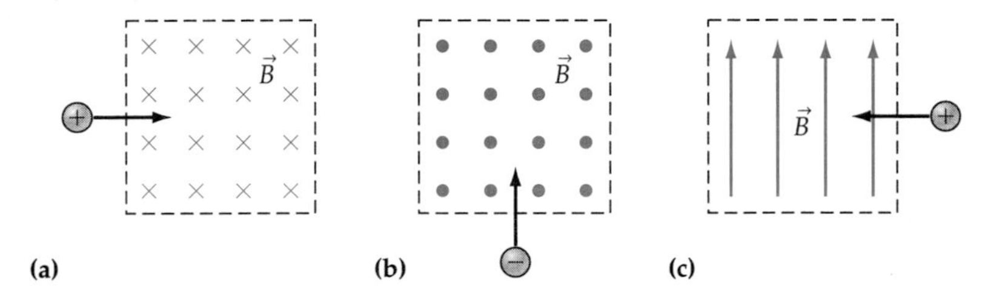

Exercise E7X.4

What are the directions of the magnetic fields causing the particles shown to experience the forces shown in each drawing? Which drawings (if any) are impossible?

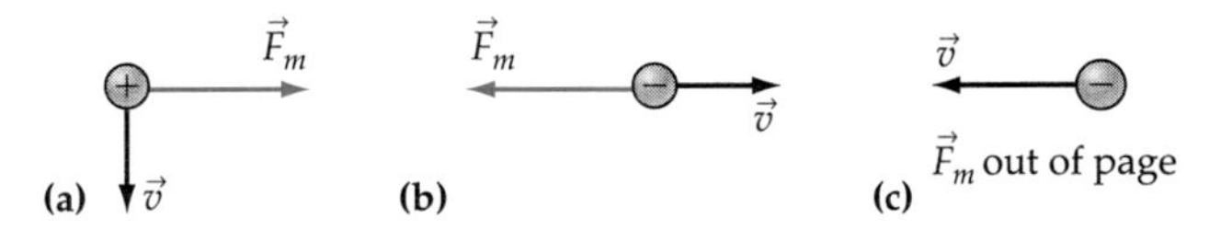

E7.4 A Free Particle in a Uniform Magnetic Field

A magnetic field contributes no k-work to a moving particle

Equation E7.6 implies that no matter how the particle moves, the magnetic field exerts a force perpendicular to the particle's velocity, so the rate at which a magnetic force directly contributes k-work to a particle is

$$\frac{[dK]}{dt} = \frac{\vec{F}_m \cdot d\vec{r}}{dt} = \vec{F}_m \cdot \vec{v} = 0 \qquad (\text{since } \vec{F}_m \perp \vec{v}) \tag{E7.7}$$

(see the discussion of k-work in chapter C8). Thus a charged particle's kinetic energy (and so its speed) is not changed by a static magnetic field.

A free particle initially moving perpendicular to $\vec{\mathbb{B}}$ moves in a circle

What the magnetic field *does* do is to change such a particle's direction of motion. Consider the case where we have uniform magnetic field (a field whose field vectors $\vec{\mathbb{B}}$ are essentially constant in magnitude and direction at all points in a region of interest) through which a particle with charge q moves with a velocity initially perpendicular to $\vec{\mathbb{B}}$ (see figure E7.9a). Since the magnetic force $\vec{F}_m$ is perpendicular to $\vec{\mathbb{B}}$, it will act in the same plane as $\vec{v}$, but will also always be perpendicular to $\vec{v}$, as shown. Under these circumstances, the magnitude of this force is

$$F_m = q\left(\frac{v}{c}\right)\mathbb{B}\sin\theta = q\left(\frac{v}{c}\right)\vec{\mathbb{B}} = \text{constant} \tag{E7.8}$$

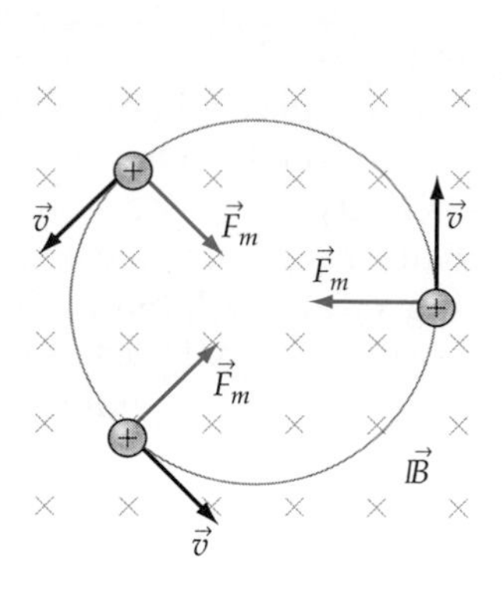

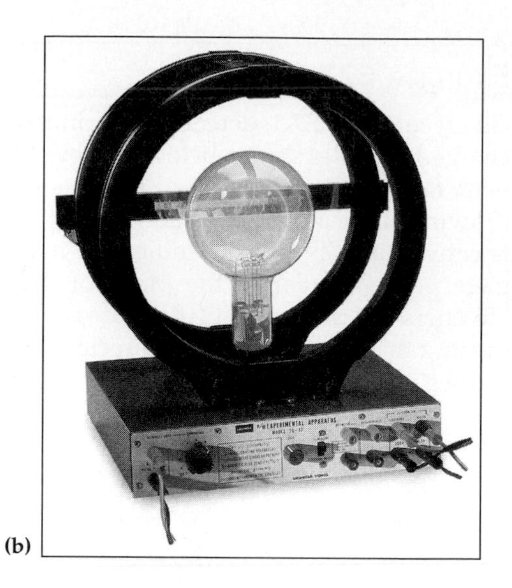

Figure E7.9

(a) The path of a charged particle whose initial velocity is perpendicular to $\vec{B}$. (b) This photograph shows an electron beam moving in the uniform magnetic field created by the large circular coils (electrons striking gas molecules in the tube causes the gas molecules to glow). Note that the trajectory of the electrons in this field is nearly a perfect circle.

since the angle θ between $\vec{v}$ and $\vec{B}$ always 90°, the value of q is fixed, the particle's speed v is constant (as we've just argued), and $B \equiv \text{mag}(\vec{B})$ is constant (since the field is uniform). If no other forces act on the particle, its *acceleration* must also be constant. One can use the Newton computer program (see chapter N4) to show that whenever a particle experiences a constant acceleration perpendicular to its motion, it will necessarily move in a *circle*, as shown in figure E7.9b (see problem E7S.13).

The radius of the particle's circular path is proportional to its momentum, and its period is independent of its speed

The acceleration of a particle moving at a constant speed v in a circle of radius R is v^2/R. Newton's second law therefore implies that in this case

$$|q|\left(\frac{v}{c}\right)B = F_m = ma = m\frac{v^2}{R} \quad \Rightarrow \quad R = \frac{mvc}{|q|B} = \frac{pc}{|q|B} \tag{E7.9}$$

This important equation says that the radius R of the circular motion of a particle with a given charge in a given field is proportional to its *momentum* p. The time T required for the particle to go once around the circle is the circle's circumference divided by the particle's speed, so

$$T = \frac{2\pi R}{v} = \frac{2\pi}{v}\frac{mvc}{|q|B} = \frac{2\pi mc}{|q|B} \tag{E7.10a}$$

Note that, surprisingly, T does *not* depend on the particle's initial speed! Therefore, a batch of identical and otherwise free particles moving in a given magnetic field will cycle around in the plane perpendicular to $\vec{B}$ at exactly the same frequency $f = 1/T$, independent of their speeds (as long as that speed is not relativistic). We call this unique frequency the particles' **cyclotron frequency** in that field (after a type of particle accelerator that takes advantage of this fact).

What happens when the particle is relativistic

Equation E7.9 turns out to be correct even if the particle is relativistic ($v \approx c$), as long as we take p to be the particle's relativistic momentum $p \equiv mv/(1 - v^2/c^2)^{1/2}$. If this is true, then equation E7.10*a* in this case becomes

$$T = \frac{2\pi R}{v} = \frac{2\pi}{v}\frac{mvc}{|q|B(1 - v^2/c^2)^{1/2}} = \frac{2\pi mc}{|q|B}\frac{1}{(1 - v^2/c^2)^{1/2}} \tag{E7.10b}$$

So we see that T is not really *exactly* independent of v, but the v dependence is negligible unless v becomes an appreciable fraction of c.

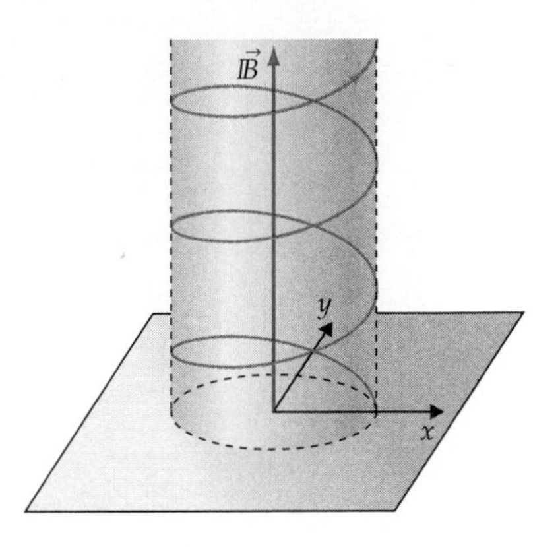

Figure E7.10
The helical path of a particle whose velocity is *not* initially perpendicular to $\vec{\mathbb{B}}$.

Exercise E7X.5

An electron beam whose electrons are traveling at a speed of 3.0×10^7 m/s perpendicular to a uniform magnetic field is observed to travel in a circle with a radius of 12.0 cm. What is the magnitude of the magnetic field? What is the cyclotron frequency of these electrons? (You can ignore relativistic effects, which change the answer by less than 1 percent.)

Now consider the more general case where the particle is *not* initially moving entirely perpendicular to $\vec{\mathbb{B}}$. Since the magnetic force $\vec{F}_m$ is always perpendicular to $\vec{\mathbb{B}}$, it has no component in the field direction, so the component of the particle's momentum parallel to $\vec{\mathbb{B}}$ is conserved. The projection of the particle's motion in a plane perpendicular to $\vec{\mathbb{B}}$ is still circular, and equation E7.9 still yields the correct radius for this circular projection as long as we replace p by $p \sin\theta$, where θ is the angle between $\vec{p}$ and $\vec{\mathbb{B}}$ (see problem E7S.14). In this situation, then, the particle moves in a helical path around and along the direction of $\vec{\mathbb{B}}$ (see figure E7.10).

Therefore the most general equations describing the motion of a free particle in a uniform magnetic field are

Equations describing the motion of a free particle in a uniform magnetic field

$$R = \frac{pc \sin\theta}{|q|\mathbb{B}} = \frac{p \sin\theta}{|q|B} \tag{E7.11a}$$

$$T = \frac{2\pi mc}{|q|\mathbb{B}\sqrt{1 - v^2/c^2}} = \frac{2\pi m}{|q|B\sqrt{1 - v^2/c^2}} \tag{E7.11b}$$

$$p_{\parallel} = \text{constant} \tag{E7.11c}$$

Purpose: These equations describe features of the helical motion of a particle with charge q and (possibly relativistic) momentum $\vec{p}$ moving in a reasonably uniform magnetic field described by $\vec{\mathbb{B}}$ or $\vec{B}$.
Symbols: R is the radius of the circular part of the helical motion, T is the time required to go once around the helix, θ is the angle between $\vec{p}$ and $\vec{\mathbb{B}}$, v is the particle's speed, c is the speed of light, $p_{\parallel}$ is the component of the particle's momentum parallel to $\vec{\mathbb{B}}$, and $B \equiv \mathbb{B}/c$ is the magnetic field strength expressed in teslas.
Limitations: The magnetic field must be nearly uniform over the region spanned by one cycle of the helix.
Note: The period T is essentially independent of v for $v \ll c$.

Applications in particle physics

These equations have a number of applications in technology and research. The most powerful particle accelerators use magnets to constrain particles to follow a circular trajectory around a closed ring. Because particles travel many times around the ring as they are being accelerated, a circular accelerator can do the same job as a much longer (and thus more expensive) linear accelerator.

After the accelerated particles reach their final speeds, they are deflected by magnetic fields into a target surrounded by a particle detector. Physicists usually place such detectors in a magnetic field because one can

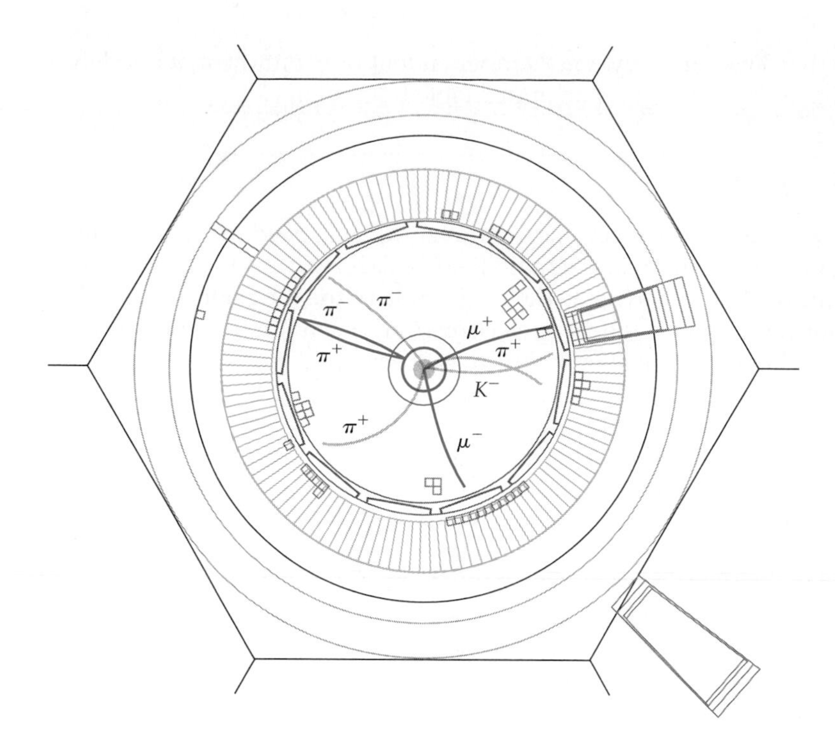

Figure E7.11
A computer reconstruction of the trajectories of particles in a particle physics experiment. Note how each particle curves to the right or the left depending on the sign of its charge. We can also determine a particle's relativistic momentum from the radius of its trajectory.

determine the particle's momentum from the radius of its trajectory and the sign of its charge from the direction of the trajectory's curvature (see figure E7.11).

Applications to the aurora

The fact that a charged particle will follow a helical path that tracks the local direction of the magnetic field is important for understanding the phenomenon of the **aurora** (commonly known as the *northern lights,* though they are visible near the south pole as well). On clear nights, people near the earth's magnetic poles often see ghostly curtains of light hanging in the sky. This is caused by charged particles from the sun that get captured by the earth's magnetic field. Because these particles then follow helical paths along the local direction of the earth's magnetic field, they get channeled into the upper atmosphere where the magnetic field lines are most vertical, that is, near the earth's magnetic poles. Collisions between these energetic particles and atoms in the upper atmosphere cause the atoms to glow, creating the auroral display (see figure E7.12).

E7.5 The Magnetic Force on a Wire

If charge is *really* moving in a current-carrying wire (as claimed in chapter E4), then these moving charges ought to respond to a magnetic field. According to equation E7.6, the magnetic force on a single electron of charge q moving through this wire with drift velocity $\vec{v}_d$ is

$$\vec{F}_m = q\frac{\vec{v}_d}{c} \times \vec{\mathbb{B}} \tag{E7.12}$$

Any impulse that this interaction delivers to electrons in a wire is almost instantly transmitted to the wire itself when the electrons next collide with

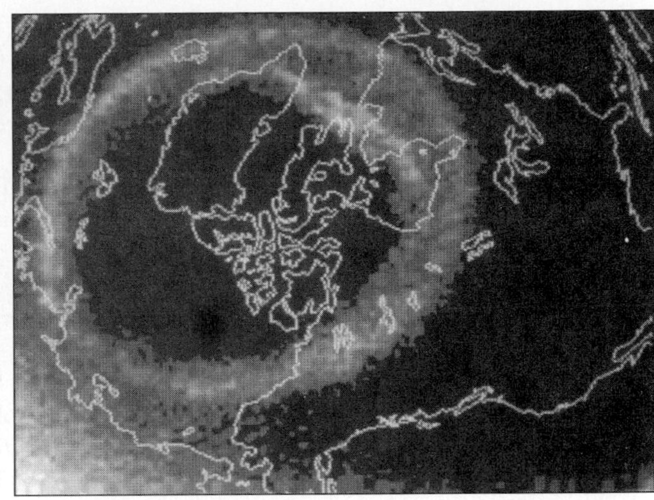

Figure E7.12
(a) The aurora as viewed from the ground. (b) The aurora as viewed from space. This ultraviolet photograph taken by a satellite shows a circular glowing region around the earth's magnetic pole where charged particles from the sun are entering the earth's upper atmosphere.

lattice atoms. This means that the total magnetic force exerted on a straight wire segment is equal to the force given in equation E7.12 multiplied by the number of flowing electrons in the wire segment (assuming that the electrons all have essentially the same drift velocity). A segment of wire of length L and cross-sectional area A has a volume LA; so if n is the number of free electrons per unit volume in the wire, the total number of flowing electrons in the wire segment is nLA. Thus the total magnetic force on the segment is

$$\vec{F}_{m,\text{seg}} = nLAq\frac{\vec{v}_d}{c} \times \vec{\mathbb{B}} \qquad \text{(E7.13)}$$

Since $\vec{I} = nqA\vec{v}_d$ (see equation E4.10), this becomes

The magnetic force on a wire segment

$$\vec{F}_{m,\text{seg}} = \frac{L}{c}\vec{I} \times \vec{\mathbb{B}} = L\vec{I} \times \vec{B} \qquad \text{(E7.14)}$$

Purpose: This equation describes the magnetic force $\vec{F}_{m,\text{seg}}$ exerted on a straight segment of wire of length L in a magnetic field described by $\vec{\mathbb{B}}$ or $\vec{B}$.

Symbols: $\vec{I}$ is the direction of the steady conventional current in the wire.

Limitations: The wire must be straight and thin, and L must be small enough that $\vec{\mathbb{B}}$ is nearly uniform over the segment.

If we need to compute the total magnetic force on a length of curved wire (or even an entire closed loop of wire), we can divide the wire up into sufficiently short straight segments, use equation E7.14 to calculate the force on each, and then sum.

Equation E7.14 makes a specific and quantitative prediction about the force that a magnetic field exerts on a current-carrying wire. One of the things it implies is that the magnetic force on a segment of wire acts *perpendicular* to the direction of the conventional current $\vec{I}$ (which always points along the wire) in a direction indicated by the usual right-hand rule for the cross product. The fact that wires are observed to experience a force of exactly this description is solid evidence that charges really *are* moving in a current-carrying wire.

Exercise E7X.6

Imagine that you hold a horizontal length of wire carrying 30 A of current to your right in a uniform magnetic field that is directed vertically upward and has a magnitude of 3 MN/C (0.01 T). What are the magnitude and direction of the magnetic force acting on each centimeter of the wire in the field?

Application to electric motors

Electric motors are an important application of this effect. Figure E7.13 shows an electric motor stripped down to its two most basic elements, which are displayed here in simplified form as a pair of permanent magnet poles that create an approximately uniform horizontal magnetic field between them, and a rectangular current-carrying loop of wire that is free to rotate about a horizontal axis (we'll worry about how to get current to the loop shortly). If the loop is oriented as shown in figure E7.13a and conducts current in the direction shown, the two loop legs parallel to the axis of rotation each experience a vertical magnetic force (as you can easily check with your right hand) that seeks to twist the loop *clockwise*. When the loop reaches the vertical orientation shown in figure E7.13b, these forces no longer twist the

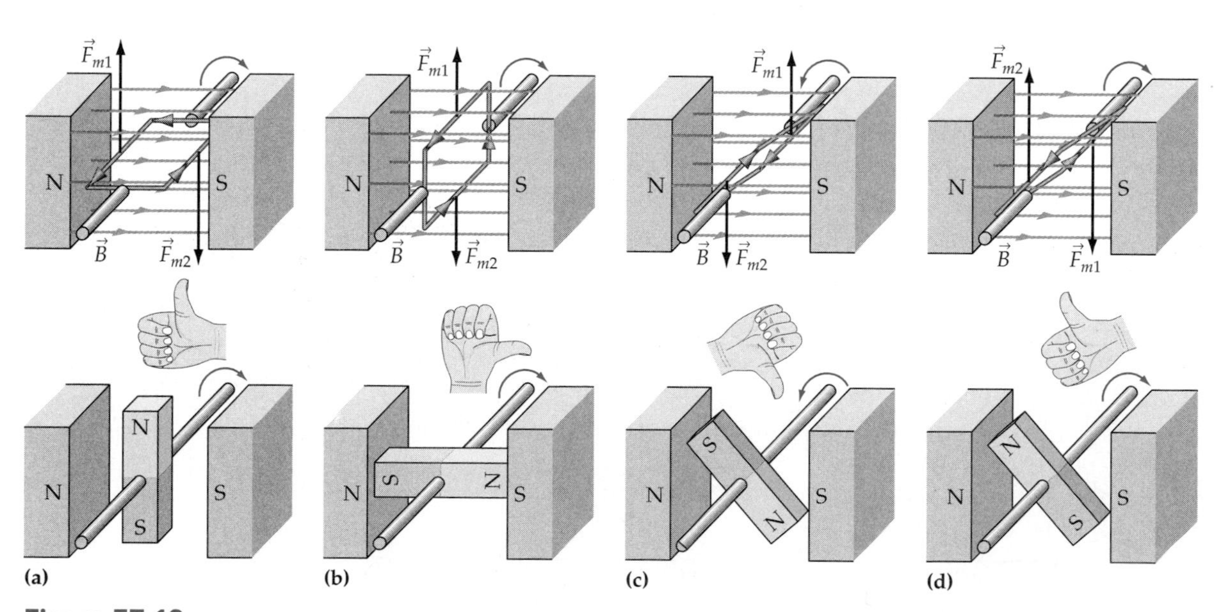

Figure E7.13
The basic elements of an electric motor are a set of permanent magnets that create a magnetic field and a current-carrying loop of wire that is free to rotate in that field. The loop behaves in the magnetic field as if it were a magnet oriented as shown below each drawing. In the last drawing, the current in the loop has been reversed.

loop but rather try to pull it apart. If the loop were to continue to rotate clockwise, the magnetic forces on these legs would seek to twist the loop counterclockwise (see figure E7.13c). The loop therefore seeks to orient itself in the direction shown in figure E7.13b.

In each of these three orientations, the loop in fact responds to the magnetic field just as the permanent magnet shown below each loop would. Note that if you curl your right fingers around the loop in the direction of current flow, your thumb points toward the north pole of the analogous permanent magnet (see figure E7.13a): we will call this the **loop-to-magnet rule.**

Exercise E7X.7

Explain why the magnetic forces on the legs perpendicular to the axis of rotation do nothing to aid or hinder rotation.

Such a loop would not make a good motor, because it gets stuck in the orientation shown in figure E7.13b. The trick to making a motor is to come up with a way to reverse the direction of the current just as the loop passes through the position shown in figure E7.13b, so that magnetic forces on the loop *continue* to twist it clockwise (see figure E7.13d) instead of counterclockwise as in figure E7.13c. If we reverse the loop's current every half rotation, the loop will then rotate endlessly.

There are a number of ways to perform this reversal. The most straightforward is illustrated in figure E7.14. In this scheme, we attach the loop's ends to two half-rings that each wrap halfway around the motor's axle (these half rings, which together are called the motor's **commutator,** are shown in an end view in figure E7.14a). A stationary **brush** touches each ring: as the

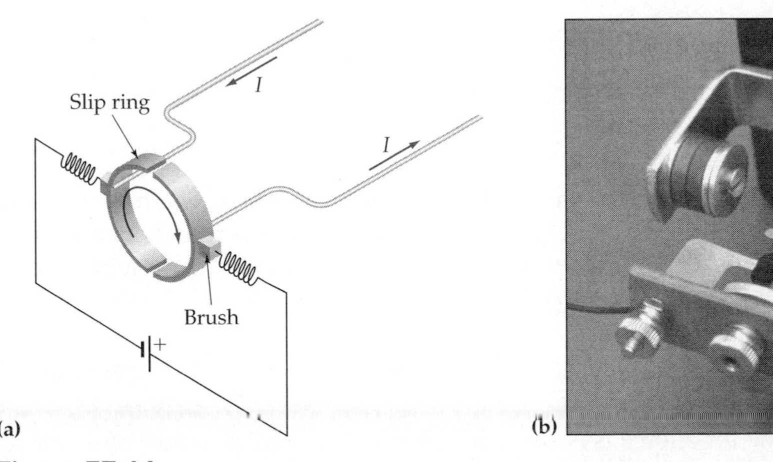

Figure E7.14
(a) A *commutator* can both feed current to a rotating loop and reverse the direction of that current each half turn so that the loop always continues to rotate clockwise.The brushes slide on the commutator's metal half rings as the axle turns but, if designed appropriately, always maintain good electrical contact with those rings, always feeding current into the loop's right leg and taking it out of the right leg. (b) An actual photograph of an electric motor. The commutator is at the lower left end of the motor's axle, and the brushes in this case touch the commutator on its top and bottom instead of on its sides. The loop here consists of many turns of wire wound around an iron bar.

axle turns, the half-rings rotate under the brushes, but maintain good electrical contact with them. Note that if we connect the two brushes to the two terminals of a battery as shown, current will always flow into whichever leg is rightmost and out whichever leg is leftmost, ensuring that the magnetic forces on the loop always seek to turn it clockwise.

We can make our motor more powerful by wrapping N sequential turns of wire around the loop instead of a single turn. In such a case, the total magnetic force on the N strands of wire in each leg will be N times greater than on a single-strand leg carrying the same current. Figure E7.14b shows an actual electric motor with a commutator and a rotating loop that has multiple turns of wire.

Example E7.1

Problem Consider a motor whose rotating loop consists of N turns of wire wound in a rectangular loop of length L and width W. If the loop carries current I as it rotates in a magnetic field of strength $\vec{B}$, how much energy does it produce per revolution?

Translation Figure E7.15 shows the situation.

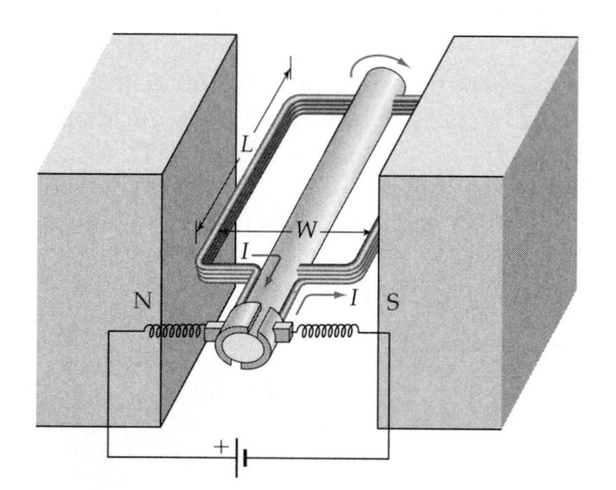

Figure E7.15
A picture of the loop described in example E7.1, with some useful symbols defined.

Model Note that no matter how the loop is oriented, the current flowing in the two legs parallel to the axis of rotation is perpendicular to the magnetic field. According to equation E7.14, the magnitude of the magnetic force on each of N strands in each of these legs is therefore

$$F_m = \frac{L}{c}\text{mag}(\vec{I} \times \vec{B}) = \frac{L}{c} I B \sin 90° = \frac{LI}{c} \vec{B} \qquad \text{(per strand)} \qquad \text{(E7.15)}$$

The total force is N times this. Now, as the loop rotates one-half revolution, each leg moves a total displacement $\Delta\vec{r}$ that is parallel to the force and whose magnitude is equal to the loop's width W. If the magnetic field is uniform over the size of the loop, the magnetic force on the legs will be constant over this displacement, so the total k-work flowing into each leg during the displacement will be

$$[\Delta K] = \sum_{\text{all steps}} \vec{F}_m \cdot d\vec{r} = \vec{F}_m \cdot \left(\sum_{\text{all steps}} d\vec{r} \right)$$
$$= \vec{F}_m \cdot \Delta\vec{r} = F_m W \cos\theta = F_m W \qquad \text{(E7.16)}$$

since the magnetic force is parallel to the total displacement $\Delta\vec{r}$. The k-work contributed during a full revolution is simply double the amount contributed per one-half revolution.

Solution Therefore, the total energy contributed to the loop per revolution is

$$[\Delta K] = 2F_{m,\text{tot}}W = 2\frac{2NLI}{c}\mathbb{B}W = 4NAI\frac{\mathbb{B}}{c} = 4NAIB \qquad \text{(E7.17)}$$

where $A = LW$ is the loop's area.

Evaluation The result is proportional to the number of turns, the current, and the magnetic field strength, which all makes sense. The result also has SI units of (m²)(C/s)(N/C)(s/m) = N · m = J, which is appropriate.

Exercise E7X.8

Imagine that the loop in a certain motor consists of a coil with 150 turns spanning an area of 2.0 cm². Imagine that in operation, the loop carries an average current of about 1.0 A and rotates at 1500 rev/min in a magnetic field of 200 MN/C. How much mechanical power does the motor provide?

TWO-MINUTE PROBLEMS

E7T.1 Which of the following statements describes an observation that represents strong physical evidence that magnetic poles are *not* the same as electric charges?

A. Poles are described as being *north* and *south* while charges are described as being *positive* and *negative.*
B. A magnet does not exert a force on a motionless charge.
C. Magnetic poles always come in pairs.
D. A and B.
E B and C.
F. A, B, and C.
T. Magnetic poles *are* the same as electric charges.

In each of problems E7T.2 through E7T.4, a particle whose charge has the specified sign enters a region of space where the magnetic field has the direction shown. What is the initial direction of the magnetic force on the particle?

In each of problems E7T.5 through E7T.7, determine the direction of the magnetic field that is causing the charged particle to experience the magnetic force shown.

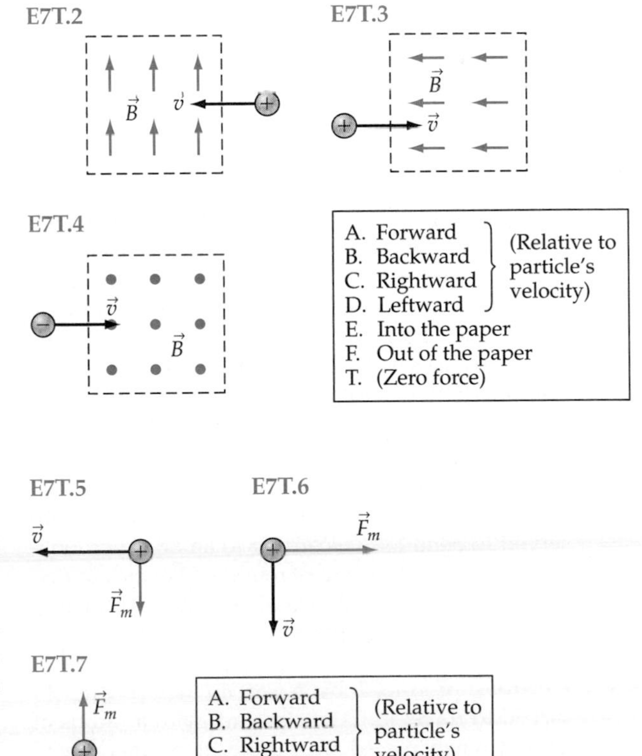

E7T.8 Imagine that you place a magnet in a *uniform* magnetic field (a field whose vectors have the same magnitude and direction at all points). Will the magnetic field exert a net force on the magnet? If so,

what is the direction of this force? (*Hint*: Use the dipole analogy.)
A. Yes, in the direction of the magnetic field.
B. Yes, opposite to the direction magnetic field.
C. Yes, but the direction depends on magnet's orientation.
D. No, the net force on the magnet is zero.

E7T.9 Imagine that you are looking at the face of a CRT. The bright spot indicating where the electron beam hits the face is exactly in the center of the screen. You bring a permanent magnet toward the CRT vertically from above. The magnet is oriented vertically with its north pole downward. Which direction will the spot deflect?
A. Up
B. Down
C. The spot does not deflect
D. Right
E. Left
F. Other (specify)

E7T.10 In a magnetic field oriented vertically downward, an otherwise free particle is observed to move counterclockwise in a horizontal circle when viewed from above. This particle has a
A. Positive charge
B. Negative charge
C. Zero charge
D. One cannot determine the sign of the charge from the information given

E7T.11 Imagine that you hold a short segment of wire vertically in a magnetic field that is directed horizontally toward you. If the segment carries an upward conventional current, what is the direction of the magnetic force exerted on the wire?
A. Right
B. Left
C. Up
D. Down
E. Toward you
F. Away from you

E7T.12 A loop facing east carries a counterclockwise current when viewed from the east. It is placed in a magnetic field whose field vectors point vertically upward. The loop is free to rotate around any axis. How will the loop respond to the magnetic field? (*Hint:* Find the orientation of this loop's equivalent magnet.)
A. When viewed from above, it will twist 90° clockwise to face south.
B. When viewed from above, it will twist 90° counterclockwise to face north.
C. When viewed from the south, it will twist 90° clockwise to face downward.
D. When viewed from the south, it will twist 90° clockwise to face upward.
E. The loop will be unaffected by the magnetic field.
F. It will respond in some other way (specify).

HOMEWORK PROBLEMS

Basic Skills

E7B.1 Imagine that you are looking at the face of a CRT. The bright spot indicating where the electron beam hits the face is exactly in the center of the screen. You bring a permanent magnet toward the middle of the CRT horizontally from the right with its south pole closest to the CRT. Which direction will the spot deflect? Explain.

E7B.2 A sample of an unknown radioactive substance is observed to emit particles. In a magnetic field oriented vertically upward, these particles are observed to bend left (according to an observer looking along their direction of motion). Are the emitted particles positively or negatively charged, or is it impossible to tell? Explain.

E7B.3 Electrons produced by a certain device are observed to travel in a circular path with a radius of 2.0 cm when placed in a uniform magnetic field whose strength is 10 MN/C. What is the speed of the electrons emitted by this device? (You may express this speed as a fraction of c.)

E7B.4 Protons produced by a certain device are observed to travel in a circular path with a radius of 1.0 m when placed in a magnetic field whose strength is 0.67 T ($\vec{B} = 200$ MN/C). What is the speed of the protons emitted by this device? (You may express this speed as a fraction of c.)

E7B.5 Electrons moving perpendicular to the direction of a magnetic field are observed to circulate about the field direction at a frequency of 150 MHz. What is the strength of the magnetic field in newtons per coulomb, in teslas, and in gauss?

E7B.6 How long would an electron moving perpendicular to the earth's magnetic field take to complete one orbit around the magnetic field direction (assuming that the electron is traveling at a nonrelativistic speed)?

E7B.7 A wire carries a current of 300 A. What is the maximum possible force per unit length that could be exerted on the wire by the earth's magnetic field? Why is what you found a *maximum* possible value (what might make the actual force smaller)?

E7B.8 A wire has a mass of 10 g/m. What is the *minimum* amount of current it would have to carry if the magnetic force exerted by a horizontal 30 MN/C magnetic field on a horizontal length of this wire were able to lift the wire off the ground?

E7S.9 Imagine that the magnetic field in a particle detector chamber points vertically downward and has a magnitude of 500 MN/C. Imagine that a collision in the target produces (among other things) a subatomic particle whose track bends to its right as we look along the direction of its motion. What is the sign of the particle's charge? If the radius of curvature of this particle's trajectory is $R = 85$ cm and we assume that the *magnitude* $|q|$ of the particle's charge is $e = 1.6 \times 10^{-19}$ C, what is the particle's relativistic momentum p?

E7B.10 A square loop 1.0 cm on a side conducts a current of 3.0 A clockwise when viewed from the east. The loop is placed in a vertically downward magnetic field having a strength of 20 MN/C. How will the loop seek to orient itself in this magnetic field? How much energy does it take to flip the loop 180° from this preferred orientation?

E7B.11 A circular coil consisting of 60 turns of wire that carry a current of 1.5 A is placed in a horizontal magnetic field with strength 50 MN/C. After experimenting with various orientations of the loop, we find that it takes 2.0 J of energy to turn it 180° from the loop's preferred orientation. What is this loop's radius?

Synthetic

E7S.1 Imagine you are looking at the face of a CRT in an airplane. The plane is initially flying south at the equator, and thus approximately opposite to the earth's magnetic field. The electron beam moves in the same direction as the plane, and the electron beam hits the CRT face exactly in the center of the screen. The plane then turns 90° to the east. In what direction will you see the spot shift due to the effects of the earth's magnetic field? Explain.

E7S.2 Your company is designing a desktop-sized proton accelerator, using a 500 MN/C superconducting magnet to hold the protons in a circular path. Estimate the maximum kinetic energy that your accelerator can give protons. Express your answer in the conventional particle physics unit of *electron-volts*, where 1 eV $= 1.6 \times 10^{-19}$ J. For comparison, the accelerator at the Fermi National Laboratory can accelerate protons to energies greater than 1 TeV. (*Hint:* Assume that the protons are nonrelativistic, do the calculation, and then check that assumption.)

E7S.3 The aurora is caused by electrons and protons from the sun spiraling in along the direction of the earth's magnetic field. The typical speed of the electrons involved is about $0.003c$, and the speed of protons is about $1.5 \times 10^{-4}c$. What are the *maximum* radii of the circular part of these particles' motions around the field direction as they spiral into the atmosphere? What are their orbital frequencies? The magnitude of near the earth is approximately 0.5 gauss.

E7S.4 It is possible to use crossed electric and magnetic fields to construct a *velocity selector*, a device that only passes charged particles having a certain velocity. One way of constructing such a velocity selector is shown in figure E7.16: ions are sent through a region of space where there is a uniform magnetic field (directed upward out of the page) and a uniform electric field acting in the direction shown (upward in the plane of the page). Only ions having a specific speed will be able to travel in a straight line through this apparatus. If the magnetic field strength is $\mathbb{B} = 3.0$ MN/C, what electric field strength would you want to select ions having a speed of exactly $0.01c$? Does your answer depend on the mass of the ion? Does it depend on the sign of the ion's charge? Does it depend on the magnitude of the ion's charge? (Ignore gravity.)

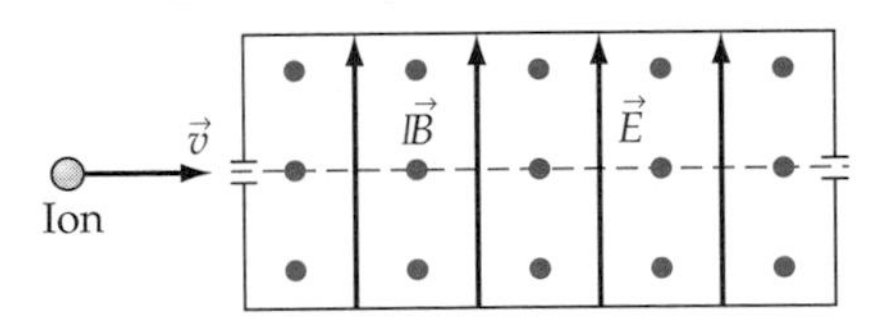

Figure E7.16
A schematic diagram of a velocity selector (see problem E7S.4).

E7S.5 A *mass spectrometer* is a device that uses a magnetic field to sort atoms by mass. In this device, atoms whose masses are to be determined are *ionized* by stripping off one electron. They are then sent through a velocity selector (see problem E7S.4) that selects ions only with a very specific speed v, and these ions are sent into a region of space filled with a known uniform magnetic field $\vec{\mathbb{B}}$ perpendicular to $\vec{v}$. The field causes each atom to follow a circular path whose radius is proportional to the atom's mass. Atoms with different masses will therefore follow somewhat different circular paths and thus end up at different places on a photographic plate, as shown in figure E7.17. Imagine that we give N_2^+ ions, O_2^+ ions, and NO^+ ions the same velocity of 30.0 km/s and then send them into a mass spectrometer where the magnetic field strength is $\mathbb{B} = 8.5$ MN/C. How far would the spot on the photographic plate be from the entry point for each ion, assuming that each ion completes one-half of an orbit, as shown in the figure? The atomic mass of a nitrogen atom is 14.0031 amu and oxygen is 15.9949 amu, where 1 amu $= 1.6605 \times 10^{-27}$ kg.

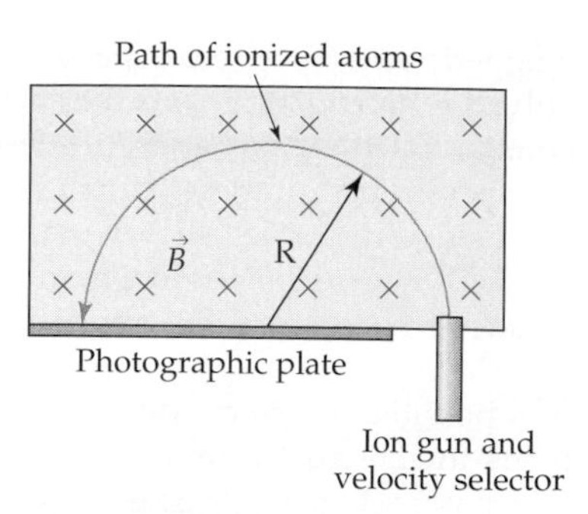

Figure E7.17
A schematic diagram of a mass spectrometer (see problem E7S.5).

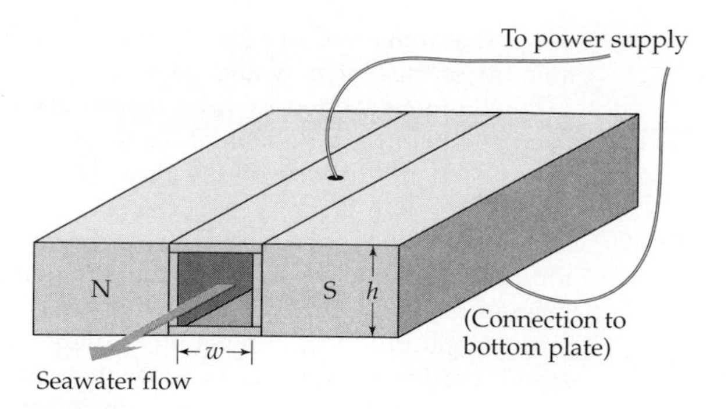

Figure E7.19
A schematic diagram of a magnetohydrodynamic boat drive (see problem E7S.8).

E7S.6 Your college physics department has a garage sale, and you pick up a particle accelerator at a *great* price. The sellers said that they *thought* that the beam produced by the accelerator was an electron beam, but it might be a proton beam. They give you an unmarked bar magnet, a piece of wire, and a battery, and they tell you to check it yourself. Carefully and completely describe how you could use these items to determine whether the particles emerging from your new accelerator are electrons or protons charged. (Assume that you can see the beam emerging from the accelerator.)

E7S.7 A long, straight wire having a mass of 0.025 kg/m is suspended by threads. When the wire carries a 20-A current, it experiences a horizontal magnetic force that deflects it to an equilibrium angle of 10° (see figure E7.18). What are the magnitude and direction of the magnetic field? (Assume that the field vectors are perpendicular to the wire.)

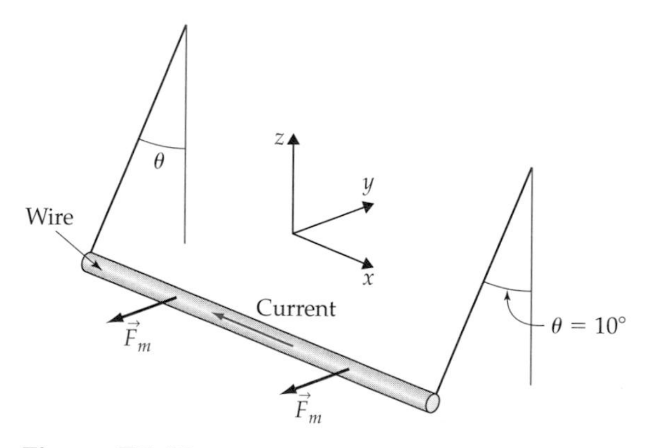

Figure E7.18
The situation described in problem E7S.7.

E7S.8 One can use the fact that seawater conducts electricity to construct a *magnetohydrodynamic* boat engine that uses magnetic forces (instead of a propeller) to thrust water backward. (This would be advantageous in creating very quiet submarine drives.) An example of such an engine is illustrated in figure E7.19. The engine consists of a channel through which seawater flows. The sidewalls of the channel are powerful permanent magnets that create a strong magnetic field perpendicular to the flow of water. Electrodes on the top and bottom faces of the channel drive a very large current through the seawater as it flows through the channel. For the sake of argument, assume that the channel has a width and height of $w = h = 50$ cm and a length of $L = 2.0$ m; the average magnetic field strength in the channel is 1 T (a *large* magnetic field); and we want to exert a total thrust force of $F_{\text{Th}} = 100$ N (22 lb) on the water flowing through the channel. Assume that the magnetic field and the current density in the water are uniform.

(a) Estimate the total current that must flow through the water to obtain this thrust.
(b) In what direction must the current flow to push the water in the direction shown?
(c) Use the data in table E4.1 to estimate the total resistance of the seawater in the channel.
(d) How much power is wasted in the form of thermal energy to get this thrust?
(e) Why do you think drives like this are not in common use?

E7S.9 A bar magnet oriented with its long axis aligned with the direction of a nonuniform magnetic field produced by another bar magnet will experience a net force toward that magnet, as shown in figure E7.20a. Explain why, using an electric dipole analogy. Also explain why a current-carrying loop oriented analogously will *also* experience a net force toward the magnet (figure E7.20b). This is an example of how current loops and bar magnets behave similarly.

E7S.10 Imagine that we construct a motor using a coil consisting of 300 turns of wire wound in the form of a square 2.5 cm on a side. The permanent magnets in the motor create a field with strength of 40 MN/C in which the coil rotates. We would like the motor

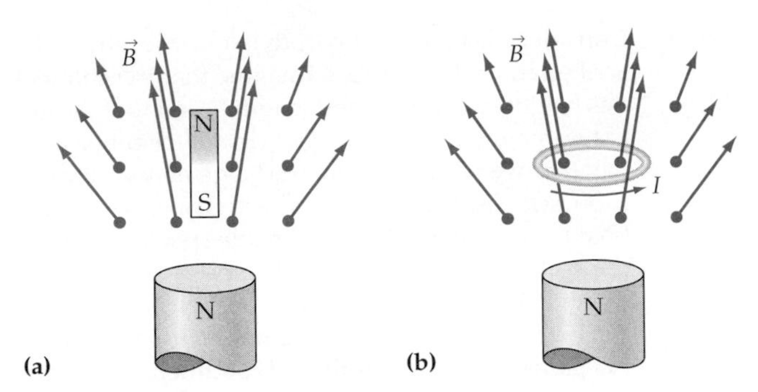

Figure E7.20
The bar magnet and the current loop both experience downward net forces here. Why? (See problem E7S.9.)

to be able produce 5.0 W of mechanical energy when it is turning at 1200 rev/min. How much current do we need to supply to the motor?

E7S.11 Imagine constructing a motor that would operate in the earth's magnetic field instead of using permanent magnets. Let's say that we were satisfied if such a motor generated a paltry 1 W of power while turning at a (very rapid) rate of 100 rev/s. Assume also that the rotor contains 1000 turns of wire, and that the wire carries a healthy 1 A of current.

(a) If the loop were a square loop, about how big would it have to be?

(b) Does such a motor seem practical to you? Discuss why or why not.

E7S.12 Consider a current-carrying rectangular loop that is free to rotate in a magnetic field, as shown in figure E7.13. For any current-carrying loop, we can define *magnetic moment* vector $\vec{\mu}$ whose direction is perpendicular to the loop's face in the direction indicated by your right thumb if you curl your right fingers in the direction of current flow (that is, the direction indicated by the loop rule) and whose magnitude is $\mu = NAI$, where N is the number of turns in the loop, A is its area, and I is the current it carries. In this problem, we will show that we can use the vector $\vec{\mu}$ to write a simple expression for the torque that a loop experiences in a magnetic field.

(a) Draw a cross-sectional side view of this loop (with the axis of rotation perpendicular to the plane of the drawing) when the loop is oriented as shown in figure E7.13d. Draw a vector representing this loop's magnetic moment vector $\vec{\mu}$, and show the angle θ that this vector makes with the magnetic field vectors $\vec{\mathbb{B}}$ in the loop's vicinity. Also draw vectors representing the magnetic force exerted on each of the loop legs that are parallel to the axis of rotation.

(b) If we apply a force $\vec{F}$ to an object at a point whose position relative to the origin is $\vec{r}$, then the torque that force exerts on the object is given by $\vec{\tau} = \vec{r} \times \vec{F}$ (see chapter C13). In our case, we will take the origin to be the axis of rotation. The magnitude of this torque is $\tau = rF\sin\phi$, where ϕ is the angle between $\vec{r}$ and $\vec{F}$. In this case, argue that $\phi = \theta$, that the magnitude of the torque exerted on each of the two loop legs parallel to the axis of rotation is

$$\tau_{\text{leg}} = \frac{1}{2}W\left(\frac{NLI}{c}\right)\mathbb{B}\sin\theta \tag{E7.18}$$

(c) By considering the directions of the torques on these legs, argue that they add, and that the total magnitude of the torque on the loop is given by the simple formula

$$\vec{\tau} = \frac{\vec{\mu}}{c} \times \vec{\mathbb{B}} = \vec{\mu} \times \vec{B} \tag{E7.19}$$

E7S.13 Set up the Newton computer program to model the situation where a particle's acceleration has magnitude proportional to its speed and perpendicular to its velocity, as in the case of a free particle moving in a uniform magnetic field. (For the sake of simplicity, let us arbitrarily choose the constant of proportionality to be 1.0 s^{-1}.) Submit printouts that demonstrate that in this situation (a) the particle moves in a circular trajectory irrespective of its speed, (b) the radius of its trajectory is proportional to its speed, and (c) the period of the particle's trajectory is independent of its speed.

E7S.14 Consider an otherwise free particle with charge q moving in a uniform magnetic field whose field vectors are $\vec{\mathbb{B}}$. For the sake of simplicity, let's orient our reference frame so that the magnetic field points entirely in the z direction.

(a) The component definition of the cross product specifies that

$$\vec{u} \times \vec{w} = \begin{bmatrix} u_y w_z - u_z w_y \\ u_z w_x - u_x w_z \\ u_x w_y - u_y w_x \end{bmatrix} \tag{E7.20}$$

Use this and Newton's second law to show that in this situation, the particle's acceleration is

$$\vec{a} = \begin{bmatrix} a_x \\ a_y \\ a_z \end{bmatrix} = \frac{q}{m}\begin{bmatrix} v_y \mathbb{B} \\ -v_x \mathbb{B} \\ 0 \end{bmatrix} \tag{E7.21}$$

(b) Show that if we integrate these equations from time $t = 0$ to some arbitrary time t, we get

$$b(v_x - v_{0x}) = y - y_0 \quad \text{or} \quad bv_x = y - (y_0 - bv_{0x}) \tag{E7.22a}$$

$$b(v_y - v_{0y}) = -x + x_0 \quad \text{or}$$

$$bv_y = -x + (x_0 + bv_{0y}) \tag{E7.22b}$$

$$v_z = v_{0z} \tag{E7.22c}$$

where x_0, y_0, and z_0 are the components of the particle's position at $t = 0$; v_{0x}, v_{0y}, and v_{0z} are components of its velocity at $t = 0$; and b is a constant whose value you should determine. Note that equation E7.22*c* shows that the particle's velocity component along the field direction is constant.

(c) Our freedom to choose the origin of our reference frame in the xy plane implies that we can find a coordinate system where $y_0 - bv_{0x}$ and $x_0 + bv_{0y}$ are zero. Draw a sketch showing how we might do this in an example case (perhaps one where either v_{0x} or v_{0y} is zero).

(d) The distance that the particle is from the z axis in this *new* coordinate system is $R = (x^2 + y^2)^{1/2}$. Show that equations E7.22 then imply that if v is constant, R is constant, implying that the projection of the particle's motion on the xy plane is a circle of radius R around the z axis.

(e) Argue that this radius is as predicted by equation E7.11*a*. [*Hint:* First argue that $(v_x^2 + v_y^2)^{1/2} = v \sin\theta$.]

Rich-Context

E7R.1 A cathode-ray tube sits on a table, oriented horizontally. The controls have been adjusted so that the electron beam should make a single spot of light exactly in the center of the screen. You observe, however, that the spot is deflected to the right. As a clever scientist, you suspect that your laboratory is in either an electric or a magnetic field, but you do not know which. Carefully describe a set of experiments that only involve reorienting your CRT that will determine (a) that you really are in a field, as opposed to having a broken CRT, and (b) whether the field is an electric field or a magnetic field.

E7R.2 In a *cyclotron* particle accelerator, a charged particle travels in a nearly circular horizontal path in a uniform vertical constant magnetic field $\vec{\mathbb{B}}$. Twice each orbit, the particle passes a place where electric fields give it a small boost in energy ε. As the particle's energy increases, its momentum increases, so its orbital radius increases (it spirals outward). Assuming that the particle is nonrelativistic and ε is small enough that each orbit remains nearly circular, find an expression for the orbital radius R as a function of time.

E7R.3 Consider the magnetohydrodynamic boat drive described in problem E7S.8. Imagine that the channel has height h, width w, and length L, and that the average magnetic field strength in the channel is B. Say that we would like to exert a given thrust force F on seawater that has conductivity σ_c while the boat is moving at some reasonable speed v through the water (assume this v is the speed of the water relative to the channel). Ignore friction.

(a) Find an expression for the rate at which the boat gains k-work from the drive's interaction with the water divided by the rate at which the drive expends thermal energy (this is a measure of the drive's efficiency).

(b) Calculate this ratio for the specific drive parameters described in problem E7S.8.

E7R.4 In section E7.4, we saw that a magnetic field cannot do any k-work on a free charged particle. In example E7.1 in section E7.5, however, we calculate the (nonzero) k-work that a magnetic field contributes to loop turning in the magnetic field. Can a magnetic field contribute k-work or not? Solve this puzzle by considering in detail what happens to charge carriers in a wire (according to the Drude model) as they are deflected by the magnetic field. Exactly how (at the microscopic level) does the magnetic field actually exert a force on a current-carrying wire? Is it really the magnetic field that does the k-work on the wire's atom, or is it some other kind of interaction? If so, what kind of interaction? Explain your responses to all these questions carefully and completely.

Advanced

E7A.1 A *synchrotron* particle accelerator is much like a cyclotron (see problem E7R.2) except that the magnitude of $\vec{\mathbb{B}}$ is increased in synchrony with the particle's energy so that the particle's orbital radius remains constant. Find an expression describing how $\vec{\mathbb{B}}$ must vary with time. Do *not* assume that the particles are nonrelativistic. [*Hint:* First show that

$$\mathbb{B}^2 = \frac{2mc^2(E^2 - m^2)}{q^2 R^2} \tag{E7.23}$$

where E is particle's total relativistic energy $= (p^2 + m^2)^{1/2} = mc^2(1 - v^2/c^2)^{-1/2}$. Take the time derivative of each side and make some appropriate approximations.]

ANSWERS TO EXERCISES

E7X.1 It is awkward to define $\vec{B}$ in terms of the magnetic force exerted on an isolated magnetic pole because we cannot in practice isolate magnetic poles the way that we can isolate electric charges. Every magnetic pole comes with a nearby opposite pole, which will complicate making measurements of the force exerted on the first pole.

E7X.2 If we multiply equation E7.3 by c/c, we get

$$\vec{F}_m = q\vec{v}\left(\frac{c}{c}\right) \times \vec{B} = q\left(\frac{\vec{v}}{c}\right) \times (c\vec{B}) \tag{E7.24}$$

If we compare this to equation E7.4, we see that $\vec{\mathbb{B}} = c\vec{B}$.

E7X.3 The directions are upward, to the left, and into the paper.

E7X.4 In the first and third cases, the directions are into the paper and upward, respectively. The second case is impossible.

E7X.5 Equation E7.9 implies that

$$\begin{aligned}\mathbb{B} &= \frac{mvc}{|q|R} \\ &= \frac{(9.11 \times 10^{-31}\ \cancel{\text{kg}})(3 \times 10^{7}\ \cancel{\text{m}}/\cancel{\text{s}})(3 \times 10^{8}\ \cancel{\text{m}}/\cancel{\text{s}})}{(1.60 \times 10^{-19}\ \text{C})(0.12\ \cancel{\text{m}})} \\ &\quad \times \left(\frac{1\ \text{N}}{1\ \cancel{\text{kg}} \cdot \cancel{\text{m}}/\cancel{\text{s}^2}}\right) \\ &= 430{,}000\ \text{N/C}\end{aligned} \tag{E7.25}$$

Equation E7.10*a* tells us that

$$\begin{aligned} f &= \frac{1}{T} = \frac{|q|\mathbb{B}}{2\pi mc} \\ &= \frac{(1.60 \times 10^{-19}\ \cancel{\text{C}})(430{,}000\ \text{N}/\cancel{\text{C}})}{2\pi(9.11 \times 10^{-31}\ \text{kg})(3 \times 10^{8}\ \text{m/s})} \\ &= 4.0 \times 10^{7}\ \frac{\cancel{\text{N}} \cdot \cancel{\text{s}}}{\cancel{\text{kg}} \cdot \cancel{\text{m}}}\left(\frac{1\ \cancel{\text{kg}} \cdot \cancel{\text{m}}/\cancel{\text{s}^2}}{1\ \cancel{\text{N}}}\right)\left(\frac{1\ \cancel{\text{Hz}}}{1\ \cancel{\text{s}^{-1}}}\right) \\ &\quad \times \left(\frac{1\ \text{MHz}}{10^{6}\ \cancel{\text{Hz}}}\right) = 40\ \text{MHz}\end{aligned} \tag{E7.26}$$

E7X.6 The right-hand rule implies that the force acts directly *toward* you. Since the current is perpendicular to the magnetic field, the sine of the angle between $\vec{I}$ and $\vec{\mathbb{B}}$ is 1, and so the magnitude of the force on a 1-cm segment is

$$\begin{aligned} F_m &= \frac{L}{c}\text{mag}(\vec{I} \times \vec{\mathbb{B}}) = \frac{L}{c}\mathbb{B} \\ &= \frac{(0.01\ \text{m})(30\ \text{A})(3 \times 10^{6}\ \text{N/C})}{3.0 \times 10^{8}\ \text{m/s}} \\ &= \frac{0.0030\ \cancel{\text{m}} \cdot \cancel{\text{A}} \cdot \text{N}}{(\cancel{\text{m}}/\cancel{\text{s}})\cancel{\text{C}}}\left(\frac{1\ \cancel{\text{C}}/\cancel{\text{s}}}{1\ \cancel{\text{A}}}\right) = 0.0030\ \text{N}\end{aligned} \tag{E7.27}$$

E7X.7 In the orientation shown in figure E7.13a, the current in these legs flows parallel to the magnetic field, so there is no force exerted on these legs at all. In the other orientations, the right-hand rule for the cross product implies that the forces on these legs are parallel to the axis of rotation. These forces therefore cannot influence the rotation at all.

E7X.8 According to equation E7.17, each revolution, the k-work done on the coil is

$$\begin{aligned} 4NAI\frac{\mathbb{B}}{c} &= 4(150)(2.0\ \text{cm}^2)(1.0\ \text{A})\frac{200 \times 10^{6}\ \text{N/C}}{3.0 \times 10^{8}\ \text{m/s}} \\ &= 80\ \cancel{\text{cm}^2} \cdot \cancel{\text{A}} \cdot \frac{\cancel{\text{N}}}{\cancel{\text{C}}(\cancel{\text{m}}/\cancel{\text{s}})}\left(\frac{1\ \cancel{\text{m}}}{100\ \cancel{\text{cm}}}\right)^2\left(\frac{1\ \cancel{\text{C}}/\cancel{\text{s}}}{1\ \cancel{\text{A}}}\right) \\ &\quad \times \left(\frac{1\ \text{J}}{1\ \cancel{\text{N}} \cdot \cancel{\text{m}}}\right) = 0.08\ \text{J}\end{aligned} \tag{E7.28}$$

If the motor coil turns at 1500 rev/min, the rate at which energy is released is

$$\frac{1500\ \cancel{\text{rev}}}{\cancel{\text{min}}}\left(\frac{1\ \cancel{\text{min}}}{60\ \cancel{\text{s}}}\right)\left(\frac{0.08\ \cancel{\text{J}}}{\cancel{\text{rev}}}\right)\left(\frac{1\ \text{W}}{1\ \cancel{\text{J}}/\cancel{\text{s}}}\right) = 2.0\ \text{W} \tag{E7.29}$$

E8

Currents and Magnets

▷ Electric Field Fundamentals
▷ Controlling Currents
▽ Magnetic Field Fundamentals
Magnetic Fields
Currents and Magnets
▷ Calculating Static Fields
▷ Dynamic Fields

Chapter Overview

Introduction

In chapter E7, we saw that electric currents respond to magnetic fields. In this chapter, we will see that *magnetic fields can create currents* in moving conductors and that *currents can create magnetic fields.*

Section E8.1: Creating Currents in Moving Conductors

The magnetic forces on charge carriers in a moving conductor can cause the ends of the conductor to become charged, creating a potential difference between the conductor's ends as if it were a battery. Such forces will also drive current through a loop moving through a *nonuniform* magnetic field (see figure E8.2) or if the loop is *rotating* in a magnetic field (see figure E8.3). We can use the latter approach to create an electric **generator.**

Section E8.2: The Magnetic Field of a Moving Charge

In 1820, Hans Oersted discovered that moving charges create magnetic fields. A moving point charge creates a magnetic field described by

$$\vec{B} = \frac{\mu_0}{4\pi}\frac{q}{r_{PC}^2}(\vec{v} \times \hat{r}_{PC}) \quad \text{or} \quad \vec{\mathbb{B}} = \frac{kq}{r_{PC}^2}\left(\frac{\vec{v}}{c} \times \hat{r}_{PC}\right) \tag{E8.4}$$

Purpose: This equation describes the magnetic field vector $\vec{\mathbb{B}}$ or $\vec{B}$ created at a point P by a charge q at point C moving with velocity $\vec{v}$.

Symbols: $\vec{r}_{PC}$ is the position of point P relative to C, $r_{PC} = \text{mag}(\vec{r}_{PC})$, $\hat{r}_{PC}$ is a directional representing the direction of $\vec{r}_{PC}$, c is the speed of light, k is the Coulomb constant, and μ_0 is an empirical constant called the **magnetic permeability:** $\mu_0 = 4\pi k/c^2 = 4\pi \times 10^{-7}\ \text{T}\cdot\text{s}\cdot\text{m/C} = 4\pi \times 10^{-7}\ \text{N}\cdot\text{s}^2/\text{C}^2$.

Limitations: Technically, $\vec{v}$ should be constant and nonrelativistic.

The second version makes it clear that this equation is essentially Coulomb's law with a velocity-dependent cross-product factor. This hints at a deep connection between electricity and magnetism that we will explore in depth in chapter E12.

The cross product implies that $\vec{\mathbb{B}}$ at P is perpendicular to the direction of the charge's velocity *and* to the position vector of P relative to the charge. Qualitatively, if you stick your right thumb in the direction of $q\vec{v}$, your fingers curl in the direction in which the magnetic field vectors go around the direction of $\vec{v}$.

Section E8.3: The Magnetic Field of a Wire Segment

Like the electric field, the magnetic field obeys the superposition principle:

$$\vec{\mathbb{B}} = \vec{\mathbb{B}}_1 + \vec{\mathbb{B}}_2 + \vec{\mathbb{B}}_3 + \cdots \tag{E8.6}$$

Purpose: This is the superposition principle for the magnetic field.

Symbols: $\vec{\mathbb{B}}$ is the total magnetic field at a specified point at a given time, and $\vec{\mathbb{B}}_1, \vec{\mathbb{B}}_2, \vec{\mathbb{B}}_3, \ldots$ are the magnetic field vectors that would be created at that same point and time by moving charges 1, 2, 3, . . . individually.
Limitations: None are known (for fields in a vacuum, anyway).

This means that we can find the magnetic field of an arbitrary wire by dividing the wire into segments of length dL that are small enough to treat as particles. Since one can show that $q\vec{v}$ for a segment carrying current $\vec{I}$ is $\vec{I}\,dL$, the field of a wire is

$$\vec{\mathbb{B}} = \frac{k}{c}\sum_{\text{all } i}\frac{dL}{r_{Pi}^2}(\vec{I}_i \times \hat{r}_{Pi}) \quad \text{or} \quad \vec{B} = \frac{\mu_0}{4\pi}\sum_{\text{all } i}\frac{dL}{r_{Pi}^2}(\vec{I}_i \times \hat{r}_{Pi}) \tag{E8.10}$$

Purpose: This equation describes the magnetic field $\vec{\mathbb{B}}$ produced at a point P by an arbitrarily shaped wire carrying a current I.
Symbols: $\vec{r}_{Pi}$ is P's position relative to the ith segment; $r_{Pi} = \text{mag}(\vec{r}_{Pi})$; $\hat{r}_{Pi} \equiv \vec{r}_{Pi}/r_{Pi}$ is a directional representing the direction of $\vec{r}_{Pi}$; dL and $\vec{I}_i$ are the segment's length and conventional current, respectively; k is the Coulomb constant; and c is the speed of light.
Limitations: This equation strictly applies only to static magnetic fields. Also, dL must be small compared to r_{Pi} and small enough that all segments are essentially straight.
Note: This is the **Biot-Savart law** (pronounced "Bee-oh Sahvahr").

Section E8.4: The Magnetic Field of a Long, Straight Wire

Using the Biot-Savart law, we can show that the magnitude of $\vec{\mathbb{B}}$ at a point a distance r from the nearest point on a long, straight wire carrying a steady current I is

$$\mathbb{B} \approx \frac{2kI}{cr} \tag{E8.15}$$

This approximation is valid if r is much smaller than the distance to any parts of the circuit that are not part of the long, straight segment. The **wire rule** says that if you point your right thumb in the direction of the wire's current, your fingers will curl in the direction of the magnetic field around the wire.

Section E8.5: The Magnetic Field of a Circular Loop

The Biot-Savart law also implies that the magnitude of $\vec{\mathbb{B}}$ at the center of a circular loop of radius R carrying a current I is

$$\mathbb{B} = \frac{2\pi kI}{cR} \tag{E8.16}$$

The magnetic field vectors at other points are harder to compute, but it turns out that far from the loop, the magnetic field looks like that of a small bar magnet aligned with the loop's axis, which in turn is like the electric field of a dipole. The **loop rule** says that if you curl your right fingers in the direction in which current flows in the loop, your thumb indicates the direction of $\vec{\mathbb{B}}$ at the loop's center and the direction from the south to the north poles of the equivalent magnet.

Section E8.6: *All* Magnets Involve Circulating Charges

Most atoms behave as little dipolelike magnets because they contain orbiting or spinning electrons that create current loops. While these atomic magnets are randomly oriented in most materials and thus create no net magnetic field, in **ferromagnetic** materials, one can easily align these atomic magnets to create a permanent magnet. Because such a magnet is simply a sum of atomic current loops, a magnet, like a loop, creates a dipolelike field and responds to an external magnetic field as a dipole would to an electric field.

E8.1 Creating Currents in Moving Conductors

In chapter E7, we saw that a magnetic field can exert magnetic forces on a conductor carrying a current. However, under the right circumstances, magnetic fields can also *create* currents in moving conductors. In this section, we will consider how this is possible.

A metal bar moving in a magnetic field

First consider a metal bar of length L moving with a velocity $\vec{v}$ in a uniform magnetic field whose field vectors at all points have the common value $\vec{\mathbb{B}}$. Assume that the length of the bar, $\vec{v}$, and $\vec{\mathbb{B}}$ are mutually perpendicular, as shown in figure E8.1. As the bar moves, it carries its conduction electrons with it. Therefore, these electrons are moving in the magnetic field, and thus they will experience a magnetic force $\vec{F}_m$ toward the bar's left end whose magnitude is

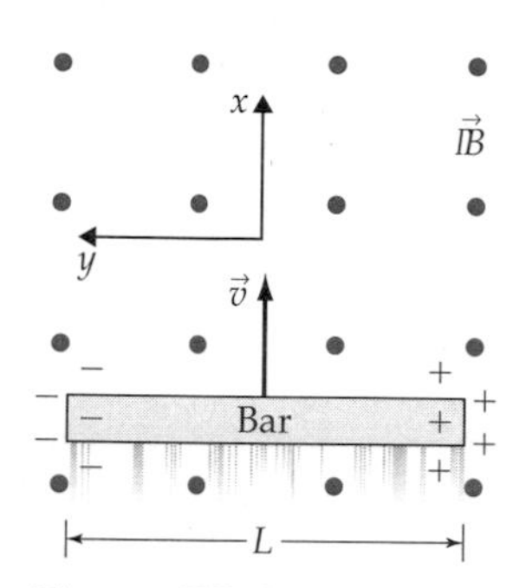

Figure E8.1
The ends of a conducting bar moving perpendicular to a magnetic field become oppositely charged.

$$F_m = \frac{ev\mathbb{B}}{c} = evB \tag{E8.1}$$

where e is the magnitude of an electron's charge.

Exercise E8X.1

Check with your right hand that the electrons are indeed pushed toward the bar's left end.

This magnetic force will cause electrons to pile up on the bar's left end, leaving a deficit of electrons on its right end, as shown in the diagram. This will continue until the charges reach a static equilibrium configuration where they create an electric field that exerts an electrostatic force on every charge in the bar that is equal in magnitude but opposite in direction to the magnetic force. The magnitude of this field (by the definition of $\vec{E}$) will then be $E = F_e/|q| = F_m/e = v\mathbb{B}/c$ everywhere in the bar. According to equation E3.11, this implies that there will be a potential difference between the bar's negative and positive ends of

$$\Delta\phi = \sum_{\text{all steps}} -\vec{E}\cdot d\vec{r} = -\vec{E}\cdot\left(\sum_{\text{all steps}} d\vec{r}\right) = -\vec{E}\cdot\Delta\vec{r} = \frac{v\mathbb{B}}{c}L = vBL \tag{E8.2}$$

Exercise E8X.2

An airplane with a wingspan of 80 m flies at 280 m/s due east in a region where the earth's magnetic field has a magnitude of 0.5 gauss and points nearly directly downward. What is the potential difference between the ends of the plane's wings? Which wing becomes positively charged?

The ends of such a bar therefore become charged with a characteristic potential difference between them, just as the terminals of a battery do. Can we exploit this magnetic battery to drive a current around a circuit? Figure E8.2a illustrates what happens when we attempt to complete the circuit by closing the loop. The problem is that *both* horizontal legs in this circuit act as opposing magnetic batteries, *both* driving electrons toward the circuit's left side. (Note that charges on the two legs parallel to the motion are driven to the left sides of these legs but not along their length, so these two legs neither oppose nor aid the flow of current.) The end result, as illustrated in the

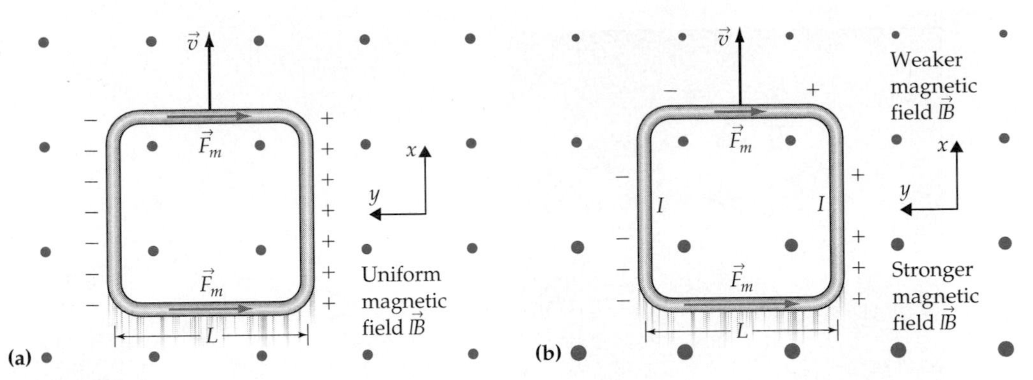

Figure E8.2
(a) A loop moving through a *uniform* magnetic field with its face perpendicular to the field. No current flows in the loop in this case, because electrons experience equal magnetic forces in the front and rear legs. (The gray arrows show the direction in which the magnetic force would act on a positive charge.) (b) If the loop moves in a *nonuniform* field, the stronger magnetic force in one leg pushes conventional current counterclockwise around the loop (colored arrows).

diagram, is that the circuit's left side becomes negatively charged and its right side positively charged, but no net current flows.

Induction in a loop moving in a nonuniform magnetic field

What if the magnetic field is *not* uniform, but decreases in strength in the $+x$ direction as illustrated in figure E8.2b? In this case, the magnetic field's magnitude $\mathbb{B}_r$ at the position of the rear leg is stronger than its magnitude $\mathbb{B}_f$ at the position of the front leg. The unbalanced magnetic forces exerted on the electrons in these legs *will* cause electrons to flow clockwise around the loop, even though electrons are pushed *against* the magnetic force in the front leg. How is this possible? The leftward magnetic force on electrons in the rear leg packs excess electrons onto the surface of the loop's *left* leg. This surface charge pushes electrons forward into the front leg. At first, they are driven back by the opposing magnetic force, but as electrons pile up behind them in the left leg (driven there by the stronger magnetic force exerted on electrons in the rear leg), there is eventually a sufficiently strong electrostatic repulsion from the surface charges on the left leg to push electrons through the front leg against the magnetic force. A steady-state situation is soon established where electrons flow clockwise at a steady rate through the loop. You can show that the net emf (energy per unit charge) contributed by the magnetic force to a charge q flowing around the loop is

$$|\mathscr{E}_{\text{loop}}| \equiv \left| \frac{1}{q} \sum_{\text{loop}} \vec{F}_m \cdot d\vec{r} \right| = \frac{v}{c} |\mathbb{B}_r - \mathbb{B}_f| L = vL|B_r - B_f| \qquad \text{(E8.3)}$$

We say in such a situation that the magnetic field creates an emf that **induces** a current in the loop. Note that current will *only* flow if the field through which the loop moves is *not* uniform, so that $\mathbb{B}_r \neq \mathbb{B}_f$.

Exercise E8X.3

Verify that equation E8.3 is correct.

Generators

Magnetic forces on electrons in a loop *rotating* in a uniform magnetic field can also induce a current to flow through the loop. Imagine that we take the electric motor shown in figure E7.14, connect its brushes to a resistive load of some type, and then compel the loop to rotate clockwise in the magnetic field,

(a)

(b)

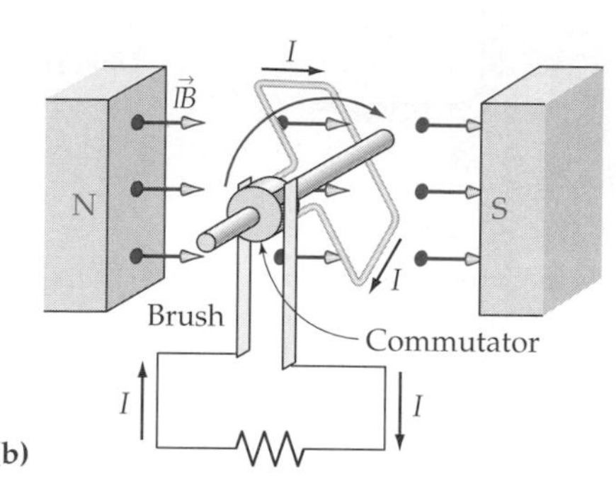

Figure E8.3
A (a) photo and (b) a schematic diagram of an electric generator.

as shown in figure E8.3. Note that as the left leg of the loop moves upward in the magnetic field, the magnetic forces on charge carriers in that leg will drive conventional current away from us. Similarly, as the right leg moves downward, magnetic forces drive current toward us. Thus the left brush will become negative and the right brush positive, and current will flow leftward through the resistor, as if it were connected to a battery. Since the right brush is always connected by the commutator to whichever leg of the loop is rightward (and the left brush likewise to the leftward leg) as the loop turns, current will *always* flow leftward through the resistor. Such a device constitutes a simplified electric **generator.** Generators used by power companies throughout the world employ this general principle to create currents.

We see that magnetic fields can actually create electric currents under suitable conditions. The converse is also true: *electric currents can create magnetic fields.* We will be exploring this idea in the remaining sections of this chapter.

E8.2 The Magnetic Field of a Moving Charge

Oersted's discovery

In the winter of 1819–1820, Hans Christian Oersted, a professor at the university in Copenhagen, gave a course in electrostatics and magnetism. At the time, the latter simply entailed the study of lodestones and other natural magnets, as no one yet knew anything about how magnets affected moving charges. Even so, Oersted had a metaphysical belief in the "unity of all forces" and particularly felt that magnetism might possibly be a variant form of electricity. One day, searching for some way to demonstrate this, Oersted tried before his class an experiment in which he placed a wire above a compass and ran a current through it in a direction perpendicular to the compass needle. He intuitively expected to see the compass needle turn to be more parallel to the flowing current, but (as is the case with so many purportedly great physics demonstrations) nothing happened. After class, it occurred to him to try again with the compass needle initially oriented parallel to the wire. This time (after everyone had gone home, of course) the needle swung decisively away from the current. Oersted's article (July 1820) describing this experiment was the first published indication of the link between electricity and magnetism.

This discovery opened the floodgates for research

The study of magnetism had not advanced at all for more than 200 years (since the publication of Gilbert's *De Magnete* in 1600) at the time of Oersted's discovery. But within weeks of learning of Oersted's discovery in September of 1820 at the Paris Academy of Sciences, Jean Biot (rhymes with "Cleo") and Felix Savart (rhymes with "bazaar") had quantitatively measured the magnitude of the magnetic field strength in the vicinity of a current-carrying wire (using the "quivering compass" method of chapter E7) and discovered that the field at a point near the wire depended on $1/R$, where R is the perpendicular distance between the point and the wire. Within a few short years, Andre Marie Ampère, Michael Faraday, and others had provided to the physics community a fairly complete description of the magnetic fields generated by currents. Within 12 years of Oersted's discovery, Faraday had discovered magnetic induction; within 60 years, James Clerk Maxwell was able to provide a complete description of the behavior of electromagnetic fields, a theory that rapidly led to the discovery of radio waves and prompted Einstein to develop the theory of special relativity. Oersted's discovery thus turned out to be the impetus needed to open the floodgates of research, research which ultimately vindicated his intuition about the unity of electricity and magnetism.

The wire rule

Biot and Savart's empirical measurements of the magnetic field near a long, straight current-carrying wire showed that the magnetic field is directed circularly around the wire, as shown in figure E8.4. Note that in this figure the wire runs perpendicular to the plane of the page: the × indicates that the current it carries flows into the plane of the diagram. The direction of the magnetic field can be determined by a simple right-hand rule: if you point your right thumb in the direction of the conventional current carried by the wire and then curl your fingers around the wire, your fingers will curl in the direction of the magnetic field. Thus if the current is flowing *into* the plane of the paper, as shown in figure E8.4, the magnetic field will curl clockwise; if the current were to flow out of the plane of the paper, the field would curl counterclockwise.

The right-hand rule just described is not the same as the right-hand rule that defines the cross product or the right-hand rule used in chapter E7 to indicate the direction of the north pole of a magnet equivalent to a loop. To help us distinguish these rules, let us call the rule described in figure E8.4 the **wire rule** and the others the **cross-product rule** and the **loop rule.** We will see later that the wire and loop rules, are really just convenient variants of the cross-product rule, which is the fundamental rule behind them all.

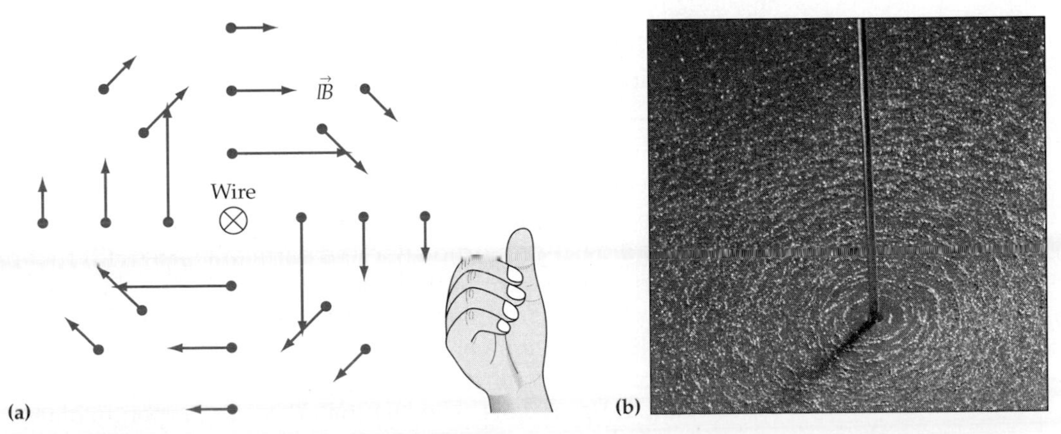

Figure E8.4
(a) The magnetic field around a long, straight wire. The hand illustrates the wire rule linking the directions of the magnetic field and current. (b) Iron filings near a current-carrying wire align themselves with the wire's magnetic field.

Exercise E8X.4

Explain why Oersted's class demonstration failed but his after-lecture experiment succeeded in displaying the magnetic effect of a current.

We also know that currents consist of moving charges. Does it then follow that *any* moving charge will create a magnetic field? For example, will a beam of electrons moving through a vacuum create a magnetic field? How about a static charge on a rotating disk?

Rowland shows that a *static* charge on a moving object can create a magnetic field

In 1875, Henry Rowland at The Johns Hopkins University performed a clever experiment to measure the magnetic field produced by a static charge placed on the perimeter of a rubber disk rotating at 60 turns per second. This was very difficult, because the sensitive compasses used to register the rotating charge's tiny magnetic field ($\approx 10^{-5}$ of the earth's magnetic field) had to be completely shielded from electrostatic effects due to that charge. Even so, Rowland was able to show that the magnetic field produced by the rotating charge on the edge of the disk was *exactly* the same as that produced by the equivalent current in a ring of the same dimensions. This makes it clear that *magnetic fields are indeed generated by charges in motion.* Subsequent experiments with electron beams and the like have underlined the truth of this statement.

Figure E8.4 suggests that a moving charge produces a magnetic field that goes around the direction of motion in a plane perpendicular to the velocity. Careful measurements indicate that the strength of the magnetic field generated by a moving point charge is proportional to the charge's speed and (like the electric field of a point charge) inversely proportional to the square of the distance that one is from the charge:

The magnetic field created by a moving point charge

$$\vec{B} = \frac{\mu_0}{4\pi}\frac{q}{r_{PC}^2}(\vec{v} \times \hat{r}_{PC}) \quad \text{or} \quad \vec{\mathbb{B}} = \frac{\mu_0 c^2}{4\pi}\frac{q}{r_{PC}^2}\left(\frac{\vec{v}}{c} \times \hat{r}_{PC}\right) \qquad \text{(E8.4}a\text{)}$$

Purpose: This equation describes the magnetic field vector $\vec{\mathbb{B}}$ or $\vec{B}$ created at a point P by a charge q at point C moving with velocity $\vec{v}$.

Symbols: $\vec{r}_{PC}$ is the position of point P relative to C; $r_{PC} = \text{mag}(\vec{r}_{PC})$; is a directional representing the direction of $\vec{r}_{PC}$; c is the speed of light; and μ_0 is an empirical constant called the **magnetic permeability:** $\mu_0 = 4\pi \times 10^{-7}\ \text{T}\cdot\text{s}\cdot\text{m/C} = 4\pi \times 10^{-7}\ \text{N}\cdot\text{s}^2/\text{C}^2$.

Limitations: Technically, $\vec{v}$ should be constant and nonrelativistic.

The cross product in parentheses may look daunting, but all that it does is to compactly express three important ideas: (1) that the magnetic field strength is indeed proportional to the particle's speed, (2) that the direction of $\vec{\mathbb{B}}$ is perpendicular to both $\vec{v}$ and $\vec{r}_{PC}$, and (3) that the magnitude of $\vec{\mathbb{B}}$ is also proportional to the sine of the angle between $\vec{v}$ and $\vec{r}_{PC}$. The fact that $\vec{\mathbb{B}}$ is perpendicular to $\vec{v}$ means that the field at a point P, for example, must lie somewhere in a plane that is perpendicular to $\vec{v}$ and contains P: this is illustrated in figure E8.4. The fact that $\vec{\mathbb{B}}$ is *also* perpendicular to $\vec{r}_{PC}$ means that $\vec{\mathbb{B}}$ must be perpendicular to the projection of $\vec{r}_{PC}$ on that plane, which in turn means that $\vec{\mathbb{B}}$ points tangent to a circle going around the charge's direction of motion (consistent with the wire rule), as illustrated in figure E8.5.

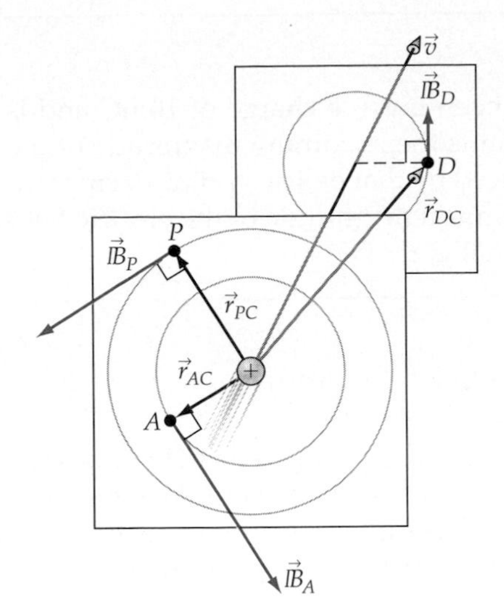

Figure E8.5
The magnetic field vectors created at selected points by a moving charge.

Exercise E8X.5

Convince yourself that the directions of $\vec{\mathbb{B}}$ shown in figure E8.5 are consistent with not only the cross-product rule but also the wire rule. (This shows that the wire rule is really just the cross-product rule in disguise.)

Expressed in the right units, this equation looks eerily like Coulomb's law!

Note that except for the term in parentheses, equation E8.4 is very similar to Coulomb's law for the electric field of a point particle:

$$\vec{E} = \frac{kq}{r_{PC}^2}\hat{r}_{PC} \tag{E8.5}$$

The empirical constant of proportionality $\mu_0/4\pi$ in equation E8.4 is to the magnetic field what the empirical Coulomb constant k is to the electric field in equation E8.5: each expresses how strong a field a given amount of charge creates.

The relationship between μ_0 and k

Now here is something that will really make the little hairs on the back of your neck stand up: if you compute the value of the constant $\mu_0 c^2/4\pi$, you will find that it has exactly the same *value and units as the Coulomb constant* k! Therefore, we could have written the second version of equation E8.4 as

$$\vec{\mathbb{B}} = \frac{kq}{r_{PC}^2}\left(\frac{\vec{v}}{c} \times \hat{r}_{PC}\right) \tag{E8.4b}$$

There is something very deep and eerie going on here: when we express the equation for the magnetic field of a point particle in the right units, we see that it is just a strangely restated velocity-dependent version of Coulomb's law! This is a tantalizing moonlit glimpse of a deep substructure connecting $\vec{E}$ and $\vec{\mathbb{B}}$, a substructure whose bold lines we will see in the broad light of day in chapters E12 and E13. (Note that this vital clue is almost completely obscured when we use the historical SI units for the magnetic field.)

Exercise E8X.6

Show that $k = \mu_0 c^2/4\pi$. (Note that this and $\vec{\mathbb{B}} = c\vec{B}$ make it easy to convert any equation in this chapter involving k and $\vec{\mathbb{B}}$ to one involving μ_0 and $\vec{B}$.)

Exercise E8X.7

Imagine that a tiny Styrofoam ball has been given a charge of 10 nC and is moving in a straight line at 30 m/s. What is the maximum magnitude of the magnetic field quantities $\vec{B}$ and $\vec{\mathbb{B}}$ produced by that ball at a point 10 cm from its path as it passes by? What is the maximum magnitude of the electric field $\vec{E}$ produced by the ball at that point? What is $E / \mathbb{B}$?

E8.3 The Magnetic Field of a Wire Segment

Experiments indicate that, like the electric field, the magnetic field obeys the superposition principle: the total magnetic field vector at a point near a set of *moving* charges $q_1, q_2, q_3, \ldots$ is the simple vector sum of the magnetic field vectors that would be produced at that point by each individual moving charge acting alone:

The superposition principle for the magnetic field

$$\vec{\mathbb{B}} = \vec{\mathbb{B}}_1 + \vec{\mathbb{B}}_2 + \vec{\mathbb{B}}_3 + \cdots \tag{E8.6}$$

Purpose: This equation states that the magnetic field obeys the superposition principle.
Symbols: $\vec{\mathbb{B}}$ is the total magnetic field at a specified point at a given time, and $\vec{\mathbb{B}}_1, \vec{\mathbb{B}}_2, \vec{\mathbb{B}}_3, \ldots$ are the magnetic field vectors that would be created at that same point and time by moving charges 1, 2, 3, . . . individually.
Limitations: None are known.

This provides a means (at least in principle) of calculating the magnetic field produced by a current flowing in an arbitrarily shaped wire. Imagine that we want to know the magnetic field at a point P near a wire of cross-sectional area A carrying a current I, as shown in figure E8.6. The basic trick is to divide the curved wire into tiny segments, each so small that the segment's current can be considered a moving point charge. We then use equation E8.4 to compute the field at P produced by each segment, and we sum over all the segments. In principle, we should be able to calculate the magnetic field of *any* curved wire in this way.

Let's see more carefully how we might do this in practice. Consider dividing the wire into infinitesimal segments of length dL. The volume of each segment is $A\,dL$, so the total amount of moving charge in a segment is

$$q_{\text{seg}} = \rho A\,dL \tag{E8.7}$$

where ρ is the density of moving charge in the wire. The current flowing in this segment of wire (as we saw in chapter E4) is

$$\vec{I} = \vec{J}A = \rho \vec{v}_d A \tag{E8.8}$$

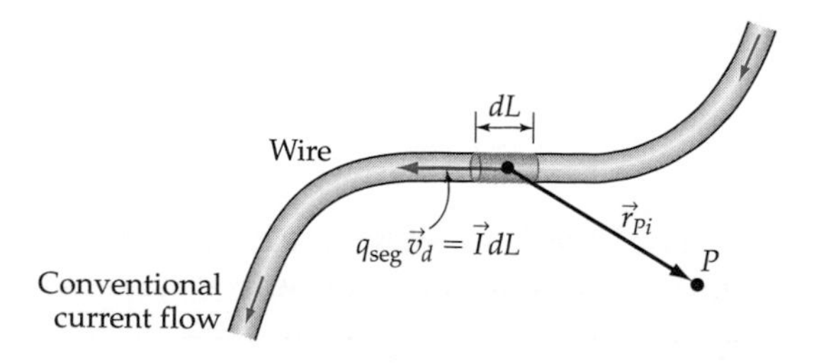

Figure E8.6
To find the magnetic field at point P due to a current flowing in a wire, divide the wire into segments so small that they can be considered point charges, calculate each segment's contribution to the field at P, and sum.

where $\vec{v}_d$ is the drift velocity of the charge carriers. This means that

$$q_{\text{seg}}\vec{v}_d = \rho A\, dL\, \vec{v}_d = (\rho A\vec{v}_d)\, dL = \vec{I}\, dL \tag{E8.9}$$

The magnetic field of an infinitesimal wire segment

This will be true for all segments, so plugging this into the fundamental equation E8.4 and adding over all segments, we find that the total magnetic field vector produced by the wire at point P is

The Biot-Savart law

$$\vec{\mathbb{B}} = \frac{k}{c}\sum_{\text{all } i}\frac{dL}{r_{Pi}^2}(\vec{I}_i \times \hat{r}_{Pi}) \quad \text{or} \quad \vec{B} = \frac{\mu_0}{4\pi}\sum_{\text{all } i}\frac{dL}{r_{Pi}^2}(\vec{I}_i \times \hat{r}_{Pi}) \tag{E8.10}$$

Purpose: This equation describes the magnetic field $\vec{\mathbb{B}}$ produced at a point P by an arbitrarily shaped wire carrying a current I.

Symbols: $\vec{r}_{Pi}$ is P's position relative to the ith segment; $r_{Pi} = \text{mag}(\vec{r}_{Pi})$; $\hat{r}_{Pi} \equiv \vec{r}_{Pi}/r_{Pi}$ is directional representing the direction of $\vec{r}_{Pi}$; dL and $\vec{I}_i$ are the segment's length and conventional current, respectively; k is the Coulomb constant; and c is the speed of light.

Limitations: This equation strictly applies only to static magnetic fields. Also, dL must be small enough that all segments are essentially straight.

Note: This is called the **Biot-Savart law.**

In principle, we can apply this formula to any wire or loop carrying a constant current. In practice, this is difficult except in the simplest cases. Even so, we can extract some useful qualitative principles from this formula. The effect of the ith segment at point P is proportional to $1/r_{Pi}^2$: this means that the magnetic field at point P will primarily be determined by the parts of the wire that are closest to P. Moreover, while information about the *direction* of the field implied in equation E8.10 looks tough to decipher, remember that this cross product turns out to be consistent with the wire rule (which is generally easier to remember and use).

Exercise E8X.8

What is the approximate magnitude of the magnetic fields at points B and C compared to that at point A?

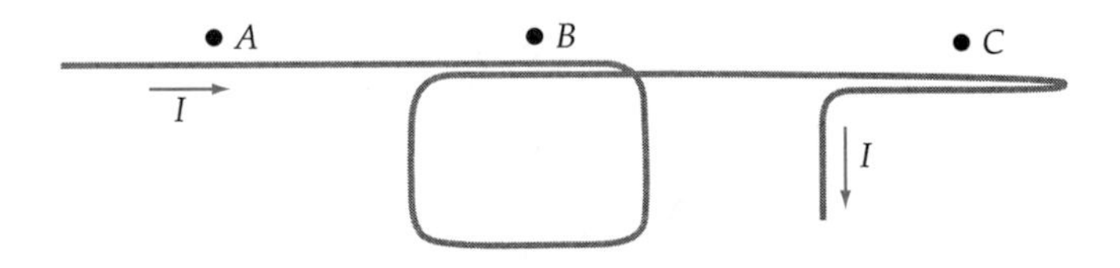

E8.4 The Magnetic Field of a Long, Straight Wire

Application to a long, straight wire

One of the easiest applications of equation E8.10 is to the problem of a *long, straight wire.* Consider a point P a distance r from such a wire. Let us choose our x axis to coincide with the wire (with the $+x$ direction pointing in the same direction as the current), and let $x = 0$ be the point on the wire closest to P. We will divide the wire into infinitesimal segments of length dx. Figure E8.7 illustrates the situation.

Note that for such a wire, both the magnitude and the direction of the current are the same in every segment: $\vec{I}_i = \vec{I}$ for all i. Now, equation E8.10 implies that the magnetic field vector contributed by a given segment will

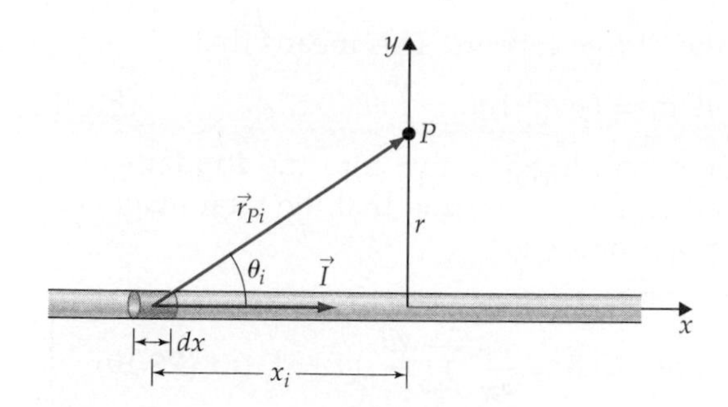

Figure E8.7
Calculating the magnetic field at a point P near a long, straight wire.

point in the same direction as $\vec{I} \times \hat{r}_{Pi}$. Since both $\vec{I}$ and $\hat{r}_{Pi}$ lie in the plane of the figure, the vector $\vec{\mathbb{B}}_i$ contributed by each segment at P must point perpendicular to the plane of the figure (in fact, toward the viewer for the point P shown). Since the $\vec{\mathbb{B}}_i$ vectors contributed by all segments are parallel, the magnitude of the total magnetic field vector $\vec{\mathbb{B}}$ at P is simply the sum of the magnitudes of the $\vec{\mathbb{B}}_i$ vectors contributed by the segments. So in this particular case, equation E8.10 becomes

$$\mathbb{B} = \frac{k}{c} \sum_{\text{all } i} \frac{dx}{r_{Pi}^2} \text{mag}(\vec{I}_i \times \hat{r}_{Pi}) = \frac{k}{c} \sum_{\text{all } i} \frac{dx}{r_{Pi}^2} I \sin\theta_i \tag{E8.11}$$

where θ_i is the angle between $\vec{r}_{Pi}$ and $\vec{I}$ (remember that the directional $\hat{r}_{Pi} \equiv \vec{r}_{Pi}/r_{Pi}$ has a magnitude of 1). Since we can see from figure E8.7 that $\sin\theta_i = r/r_{Pi}$ and that $r_{Pi}^2 = r^2 + x_i^2$, this becomes

$$\mathbb{B} = \frac{kI}{c} \sum_{\text{all } i} \frac{dx}{r_{Pi}^2} \frac{r}{r_{Pi}} = \frac{kIr}{c} \sum_{\text{all } i} \frac{dx}{r_{Pi}^3} = \frac{kIr}{c} \sum_{\text{all } i} \frac{dx}{(r^2 + x_i^2)^{3/2}} \tag{E8.12}$$

In the limit that dx becomes infinitesimal, the sum becomes an integral

$$\mathbb{B} = \frac{kIr}{c} \int_{x_1}^{x_2} \frac{dx}{(r^2 + x^2)^{3/2}} \tag{E8.13}$$

We have seen this integral before (see the inside front cover and example E2.3): the result is

$$\mathbb{B} = \frac{kIr}{c} \left[\frac{x}{r^2\sqrt{r^2 + x^2}} \right]_{x_1}^{x_2} = \frac{kI}{cr} \left(\frac{x_2}{\sqrt{r^2 + x_2^2}} - \frac{x_1}{\sqrt{r^2 + x_1^2}} \right) \tag{E8.14}$$

If the ends of the straight segment and all other parts of the circuit are very far away from P compared to r ($x_1^2 \gg r^2$ and $x_2^2 \gg r^2$, with x_1 negative and x_2 positive), then this becomes

The magnetic field of a long, straight wire

$$\mathbb{B} \approx \frac{kI}{cr} \left(\frac{x_2}{\sqrt{x_2^2}} - \frac{x_1}{\sqrt{x_1^2}} \right) = \frac{kI}{cr}[1 - (-1)] = \frac{2kI}{cr} \tag{E8.15}$$

This result can serve as a useful approximation in a number of situations. Note that this result is consistent with the empirical results described by Biot and Savart, as discussed in section E8.2. We will refer to this result again in chapters E11 and E12.

Exercise E8X.9

How much current would a long, straight wire have to carry to produce a magnetic field that at a distance of 10 cm from the wire has a magnitude comparable to that of the earth's magnetic field (15 kN/C)?

Exercise E8X.10

What are the magnitude (in terms of k, I, c, and r) and the direction of the magnetic field at the point P shown? Assume that the wire continues to essentially infinity in both the horizontal and vertical directions. (*Hint:* Argue that the horizontal half of the wire contributes *nothing* to the magnetic field at point P. Why?)

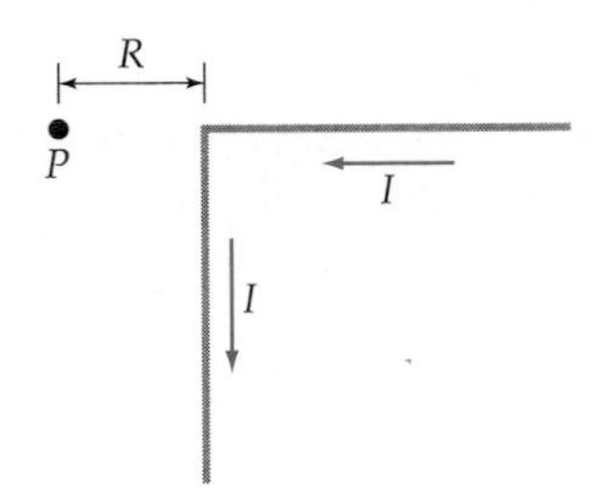

E8.5 The Magnetic Field of a Circular Loop

In chapter E7, we saw that a current-carrying loop of wire behaves in a magnetic field as if it were a bar magnet. Our discussion so far in this chapter makes it clear that a current-carrying loop will also produce a magnetic field. What does this magnetic field look like?

Example E8.1

Problem Calculate the magnitude of the magnetic field at the center of a circular loop of radius R carrying a current I.

Translation Figure E8.8 shows the situation and defines some useful symbols.

Model Note that all segments on the loop are the same distance R from the loop. Moreover, note that since both $\vec{I}$ and $\vec{r}_{Pi}$ lie in the plane of the drawing for each segment, each segment will contribute a magnetic field vector $\vec{\mathbb{B}}_i$ that is perpendicular to the plane of the loop (as illustrated in figure E8.8). Therefore, as in the case of the long wire, we can find the magnitude of $\vec{\mathbb{B}}$ at the center by summing the *magnitudes* of the vectors $\vec{\mathbb{B}}_i$ contributed by the segments. Finally, note that $\vec{I}_i$ is perpendicular to $\vec{r}_{Pi}$ for all segments, so $\text{mag}(\vec{I}_i \times \vec{r}_{Pi}) = I \sin 90° = I$ for all segments.

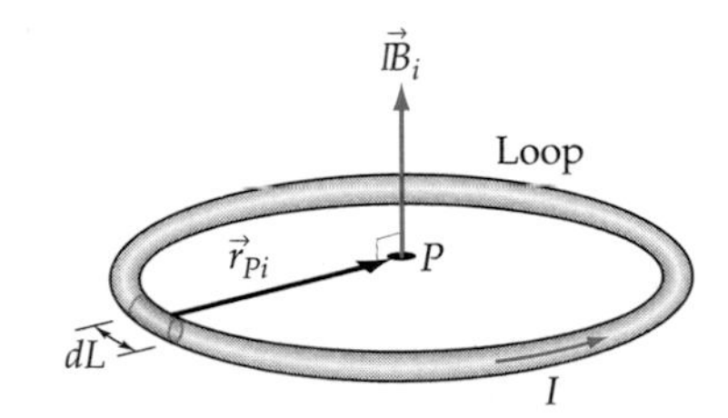

Figure E8.8
This diagram shows the part of the magnetic field contributed at the center of a circular loop by the loop's *i*th segment.

Solution Therefore, the total magnetic field at the loop's center is

$$\mathbb{B} = \frac{k}{c}\sum_{\text{all } i}\frac{dL}{r_{Pi}^2}\,\text{mag}(\vec{I}_i \times \hat{r}_{Pi}) = \frac{k}{c}\sum_{\text{all } i}\frac{dL}{R^2}I = \frac{kI}{cR^2}\sum_{\text{all } i} dL = \frac{2\pi kI}{cR} \qquad \text{(E8.16)}$$

Evaluation Note that the units of this expression are $(\text{N}\cdot\text{m}^2/\text{C}^2)(\text{C/s})(\text{s/m})/\text{m} = \text{N/C}$, which are the correct units for $\vec{\mathbb{B}}$.

The magnetic field vector at any point *other* than the loop's center is more difficult to calculate. Computing the magnetic field at a point along the loop's axis is not so bad (see problem E8S.9), but doing so at any point off the axis presents a real mathematical challenge. However, one can pretty easily write a computer program that performs the sum given by equation E8.10 directly for any point P. When one does this, one finds that the magnetic field vectors created by a loop at points close to the loop are as shown in figure E8.9a. Figure E8.9b indicates that at points far from the loop compared to its radius, the magnetic field vectors are essentially the same as those for a small bar magnet, which in turn are analogous to the electric field vectors produced by a dipole (see problems E8S.9, E8S.10, and E8A.1 for more discussion of this issue).

The loop-to-magnet rule introduced in section E7.5 tells us that if you wrap your right fingers in the direction of current flow in a loop, your right thumb indicates the direction of the north pole of the bar magnet that *responds*

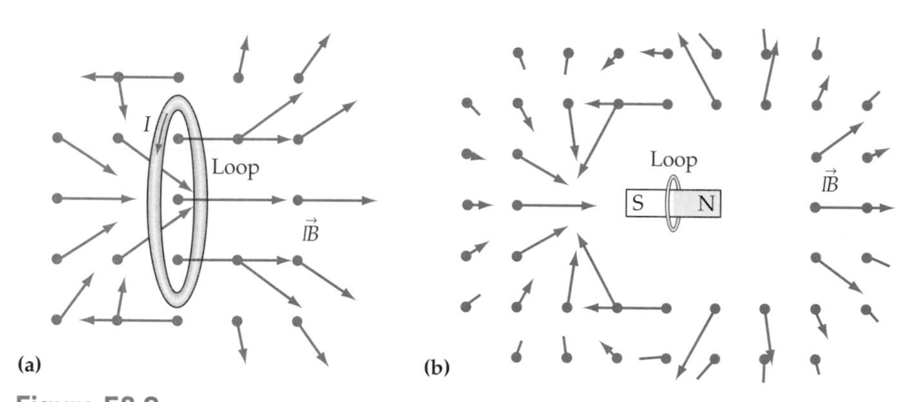

Figure E8.9
(a) The magnetic field close to a circular current-carrying loop is complicated and not particularly dipolelike. (b) However, the magnetic field far from the loop is equivalent to the electric field produced by a dipole, and thus (according to the dipole model) to the magnetic field produced by a permanent bar magnet in the orientation shown.

Table E8.1 Right-hand rules

Rule	Fingers	Thumb	Figure
Cross-product rule	Point index finger in the direction of the first vector in the product, and second finger in the direction of the second vector.	Indicates the direction of the cross product	$\vec{u} \times \vec{v}$, $\vec{u}$, $\vec{v}$ E8A.11a
Loop rule (= loop-to-magnet rule)	Curl fingers in the direction of the current flowing around the loop.	Indicates (1) the S-to-N direction of the analogous bar magnet and (2) the direction of the magnetic field at the loop's center	$\vec{B}$, I, =, N, S E8A.11b
Wire rule	Fingers curl in the direction of the magnetic field that the wire creates.	Point thumb in the direction of the conventional current flowing in the wire.	$\vec{B}$, $\vec{I}$ E8A.11c

to a magnetic field in the same way as a loop does. Figure E8.9 makes it clear that this rule also describes the orientation of the bar magnet that *creates* a magnetic field that is the same as the loop's magnetic field at large distances. It is also useful to note that in this context, if you curl your right fingers in the direction of the current flowing in the loop, your right thumb indicates the direction of the magnetic field vectors $\vec{B}$ at points *inside* the loop.

Since the field vectors shown in figure E8.9 ultimately come from a calculation involving equation E8.10, which employs the cross-product rule to determine the direction of $\vec{B}$, we see that the loop rule is actually just another application of the cross-product rule. But it is actually easier, in my opinion, to memorize the loop and wire rules separately than always to work everything out using the cross-product rule.Table E8.1 summarizes the right-hand rules we have introduced so far.

E8.6 *All* Magnets Involve Circulating Charges

What creates a permanent magnet's magnetic field?

So a current-carrying loop not only *responds* to a magnetic field as a small permanent bar magnet would (which is analogous to how an electric dipole would respond to an analogous electric field), but it also *creates* a magnetic field like the distant field of a permanent magnet (which is analogous to the electric field of an electric dipole). But what creates the field of a permanent magnet?

Orbiting electrons in atoms act as current loops

Ampère was perhaps the first physicist to intuit (in the early 1800s) that a permanent magnet's field might be produced by microscopic current loops. However, how these currents could circulate indefinitely remained a mystery for a long time.

We now know that atoms are constructed of electrons circulating around an atomic nucleus. A classical model of the atom has the electrons orbiting the nucleus as planets orbit the sun. Quantum mechanics teaches us to be more careful about this description, but even in quantum mechanics, an electron can have nonzero orbital angular momentum, which is just a quantum-mechanical way of saying that the electron is indeed circulating around the nucleus.

Such circulating electrons therefore are essentially tiny current loops. Most atomic electrons are members of pairs that circulate in opposite directions (thus yielding zero net circulating current), but many atoms have at least one unpaired electron with nonzero orbital angular momentum that represents a tiny permanent current loop. Unlike a macroscopic current flowing in a wire, the circulating electrons in an atom flow in a vacuum without losing energy and thus do not require a power supply.

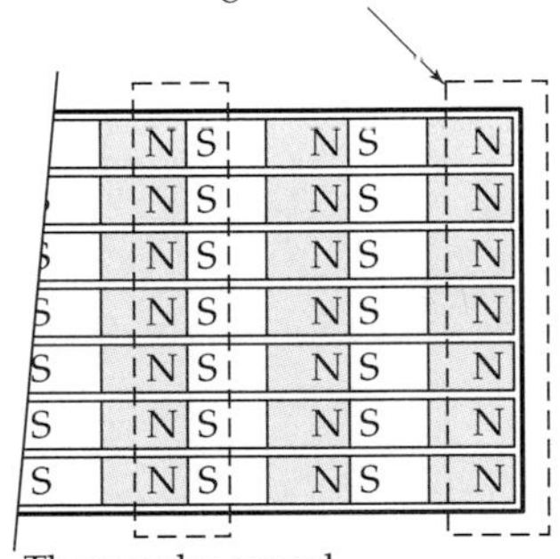

Figure E8.10
This close-up side view of one end of a bar magnet shows (schematically) how the tiny dipoles inside a permanent magnet essentially add up to poles at the magnet's ends.

The magnetic fields of these atomic current loops can combine to create a stronger magnetic field if the loops are aligned in the same direction. If we could align all the atomic current loops in a cylindrical bar in the same direction, we would create a powerful permanent bar magnet whose total field would essentially be the sum of atomic dipolelike fields. Figure E8.10 illustrates how microscopic atomic magnets can combine to create what looks like a single north pole at one end of the magnet and a single south pole at the other end.

In ferromagnetic materials, these atomic current loops tend to align

In most substances, however, these atomic magnets are oriented randomly, and so the net magnetic field produced by the substance is essentially zero. However, in a very few substances (**ferromagnetic** substances such as iron, nickel, and cobalt), it turns out to be energetically favorable for adjacent atoms to align their effective current loops in the *same* direction. (This is a bit strange, since if you put two bar magnets side-by-side and pointing in the

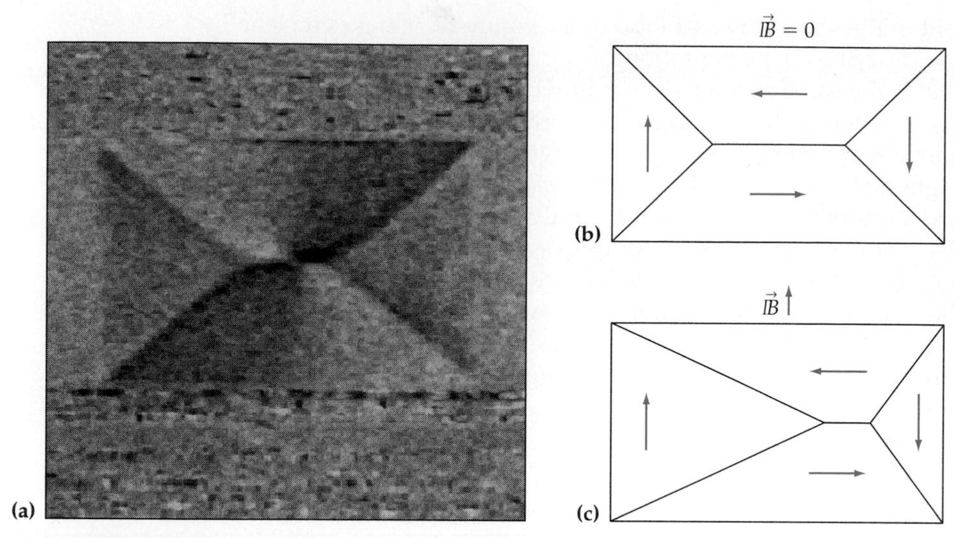

Figure E8.11
(a) A photograph showing the magnetic domains in a tiny rectangle of permalloy (80 percent nickel, 20 percent iron). (b) A schematic diagram of the domains in such a ferromagnetic substance. The arrows show the orientation of the atomic magnets inside the domains. (c) When the iron is placed in an external magnetic field, domains most nearly aligned with the field grow, while others shrink.

same direction, their adjacent like poles repel each other, exerting forces that tend to reorient the magnets in *opposite* directions. Clearly, some nonmagnetic effect must overcome this tendency.) In a typical piece of iron, the iron atoms tend to form small regional "clubs" called **domains,** within which all atoms agree on a given orientation (see figure E8.11). In an unmagnetized piece of iron, the alignments of the various domains are random, and thus the net magnetic field is still zero. But if one puts a piece of iron in a strong external magnetic field, domains whose orientations are basically aligned with the field tend to grow (by "recruiting" along their boundaries), and domains that are not aligned with the field shrink. The result is that the iron develops its own net magnetic field whose average direction is the same as the direction of the applied field.

This is why a coil of wire wrapped around a piece of iron creates a stronger magnetic field than the same coil wrapped around a piece of wood. In the first case, the coil's field is augmented by the net magnetic field of the iron atoms as they line up with coil's field.

When you remove the external magnetic field, most ferromagnetic materials retain a residual magnetic field (some materials better than others), because once the atoms are aligned in a given direction, they tend to stay oriented that way. So an unmagnetized piece of iron, once exposed to a strong external magnetic field, will remain permanently magnetized thereafter: we have created a *permanent magnet.* The atoms will remain aligned unless the temperature becomes too large (the jostling of atoms due to thermal effects can destroy the alignment) or a strong magnetic field in a different direction is imposed.

All magnetic fields are caused by moving charges

The fundamental point of this discussion is that the magnetic field of a permanent magnet, like that of a current loop, is ultimately produced by charges in motion. *Moving or spinning*† *charges are the only known source of*

†Spinning charges will be discussed in greater depth in unit Q.

magnetic fields. Since each atomic current loop in a permanent magnet both *responds* to an external magnetic field as an electric dipole responds to an external electric field (see section E8.2) and *creates* a magnetic field that (far from the atom) is like an electric dipole's electric field (see section E8.5), a permanent magnet comprised of such atomic current loops will *also* respond to a field and create a field in a similar way (the sum of little dipolelike magnets is just a big dipolelike magnet). This is fundamentally why the dipole model for permanent magnets introduced in chapter E7 works so well in practical situations.

TWO-MINUTE PROBLEMS

E8T.1 A jet flies in a region of the earth where the magnetic field points nearly straight upward. Which part of the airplane becomes negatively charged?

A. Its nose
B. Its tail
C. Its top
D. Its bottom
E. Its left wing
F. Its right wing
T. No part becomes charged

E8T.2 A large bar magnet is embedded in a table so that its north pole sticks a certain distance above the table and its south pole sticks out an equal distance below the table. A ring of wire lying flat on the table is pushed so that it slides directly toward the magnet. What is the direction of the current induced in the ring when viewed from above? (Assume the magnet's field is not affected by the table.)
A. Clockwise
B. Counterclockwise
C. There is no induced current
D. Insufficient information

E8T.3 Imagine placing two long wires parallel to each other. If currents flow through the wires in the directions shown, will the wires attract, repel, or exert no forces on each other?

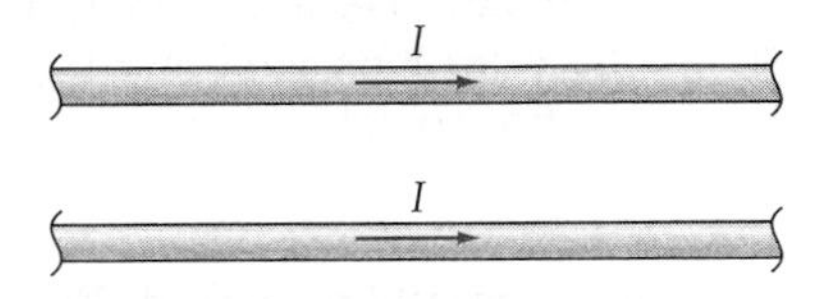

A. Attract
B. Repel
C. Exert no forces

E8T.4 Imagine placing two long wires parallel to each other. If currents flow through the wires in the directions shown, will the wires attract, repel, or exert no forces on each other?

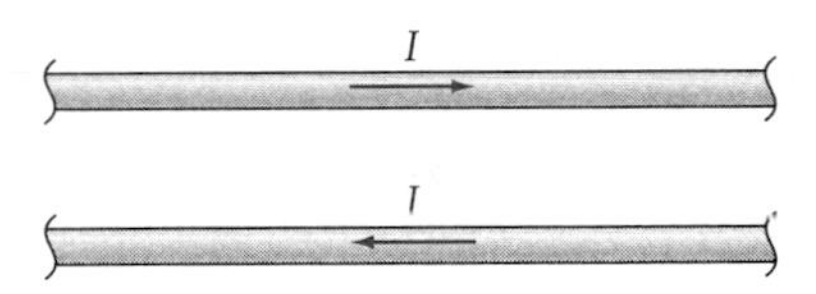

A. Attract
B. Repel
C. Exert no forces

E8T.5 Consider the loop of flexible wire shown. If a current is passed through the loop, will it tend to expand (becoming more circular), scrunch up even more tightly, or do nothing?

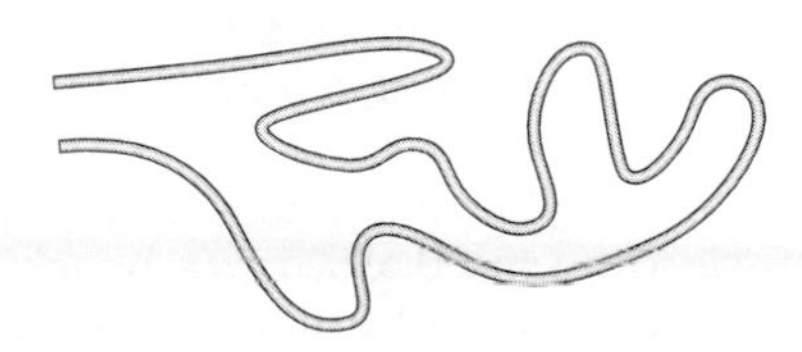

A. Expand
B. Scrunch up
C. Do nothing
D. The answer depends on the current's direction.

E8T.6 The picture next shows an end view of a loop conducting a current whose conventional direction is shown. The magnetic field at the center of the loop points

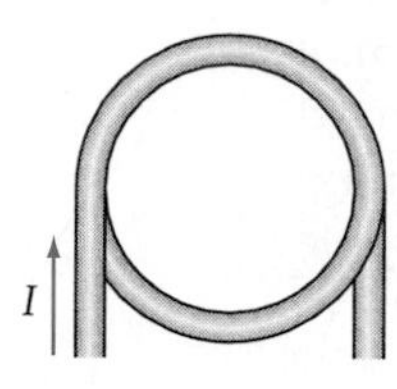

A. Toward us
B. Away from us
C. To the left
D. To the right
E. Nowhere, because there is no field
F. In a direction we don't have enough information to determine

E8T.7 A wire bent in the shape shown below carries a current in the direction marked. What is the magnitude $\mathbb{B}$ of the magnetic field at point a due to the *horizontal segments* of the wire?

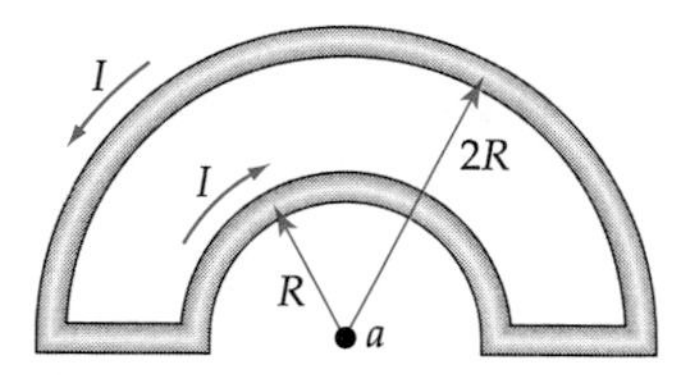

A. kI/cR
B. $\pi kI/cR$
C. $\pi kI/2cR$
D. $2\pi kI/cR$
E. kI/c
F. 0
T. Other (specify)

E8T.8 A wire bent in the shape shown in problem E8T.7 carries a current in the direction marked. What is the magnitude of the magnetic field at the point a due to the entire loop? (Select from the answers provided for problem E8T.7.)

E8T.9 Consider a wire bent in the hairpin shape shown below. The wire carries a current I. What is the approximate magnitude $\mathbb{B}$ of the magnetic field at point a?

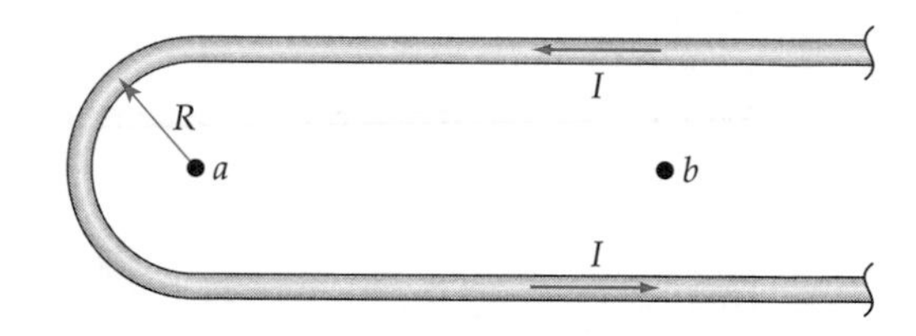

A. kI/cR
B. $2kI/cR$
C. $2\pi Ik/cR$
D. $(2+\pi)kI/cR$
E. 0
F. Other (specify)

E8T.10 Consider a wire bent in the hairpin shape shown in problem E8T.9. The wire carries a current I. What is the approximate magnitude of the magnetic field at point b? (Select from the answers provided for problem E8T.9.)

HOMEWORK PROBLEMS

Basic Skills

E8B.1 An airplane develops a 0.25-V potential difference between the tips of its wings when flying in a region where the earth's magnetic field points essentially vertically upward. If the plane's wingspan is 35 m, how fast is it flying?

E8B.2 A bullet with a charge of +10 nC is fired along the x axis with a speed of 520 m/s. Imagine that you sit at the position $y = 1.0$ m, $x = 0$. What are the magnitude and direction of the *maximum* magnetic field that you feel as a result of the bullet's motion? Where is the bullet when its field at your position is maximum?

E8B.3 An electron travels at $0.1c$ along the x axis. An atom sits along the y axis 2.5 nm away from the x axis. What are the magnitude and direction of the *maximum* magnetic field that the atom feels as a result of the electron? Where is the electron when it exerts this maximum effect?

E8B.4 How much current would a long, straight wire have to carry to create a magnetic field whose strength is $\mathbb{B} = 100$ MN/C at a point 1 cm from the wire?

E8B.5 A long, straight wire carries a current of 200 A. What is the magnetic field strength at a point 3 cm from the wire?

E8B.6 A circular loop of wire has a radius of 12 cm and carries a current of 0.5 A. What is the magnitude of the magnetic field at its center? How does this compare to the magnitude of the earth's magnetic field?

E8B.7 Imagine that you have a loop of wire 10 cm in diameter. How much current would have to flow through the wire to create a field at the loop's center that is twice as strong as the earth's magnetic field (≈ 0.5 gauss)?

E8B.8 Imagine winding N turns of wire in a circular loop of radius R, and assume the wire is so thin that the width of a bundle of N wires is much less than R. Each turn in this coil acts as a loop of radius R, and so the magnetic field at the coil's center is N times larger than that for a single loop of the same size. If the wire can safely carry 1.0 A of current, how many turns are needed to produce a magnetic field with strength 0.0033 T in the center of a coil 10 cm in diameter?

E8B.9 How does the model of a permanent magnet described in section E8.6 explain why we cannot isolate a magnet's poles by breaking a permanent magnet in half?

Synthetic

E8S.1 Imagine a space station module shaped like a cylinder 30 m long and 5 m in diameter orbiting close to the earth above the earth's equator. What would be the maximum potential difference that could develop between the ends of the module? How would the module have to be oriented to get this maximum voltage difference? Could this voltage difference be used to supply the station with electric power? If so, how? If not, why not?

E8S.2 Figure E8.12 shows a way to take advantage of the way that a conducting bar moving in a magnetic field becomes a magnetic battery. Imagine that we have two metal rails 20 cm apart connected by a lightbulb. The rails are placed in a 1.0-T magnetic field oriented vertically downward. We then constrain a bar to slide on the rails toward the lightbulb in this magnetic field. How fast do we have to move the bar to put a voltage difference of 3.0 V across the bulb? In what direction does the current flow?

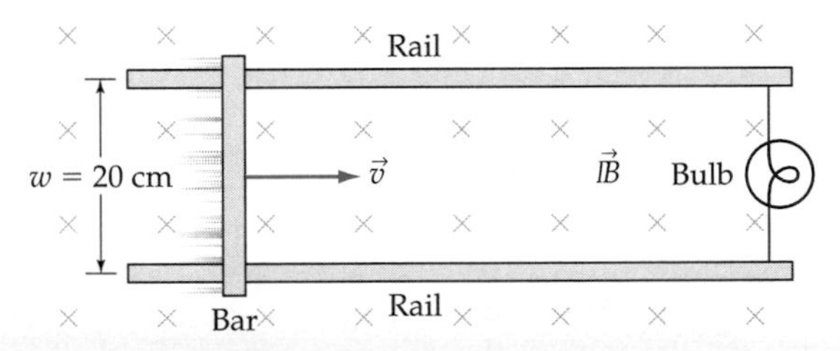

Figure E8.12
How to create a current flow by using a moving bar in a magnetic field (see problem E8S.2).

E8S.3 Imagine we have a conductor with a rectangular cross section whose height in the z direction is h and whose width in the x direction is w. This conductor carries a conventional current $\vec{I}$ in the $+x$ direction along its length in a magnetic field $\vec{\mathbb{B}}$ directed in the $+y$ direction, as illustrated in figure E8.13.

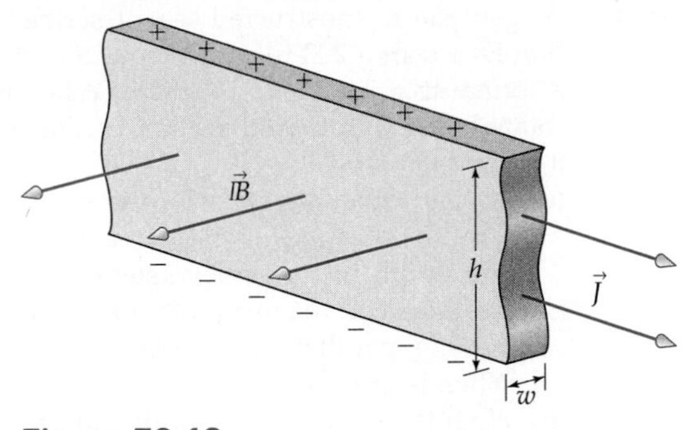

Figure E8.13
The edges of a rectangular conductor carrying a current perpendicular to a magnetic field become oppositely charged due to the Hall effect (see problem E8S.3).

(a) The moving charge carriers will experience magnetic forces that will push flowing charges either upward or downward, depending on their charge and their direction of motion. Yet when a steady-state current flow is established, the charges must move directly along the conductor. Argue that whatever the charge q of the charge carriers, this only becomes possible once surface charges have been established on the top and bottom edges that create an electrostatic field such that

$$\vec{E} = -\frac{\vec{v}}{c} \times \vec{\mathbb{B}} \tag{E8.17}$$

(b) Argue that this means that the potential difference between the conductor's bottom surface and its top surface is given by

$$\Delta\phi = \frac{\mathbb{B} I}{nq\,wc} \tag{E8.18}$$

where n is the number density (number per unit volume) of charge carriers.

(c) The fact that a current-carrying conductor in a magnetic field develops a potential difference between its top and bottom edges is called the *Hall effect*. The Hall effect is useful (1) because we can use it to measure the strength of a magnetic field, using only simple measurements of current and potential difference, and (2) because the sign of this potential difference depends on the actual sign of the charge carriers, implying that we can *determine* whether it is electrons or positive charges that are actually flowing in a substance by measuring the Hall potential difference. If it is really electrons that flow in a copper conductor, and we have a conductor of 1-mm width carrying a current of 10 A in a magnetic field of 100 MN/C, what would we predict the Hall-effect potential difference between the conductor's top and bottom to be?

E8S.4 A generator constructed as described in section E8.1 uses a 200-turn coil wound in the form of a square that is 5.0 cm on a side. When the coil is rotated at 3600 rev/min, its emf oscillates between 0 V and +30 V.

(a) Explain qualitatively why the emf is not constant for such a generator, and sketch a qualitative graph of the emf versus time. (*Hint:* The emf is zero at the instant that the loop's legs are moving parallel to the magnetic field. Why? When is the emf maximum?)

(b) What is the magnitude of the magnetic field in this generator?

E8S.5 Consider a generator constructed as shown in figure E8.14. The loop in this generator rotates at an angular rate of ω. The slip rings in this design ensure that the same leg of the rotating loop is always connected to the same output terminal. Show that the potential difference between the output terminals is given by $\mathscr{E} = \mathscr{E}_0 \cos\theta$ (where $\theta = \omega t$ is the angle that the plane of the loop makes with the magnetic field direction at time t), and determine $\mathscr{E}_0$ in terms of the loop's length L, width W, the magnetic field strength $\mathbb{B}$, and the loop's angular rotation rate ω.

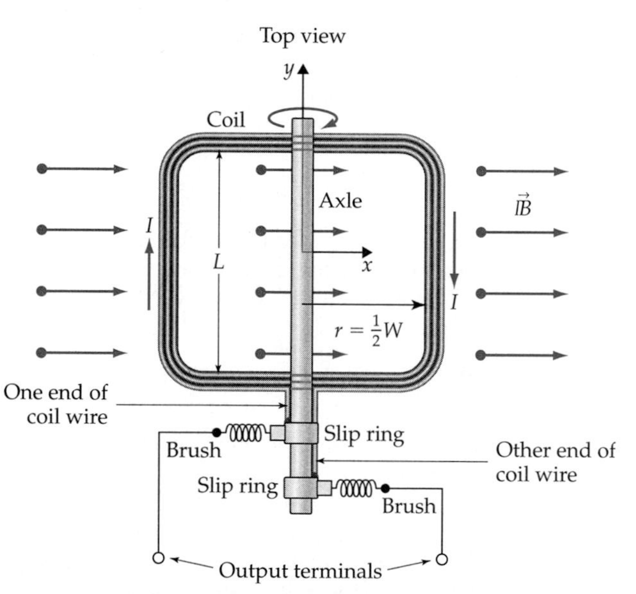

Figure E8.14
A generator design that creates a cosinusoidally varying emf between its terminals (see problem E8S.5).

E8S.6 A device called a *homopolar generator* is shown in figure E8.15. A metal disk rotates in the presence of a magnetic field perpendicular to the disk. Sliding contacts connect the axle and the outer surface of the disk to an external circuit.

(a) Explain why this device produces an emf between the two contacts. Is this emf oscillating or constant?

(b) Determine the magnitude of this emf in terms of the disk's radius R, the magnetic field strength $\mathbb{B}$, and the angular rate of rotation ω. (*Hint:* Divide the disk up into rings of infinitesimal thickness dr, compute the emf created across each ring, then sum.)

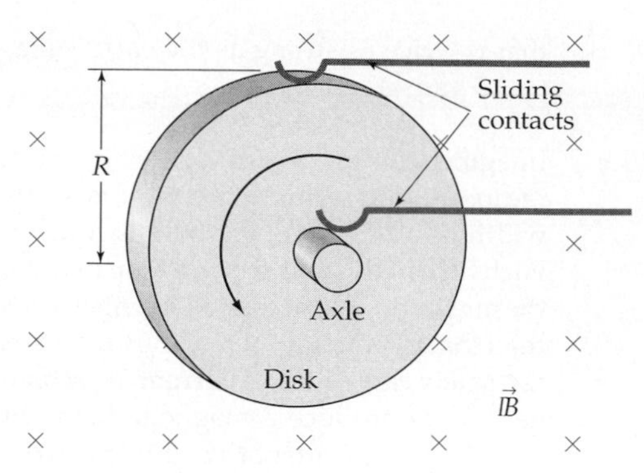

Figure E8.15
A homopolar generator (see problem E8S.6).

E8S.7 Use the Biot-Savart law to calculate the field in the center of the loop shown in the diagram for problem E8T.7, except assume that the radius of the larger loop is $3R/2$.

E8S.8 Imagine that we have a circular plastic disk with 20-cm radius. By rubbing its edge with fur, we create a total negative charge of about 30 nC essentially uniformly distributed around the disk's edge. We then spin the disk at a rate of 60 turns per second. What is the approximate magnitude of the magnetic field that rotating charge creates at the disk's center? How does this compare to the magnitude of the earth's magnetic field?

E8S.9 Consider a circular loop of wire of radius R that carries a current I. Let us set up our coordinate system so that the loop lies in the yz plane and the x axis goes through the loop's center (the x axis will then correspond to the loop's central axis).

(a) Use the Biot-Savart law to show that the magnetic field strength $\mathbb{B}$ at a point along the x axis is given by

$$\mathbb{B} = \frac{2\pi k I R^2}{c(x^2 + R^2)^{3/2}} \tag{E8.19}$$

(b) Argue that at large distances r from the loop's center, the loop's magnetic field $\vec{\mathbb{B}}$ at points along the loop's central axis is exactly analogous to the electric field $\vec{E}$ at distant points along an electric dipole's central axis (see equation E2.16) if we substitute qd for $\pi R^2 I/c$, where q is the absolute value of the electric charge at either end of a dipole of length d.

E8S.10 Consider a square loop of wire whose sides have length L and which carries current I. Let us set up our coordinate system so that the loop lies in the yz plane and the x axis goes through the loop's center (the x axis will then correspond to the loop's central axis).

(a) Use the Biot-Savart law and the small-angle approximation $\sin\phi \approx \tan\phi \approx \phi$ when $\phi \ll 1$ to compute the total magnetic field at a point P along the x axis that is a distance $x \gg L$. (*Hints:* Since each leg is very small compared to the distance x, we can treat each leg as a single segment in the sum in the Biot-Savart law. Draw a picture of the situation looking at the loop from above the xy plane, and argue that the $\mathbb{B}_y$ and $\mathbb{B}_z$ components contributed by the four legs cancel out in pairs, but that the $\mathbb{B}_x$ components contributed by all four legs are identical and add.)

(b) Show that the formula for the magnitude of the magnetic field created by the square loop at a point along its central axis such that $x \gg L$ is the same as that created by a *circular* loop at a point along its central axis such that $x \gg R$ (see equation E8.19) if IL^2 for the square loop is the same as $I\pi R^2$ for the circular loop. (This is a special case of a general result that a loop's magnetic field in the large-distance limit does not depend on the loop's shape, but only on the product of its area and the current it carries.)

E8S.11 Imagine that we have two parallel current-carrying coils of radius R that are perpendicular to and centered on the x axis, with one coil's center at $x = \frac{1}{2}D$ and the other at $x = -\frac{1}{2}D$ (so the two coils are a distance D apart). These coils both carry the same magnitude of current I in the same direction (see figure E8.16). We would like to adjust the value of D so that the magnetic field created by the coils is as constant as possible along the x axis in the neighborhood of $x = 0$.

(a) Use equation E8.19 to show that the first derivative of $\mathbb{B}_x$ with respect to x is zero for all values of D just because the arrangement of coils is symmetric about the origin. (*Hint:* Argue that x in equation E8.19 should be replaced by $x \pm \frac{1}{2}D$ in this problem, where the sign in front of $\frac{1}{2}D$ depends on which of the two coils we are talking about.)

(b) Show that if we place the coils a certain distance D apart, the second derivative of $\mathbb{B}_x$ will also be zero at $x = 0$. Find this distance in terms of R. (This optimal spacing gives the most constant field possible with two coils: coils arranged with this optimal spacing are called *Helmholtz coils*.)

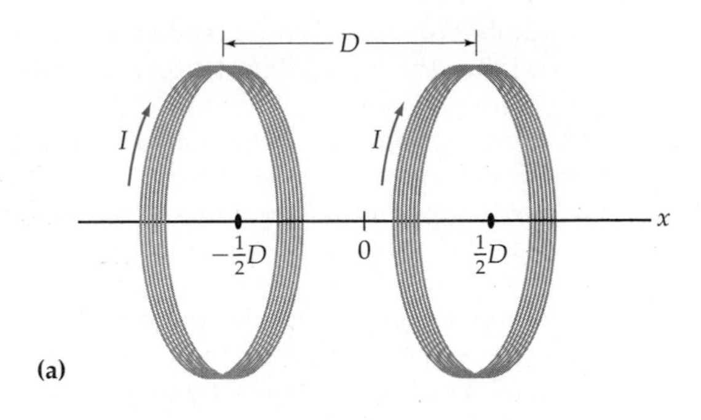

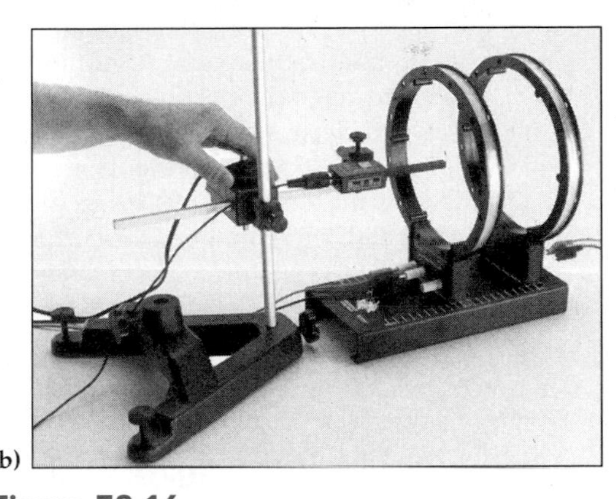

Figure E8.16
(a) A schematic diagram of a pair of Helmholtz coils (see problem E8S.11). (b) A photograph showing Helmholtz coils being used in an experiment.

Rich-Context

E8R.1 A particle accelerator injects $N = 10^{15}$ protons traveling at speed $v = 0.2c$ into a region where there is a uniform magnetic field of strength $\mathbb{B}_{ext} =$ 300 MN/C ($B_{ext} = 1.0$ T). As discussed in chapter E7, the protons will travel in a circle in this region. They will therefore act as a current loop that creates its own magnetic field $\vec{\mathbb{B}}_{pro}$.

(a) Will this field reinforce or oppose $\vec{\mathbb{B}}_{ext}$?

(b) *Assuming* that the magnitude of $\vec{\mathbb{B}}_{pro}$ is small compared to $\mathbb{B}_{ext}$, derive a *symbolic* expression for $\mathbb{B}_{pro}$ at the center of the proton loop in terms of k, c, N, $\mathbb{B}_{ext}$, and the proton's charge e and mass m.

(c) Show that in this case, $\mathbb{B}_{pro}/\mathbb{B}_{ext} \approx \frac{1}{400}$ at the loop's center, justifying the initial assumption.

(d) If v were smaller, would $\mathbb{B}_{pro}/\mathbb{B}_{ext}$ get smaller or larger? Explain your answer *physically* (not by just appealing to the equation).

E8R.2 Imagine that you are designing a circular coil that must have a radius of 10 cm, create a magnetic field of magnitude 0.01 T at its center, have a cross-sectional area of less than 0.25 cm^2, and use as little power as possible. The question is, Should you use many turns of thin wire in this coil or one turn of wire with a cross-sectional area 0.25 cm^2? Assume that you are using copper wire.

(a) Show that, to a first approximation anyway, the power this coil will use is fixed by the design requirements, and the number of turns of wire is therefore irrelevant for minimizing the power. What assumptions go into this result?

(b) Since the number of turns N does not affect the power much, we can choose N to satisfy other constraints. What would be the appropriate number of turns if you intended to connect this coil directly to a 12-V car battery, and why?

E8R.3 Just how strong can a permanent magnet be? Specifically, imagine a small cylindrical permanent magnet that is 2.0 cm long and 1.0 cm in diameter. What is the maximum magnetic field we could reasonably expect such a magnet to create at a point 10 cm from its center along its long axis? This might seem impossible to guess, but as a matter of fact, we can make a pretty good guess by making only some basic assumptions.

(a) Consider an electron of mass m traveling at speed v in a circular orbit of radius r about an atom. The magnitude of the angular momentum of such an electron is $L = mvr$. The current that this orbiting electron represents is its charge divided by the time it takes to go around once. Show that the magnitude of the current times the area of this atomic current loop can be written as

$$IA = \frac{eL}{2m} \tag{E8.20}$$

The product of a loop's current and its area is called the loop's **magnetic moment** μ.

(b) Quantum mechanics tells us that the electron's orbital angular momentum L is quantized in steps of $\hbar = h/2\pi = 1.054 \times 10^{-34}$ J · s (h is Planck's constant). Show that this means that an atom's magnetic moment must be quantized in steps of $\mu_B = 9.26 \times 10^{-24}$ A · m^2 (this quantity is called the *Bohr magneton*).

(c) Iron has a molar mass of about 55 g/mol. Most of the other ferromagnetic elements are close to iron on the periodic table, so ferromagnetic alloys will have very roughly this molar mass. Iron also has a density of about 7900 kg/m^3. Roughly how many atoms will there be in our magnet?

(d) In a typical atom, most electrons orbit in pairs having opposite angular momenta, so they contribute nothing to the atom's magnetic moment. The electron also has a spin angular momentum whose magnetic moment is quantized in roughly the same-size steps as the orbital magnetic moment. So we will likely have one or at most two electrons contributing to the magnetic moment per atom. Let's be very generous and imagine that the total magnetic moment per atom is $4\mu_B$ ($1\mu_B$ is probably more realistic). Also let's assume that *all* the atomic magnetic moments in our magnet are perfectly aligned. Use this information and equation E8.19 to compute the magnetic field of our magnet at a point P that is 10 cm from its center along its axis. (*Hints:* All the atomic loops will be very small compared to their separation from point P. Note also that the loops are all roughly the same distance from P and that P lies pretty close to the central axis of each.)

(e) Analyze the assumptions that went into this calculation. What would we have to assume to get a magnet of the same volume that is, say, 10 times stronger? Could using metals higher on the periodic table help? Is it realistic to assume higher total magnetic moment per atom? Why or why not?

Advanced

E8A.1 Calculate the magnetic field $\vec{\mathbb{B}}$ at a point P a displacement $\vec{r}_{PQ}$ away from a square loop with side L centered at point Q as a function of the angle θ between $\vec{r}_{PQ}$ and the loop's central axis, assuming that $\text{mag}(\vec{r}_{PQ}) \equiv r \gg L$. Show that the formulas for the radial and perpendicular components $\mathbb{B}_r$ and $\mathbb{B}_\perp$ are the same as those for E_r and $E_\perp$ given in problem E2A.1 if we substitute qd for IL^2/c. (*Hints:* To find the radial and perpendicular components $\mathbb{B}_r$ and $\mathbb{B}_\perp$ of the magnetic field, I think that it is easiest to orient the coordinate system so that the x axis always goes from Q to P, and so that two of the loop's legs are parallel to the z axis. Then the loop's central axis will lie in the xy plane, and we will have $\mathbb{B}_r = \mathbb{B}_x$ and $\mathbb{B}_\perp = \mathbb{B}_y$. In this coordinate system, evaluate $\vec{I}_i \times \hat{r}_{PQ}$ for all four legs, using the *component* form of the cross product:

$$\vec{u} \times \vec{w} = \begin{bmatrix} u_y w_z - u_z w_y \\ u_z w_x - u_x w_z \\ u_x w_y - u_y w_x \end{bmatrix} \tag{E8.21}$$

and use the binomial approximation (if appropriate).

ANSWERS TO EXERCISES

E8X.1 Figure E8.17 shows how to use your right hand. Note that electrons have a negative charge, so the direction of $q\vec{v}$ points opposite to the direction of $\vec{v}$.

E8X.2 The right-hand rule implies that positive charges will move to the north, so the plane's left wing becomes positively charged. Plugging the given values into equation E8.2 yields

$$\Delta\phi = \frac{v\mathbb{B}L}{c} = \frac{(280\ \cancel{\text{m/s}})(0.5\ \text{gauss})(80\ \text{m})}{3.0 \times 10^8\ \cancel{\text{m/s}}}$$

$$= 3.73 \times 10^{-5}\ \text{gauss} \cdot \cancel{\text{m}} \left(\frac{30{,}000\ \cancel{\text{N}}/\cancel{\text{C}}}{1\ \text{gauss}}\right)\left(\frac{1\ \cancel{\text{J}}}{1\ \cancel{\text{N}} \cdot \cancel{\text{m}}}\right)$$

$$\times \left(\frac{1\ \text{V}}{1\ \cancel{\text{J}}/\cancel{\text{C}}}\right) \approx 1.1\ \text{V} \tag{E8.22}$$

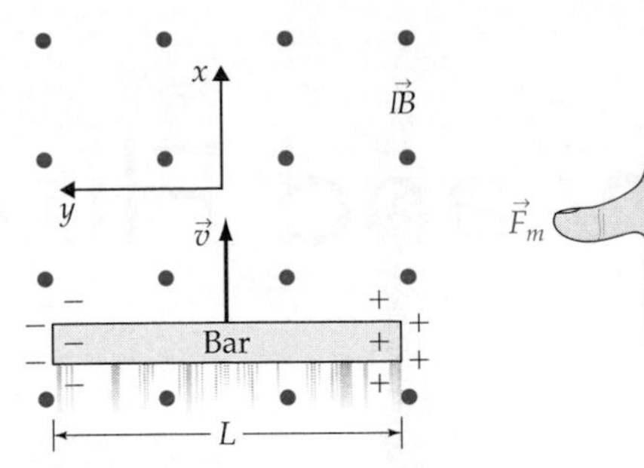

Figure E8.17
The answer to exercise E8X.1.

E8X.3 The magnetic force vectors on (hypothetical) positive charge carriers all point to the right in figure E8.2b. Let us divide the loop up into a sequence of infinitesimal steps $d\vec{r}$ that go counterclockwise around the loop. These displacement vectors will be parallel to the magnetic force in the rear leg, opposite to the magnetic force in the front leg, and perpendicular to the magnetic force in the other legs. Equation E8.1 implies that the magnitude of the magnetic force at any point is $F_m = qv\mathbb{B}/c$. This has the same value at all points on the rear leg, so by the definition of the dot product, the k-work per unit charge contributed by this force is

$$\frac{1}{q}\sum_{\text{rear leg}} \vec{F}_m \cdot d\vec{r} = \frac{1}{q}\sum_{\text{rear leg}} \frac{qv\mathbb{B}_r}{c}\, dr\cos(0^\circ) = \frac{v\mathbb{B}_r}{c}\sum_{\text{rear leg}} dr = \frac{v\mathbb{B}_r}{c}L \tag{E8.23}$$

For the front leg, $\vec{F}_m$ and $d\vec{r}$ are opposite, so

$$\frac{1}{q}\sum_{\text{front leg}} \vec{F}_m \cdot d\vec{r} = \frac{1}{q}\sum_{\text{front leg}} \frac{qv\mathbb{B}_f}{c}\, dr\cos(180^\circ) = -\frac{v\mathbb{B}_f}{c}\sum_{\text{rear leg}} dr = -\frac{v\mathbb{B}_f}{c}L \tag{E8.24}$$

For either of the other legs

$$\frac{1}{q}\sum_{\text{front leg}} \vec{F}_m \cdot d\vec{r} = \frac{1}{q}\sum_{\text{front leg}} F_m\, dr\cos(90^\circ) = 0 \tag{E8.25}$$

Putting all these sums together, we get

$$\mathscr{E}_{\text{loop}} = \frac{1}{q}\sum_{\text{loop}} \vec{F}_m \cdot d\vec{r} = \frac{qv\mathbb{B}_r L}{c} - \frac{qv\mathbb{B}_f L}{c} = \frac{qvL}{c}(\mathbb{B}_r - \mathbb{B}_f) \tag{E8.26}$$

Taking the absolute value yields equation E8.3.

E8X.4 The compass needle will align itself with the wire's magnetic field. A needle initially perpendicular to the wire is *already* aligned with its field and thus will not move when the current is turned on, but one initially *parallel* will visibly swing to align itself with the field.

E8X.5 In applying the cross-product rule, point your right index finger in the direction of $q\vec{v}$ and your long finger in the direction of $\hat{r}_{PC}$ for the point P in question. When you apply the wire rule, point your thumb parallel to $q\vec{v}$.

E8X.6 According to the value of μ_0 given,

$$\frac{\mu_0 c^2}{4\pi} = \frac{(4\pi \times 10^{-7}\ \text{T}\cdot\text{s}\cdot\text{m/C})(2.998 \times 10^8\ \text{m/s})^2}{4\pi}$$
$$= 8.99 \times 10^9 \frac{\cancel{\text{T}}\cdot\text{m}^3}{\cancel{\text{s}}\cdot\text{C}}\left(\frac{1\,(\text{N/C})(\cancel{\text{s}}/\text{m})}{1\,\cancel{\text{T}}}\right)$$
$$= 8.99 \times 10^9\,\frac{\text{N}\cdot\text{m}^2}{\text{C}^2} = k \tag{E8.27}$$

E8X.7 Note that the maximum value of mag$(\vec{v} \times \hat{r})$ is v. Plugging in the numbers, we get

$$E_{\max} = \frac{kq}{r^2_{\min}} = \frac{(8.99 \times 10^9\ \text{N}\cdot\cancel{\text{m}^2}\cdot\text{C}^{-\cancel{2}})(10 \times 10^{-9}\ \cancel{\text{C}})}{(0.1\ \cancel{\text{m}})^2}$$
$$= 9000\ \text{N/C} \tag{E8.28}$$

$$B_{\max} = \frac{\mu_0 qv}{4\pi r^2_{\min}} = \frac{(4\pi \times 10^{-7}\ \text{T}\cdot\cancel{\text{m}}\cdot\cancel{\text{s}}/\cancel{\text{C}})(10 \times 10^{-9}\ \cancel{\text{C}})(30\ \cancel{\text{m}}/\cancel{\text{s}})}{4\pi(0.1\ \cancel{\text{m}})^2}$$
$$= 3.0 \times 10^{-12}\ \text{T} \tag{E8.29}$$

$$\mathbb{B}_{\max} = \frac{kq}{r^2_{\min}}\frac{v}{c} = E_{\max}\frac{v}{c} = 9000\,\frac{\text{N}}{\text{C}}\left(\frac{30\ \cancel{\text{m/s}}}{3.0 \times 10^8\ \cancel{\text{m/s}}}\right)$$
$$= 9.0 \times 10^{-4}\ \text{N/C} \tag{E8.30}$$

Note that $\mathbb{B}/E = v/c = 10^{-7}$ here.

E8X.8 Since only the segments of the wire closest to the point really affect the field at the point, we can guess that $\mathbb{B}_B \approx 2\mathbb{B}_A \gg \mathbb{B}_C$.

E8X.9 Solving equation E8.15 for I and plugging in the numbers yield

$$I = \frac{\mathbb{B}cr}{2k} = \frac{(15{,}000\ \cancel{\text{N}}/\cancel{\text{C}})(3.0 \times 10^8\ \cancel{\text{m}}/\text{s})(0.1\ \cancel{\text{m}})}{2(8.99 \times 10^9\ \cancel{\text{N}}\cdot\cancel{\text{m}^2}/\text{C}^{\cancel{2}})}$$
$$= 25\ \text{C/s} = 25\ \text{A} \tag{E8.31}$$

E8X.10 The magnetic field at the point in question is $\mathbb{B} = kI/cR$ (one-half of full line). The horizontal segment of the wire contributes nothing because $\vec{I}$ is parallel to $\hat{r}_{Pi}$ for all segments along the wire, so $\vec{I} \times \hat{r}_{Pi} = 0$ for that part of the wire.

E9 Symmetry and Flux

▷ Electric Field Fundamentals
▷ Controlling Currents
▷ Magnetic Field Fundamentals
▽ Calculating Static Fields
Symmetry and Flux
Gauss's Law
Ampere's Law
▷ Dynamic Fields

Chapter Overview

Introduction

This chapter opens a three-chapter unit subdivision on more advanced methods of calculating static electric and magnetic fields. This chapter focuses on developing two important new mathematical tools.

Section E9.1: Symmetry Arguments

A **symmetry argument** is an argument that uses the following principle to determine characteristics of the field of a symmetric charge or current distribution:

> If a charge or current distribution is unchanged by some kind of transformation (such as a rotation around an axis or sliding along an axis), then the distribution's *field* is also unchanged by that same transformation.

This applies to all transformations except for mirror reflections of magnetic fields.

Section E9.2: The Electric Fields of Symmetric Objects

Examples in this section illustrate how we can apply this principle to determine characteristics of the electric fields created by symmetric charge distributions. The results are summarized in table E9.1. We usually use a transformation that keeps a vector's location fixed to determine the vector's direction, but one that rotates or slides the vector to a new position to constrain how its magnitude might vary.

Section E9.3: The Mirror Rule for Magnetic Fields

Magnetic fields obey the symmetry principle stated in section E9.1 for all transformations except for mirror reflections, which are described instead by the **mirror rule:**

> If we can slice a current distribution with a mirror in such a way that the distribution (taking account of current directions) looks exactly the same after we insert the mirror as before, then a nonzero magnetic field vector at any point on the mirror's surface will be perpendicular to that surface.

This rule is the easiest way to determine the directions of magnetic field vectors.

Section E9.4: The Magnetic Fields of Symmetric Current Distributions

Examples in this section illustrate how to apply the symmetry principle and the mirror rule to current distributions. The results are summarized in table E9.1.

Section E9.5: The Flux Through a Surface

An imaginary surface in space is a **closed surface** if it completely surrounds a certain volume of space, or a **bounded** surface if it is completely surrounded by a closed curve having a **circulation direction**. We can divide any such surface into a set of infinitesimal patches of area called **tiles**. We describe a tile mathematically by using a **tile vector** $d\vec{A}$ whose magnitude is the tile's area and whose direction is perpendicular to the tile's surface toward the surface's exterior (if the tile is on a closed surface) or as indicated by the loop rule (if it is on a bounded surface). The flux of a field through such a tile is

$$d\Phi_E \equiv \vec{E} \cdot d\vec{A} = E\, dA \cos\theta \quad \text{or} \quad d\Phi_{\mathbb{B}} \equiv \vec{\mathbb{B}} \cdot d\vec{A} = \mathbb{B}\, dA \cos\theta \quad \text{(E9.1)}$$

Purpose: These equations define the **flux** of a field through a tile.

Symbols: Φ is a capital Greek phi (for "phlux" I guess), the d in front reminds us that the flux through an infinitesimally small tile will be an infinitesimally small number, and the E or $\mathbb{B}$ subscript specifies whether we are talking about the flux of the electric or magnetic field through the tile. The $\vec{E}$ and $\vec{\mathbb{B}}$ are the electric and magnetic field vectors, respectively, at the tile; $d\vec{A}$ is the tile's tile vector; and θ is the angle between the field and tile vectors.

Limitations: The tile must be small enough that it is both approximately flat and that $\vec{E}$ or $\vec{\mathbb{B}}$ is approximately uniform over its surface.

The flux through a tile can be positive, negative, or zero, depending on the value of θ. The **net flux** through a surface is the sum of $d\Phi_E$ or $d\Phi_{\mathbb{B}}$ for all tiles on a surface.

Table E9.1 The fields of symmetric objects

Type of Distribution	Unchanged by:	Field Characteristics
Spherically symmetric charge distribution; $\rho = \rho(r)$	• Rotations around any axis • Reflections across any plane through the distribution's center	• $\vec{E}$ points radially away from or toward the distribution's center • mag($\vec{E}$) depends only on the distance r from the center
Infinite cylindrically symmetric charge distribution; Axis; $\rho = \rho(r)$	• Rotations around central axis • Slides along the central axis • Reflections across any plane containing the central axis • End-for-end flips around any axis perpendicular to the central axis	• $\vec{E}$ points radially away from or toward the nearest point on the distribution's central axis • mag($\vec{E}$) depends only on the distance r from the axis
Infinite, uniformly charged planar slab; z, y, x; ρ = const.	• Slides parallel to the slab's central plane • Reflections across the central plane or planes perpendicular to that plane • End-for-end flips around any axis lying in the central plane • Rotations around any axis perpendicular to the central plane	• $\vec{E}$ points perpendicular to the slab's central plane • mag($\vec{E}$) depends only on the distance from the central plane • $\vec{E} = 0$ on the central plane
Axially symmetric current distribution; $\vec{J}$; Axis; $J = J(r)$	• Rotations around the central axis • Slides along the central axis • Reflections across any plane containing the central axis	• $\vec{\mathbb{B}}$ is tangent to a circle going around the central axis • mag($\vec{\mathbb{B}}$) depends only on the distance r from the central axis
Infinite solenoidal current distribution; $\vec{J}$; Axis; $J = J(r)$	• Rotations about the central axis • Slides along the central axis • Reflections across any plane perpendicular to the central axis	• $\vec{\mathbb{B}}$ is parallel to the central axis • mag($\vec{\mathbb{B}}$) depends only on the distance r from the central axis
Infinite planar slab carrying a uniform $\vec{J}$ parallel to its surface; z, y, x, $\vec{J}$	• Slides parallel to the slab's central plane • Reflections across the central plane • Reflections across any plane perpendicular to the central plane but parallel to the current direction	• $\vec{\mathbb{B}}$ is parallel to the surface and perpendicular to the current • mag($\vec{\mathbb{B}}$) depends only on the distance from the central plane • $\vec{\mathbb{B}} = 0$ on the central plane

E9.1 Symmetry Arguments

Introductory comments

The methods that we learned in chapters E2 and E3 for calculating electric fields and in chapter E8 for calculating magnetic fields are often difficult to apply to even simple extended objects. This chapter opens a unit subdivision that introduces a set of powerful techniques that (in certain situations) allow us to calculate electric and magnetic fields much more easily than we could by using Coulomb's law or the Biot-Savart law. Moreover, these techniques provide a useful first step toward Maxwell's equations, which we will explore in greater detail in chapter E13.

What is a symmetry argument?

These new methods are simpler in such cases partly because they take full advantage of what physicists call *symmetry arguments.* A **symmetry argument** is a technique for determining, by sheer logic, useful characteristics of the field created by a symmetric charge or current distribution. In electricity and magnetism, symmetry arguments are based on the following symmetry principle:

> If a charge or current distribution is unchanged by some kind of transformation (such as a rotation around an axis or sliding along an axis), then the distribution's *field* is also unchanged by that same transformation.

For example, if a ball's final charge distribution after rotating around some axis is indistinguishable from its inital charge distribution, then it makes sense that the electric fields created by the ball in its initial and final orientations will likewise be indistinguishable.

The principle in this simple form applies to all transformations *except* for mirror reflections of magnetic fields (which we will study in section E9.3). A careful logical argument based on this principle can determine, without involving any calculations, an astonishing amount about the field created by a suitably symmetric object, which greatly reduces the remaining work we must do to determine the field completely.

Visualizing what a transformation does to a field

It helps me when I construct symmetry arguments if I visualize the field vectors created by an object to be rigidly attached to it as bristles are attached to the body of a brush. Thus if we rotate or slide a charge distribution, its field vectors rotate or slide along with it, just as a brush's bristles would rotate or slide along with a brush. This visually expresses the idea that whatever transformation we apply to the object, we also apply to the field.

E9.2 The Electric Fields of Symmetric Objects

The easiest way to illustrate how a symmetry argument works is to work some concrete examples. In this section, we will use symmetry arguments to determine important characteristics of the electric fields of three important types of symmetric charge distributions.

Example E9.1 A Spherical Charge Distribution

Problem Use symmetry arguments to show that an electric field vector created by *any* **spherically symmetric** charge distribution at a point P inside or outside that distribution must (1) point radially toward or away from the distribution's center and (2) have a magnitude that depends at most on the distance P from that center.

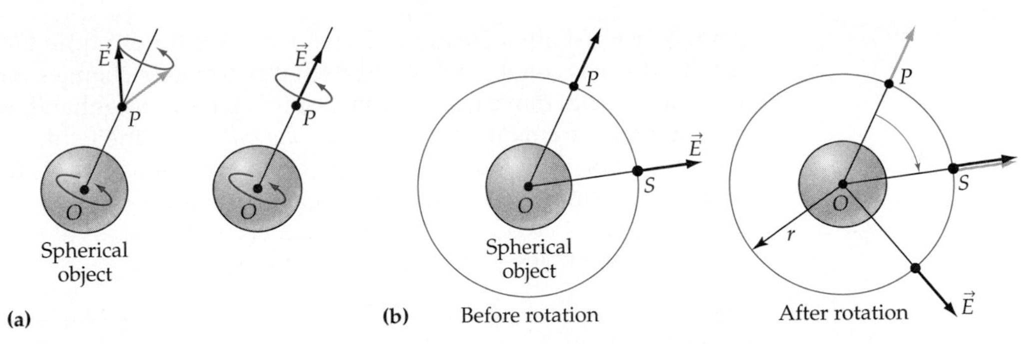

Figure E9.1
(a) A rotation about the axis *OP* changes the direction of the field vector at *P* unless it points directly away from or toward *O*. (b) Rotation around an axis perpendicular to *OP* moves the field vector at *P* to a new position *S* the same distance *r* from *O*. If the field is to remain unchanged by this rotation, the vectors originally at *P* and *S* must have the same magnitude.

Solution A distribution is *spherically symmetric* if we can rotate it about *any* axis through its center without changing it. A spherical object (like a ball or shell) whose charge density either is uniform or depends at most on the radial distance r from its center qualifies as a spherically symmetric charge distribution. The symmetry principle puts some significant constraints on the electric field that such a distribution can create.

For example, imagine that the electric field vector $\vec{E}$ at some point P near the charge distribution does *not* point along the line connecting the object's center O to P. Then rotating the object around this axis rotates the field vector at P to point in a new direction, as shown in figure E9.1a. But this rotation is not supposed to change the vector! We can avoid this logical contradiction *if and only if* the field vector at P points either directly toward or away from O (see the right drawing in figure E9.1a): a field vector pointing in either of these directions will be unchanged by the rotation. (Note that while figure E9.1 illustrates the idea using a point *outside* the charge distribution, the argument applies just as well to a point *inside* the distribution.)

Now consider rotating the object around any axis perpendicular to the line connecting O and P. This rotates the field vector that was at point P to a new position S the same distance r from point O (see figure E9.1b). Because this rotation does not affect the charge distribution, it cannot affect the field. If the field is not to change, then the length of the field vector that was at P and is now at S must have the same length as the vector *originally* at S, so that the replacement exactly matches the original. Since both the magnitude and the axis of the rotation are completely arbitrary, it follows that *all* electric field vectors that are the same distance r from the object's center O *must* have the same magnitude. This means that the magnitude of the electric field vector can at most depend on the distance r.

Evaluation This is what we set out to prove. Note that these simple statements would have been *very* hard to prove by summing contributions to the field at an arbitrary point P produced by all bits of charge in the distribution.

The structure of such an argument

Example E9.1 actually illustrates two different kinds of symmetry arguments. When making an argument regarding the field's *direction* at a given point P, we usually consider a transformation (in example E9.1, a rotation around the axis connecting O and P) that potentially changes the field vector at P but does not move it *away* from point P. On the other hand, when making a symmetry argument about how the *magnitude* of the field might or might not depend on position, we consider transformations that *do* transport field vectors from one place to another (in example E9.1, a rotation that moved the vector at point P to point S), so that field vectors at different positions can be compared. We will see that this general pattern recurs.

Exercise E9X.1

Consider a thin, uniformly charged disk. **(a)** Argue, using the approach illustrated in figure E9.1a, that the electric field vector $\vec{E}$ produced by this disk at any point along its central axis must point along that axis. **(b)** Use the approach shown in figure E9.1b to argue that the *magnitude* of the field vector $\vec{E}$ created by the disk at an *arbitrary* point P on the disk's surface can only depend on the distance r that P is from the central axis.

Example E9.2 An Infinite Cylindrical Charge Distribution

Problem Consider an infinitely long **cylindrically symmetric** charge distribution that is unchanged by rotating it *around* some central axis or sliding it any distance *along* that axis. (An infinitely long, straight cylinder, pipe, or wire has this symmetry.) Use symmetry arguments to show that an electric field vector created by *any* such distribution at an arbitrary point P inside or outside the distribution must (1) point radially toward or away from the nearest point on the distribution's central axis and (2) have a magnitude that depends at most on the distance r that P is away from that axis.

Solution Let us define our x axis to coincide with the distribution's central axis. Let O be the point on the axis closest to our arbitrary point P. If the distribution is really infinite in length, it is unchanged by flipping it end for end around the OP axis, so this flip should also not change the field vector at point P. As figure E9.2a illustrates, the field vector at P is unchanged by this

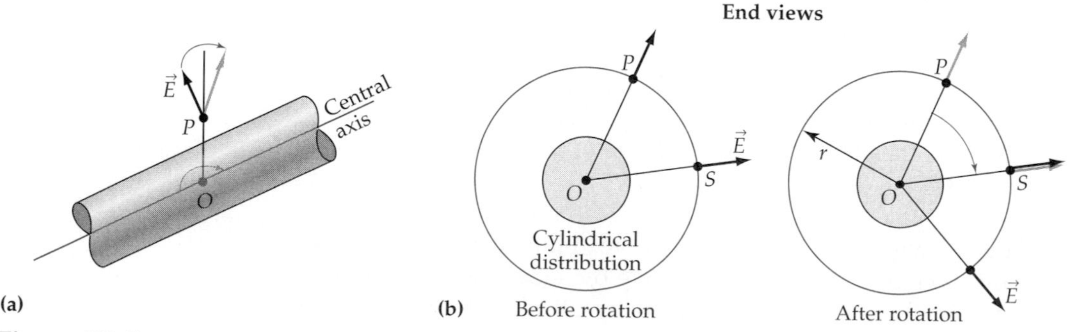

Figure E9.2
(a) A 180° flip around the axis OP does not change a cylindrical charge distribution but does change the direction of the field vector at P unless it points directly away from or toward O. (b) Rotation around the central axis moves the field vector at P to a new position S the same distance r from O. If the field is to remain unchanged by this rotation, the vectors originally at P and S must have the same magnitude.

transformation only if it points directly toward or away from the nearest point O on the x axis.

This infinite distribution is also unchanged by sliding it along the x axis or rotating it around that axis. Either transformation moves the field vector at point P to a point S the same distance r from the x axis but with a different x coordinate (see figure E9.2b). Since the field vector originally at P must be identical to the one originally at S if this transformation is not to change the field, the magnitude of the field vector must be independent of x. Indeed, a combination of a slide and a rotation can move the field vector at P to *any* other arbitrary point the same distance r from the axis, so the field vectors at *all* points a given distance r from the axis must have the same magnitude. Therefore, this magnitude can at most depend on r.

Example E9.3 An Infinite Planar Slab of Charge

Problem Consider an infinite planar slab that has a finite fixed thickness in, say, the x direction but stretches infinitely in the y and z directions. We define the slab's **central plane** to be the plane halfway between the slab's two surfaces. Imagine that we distribute charge uniformly throughout the slab's volume. Use symmetry arguments to argue that an electric field vector created by the slab at any point P inside or outside the slab must (1) point perpendicular to the slab's central plane, (2) be zero if the point lies on the slab's central plane, and (3) have a magnitude that depends at most on the distance x between P and the nearest point on the central plane.

Solution Let us define coordinates so that the slab's central plane coincides with the yz plane. Consider an arbitrary point P with a nonzero x coordinate, and let O be the point on the central plane closest to P. If the distribution is really infinite, it is unchanged by any rotation about the OP axis, so this rotation should also not change the field vector at point P. As figure E9.3a illustrates, the field vector is unchanged by this transformation only if the

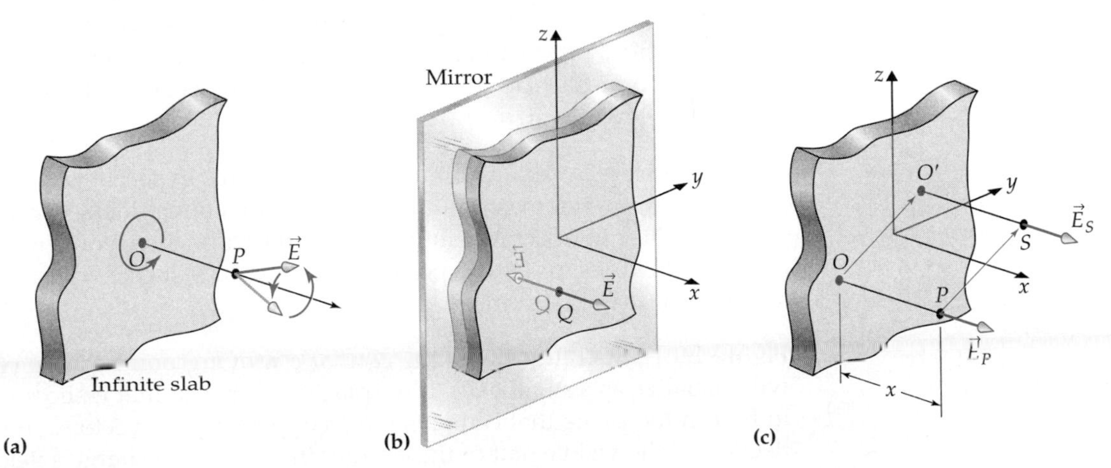

Figure E9.3
(a) A rotation around the axis OP does not change an infinite slab but does change the direction of the field vector at P unless it points directly away from or toward O. (b) A mirror reflection across the slab's central plane does not change the slab, but reverses the electric field at any point on the central plane unless the electric field is zero there. (c) Since sliding the slab an arbitrary displacement parallel to the xy plane does not affect the slab, all field vectors at points with the same x coordinate must be the same.

field vector is perpendicular to the slab. Since P was arbitrary, this applies to all points.

Now consider a point Q that lies on the slab's central plane, and let's assume for the sake of argument that the field vector has a nonzero magnitude. By the argument we have just given, this field vector must point perpendicular to the slab's central plane. But the slab is also unchanged by a mirror reflection across this central plane, so the field vector at Q should be unchanged also. But as figure E9.3b illustrates, such a mirror reflection will *reverse* the vector at Q. A field vector of zero length, however, can be reversed without changing it ($E_x = -E_x$ if $E_x = 0$). Therefore, a field vector at any point on the slab's central plane must be zero.

Finally, consider sliding the slab in the y and/or z direction. Such a transformation can carry the field vector originally at P to any arbitrary point S that is the same distance x from the slab's central plane, as shown in figure E9.3c. Since this transformation does not change the charge distribution by hypothesis, it cannot change the field, so the field vector originally at S must be the same magnitude as the field vector originally at P. Since the sliding transformation was completely arbitrary, field vectors at *all* points a given distance x from the central plane must have the same magnitude, so magnitude can depend at most on x.

E9.3 The Mirror Rule for Magnetic Fields

Two differences between electric and magnetic symmetry arguments

We can use similar symmetry arguments to determine the characteristics of the magnetic fields produced by symmetric current distributions. There are, however, two crucial differences between symmetry arguments involving electric fields and those involving magnetic fields.

The first difference is that current is a *vector* having a direction, whereas charge is a *scalar* with no direction. We can only say that a transformation leaves a *current* distribution unchanged if that transformation changes neither the shape *nor* the direction of the current (in the case of charge distributions, we only needed to worry about the shape).

The second difference is that while magnetic fields obey the simple version of the symmetry principle for transformations involving rotations or slides, the cross product that appears in the basic equation for the magnetic field of a point particle requires that mirror reflections of magnetic fields obey the following **mirror rule**†:

A statement of the mirror rule

> If we slice a current distribution with a mirror in such a way that the distribution (taking account of current directions) looks exactly the same after we insert the mirror as it did before, then a nonzero magnetic field vector at any point on the mirror's surface must be *perpendicular* to that surface.

Figure E9.4 illustrates this in the case of a moving point charge (which we will visualize as a small ball). If we place a mirror so that it slices the charge in half in the plane that contains the charge's velocity vector $\vec{v}$, then the reflection of the visible half of the charge (including the charge's velocity vector) looks identical to the part of the charge hidden by the mirror. We already know what the magnetic field vectors surrounding a point charge look like;

---

†Thanks to my Pomona colleague David Tanenbaum for formulating this rule.

we can see that each field vector on the mirror's surface is indeed perpendicular to the mirror's surface.

Note that if magnetic fields obeyed the simple version of the symmetry principle, then the field vectors on the mirror's surface would be *unchanged* by the reflection transformation, which would imply that the field vectors at points on the mirror's surface would have to lie *in the plane* of the surface. The mirror rule, in contrast, implies that magnetic field vectors at points on the mirror's surface (since they are perpendicular to that surface) are *exactly reversed* by the reflection transformation instead of being unchanged.

Once you get used to it, you will find that the mirror rule is generally the most effective way to determine the directions of magnetic field vectors created by a current distribution. We will generally use symmetry arguments involving rotation, flipping, and/or sliding to determine constraints on the *magnitudes* of magnetic fields.

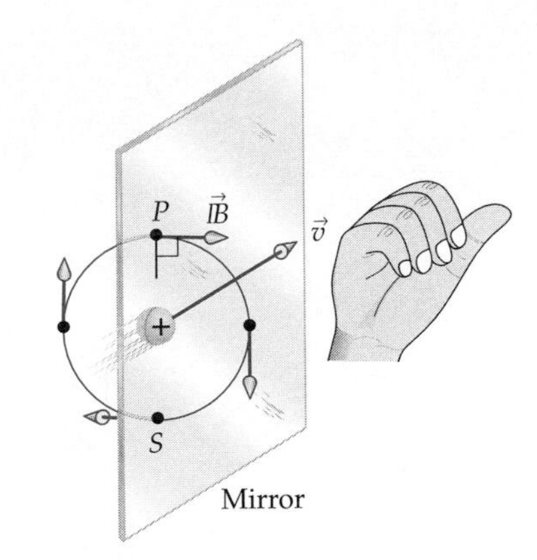

Figure E9.4
The current distribution represented by a moving charge is unaffected if we bisect it with a mirror containing its velocity vector. Note that the magnetic field produced by this charge is always perpendicular to the mirror at points on the plane of the mirror.

Exercise E9X.2

In the moving point charge example, imagine placing the mirror so that it slices the charge in half in the plane *perpendicular* to the charge's velocity. Explain why this does not satisfy the condition on mirror placement specified in the mirror rule.

E9.4 The Magnetic Fields of Symmetric Current Distributions

Examples E9.4 to E9.6 in this section illustrate the application of the mirror rule to some important types of current distributions.

Example E9.4 An Axial Current Distribution

Problem Consider a cylindrically symmetric current distribution (one that is unchanged by rotating the distribution about its central axis or sliding along that axis) that involves currents that flow only parallel to its central axis. We call such a distribution an **axial current distribution.** An infinite cylindrical wire or pipe carrying current parallel to its length is an example of such distributions. Argue that the magnetic field vector created at a given point P by such a distribution must point tangent to a circle perpendicular to the axis but centered on it, and that its magnitude can at most depend on the distance r that P is from the axis.

Solution Let us see whether we can use the mirror rule to determine the direction of the field vectors in this case. How should we place the mirror? Figure E9.5a shows that if we place a mirror in any plane perpendicular to the current distribution's central axis, the distribution's mirror image *looks* exactly like the part of the distribution behind the mirror except that the current direction is reversed. The mirror rule, therefore, does *not* apply to such a mirror. However, figure E9.5b shows that if we place the mirror so that its plane contains the central axis and the arbitrary point P where we would like to evaluate the field, the current distribution's mirror image (including the current) looks *exactly* like the part behind the mirror. The mirror rule therefore implies that the magnetic field vector P must point perpendicular to the

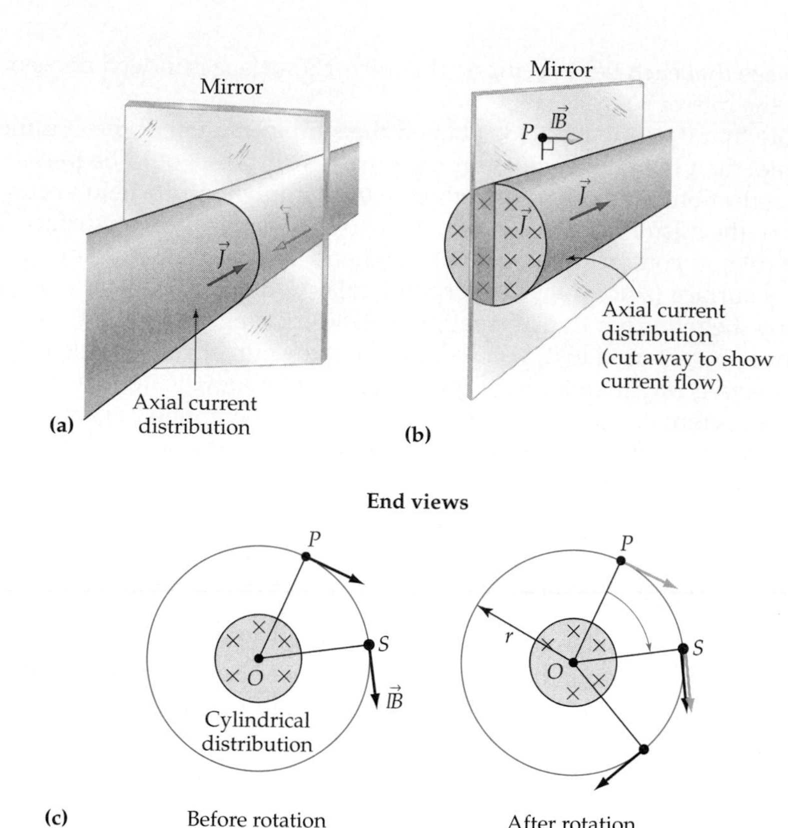

Figure E9.5

(a) If we place a mirror perpendicular to the central axis of an axial current distribution, the distribution looks the same *except* that the current is reversed in the mirror image. (b) If we place a mirror so that it bisects the distribution in a plane that contains its central axis, the distribution looks exactly the same. Therefore its magnetic field is perpendicular to the mirror at points on the plane of the mirror. (The distribution goes on indefinitely in the direction toward the viewer, but I have sliced it to display the mirror image more clearly.) (c) Rotation about the distribution's central axis moves the field vector at P to a new position S the same distance r from O. If the field is to remain unchanged by this rotation, the vectors originally at P and S must have the same magnitude.

mirror. But as figure E9.5b shows, such a field vector is indeed tangent to a circle going around the axis in a plane perpendicular to the axis. Since P was arbitrary, this applies to all points inside and outside the distribution.

Now consider rotating the current distribution an arbitrary amount about the central axis. As figure E9.5c shows, this rotation carries the field vector originally at point P to a point S that is the same distance r from the central axis. The current distribution is unchanged by such a rotation (by hypothesis), so the magnetic field should be unchanged. This will be true only if the field vector originally at P has the same magnitude as the field vector that was originally at S. Sliding the distribution along the axis also leaves it unchanged, and a combination of a rotation and a slide will carry the field vector at P to any arbitrary point that is the same distance r from the axis. Therefore the field vectors at *all* points a given distance r from the axis must have the same magnitude, implying that the magnitude of the magnetic field at a point can *at most* depend on r.

Example E9.5 A Solenoidal Current Distribution

Problem A **solenoid** is a coil of wire closely wound around a cylindrical form. Consider an infinite solenoid wrapped around a cylindrical pipe of uniform radius and infinite length. Such a coil (like the distribution in example E9.4) is essentially unchanged by rotating it about its central axis or sliding it along that axis, but the currents in this case flow *perpendicular* to the axis (contrary to the situation in example E9.4). We call such a distribution a **solenoidal** current distribution. Argue that the magnetic field created by such a solenoidal distribution is parallel to the distribution's central axis at all points, and has a magnitude at a given point P that can, at most, depend the distance that P is from the axis.

Solution Figure E9.6a shows that if we place a mirror perpendicular to the solenoid's central axis, the reflected image looks exactly the same as the part of the solenoid behind the mirror. Since we can slide the solenoid along the axis an arbitrary amount without changing it, we can slide it until any arbitrary point P lies in the plane of such a mirror. The mirror rule then implies that the magnetic field at this arbitrary point P is perpendicular to the mirror and thus is parallel to the axis.

As in example E9.4, the solenoid's current distribution is unaffected by rotating it about the axis or sliding it along the axis. Since the combination of a rotation and a slide can carry the magnetic field vector at point P to any arbitrary point S that is the same distance r from the solenoid's central axis, the field vector originally at P must have the same magnitude as that originally at point S if the field is to be unchanged by such a transformation. Therefore, the magnitude of the magnetic field vector at a given point P can at most depend on the distance r that P is from the solenoid's central axis.

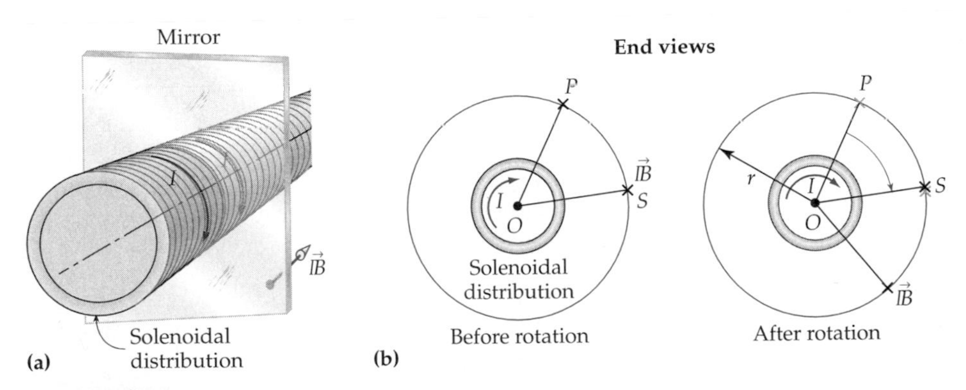

Figure E9.6
(a) In this case of a solenoidal distribution, we have to place the mirror perpendicular to the distribution's central axis to preserve the current direction in the mirror image. Therefore the magnetic field in this case is parallel to the axis. (b) Rotation about the distribution's central axis moves the field vector at P to a new position S the same distance r from O. If the field is to remain unchanged by this rotation, the vectors originally at P and S must have the same magnitude.

Exercise E9X.3

Explain why placing a mirror so that its plane contains the solenoid's central axis (as we did in example E9.4) does *not* help us determine anything about the field in this case.

Example E9.6 A Planar Current Distribution

Problem Consider an infinite conducting planar slab which (for the sake of argument) has a fixed finite width in the x direction but is infinite in the y and z directions and which is uniformly filled with current flowing in the $-y$ direction. Argue (1) that the magnetic field vector created by this distribution at *any* point P must point in the $\pm z$ direction, (2) that its magnitude can depend at most on the distance x that P is from the slab's central plane, and (3) that it must be zero if P is on the slab's central plane.

Solution For the sake of concreteness, let us define coordinates so that the slab's central plane is the yz plane. Figure E9.7a shows that if we place a mirror in the xy plane, the slab's reflected image looks exactly the same as the part of the slab behind the mirror. Since we can slide an infinite slab an arbitrary amount in the z direction without changing it, we can slide it until any arbitrary point P lies in the plane of such a mirror. The mirror rule then implies that the magnetic field at this arbitrary point P is perpendicular to the mirror and so points in the $\pm z$ direction.

Figure E9.7b illustrates that a slide parallel to the y axis combined with a slide parallel to the z axis can carry the field vector originally at any given point P to an arbitrary point S having the same x coordinate. If the field is not to be changed by such a transformation, the field vector originally at P must have the same magnitude as the field vector originally at S. Therefore, the field vectors at *all* points with the same x coordinate have the same magnitude, implying that the magnitude of such a field can at *most* depend on x.

Figure E9.7c shows that if we place a mirror in the yz plane (the slab's central plane), the reflected image of the slab *still* looks exactly like the half behind the mirror. The mirror rule tells us that nonzero magnetic field

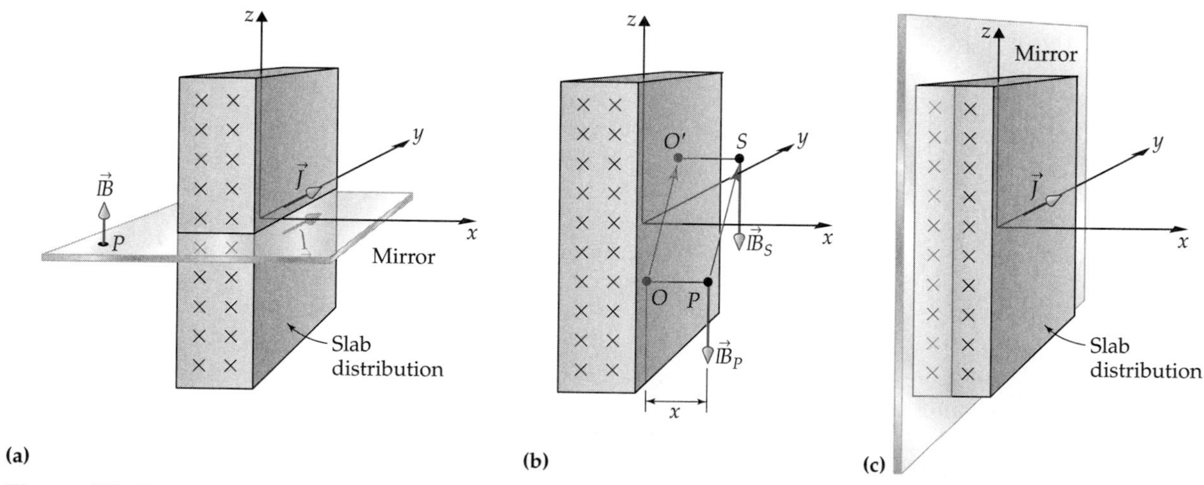

Figure E9.7
In these drawings, the slab goes indefinitely in the $\pm y$ and $\pm z$ directions, but I have sliced out a rectangular portion to make things easier to see. (a) If we place a mirror parallel to the xy plane, the mirror image of the slab looks the same as the part behind the slab. The mirror rule then implies that the magnetic field must point in the $\pm z$ direction. (b) Since sliding the slab an arbitrary displacement parallel to the xy plane does not affect the slab, all field vectors at points with the same x coordinate must be the same. (c) A mirror in the yz plane (the slab's central plane) also preserves the distribution, so vectors on the central plane must be simultaneously horizontal and vertical. This is absurd, so the magnetic field must be zero on the central plane.

vectors at points on the central plane must point perpendicular to that plane, that is, purely in the $\pm x$ direction. But we have just seen that all nonzero field vectors created by this current distribution must point purely in the $\pm z$ direction! We can avoid this contradiction only if the field vectors at all points on the slab's central plane are zero.

Exercise E9X.4

Explain why the last argument implies only that field vectors at points on the slab's central plane (as opposed to everywhere) must be zero.

Some general comments about infinite distributions

Even though infinite charge or current distributions cannot exist in the real world, they are worth discussing for three reasons: (1) Because these infinite objects have simpler symmetries than most realistic objects, they are useful when displaying how symmetry arguments work. (2) We will see that such simple distributions also provide simple example applications of equations we will develop in chapters E12 and E13. (3) These infinite distributions are actually useful as *approximations* to real objects whose actual fields would be much more difficult to calculate.

For these highly symmetric infinite objects, symmetry arguments *alone* determine everything about their fields *except* for exactly how the magnitudes of their field vectors depend on some *single* variable such as r or x. For *finite* charge or current distributions, we can still use symmetry arguments to determine the field vector directions at certain sets of points, and/or put some constraints on what the field vector magnitudes can depend on, but we cannot determine the field quite so completely (with the sole exception of the spherically symmetric charge distribution).

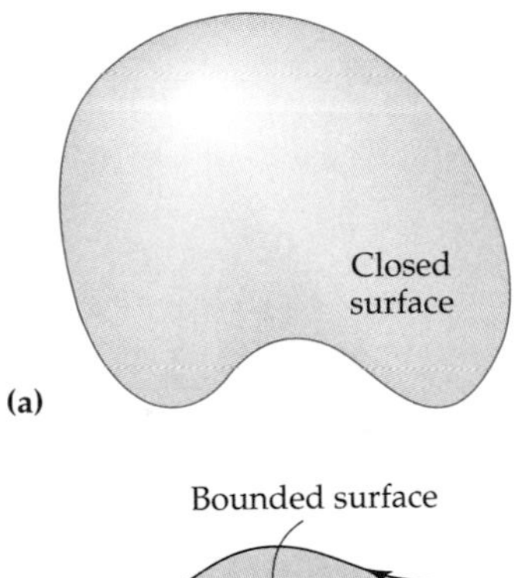

Figure E9.8
(a) An example of a closed surface. (b) An example of a bounded surface whose boundary has a defined circulation direction.

E9.5 The Flux Through a Surface

In addition to symmetry arguments, we need one more tool before diving into the next few chapters. This tool is the mathematical concept of the *flux of a field through a surface*.

Definition of some important terms

Consider a smooth imaginary two-dimensional surface in space. This surface need not be flat or correspond to the physical surface of any object. (I like to visualize such a surface instead as being like an insubstantial and permeable membrane floating motionless in space.) We call a surface a **closed surface** if it completely surrounds a certain three-dimensional volume of space. A spherical shell or cylindrical can is an example of a closed surface. Otherwise, we will usually be interested in two-dimensional **bounded surfaces** whose boundary is some **closed curve** (such as a circle or rectangle). In the applications we will consider, we will define an associated **circulation direction** for a closed curve that gives the curve a direction. Figure E9.8 illustrates these concepts. (We will see later how to define *useful* circulation directions for such boundary curves: for now, just assume that we have defined such a direction.)

Tiling a surface

We can divide any surface into tiny patches of area (which I will call **tiles**), each small enough to be essentially flat. We can compactly describe the orientation and area of such a tile by using a **tile vector** $d\vec{A}$ whose magnitude is the tile's area and whose direction is perpendicular to the tile's face (see figure E9.9). If the surface in question is *closed*, we conventionally define

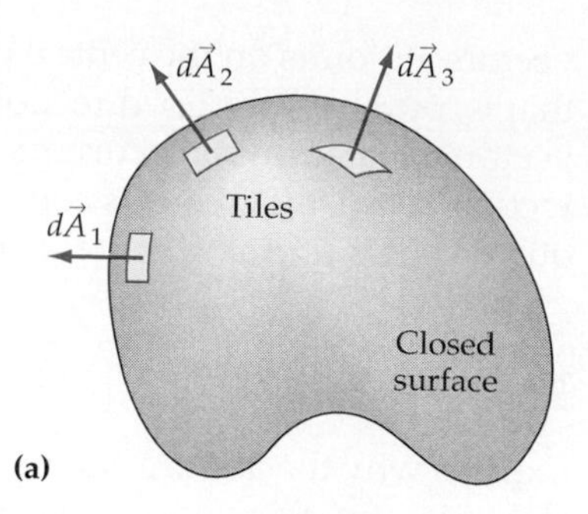

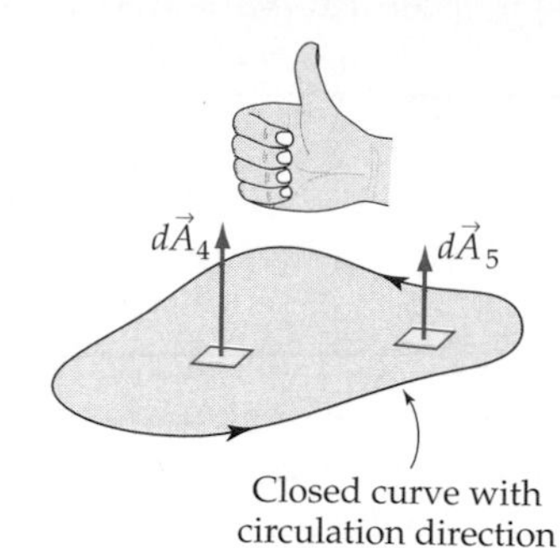

Figure E9.9
(a) Some tiles on a closed surface with their associated vectors. Such vectors conventionally point toward the outside of such a surface. (b) Some tiles on a bounded surface and their associated tile vectors. Such vectors conventionally point in the direction specified by the loop rule (if you curl your right fingers in the direction of the circulation, your thumb indicates the direction of the tile vector).

the tile vector to point away from the surface's interior. If the surface in question is *bounded*, we conventionally define the tile vector to point in the direction indicated by the loop rule: if you curl your right fingers in the direction of the boundary curve's circulation, your right thumb indicates the direction of the tile vector.

Do not stay up late puzzling over the idea that we are describing a piece of surface by a *vector*! A tile vector is simply a clever device for compactly encoding some useful information about the tile, nothing more. If when you are given a tile, you can determine its $d\vec{A}$ vector, or if when you are given a $d\vec{A}$ vector, you can imagine the size and orientation of the corresponding tile, then you have understood exactly what you need to understand.

Now imagine that our surface is floating in a region of space where there is an electric or magnetic field. We will assume in what follows that every tile is "small enough" that (1) it is essentially flat and (2) the field vectors at all points on its surface are close enough to being the same that we can adequately describe them by a single vector $\vec{E}$ (or $\vec{\mathbb{B}}$). (Note that this criterion does *not* require that the tile be infinitesimally small if the field is uniform and a large region of the surface is flat.) Given an appropriately small tile, we then *define* the **flux of a field through the tile** to be the quantity

Definition of the flux of a field through a tile

$$d\Phi_E \equiv \vec{E} \cdot d\vec{A} = E\,dA\cos\theta \quad \text{or} \quad d\Phi_{\mathbb{B}} \equiv \vec{\mathbb{B}} \cdot d\vec{A} = \mathbb{B}\,dA\cos\theta \quad \text{(E9.1)}$$

Purpose: These equations define the flux of a field through a tile.
Symbols: Φ is a capital Greek phi (for "phlux" I guess), the d in front reminds us that the flux through an infinitesimally small tile will be an infinitesimally small number, and the E or $\mathbb{B}$ subscript specifies whether we are talking about the flux of the electric or magnetic field through the tile. The quantities $\vec{E}$ and $\vec{\mathbb{B}}$ are the electric and magnetic field vectors, respectively, at the tile, $d\vec{A}$ is the tile's tile vector, and θ is the angle between the field and tile vectors.

Limitations: The tile must be small enough that it is approximately flat and that $\vec{E}$ or $\vec{\mathbb{B}}$ is approximately uniform over its surface.

How to determine the sign of the flux

Note that the flux through a tile can be positive, negative, or zero depending on the angle θ between the field and tile vectors. For example, if we are considering the flux of an electric field through tiles on a closed surface, then

$d\Phi_E > 0$	$\Leftrightarrow$	$\theta < 90^\circ$	$\Leftrightarrow$	$\vec{E}$ has a component toward surface's exterior
$d\Phi_E = 0$	$\Leftrightarrow$	$\theta = 90^\circ$	$\Leftrightarrow$	$\vec{E}$ is perpendicular to tile and thus parallel to surface (or $\vec{E} = 0$)
$d\Phi_E < 0$	$\Leftrightarrow$	$\theta > 90^\circ$	$\Leftrightarrow$	$\vec{E}$ has a component toward surface's interior

Exercise E9X.5

Imagine that in a certain region of space, the electric field $\vec{E}$ has a magnitude of 200 N/C and points in the $+x$ direction. **(a)** What is the flux through a tile whose area is 1.0 cm^2 and whose tile vector points at a 30° angle with the $-x$ axis? **(b)** What is the flux through a tile with the same area whose vector points in the $+z$ direction?

Calculating the net flux through a surface

We define the **net flux of the field through a surface** to be the sum of the fluxes through each of its tiles in the limit that the tiles become infinitesimal:

$$\Phi_E \equiv \lim_{dA\to 0} \sum_{\text{all tiles}} \vec{E}\cdot d\vec{A} \equiv \int \vec{E}\cdot d\vec{A} \quad \text{or} \quad \Phi_{\mathbb{B}} \equiv \lim_{dA\to 0} \sum_{\text{all tiles}} \vec{\mathbb{B}}\cdot d\vec{A} \equiv \int \vec{\mathbb{B}}\cdot d\vec{A} \tag{E9.2}$$

Do not let the integral symbols frighten you. These symbols are simply a shorthand way of saying, "Sum the following infinitesimal quantities." (One sometimes might use an actual integral as a tool to *evaluate* such a sum, but we will avoid such cases whenever possible. Thinking of a flux "integral" as being simply a sum is almost always the right thing to do.)

We will find it helpful to distinguish between a sum done over a closed surface and one done over a bounded surface. I will do this in what follows, using the following notation:

$$(\Phi_E) = \oint \vec{E}\cdot d\vec{A} \equiv \text{ net electric flux through a closed surface} \tag{E9.3a}$$

$$\Phi_E = \int \vec{E}\cdot d\vec{A} \equiv \text{ net electric flux through a bounded surface} \tag{E9.3b}$$

and similarly for magnetic fluxes. The parentheses around the flux symbol and the circle in the integral sign both indicate that the flux in question is calculated for a *closed* surface. I will generally use the circled integral symbol for flux through a closed surface, but the Φ symbol for flux through a bounded surface (for reasons that will become clearer later).

Example E9.7 Computing the Net Flux

Problem Consider a point particle with positive charge q. Compute the net flux of this particle's electric field through a spherical surface of radius r centered on the particle.

Model The tile vector $d\vec{A}$ of any arbitrary tile on this surface points radially outward from the point, as does the electric field vector $\vec{E}$ at that tile. So the flux through any given tile in this case is simply $d\Phi_E = \vec{E} \cdot d\vec{A} = E\,dA \cos 0° = E\,dA$. Since $E = kq/r^2$ depends only on r, it has the same magnitude for all tiles on the surface.

Solution Therefore, the net flux is

$$\oint \vec{E} \cdot d\vec{A} \equiv \sum_{\text{all tiles}} \vec{E} \cdot d\vec{A} = E \sum_{\text{all tiles}} dA = E(4\pi r^2) = \frac{kq}{r^2} 4\pi r^2 = 4\pi kq \quad \text{(E9.3)}$$

Do not worry if you don't have an intuitive concept of what flux "really means." The idea of flux has its origins in fluid mechanics, since one can calculate the rate at which a fluid flows through a surface using similar mathematics (see problem E9S.7). An electric or magnetic field is not actually "flowing," of course, but the image of fluid flowing through a surface might be a helpful metaphor.

For our purposes, though, flux is simply a mathematical tool, like the cross product, that makes writing certain future equations easier. As long as you know exactly what to *do* when you are asked to calculate the flux or net flux in a given situation, you have understood what you need to understand. We will see what this tool is good for in chapter E10.

TWO-MINUTE PROBLEMS

E9T.1 Consider a uniformly charged circular ring. Which of the following transformations leave the charge distribution unchanged?

A. An arbitrary rotation about the ring's central axis
B. An arbitrary rotation about an axis in the plane of the ring
C. A reflection across any plane containing the ring's central axis
D. Sliding the ring parallel to its central axis
E. All the above
F. A, B, and C
T. A and C

E9T.2 Consider a *finite* square planar slab that is uniformly filled with charge. Assume that the slab's central plane coincides with the yz plane. The following statements are true about the electric field of an infinite planar slab; which of them remain true for the *finite* slab?

A. The electric field's direction at any point is perpendicular to the slab.
B. The field vector's magnitude at a point can at most depend on the x coordinate of that point.
C. The electric field is zero at any point on the slab's central plane.
D. All the above remain true.
E. B and C remain true.
F. All these statements are false for the finite slab.

E9T.3 Consider a uniformly charged plastic hemispherical bowl (half a spherical shell). Let us define the bowl's "center" to be the point at the center of the circle defined by the bow's rim (this would be the center of the complete sphere formed by two such bowls), and its "central axis" to be a line going through the center perpendicular to the plane defined by the bowl's rim. Which of the following statements follow logically from the bowl's symmetry?

A. All electric field vectors point radially away from or toward the bowl's center.

B. The electric field vectors at all points on the plane defined by the bowl's rim are perpendicular to that plane.
C. The magnitude of the electric field at an arbitrary point depends at most on the distance that point is from the bowl's central axis.
D. The electric field vector at any point must lie in a plane containing that point and the bowl's central axis.
E. All these statements logically follow.
F. None of these statements logically follow.
T. Some combination of these statements follows. (Specify which.)

E9T.4 Consider a circular loop that carries a uniform current. Define the loop's central plane to be the plane that slices the loop exactly in half as one would slice a bagel. Which of the following ways to place a mirror satisfy the conditions of the mirror rule?
A. Placing a mirror in any plane containing the loop's central axis
B. Placing a mirror in the loop's central plane (dividing the loop in half like a bagel)
C. Placing a mirror in any plane parallel to the loop's central plane
D. All the above
E. A and B
F. None of the above

E9T.5 Imagine that the charge density inside an infinite slab is not uniform but is proportional to some function $f(|x|)$ of the distance $|x|$ that one is from the slab's central plane. The following statements are true for the infinite uniformly charged slab. Which of these statements remain true for the nonuniformly charged slab?
A. The electric field at all points is perpendicular to the slab.
B. The electric field has a magnitude at a point that at most depends on the distance $|x|$ the point is from the slab's central plane.
C. The electric field is zero at the center of the slab.
D. All these statements remain true.
E. None of these statements remain true.
F. A pair of these statements remains true (specify).

E9T.6 Consider a conducting loop that carries current in a counterclockwise direction, and let the loop define the boundary of a bounded surface, with the current direction specifying the circulation direction of the boundary. The magnetic flux through this surface is
A. Positive
B. Negative
C. Zero
D. Some quantity whose sign we could determine if given more information

E9T.7 Imagine that in a certain region of space, the electric field is essentially $\vec{E} = 200$ N/C in the vertical direction, as shown in figure E9.10. Consider also the cubic closed surface shown in that figure. Find the electric field flux through the
(a) Top horizontal face
(b) Bottom horizontal face
(c) Left vertical face

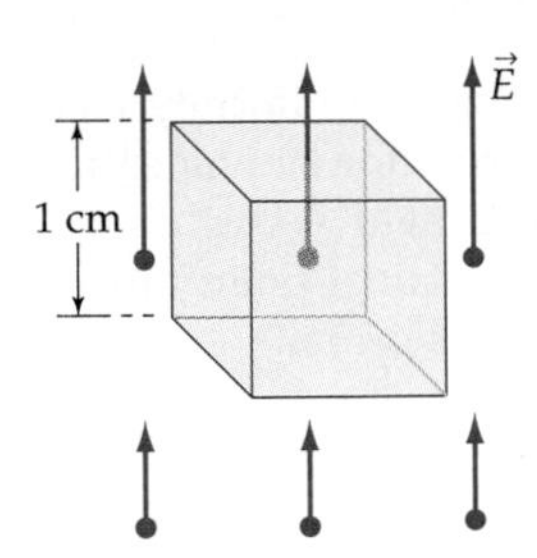

Figure E9.10
What is the net electric field flux through this cubic surface? (See problem E9T.7.)

A. $+200\ \text{N}\cdot\text{m}^2/\text{C}$
B. $-200\ \text{N}\cdot\text{m}^2/\text{C}$
C. $+2\ \text{N}\cdot\text{m}^2/\text{C}$
D. $-2\ \text{N}\cdot\text{m}^2/\text{C}$
E. 0
F. Some other result (specify)

E9T.8 In chapter E8, we determined that the magnetic field of an infinite straight wire carrying current I points circularly around the wire and, at a point a distance r from the wire, has a magnitude of $\mathbb{B} = 2\pi kI/cr$. Imagine that we surround the wire with the cylindrical can-shaped closed surface, illustrated in figure E9.11. The net flux through this surface is

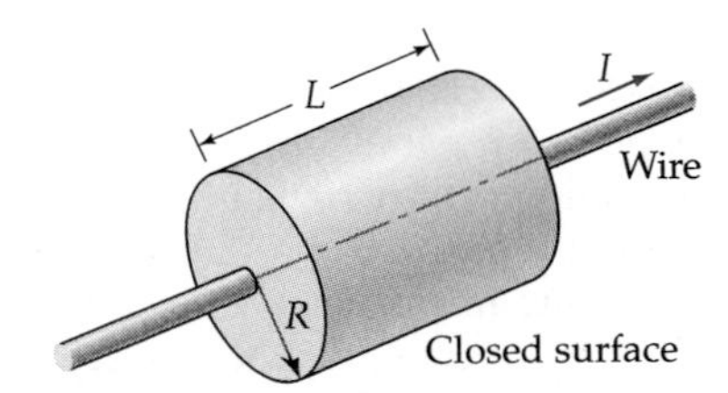

Figure E9.11
What is the net magnetic field flux through this can-shaped surface? (See problems E9T.8 and E9S.9.)

A. Positive
B. Negative
C. 0
D. Some quantity whose sign we could determine if given more information

HOMEWORK PROBLEMS

Basic Skills

E9B.1 Consider a uniformly charged, circular thin ring. Use a rotation symmetry argument to prove that the electric field vector created by this ring at any point along its central axis must point parallel to that axis.

E9B.2 Consider a uniformly charged, circular thin ring. Use a rotation symmetry argument to prove that the electric field vector at any point in the plane of the ring must point directly toward or away from the ring's center.

E9B.3 Consider a uniformly charged, circular thin ring. Use a rotation symmetry argument to prove that the magnitude of the electric field vector at any point in the plane of the ring can depend at most on the distance r that the point is from the ring's center.

E9B.4 Consider the infinite planar current distribution described in example E9.6. Use a rotation argument to prove that if the magnetic field points in the $+z$ direction at points on one side of the slab's central plane, it must point in the $-z$ direction at points on the other side.

E9B.5 Consider a circular loop of wire that carries a uniform current. Use a rotation argument to prove that the magnetic field vector at any point along the loop's central axis must point parallel to that axis.

E9B.6 In example E9.5, we used a mirror placed perpendicular to the solenoid's axis to determine the direction of the magnetic field. Explain why this is the only possible choice for mirror placement that satisfies the conditions of the mirror rule in the case of an infinite solenoid.

E9B.7 In example E9.6, we used mirrors in different orientations but parallel to the current to determine the direction of the magnetic field and argue that it was zero at points on the slab's central plane. Explain why any mirror whose plane is *not* parallel to the current direction cannot satisfy the conditions of the mirror rule in the case of the infinite plane.

E9B.8 A tile in the horizontal plane xy plane has an area of 2.0 cm^2. Assume that the circulation direction for its boundary is counterclockwise when viewed from above.

(a) What is the flux of the magnetic field through this tile if in the vicinity of this tile, the magnetic field is approximately $\vec{\mathbb{B}} = 300$ MN/C in a direction 37° up from the horizontal?

(b) What is the flux if the magnetic field has a magnitude of 100 MN/C in the $+y$ direction?

E9B.9 The electric field at a certain tile on a surface is 1500 N/C in the $+x$ direction.

(a) What is the flux of the electric field through the tile if $d\vec{A} = 2.0 \times 10^{-5}$ m^2 60° from the $+x$ direction?

(b) What is the flux if $d\vec{A} = 5.0 \times 10^{-5}$ m^2 in the $+y$ direction?

Synthetic

E9S.1 Consider a uniformly charged, circular thin disk. In exercise E9X.1, we saw that a rotation argument shows that the electric field vectors at a point along the disk's axis must point parallel to that axis. Find another kind of symmetry argument that leads to the same conclusion.

E9S.2 Use a symmetry argument to prove that the electric field must be zero at the center of any spherically symmetric charge distribution.

E9S.3 Use a symmetry argument to prove that the electric field must be zero at points along the central axis of a cylindrically symmetric charge distribution.

E9S.4 Use a symmetry argument to prove that the magnetic field must be zero at points along the central axis of an axially symmetric current distribution.

E9S.5 Consider the infinite planar current distribution described in example E9.6. In the example, we used a reflection argument to prove that the magnetic field everywhere must point in the $\pm z$ direction. We then used another reflection argument to prove that the field vectors at points on the slab's central plane must be zero. Find a different way to prove the second conclusion, given that the field vectors everywhere must point in the $\pm z$ direction.

E9S.6 A toroidal coil (see figure E9.12) is a coil of wire wound around a form shaped like a donut, a shape that mathematicians call a *torus*. (One can also think

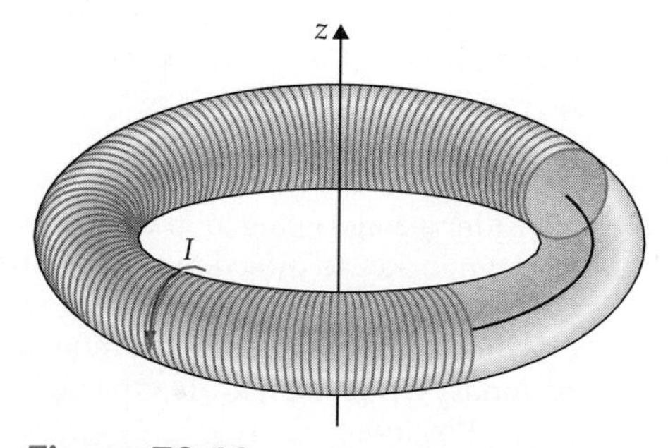

Figure E9.12
What are the characteristics of the magnetic field produced by this toroidal coil? (See problem E9S.6.)

of a toroidal coil as being a solenoid that has been bent into a circle.) Let us define coordinates so that the z axis coincides with the toroid's central axis, and let r be the distance that a given point is from that central axis.

(a) Use the mirror rule to prove that at any point inside or outside the coil the magnetic field of such a coil points tangent to a circle drawn around the toroid's center (the center of the donut hole).

(b) Use a symmetry argument to prove that the magnitude of this field can at most depend on the values of r and z for that point.

E9S.7 Consider a body of flowing water, such as a river. We can describe the flow of river water mathematically by a vector field that describes the water velocity $\vec{v}$ at every point in the river at an instant of time. Now imagine a small, flat surface submerged in the river. (If you want to visualize something specific, imagine a square grate or a picture frame or something analogous that clearly defines an area but that water can flow through easily.) Assume that the surface is small enough that $\vec{v}$ is roughly constant over its surface, and that we describe this surface by the tile vector $d\vec{A}$.

(a) First, imagine that the surface is perpendicular to the water flow in its vicinity, so that the flux of water through the surface is $d\Phi_v \equiv \vec{v} \cdot d\vec{A} = +v\, dA$. Argue that the volume of water that flows through this surface per unit time is equal to the flux. (*Hint:* Argue that the volume of water that will flow through the surface in a time interval Δt is $v \Delta t\, dA$.)

(b) Now imagine that the surface has an arbitrary orientation with respect to the flow. By extending your argument for part (a), argue that no matter how you orient the surface, the volume of water per unit time that flows through the surface is equal to the flux $d\Phi_v \equiv \vec{v} \cdot d\vec{A}$. (*Hint:* What cross-sectional area does the surface present to the flow?)

E9S.8 In chapter E2, we found that the magnitude of the electric field at a point a distance r from an infinite straight wire with a uniformly distributed positive charge is $E = 2k\lambda/r$, where λ is the charge per unit length on the wire. Imagine that we surround a portion of such a wire with a closed surface shaped like a cylindrical can, as shown in figure E9.13. What is the total flux of the electric field through this surface? (*Hint:* Calculate first the flux through the two end caps, then the flux through the remainder, and sum.)

E9S.9 In chapter E8, we determined that the magnetic field of an infinite straight wire carrying current I points circularly around the wire and, at a point a distance r from the wire, has a magnitude of $\mathbb{B} = 2\pi k I/cr$. Imagine that we surround the wire with the cylindrical can-shaped closed surface,

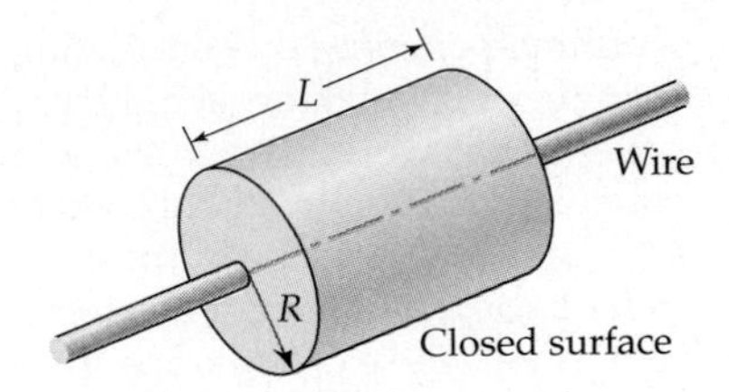

Figure E9.13
What is the net electric field flux through this can-shaped surface? (See problem E9S.8.)

illustrated in figure E9.11 on page 187. Calculate the net flux of the magnetic field through this surface.

Rich-Context

E9R.1 Consider a uniformly charged infinite slab moving with velocity $\vec{v}$ perpendicular to its surface (see figure E9.14). Since any moving charge constitutes a current, this moving slab represents a *current* distribution. In this situation, it turns out that symmetry arguments *completely determine* the magnetic field at all points inside and outside the slab. Describe the magnetic field of this slab and provide the symmetry arguments that support your description.

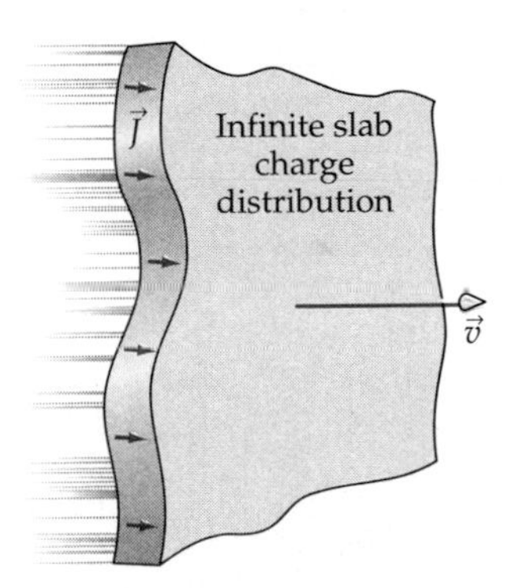

Figure E9.14
What is the magnetic field produced by this moving infinite slab? (See problem E9R.1.)

E9R.2 The government has discovered that unknown alien entities have placed a wormhole in deep waters in the middle of the Pacific Ocean. The surface of the wormhole opening looks like a sphere that appears to be about 12 m in diameter; any water that crosses the boundary of this sphere is teleported somewhere, where (according to one CIA theory) the aliens are using the water to supply an outpost in the solar system. Assume that the wormhole is suspended in the ocean far enough below the ocean's surface and far enough above the bottom that we can adequately model the behavior of the water within a few hundred meters of the wormhole opening by considering the ocean to be infinite when making symmetry arguments. The government would like to know the rate at which water is being removed: this will help investigators estimate the size of the outpost and therefore what to look for in surveys of the solar system. You are sent in a deep-sea submarine to investigate. When you are 50 m from the wormhole opening, you find that you are just barely able to keep the submarine at rest against the inward current. Your submarine's top speed in still water is 5 mi/h. Use this information, a suitable symmetry argument, and the result of problem E9S.7 to estimate the rate at which the wormhole is removing water. (*Hint:* Consider the total flux of water through an imaginary spherical surface surrounding the wormhole opening.)

Advanced

E9A.1 Imagine a point particle with charge q placed a distance d above the center of a horizontal disk-shaped surface of radius R (see figure E9.15). Let's define the circulation direction of the boundary of this disk to be counterclockwise when viewed from above. Compute the flux of the charge's electric field through this disk. (*Hint:* Divide the disk's surface into rings of radius $r < R$ and width dr. Compute the flux through each ring, and then sum over all rings, using one of the integrals on the inside front cover.)

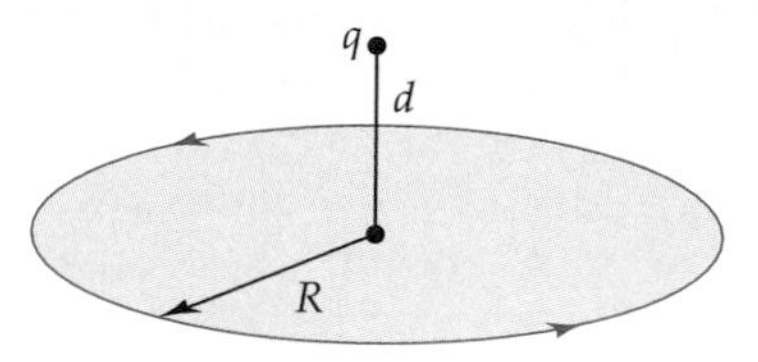

Figure E9.15
What is the net electric field flux through this disk-shaped surface? (See problem E9A.1.)

ANSWERS TO EXERCISES

E9X.1 (a) Imagine rotating the disk around its central axis. If the electric field vector at any point along that axis were *not* to point parallel to the axis, it would be changed by such a transformation, as shown in figure E9.16a. But this transformation does not change the disk and so should not change the disk's field. The only way for the electric field to be unchanged at points along the axis under a rotation about that axis is for the electric field to point parallel to the axis (either *toward* the disk's center or away from it). (b) No matter what direction the electric field has at some point P on the disk's surface, rotating the disk carries that vector to some other point S the same distance r from the disk's center, as shown in figure E9.15b. The disk is not changed by this rotation, so the field should not be changed either. For this to be true, the electric field vector originally at P must have the same magnitude as the vector originally at S, so that $\vec{E}_P$ matches $\vec{E}_S$ when the rotation carries it to the point S. Since P and S are arbitrary, this means that *all* vectors the same distance r from the disk's center must have the same magnitude.

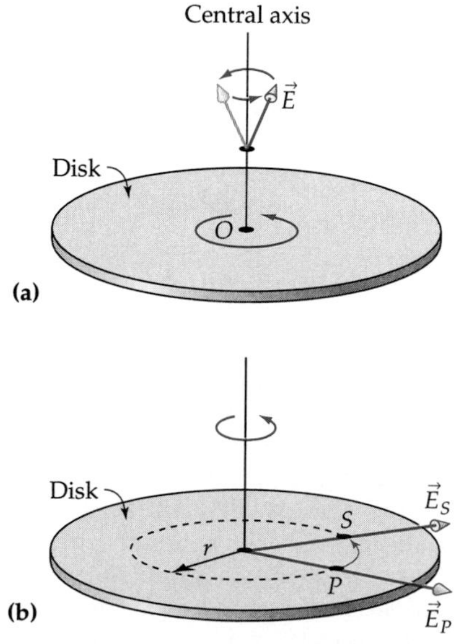

Figure E9.16
See the solution to exercise E9X.1.

E9X.2 If the mirror is placed perpendicular to the charge's velocity, the charge's velocity will appear reversed, as shown in figure E9.17. This means that the *current* distribution represented by the moving charge is *not* the same after the mirror is placed as

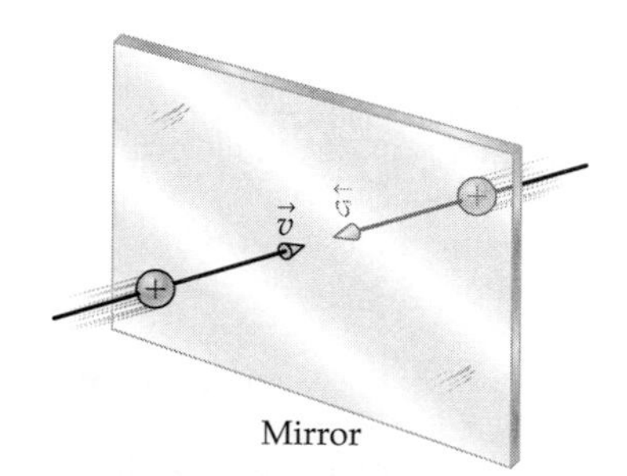

Figure E9.17
See the solution to exercise E9X.2.

it was before, even if the mirror exactly bisects the charge at an instant. Therefore, this mirror does not satisfy the conditions imposed by the mirror rule.

E9X.3 If the plane of the mirror contains the symmetry axis, figure E9.18 shows that the solenoid current distribution does *not* look the same before and after we place the mirror, because the current direction is reversed in the mirror image. Therefore, the conditions of the mirror rule do not apply, and this means that placing the mirror this way does not tell us anything useful about the current distribution.

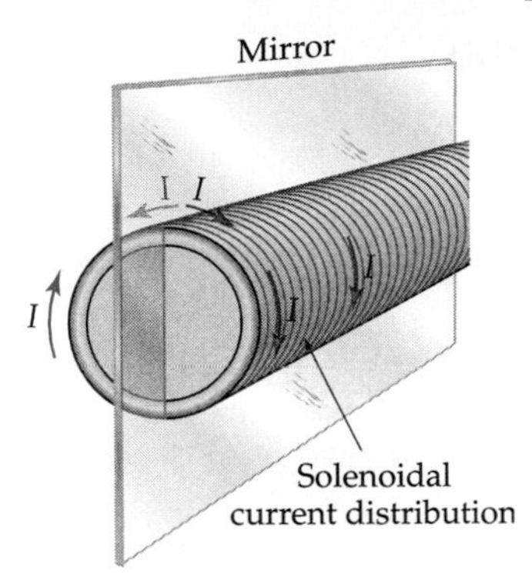

Figure E9.18
See the solution to exercise E9X.3.

E9X.4 If the plane of the mirror does not exactly coincide with the slab's central plane, then the distribution will *not* look exactly the same after the mirror was placed as before, because the real part of the slab plus the reflected image will look either thicker or thinner than the initial slab (or like two slabs or no slab). Only when the mirror exactly bisects the slab down the middle do we get something that looks exactly like what we had before, and therefore only this situation satisfies the conditions of the mirror rule.

E9X.5 The flux of the electric field through a tile is given by $\vec{E} \cdot d\vec{A} = E\,dA \cos\theta$, where $\vec{E}$ is the electric field at the tile, $d\vec{A}$ is the tile's tile vector, and θ is the angle between those vectors. (a) When $d\vec{A}$ points 30° from the $-x$ direction, which is 150° from the $+x$ direction, then $E\,dA\cos\theta = (200\text{ N/C})(1.0\text{ cm}^2) \times (\cos 150°) = -173\text{ N}\cdot\text{cm}^2/\text{C} = -0.017\text{ N}\cdot\text{m}^2/\text{C}$ (since $1.0\text{ cm}^2 = -1.0 \times 10^{-4}\text{ m}^2$). (b) In this case, the angle between the $\vec{E}$ and $d\vec{A}$ vectors is 90°, so $E\,dA\cos\theta = 0$.

E10 Gauss's Law

▷ Electric Field Fundamentals
▷ Controlling Currents
▷ Magnetic Field Fundamentals
▽ Calculating Static Fields
Symmetry and Flux
Gauss's Law
Ampere's Law
▷ Dynamic Fields

Chapter Overview

Introduction

This chapter and chapter E11 explore how we can use the tools developed in chapter E9 to state new laws for electric and magnetic fields that make it easy to calculate static fields of symmetric objects. These new electromagnetic laws have many important applications in physics, chemistry, and engineering. In particular, we will use them in chapters E12 and E13 as we construct Maxwell's equations.

Section E10.1: Gauss's Law

$$(\Phi_E) \equiv \oint \vec{E} \cdot d\vec{A} = 4\pi k q_{\text{enc}} = \frac{q_{\text{enc}}}{\varepsilon_0} \quad \text{(E10.1)}$$

Purpose: This equation, which is called **Gauss's Law**, provides an easy way to calculate the electric field of a suitably symmetric charge distribution.

Symbols: (Φ_E) denotes the net electric flux through an arbitrary closed surface, $\oint$ is simply a shorthand notation for "sum over all infinitesimal tiles on a closed surface" in this case, q_{enc} is the net charge enclosed by the particular chosen surface, $\vec{E}$ is the electric field vector at a given tile, $d\vec{A}$ is the tile's tile vector, k is Coulomb's constant, and $\varepsilon_0 = 1/4\pi k$.

Limitations: Each tile must be small enough that it is approximately flat and that $\vec{E} \approx$ uniform over its surface. The surface must completely enclose the volume (but can otherwise have an arbitrary shape).

Note: The closed surface may be freely chosen and can have any shape, and can even be completely imaginary.

In the case of *static* charges, this law is mathematically equivalent to the combination of Coulomb's law and the superposition principle. However, Gauss's law also works in cases where Coulomb's law fails, so it is actually more fundamental.

Section E10.2: Using Gauss's Law

This four-step process makes it easy to calculate electric fields using Gauss's law:

1. ***Use symmetry*** to learn as much as possible about the field's direction at all points and what its magnitude does *not* depend on.
2. ***Choose a gaussian surface*** (an imaginary closed surface) shaped so that it is easy to calculate the net flux across the surface.
3. ***Calculate the flux*** through the gaussian surface.
4. ***Apply Gauss's law*** to finish solving the problem.

Section E10.3: The Field of a Spherical Charge Distribution

Examples E10.2 and E10.3 illustrate how we can use Gauss's law to prove that the electric field of any spherically symmetric object at a given distance r from its center is as if all the charge inside the radius r were concentrated at that center and any charge outside that radius did not exist.

Section E10.4: The Field of an Infinite Cylindrical Distribution

A symmetry argument coupled with Gauss's law implies that the external field of any cylindrically symmetric charge distribution is indeed that claimed in table E3.1:

$$\vec{E} = \frac{2k\lambda}{r}\hat{r} = \frac{1}{2\pi\varepsilon_0}\frac{\lambda}{r}\hat{r} \tag{E10.9}$$

Purpose: This equation describes the external electric field $\vec{E}$ of a charge distribution that is unchanged by rotation about and/or sliding along an axis.

Symbols: r is the distance that the point where the field is evaluated is from the central axis, $\hat{r}$ is a unit vector pointing directly away from the central axis, k is Coulomb's constant, and $\varepsilon_0 = 1/4\pi k$.

Limitations: This only applies to an infinitely long charge distribution, though the result yields a pretty good approximation if the distribution is very long compared to its diameter.

Section E10.5: The Field of an Infinite Planar Slab

Consider an infinite planar slab perpendicular to the x direction with a uniform charge density ρ_0 spread throughout its thickness W. A symmetry argument coupled with Gauss's law implies that the electric field inside and outside the slab is

$$E_x = \begin{cases} +2\pi k\sigma_0 = +\dfrac{1}{2}\dfrac{\sigma_0}{\varepsilon_0} & \text{for } x > \tfrac{1}{2}W \\ (4\pi k\rho_0)x = \dfrac{\rho_0}{\varepsilon_0}x & \text{for } -\tfrac{1}{2}W \le x \le \tfrac{1}{2}W \\ -2\pi k\sigma_0 = -\dfrac{1}{2}\dfrac{\sigma_0}{\varepsilon_0} & \text{for } x < -\tfrac{1}{2}W \end{cases} \tag{E10.15}$$

$$E_y = E_z = 0 \tag{E10.16}$$

Purpose: This equation describes the electric field $\vec{E}$ of an infinite planar slab perpendicular to the x axis and having a uniform internal charge density ρ_0.

Symbols: x is the x coordinate of the point where the field is evaluated, $\hat{x}$ stands for the $+x$ direction (the direction perpendicular to the slab), W is the slab's thickness, $\sigma_0 = \rho_0 W$ is the slab's charge per unit area, k is Coulomb's constant, and $\varepsilon_0 = 1/4\pi k$.

Limitations: These results strictly only apply to an *infinite* slab.

Notes: Note that the field outside the slab is uniform. The equations for the slab's external field apply even if the slab's charge density varies with x.

Section E10.6: Gauss's Law for the Magnetic Field

$$\oint \vec{\mathbb{B}} \cdot d\vec{A} = 0 \quad \text{or} \quad \oint \vec{B} \cdot d\vec{A} = 0 \tag{E10.18}$$

Purpose: This equation, which is Gauss's law for the magnetic field, describes an intrinsic characteristic of magnetic fields.

Symbols: $\oint$ means "sum the following over all infinitesimal tiles on a closed surface," $\vec{\mathbb{B}} = c\vec{B}$ represents the magnetic field vector at a given tile on the surface, and $d\vec{A}$ is the tile's tile vector.

Limitations: Each tile must be small enough that it is approximately flat and that $\vec{\mathbb{B}} \approx$ uniform over its surface. There are no restrictions on the shape or size of the surface, or on whether it contains moving charges or not.

This law essentially means that there is no such thing as magnetic charge.

E10.1 Gauss's Law

In chapter E9, we learned how to construct symmetry arguments and how to calculate the flux of a field through a surface. It is pretty obvious why symmetry arguments might be useful, but the point of calculating flux remained obscure at the end of chapter E9. We are just about to discover why that mathematical tool is valuable.

A verbal statement of Gauss's law

Gauss's law makes the following bold and unexpected assertion about the nature of the electric field created by an arbitrary charge distribution:

> The net flux of the electric field through any arbitrary closed surface is equal to $4\pi k$ times the charge enclosed by that surface.

Mathematically, we express this law as follows:

A mathematical statement of Gauss's law

$$(\Phi_E) \equiv \oint \vec{E} \cdot d\vec{A} = 4\pi k q_{\text{enc}} = \frac{q_{\text{enc}}}{\varepsilon_0} \qquad \text{(E10.1)}$$

Purpose: This equation provides an easy way to calculate the electric field of a suitably symmetric charge distribution.

Symbols: (Φ_E) denotes the net electric flux through an arbitrary closed surface, $\oint$ is a shorthand notation for "sum over all infinitesimal tiles on a closed surface," q_{enc} is the net charge enclosed by the particular chosen surface, $\vec{E}$ is the electric field vector at a given tile, $d\vec{A}$ is the tile's tile vector, k is Coulomb's constant, and $\varepsilon_0 = 1/4\pi k$.

Limitations: Each tile must be small enough that it is approximately flat and that $\vec{E} \approx$ uniform over its surface. The surface must completely enclose the volume (but can otherwise have an arbitrary shape).

Note: The closed surface may be freely chosen and can have any shape, and can even be completely imaginary.

This is one of the most marvelous laws in physics, because (once one understands the concept of flux) it is easy to state and use, and yet (as we will see) it is also very powerful, making almost trivial calculations that are otherwise very difficult.

Gauss's law follows from Coulomb's law and superposition

However, I suspect that you may feel this law comes out of nowhere. Who could have thought of such a law, and why believe it to be true? Remember that in example E9.7, we saw that when we have a point particle with charge q at rest at the center of a spherical surface, the net electric flux through that spherical center was $4\pi kq$. One *can* show (see problem E10S.8) that because of the inverse-square nature of the electric field, we can deform the surface arbitrarily and still get the same result as long as it still encloses the charge. Moreover, one can show that the net flux through the surface due to any *external* point charge is zero. By the superposition principle, then, the net flux due to any *set* of point charges is simply equal to the sum of $4\pi kq$ for any charges inside the surface and 0 for any external charges. Therefore, at least for static fields, Gauss's law is mathematically equivalent to a combination of the superposition principle and Coulomb's law $\vec{E} = (kq/r^2)\hat{r}$ for the field of a point particle. This is basically the approach Carl Friedrich Gauss used when developing the law in the early 1800s.

Gauss's law is more fundamental than Coulomb's law

However, from a 21st-century perspective, this way of looking at Gauss's law is exactly *backward!* As we will see in chapter E12, $\vec{E} = (kq/r^2)\hat{r}$ really only applies to point particles *at rest*, while experiments show that Gauss's law is correct even for charges in relativistic motion. Therefore, *Gauss's law is more fundamental than Coulomb's law.* Though it would have made no sense as a teaching strategy, if I had wanted to start with the most fundamental principles, I should have taken Gauss's law as a given in chapter E1 and derived Coulomb's law from it!

Why charge on a conductor ends up on its surface

Example E10.1 illustrates how Gauss's law makes it simple to prove an important and useful fact that would be very difficult to prove using Coulomb's law.

Example E10.1 Excess Charge Ends Up on a Conductor's Surface

Problem Use Gauss's law to prove that any excess charge placed on any arbitrarily shaped conducting object must ultimately end up on the conductor's surface when the conductor reaches static equilibrium. (This was something I claimed without proof in section E4.4.)

Model and Solution We know from chapter E4 that when a conductor reaches static equilibrium, the electric field inside the object must be zero at all points inside the conductor's surface. Therefore, consider an arbitrary closed surface anywhere completely within the conductor. Since the electric field is zero at all points on such a surface, the net electric flux $\Phi_E \equiv \sum \vec{E} \cdot d\vec{A}$ through this surface will necessarily be zero. Gauss's law then directly implies that the net charge enclosed by *any* closed surface that lies completely inside the conductor must be zero. If we choose the surface to be just barely inside the conductor's surface, as shown in figure E10.1, we can see that since this imaginary surface must contain zero net charge, any excess charge must lie *outside* this surface, that is, on the conductor's surface. Q.E.D.

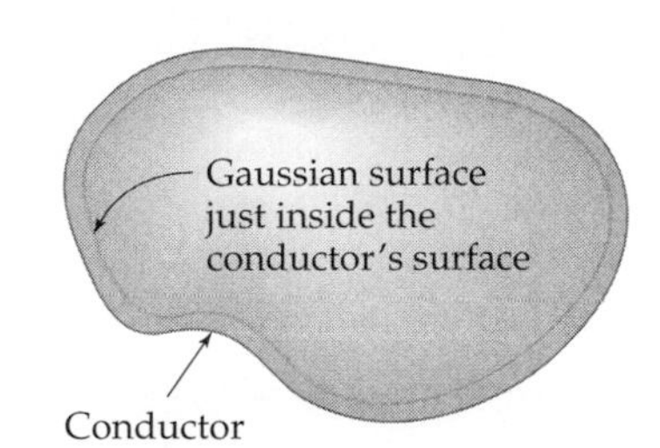

Figure E10.1
Because $\vec{E} = 0$ at all points on the gaussian surface shown, the net charge enclosed by this surface must be zero. Therefore, any net charge on the conductor must lie on its surface.

Exercise E10X.1

Choosing the closed surface to be just inside the conductor's surface only implies that the *net* charge within the conductor is zero. Is it possible that in static equilibrium, there could be a clump of positive charge hiding somewhere within the conductor that is canceled by a clump of negative charge somewhere else? Use Gauss's law to show that this is impossible: the interior of a conductor must be uniformly electrically neutral. (*Hint:* Consider a gaussian surface surrounding a region inside the conductor where you think charge might be hiding.)

E10.2 Using Gauss's Law to Calculate Electric Fields

A four-step process for using Gauss's law

In example E10.1, we used Gauss's law and knowledge about the electric field inside a conductor to determine the charge inside that conductor. More often, however, we will do the reverse; we use Gauss's law to calculate the electric field surrounding a given charge distribution.

You can often solve the latter kind of problem by following these four steps:

1. ***Use symmetry.*** You should first use symmetry and/or other arguments to determine as much as possible about the field's direction and magnitude at all points, before you attempt to use Gauss's law. Gauss's law will be easiest to apply when you know the field's precise direction at all

points and you know at least what the field's magnitude does *not* depend on.
2. ***Choose a gaussian surface.*** You should then pick a useful imaginary closed surface (which we will call a **gaussian surface**) shaped so that it is easy to calculate the net flux across the surface.
3. ***Calculate the flux.*** Then calculate the flux through this surface. If you have chosen your surface well, this should be easy.
4. ***Apply Gauss's law*** to finish determining the field.

Choosing a gaussian surface

A key part of this process is to choose the appropriate gaussian surface; if you do this well, the rest of the solution is pretty easy, but an inappropriately chosen surface can make your life miserable. The appropriate surface is not always obvious, but there are some basic guidelines that you can follow. (I will express them assuming that we are calculating electric flux, but the same rules apply to calculating magnetic flux.)

The trick is usually to construct the closed surface from pieces of surface (let us call them **surface elements**) that are shaped so that calculating the flux is as easy as possible. Calculating the flux through a surface element is *easiest* (as we found in example E10.1) when $\vec{E} = 0$ everywhere on the surface element. This means that the flux through any of its tiles is simply $d\Phi_E = \vec{E} \cdot d\vec{A} = 0 \cdot d\vec{r} = 0$, implying that the net flux through the surface element is trivially zero.

It is also very easy to calculate the net flux through a surface element chosen so that it is parallel to the field. Since the tile vector $d\vec{A}$ for any arbitrary tile on the surface element is perpendicular to that surface, this means that $d\vec{A}$ will also be perpendicular to the field vector $\vec{E}$. Since the dot product of two perpendicular vectors is zero, the flux through the tile is $d\Phi_E = \vec{E} \cdot d\vec{A} = 0$, and since this applies to all tiles on the surface element, the net flux through that surface element is zero.

Finally, it is fairly easy to calculate the flux through a surface element that is always perpendicular to the field and arranged so that the field has the same magnitude at every point on the surface. Then the field vectors and tile vectors will be parallel (or antiparallel) at all points on this surface element; so for every tile $d\Phi_E = \vec{E} \cdot d\vec{A} = E\ dA \cos\theta = \pm E\, dA$, since the angle θ between $\vec{E}$ and $d\vec{A}$ is either $0°$ or $180°$. The sign is positive if the field vectors point toward the closed surface's exterior (because then $\theta = 0°$) and negative if the field vectors point toward its interior (because then $\theta = 180°$). Since we have designed the surface portion so that $E = \text{mag}(\vec{E})$ has the same value at each tile, the total flux through this portion of surface is simply

$$\Phi_E \equiv \sum_{\text{all tiles}} \vec{E} \cdot d\vec{A} = \pm E \sum_{\text{all tiles}} dA = \pm EA \qquad \text{(E10.2)}$$

where A is the total area of the surface element.

So the scheme is that you should try to construct your closed surface entirely out of surface elements that fall into one of these three categories. Assume that some kind of symmetry argument determines the direction of the field and specifies that its magnitude depends on some single variable, say, r for the sake of argument. If at all possible, you should construct the closed surface entirely out of surface elements that each satisfy one of the following criteria:

1. The field is zero everywhere on the surface element.
2. The surface element everywhere parallel to the field. and/or
3. The surface element both is perpendicular to the field and has constant r.

Having a surface element of the *last* type is usually essential, because only when the flux through a surface element is *not* zero does Gauss's law non-trivially link the field's magnitude on that surface to the presence of charge inside the closed surface. Therefore it is often best to start by constructing a surface element of this type and then use elements of the other types to close the surface (if necessary).

In some situations, you may not be able to find a surface that satisfies *both* criteria for a surface of the last type. Under such circumstances, you should choose to satisfy whichever criterion makes evaluating the flux easiest, but doing the sum will not be nearly as straightforward as in the case shown in equation E10.2 (it will usually involve doing an actual integral).

The worked examples in sections E10.3 through E10.5 illustrate the four-step process and the art of choosing an appropriate gaussian surface.

E10.3 The Field of a Spherical Charge Distribution

Table E3.1 claimed that the electric field *outside* a uniformly charged spherical surface is as if the sphere's total charge were located at its center, and the electric field in the space *inside* a spherical surface charge is zero. We can easily use Gauss's law to prove these assertions, and indeed we can show that these descriptions apply to spherically symmetric charge distributions in general.

Example E10.2 The Field Outside a Spherical Charge Distribution

Problem Use Gauss's law to prove that the electric field at a point P *outside* of *any* spherically symmetric charge distribution is as if the distribution's total electric charge were concentrated at its center.

Use Symmetry We saw in example E9.1 that symmetry implies that the electric field vector $\vec{E}$ at any point P inside or outside *any* spherically symmetric charge distribution must point either directly away from or toward the distribution's center O. Moreover, $E = \text{mag}(\vec{E})$ can depend at most on the distance r that P is from the center.

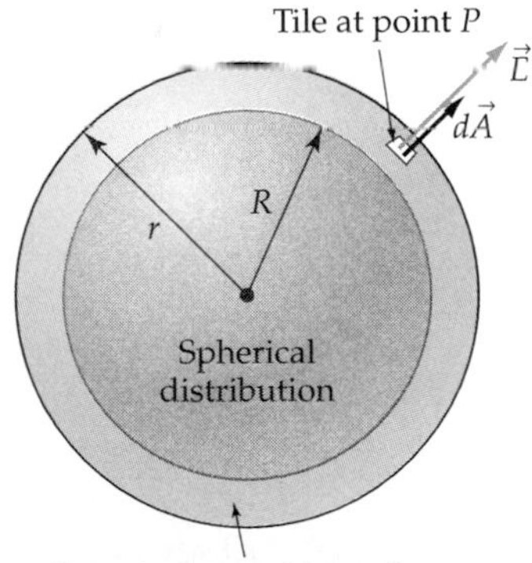

Figure E10.2
This figure shows the appropriate gaussian surface for determining the field at a point P outside a spherical charge distribution.

Choose a Gaussian Surface In this case, we can choose our gaussian surface to be a sphere whose center coincides with the distribution's center and whose radius r is the same as that of the point P where we want to evaluate the field. Note that if P is outside the distribution, then $r > R$, where R is the distribution's outer radius (see figure E10.2). The distribution's electric field $\vec{E}$, since it points radially outward or inward, is perpendicular to this surface, and since this is a surface of constant r, $E = \text{mag}(\vec{E})$ will be constant on the surface and have the same magnitude as the electric field at P. Such a surface is itself closed, so we do not need any other surface elements to construct a closed surface.

Calculate the Flux The flux through an arbitrary tile on such a surface is $\vec{E} \cdot d\vec{A} = E\,dA\cos 0° = E\,dA$ if $\vec{E}$ points outward and $E\,dA\cos 180° = -E\,dA$ if it points inward. Let us define E_r to be the component of $\vec{E}$ in the radial (outward) direction at P (and thus at all points on our gaussian surface):

$$E_r = \begin{cases} +E & \text{if } \vec{E} \text{ is outward} \\ -E & \text{if } \vec{E} \text{ is inward} \end{cases} \quad \text{and} \quad \vec{E} = E_r\hat{r} \qquad \text{(E10.3)}$$

We see then that $\vec{E} \cdot d\vec{A} = E_r\, dA$ in both cases. Since $E_r = \pm E$ is the same at all points on a surface of constant r, the net flux through all tiles on our gaussian surface is simply

$$\oint \vec{E} \cdot d\vec{A} = E_r \oint dA = 4\pi r^2 E_r \tag{E10.4}$$

since the total area of a spherical surface of radius r is $4\pi r^2$.

Apply Gauss's Law Gauss's law then implies that

$$\oint \vec{E} \cdot d\vec{A} = 4\pi k q_{\text{enc}} \quad \Rightarrow \quad 4\pi r^2 E_r = 4\pi k q_{\text{enc}} \quad \Rightarrow \quad E_r = \frac{k q_{\text{enc}}}{r^2} \tag{E10.5}$$

In this case, a gaussian surface having the same radius as our external point P encloses the distribution's net charge Q by definition, and since $\vec{E} = E_r\hat{r}$, we have

$$\vec{E} = \frac{kQ}{r^2}\hat{r} \tag{E10.6}$$

Evaluation This is indeed exactly the field we would expect at the point P if the distribution's total charge were a point charge located at the distribution's center.

Example E10.3 The Field Inside a Spherical Surface Charge

Problem Consider now a sphere that has a uniformly distributed surface charge but zero internal charge. Use Gauss's law to prove that the electric field at *any* point P *inside* the surface charge is zero.

Model The situation here is exactly the same as in example E10.2 except that the point P where we want to evaluate the field is *inside* the surface charge. Since we want to use a spherical gaussian surface that has the same radius as P, our gaussian surface will thus have a radius $r < R$, where R is the radius of the surface charge (see figure E10.3). Even so, the net flux through this gaussian surface is still $4\pi r^2 E_r$ (see equation E10.4), and Gauss's law still implies that $E_r = kq_{\text{enc}}/r^2$ (see equation E10.5). The difference is that since

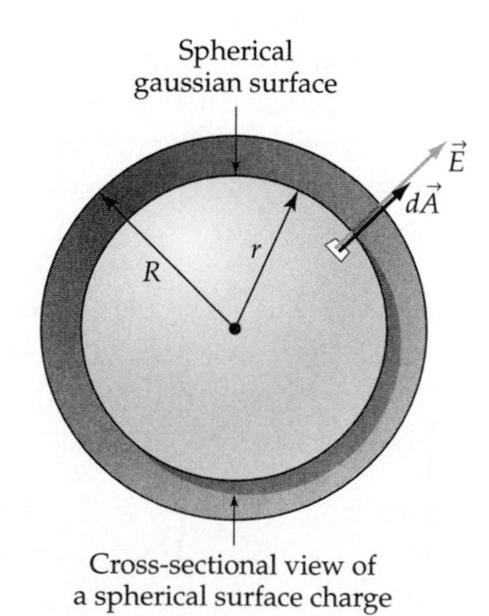

Figure E10.3
This figure shows the appropriate gaussian surface for determining the field at a point P inside a spherical surface charge distribution. (The electric field vector $\vec{E}$ in this case is hypothetical.)

there is no net charge inside the sphere by hypothesis, our gaussian sphere in this case encloses no net charge: $q_{enc} = 0$.

Solution Equation E10.5 therefore implies that $E_r = 0$. This argument applies to any arbitrary point P inside the surface, so the electric field due to the surface charge must be zero *everywhere* inside the spherical object.

Evaluation This is what we set out to show.

If you compare the relative simplicity of this argument with the complexity of the mathematical arguments in problem E3A.2, you can perhaps appreciate the value of Gauss's law! In situations with lots of symmetry, an argument based on Gauss's law is usually much simpler than a direct calculation.

Note also that if we model a point particle as being essentially a very tiny sphere, then symmetry implies that the field of such a particle at rest must be radial and equation E10.6 clearly indicates that the combination of symmetry and Gauss's law implies that its field is $\vec{E} = (kq/r^2)\hat{r}$ at a point a distance r from the particle's center. We see, therefore, that Gauss's law does indeed imply Coulomb's law for the electric field of a point particle at rest.

E10.4 The Field of an Infinite Cylindrical Distribution

In this section and section E10.5, we will be proving results that will be useful to us in chapters E12 and E13. We first consider the case of an infinitely long cylindrical charge distribution (such as a cylindrical wire or pipe).

The definition of a *cylindrical charge distribution*

Example E10.4 A Cylindrical Distribution's External Field

Problem Use Gauss's law to determine the electric field outside a cylindrically symmetric charge distribution.

Use Symmetry As discussed in example E9.2, the electric field at any point P inside or outside a cylindrically symmetric charge distribution must point radially away from or toward this axis, and the magnitude of the field can, at most, depend on the distance r that P is from the central axis.

Choose a Gaussian Surface A surface of constant r in this case looks like a cylindrical band around the distribution's axis, as shown in figure E10.4a. This surface is perpendicular to the electric field at all points, so we should be able to calculate the flux through this surface pretty easily. We can make the surface closed by adding two circular end caps, as shown in figure E10.4b. These end cap surface elements are everywhere parallel to the field. The resulting total gaussian surface looks something like a tin can.

Calculate the Flux Since the field points either radially outward or radially inward, $\vec{E}$ will be either parallel or antiparallel to $d\vec{A}$ for all tiles on the cylindrical band element of our gaussian surface, implying that $\vec{E} \cdot d\vec{A} = E_r\,dA$ for such tiles, where E_r (the radial component of $\vec{E}$) $\equiv +E$ if $\vec{E}$ points outward and $-E$ if $\vec{E}$ points inward. Since all tiles on this part of the surface are the same distance r from the axis, E will have the same value at all these tiles. This

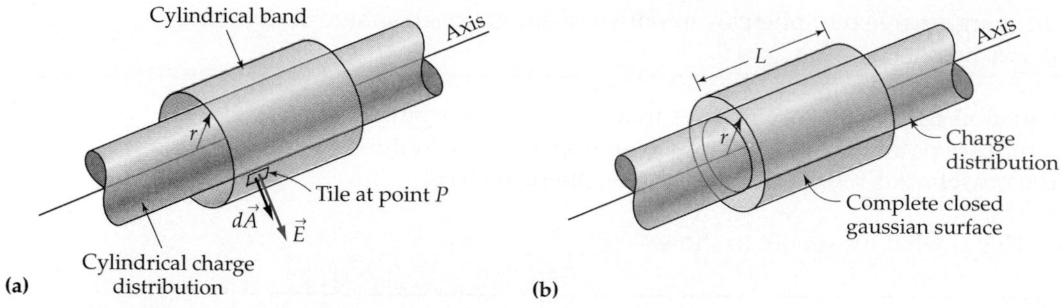

Figure E10.4

(a) A surface of constant r in the case of an infinite cylindrical distribution looks like a cylindrical band. (b) An appropriate closed gaussian surface for calculating the electric field at a point P outside a cylindrical charge distribution. The circular end caps cut completely through the distribution.

implies that since the total area A of the curved part of the surface is the band's circumference $2\pi r$ times its length L, the flux through this element of the gaussian surface must be

$$\int \vec{E} \cdot d\vec{A} = E_r \int dA = E_r A = E_r(2\pi r L) \tag{E10.7}$$

(Remember that $\int$ here simply means "sum the following infinitesimal quantities over the surface of interest.") Since the field is parallel to the end caps, $\vec{E} \cdot d\vec{A} = 0$ for all tiles on these surface elements. Therefore, the net flux through the entire gaussian surface is also $E_r(2\pi r L)$.

Apply Gauss's Law Gauss's law therefore implies that

$$E_r(2\pi r L) = \oint \vec{E} \cdot d\vec{A} = 4\pi k q_{\text{enc}} \quad \Rightarrow \quad E_r = \frac{4\pi k q_{\text{enc}}}{2\pi L} = \frac{2k}{r}\frac{q_{\text{enc}}}{L} \tag{E10.8}$$

So, if we define $\hat{r}$ to be the direction directly away from the distribution's central axis and $\lambda \equiv$ the distribution's charge per unit length $= q_{\text{enc}}/L$, then we can summarize what we have learned about this distribution's field as follows:

The external electric field of a cylindrical charge distribution

$$\vec{E} = \frac{2k\lambda}{r}\hat{r} = \frac{1}{2\pi\varepsilon_0}\frac{\lambda}{r}\hat{r} \tag{E10.9}$$

Purpose: This equation describes the external electric field $\vec{E}$ of an infinite, axially symmetric charge distribution that has a uniform charge per unit length λ.

Symbols: r is the distance that the point where the field is evaluated is from the central axis, $\hat{r}$ is a unit vector pointing directly away from the central axis, k is Coulomb's constant, and $\varepsilon_0 = 1/4\pi k$.

Limitations: This only applies to an infinitely long charge distribution, though the result yields a pretty good approximation if the distribution is very long compared to its diameter.

Evaluation This is the same result we found in section E2.6 in the infinite-wire limit (and displayed in table E3.1) except now we see that the same result describes the external electric field of *any* infinite, cylindrical charge distribution, including wires of finite diameter, hollow pipes, and so on.

Of course, creating a truly *infinite* charge distribution is physically impossible (for one thing, it would require the approval of an infinite budget). We saw in section E2.6, though, that this result represents a useful approximation for the field of a wire at points much closer to the wire than to its ends. Likewise, equation E10.9 turns out to be a good approximation to the external field of *any* cylindrical distribution at points much closer to the distribution's axis than to its ends.

Exercise E10X.2

Argue that equation E10.9 implies that the electric field points away from a distribution with positive charge and toward a distribution with negative charge.

Exercise E10X.3

Argue that Gauss's law implies that the electric field in the hollow interior of a uniformly charged pipe must be zero.

E10.5 The Field of an Infinite Planar Slab

Description of an *infinite planar slab*

Consider now the case of a planar slab parallel to the yz plane that is infinite in the y and z directions but has a finite width W in the x direction. Let us define the yz plane so that it coincides with the central plane of the slab (the plane halfway between the slab's surfaces). Assume that this slab has a uniformly distributed charge throughout its interior with a charge density (charge per unit volume) ρ_0. Such a slab exhibits the symmetry discussed in example E9.3: it is unchanged by rotations about any axis parallel to the x direction, by sliding in either the y or z direction, and by reflection across the yz plane.

Example E10.5 The External Field of an Infinite Slab

Problem Use Gauss's law to determine such a slab's electric field at an arbitrary point P whose x coordinate is positive but outside the slab ($x > \frac{1}{2}W$).

Use Symmetry According to example E9.3, the electric field at any point inside or outside this slab must be parallel to the x axis and have a magnitude that can at most depend on x. Moreover, the electric field on the slab's central plane (the $x = 0$ plane) is zero.

Choose a Gaussian Surface In this case, a surface of constant x will be a plate parallel to the yz plane. Such a surface is also perpendicular to the field, so it will be easy to evaluate the flux across such a surface. Let us put one circular end cap of area A at $x = 0$ and another at the same x position as the external point P where we want to evaluate the field. Since the electric field at all points on the former end cap is zero, the flux through that cap will be zero. We can close our gaussian surface by adding a cylindrical band surface element that connects the two circular end caps. The band surface element is everywhere parallel to the field, so the flux through this element will also be zero. The complete surface in this case looks like a can with its central axis parallel to the x axis (see figure E10.5).

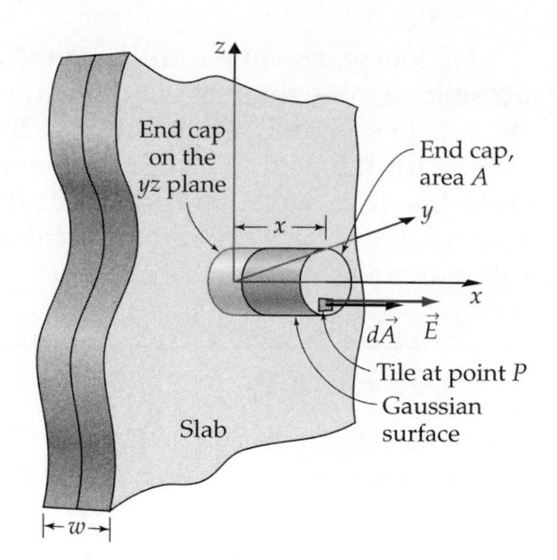

Figure E10.5
An appropriate gaussian surface for calculating the electric field at a point P outside an infinite, uniformly charged slab.

Calculate the Flux The tile vectors for each tile on the rightmost plate point in the $+x$ direction, so $d\vec{A} = [dA, 0, 0]$ and the component definition of the dot product implies that $\vec{E} \cdot d\vec{A} = E_x\, dA_x + E_y\, dA_y + E_z\, dA_z = E_x\, dA$. Since this plate is a surface of constant x, E_x has the same value for all tiles on the end cap, so the total flux through this end cap is therefore

$$\int \vec{E} \cdot d\vec{A} = E_x \int dA = E_x A \tag{E10.10}$$

where A is the end cap's area. Since the flux through each of the other surface elements is zero, this is also the net flux through our gaussian surface.

Apply Gauss's Law Our gaussian surface in this situation encloses the charge within a disklike plug of slab with area A and thickness $\frac{1}{2}W$. The total charge enclosed is

$$q_{\text{enc}} = (\text{volume})(\text{charge density}) = \tfrac{1}{2} W A \rho_0 \tag{E10.11}$$

Gauss's law in this case then implies that

$$E_x A = \oint \vec{E} \cdot d\vec{A} = 4\pi k q_{\text{enc}} = 2\pi k \rho_0 A W \quad \Rightarrow \quad E_x = \frac{2\pi k \rho_0 W \not{A}}{\not{A}} = 2\pi k \sigma_0 \tag{E10.12}$$

where $\sigma_0 \equiv \rho_0 W$ is the charge per unit area on the slab. Since $E_y = E_z = 0$ in this case, $\vec{E} = 2\pi k \sigma_0 \hat{x}$ for all points such that $x > \frac{1}{2}W$.

Evaluation This is consistent with the result stated in table E3.1 (we can think of a planar surface charge as being a very thin slab). Note that Gauss's law implies that the electric field of an infinite slab has a *constant magnitude* outside the slab, independent of how far one is from the slab.

Example E10.6 The Internal Field of an Infinite Slab

Problem Use Gauss's law to determine such a slab's electric field at an arbitrary point P whose x coordinate is positive but inside the slab ($0 \le x \le \frac{1}{2}W$).

Model and Solution The situation is exactly as described in example E10.5, except now our gaussian surface's rightmost end cap is inside the slab (see figure E10.6). The net flux through this surface is still $\langle \Phi_E \rangle = E_x A$, but the total charge enclosed by the slab is now the charge within a disklike plug of slab with area A and a thickness x (not $\frac{1}{2}W$). Therefore, in this case

$$q_{\text{enc}} = (\text{volume}) \cdot (\text{charge density}) = Ax\rho_0 \tag{E10.13}$$

and Gauss's law implies that

$$E_x A = 4\pi k q_{\text{enc}} \quad \Rightarrow \quad E_x = \frac{4\pi k(\cancel{A}\rho_0 x)}{\cancel{A}} = 4\pi k \rho_0 x \tag{E10.14}$$

Evaluation This means that *inside* the slab, E_x increases (or decreases, if $\rho_0 < 0$) linearly with increasing x, with a constant slope of $4\pi k \rho_0$. Note that the slab's field at $x = 0$ is indeed zero, as required by symmetry.

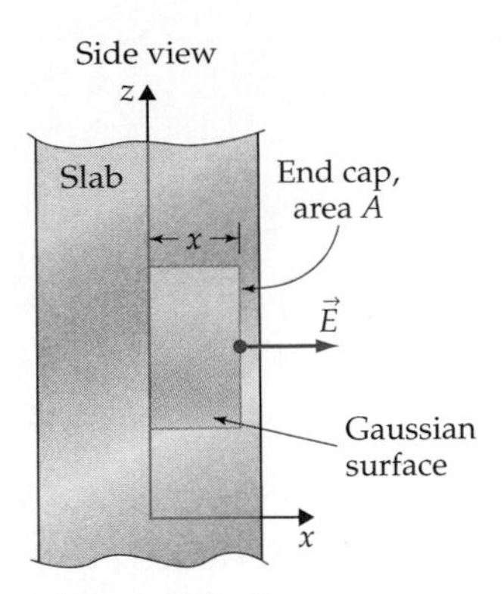

Figure E10.6
A cross-sectional view of an infinite slab and the appropriate gaussian surface for calculating the electric field at a point P inside that slab.

Exercise E10X.4

To completely determine the slab's field, we ought to consider points where $x < 0$. Use Gauss's law to argue that $E_x(-x) = -E_x(x)$.

Figure E10.7 shows a graph of E_x versus x for both sides of the plane (assuming that ρ_0 is positive). Mathematically, we can summarize the results of examples E10.5 and E10.6 and exercise E10X.4 as follows:

$$E_x = \begin{cases} +2\pi k \sigma_0 = +\dfrac{1}{2}\dfrac{\sigma_0}{\varepsilon_0} & \text{for } x > \frac{1}{2}W \\ (4\pi k \rho_0)x = \dfrac{\rho_0}{\varepsilon_0}x & \text{for } -\frac{1}{2}W \le x \le \frac{1}{2}W \\ -2\pi k \sigma_0 = -\dfrac{1}{2}\dfrac{\sigma_0}{\varepsilon_0} & \text{for } x < -\frac{1}{2}W \end{cases} \tag{E10.15}$$

$$E_y = E_z = 0 \tag{E10.16}$$

Purpose: This equation describes the electric field $\vec{E}$ of an infinite planar slab perpendicular to the x axis and having a uniform internal charge density ρ_0.

Symbols: x is the x coordinate of the point where the field is evaluated, $\hat{x}$ stands for the $+x$ direction (the direction perpendicular to the slab), W is the slab's thickness, $\sigma_0 = \rho_0 W$ is the slab's charge per unit area, k is Coulomb's constant, and $\varepsilon_0 = 1/4\pi k$.

Limitations: These results only apply to an infinite slab.

Notes: Note that the field outside the slab is uniform. The equations for the slab's external field apply even if the slab's charge density varies with x.

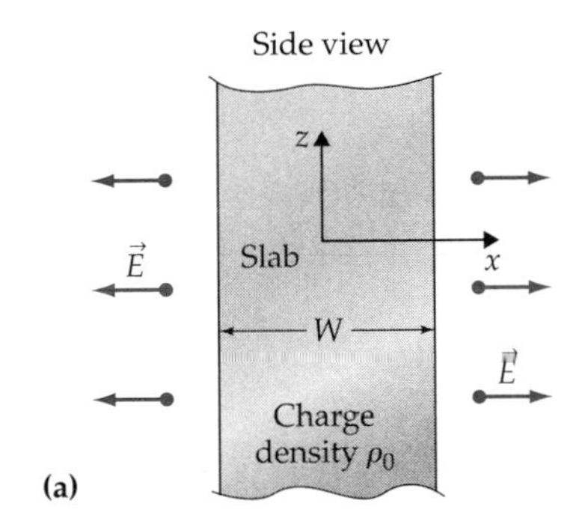

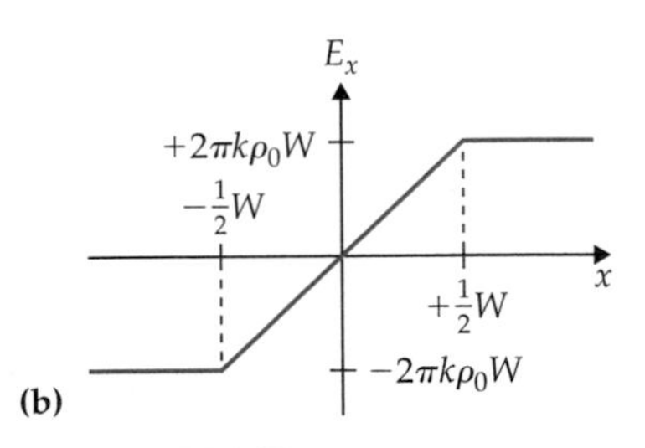

Figure E10.7
(a) A cross-sectional view of an infinite, uniformly charged slab, showing that the electric field vectors on opposite sides of the slab's central plane point in opposite directions. (b) A graph E_x as a function of x for the infinite slab.

This result will be *very* useful to us in chapters E12 and E13, where the simple character of this field will make certain theoretical arguments clearer. The "infinite slab" model is somewhat less usable in practice than the "infinite wire" model discussed in section E10.4, because the field of a finite slab can have significant E_y and E_z components, even at points that are much closer to the slab's central plane than to its ends. However, equation E10.15 remains an excellent approximation for the field's x component at points much closer to the central plane of a finite planar slab than to its ends.

E10.6 Gauss's Law for the Magnetic Field

Is there a law for magnetic fields analogous to Gauss's law? Let us consider putting a gaussian surface around a moving point charge. Equation E8.4 tells us that the magnetic field for a particle moving at a steady speed much slower than that of light is given by

$$\vec{\mathbb{B}} = \frac{kq}{r_{PC}^2}\left(\frac{\vec{v}}{c} \times \hat{r}_{PC}\right) \tag{E10.17}$$

Steps toward a magnetic version of Gauss's law

This equation implies that the magnetic field of a point particle always lies tangent to a circle centered on the axis along which the particle is moving, as shown in figure E10.8a. This applies to a circle in any plane perpendicular to the particle's direction of motion, even if the plane is far ahead of or far behind the particle.

Consider, then, a spherical gaussian surface surrounding the charge (see figure E10.8b). You can see from the drawing that since the tile vector $d\vec{A}$ for any tile on such a surface points perpendicular to the surface and since the magnetic field is *always* tangent to such a surface, the magnetic field flux $\vec{\mathbb{B}} \cdot d\vec{A}$ through any tile on the surface will be zero. This means that the *net* magnetic field flux ($\Phi_{\mathbb{B}}$) through such a surface will be zero as well, even though the surface encloses a moving charge!

If we move the charge outside the sphere, the magnetic field is still tangent to the surface of the sphere, as shown in figure E10.8c. Therefore, the net magnetic flux through the sphere centered on the particle's direction of motion is zero whether the moving charge is inside or outside the sphere. Finally, while equation E10.17 yields incorrect predictions about the *strength* of the field of a particle moving at relativistic speeds, it does correctly predict the field's direction, so the result applies to relativistic particles as well. Moreover, one can show (with some effort, see problem E10S.9) that even arbitrarily deforming the surface does not change this result!

If all this is true, it means that the net magnetic field flux of *any* moving point particle through *any* closed surface is zero. By the superposition principle, if this is true for the magnetic field of any single moving particle in a current distribution, it is also true for the total field created by that charge distribution. The following law therefore appears to describe something intrinsic about the character of *any* magnetic field:

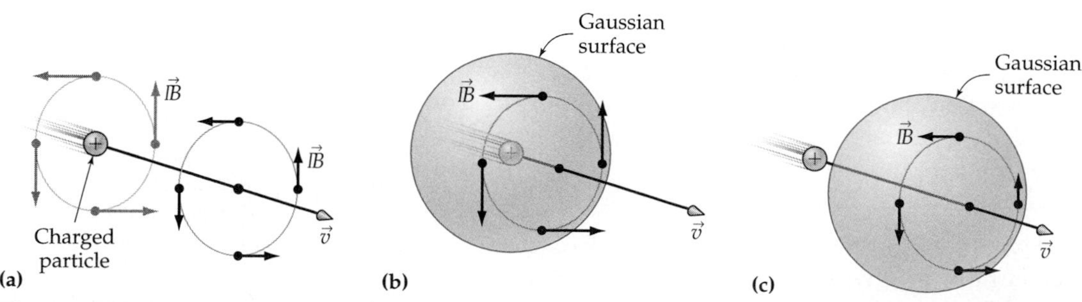

Figure E10.8
(a) Magnetic field vectors at all points around a moving charged particle are tangent to circles centered on and perpendicular to the axis defined by the particle's direction of motion. (b) This means that these field vectors all lie tangent to a spherical gaussian surface centered on the particle, implying that the net flux through such a surface is zero. (c) The same thing is true even if the gaussian sphere does not contain the particle, as long as the axis of the particle's motion passes through the sphere's center.

Gauss's law for the magnetic field

$$\oint \vec{\mathbb{B}} \cdot d\vec{A} = 0 \quad \text{or} \quad \oint \vec{B} \cdot d\vec{A} = 0 \qquad \text{(E10.18)}$$

Purpose: This equation describes an intrinsic characteristic of magnetic fields.

Symbols: $\oint$ means "sum the following over all infinitesimal tiles on a closed surface," $\vec{\mathbb{B}} = c\vec{B}$ represents the magnetic field vector at a given tile on the surface, and $d\vec{A}$ is the tile's tile vector.

Limitations: Each tile must be small enough that it is approximately flat and that $\vec{B} \approx$ uniform over its surface. There are no restrictions on the shape or size of the surface, or on whether it contains moving charges or not.

This law means that magnetic charge does not exist

If you compare this with Gauss's law $(\Phi_E) \equiv \oint \vec{E} \cdot d\vec{A} = 4\pi k q_{\text{enc}}$ for the electric field, you can see that this law tells us that there is no such thing as "magnetic charge": nothing that plays quite the same role for the magnetic field that charge plays for the electric field. A positively charged particle creates an electric field that everywhere points away from that charge, creating a net positive flux through a small gaussian surface surrounding the charge (see figure E10.9a). However, the right side of a current loop is equivalent to a north magnetic pole, and if we draw a gaussian surface around the region to the right of a loop, we see that magnetic field vectors flow outward through much of the surface, but magnetic field vectors also flow into the *left* side of the surface (see figure E10.9b). Equation E10.18 indeed implies that the inward magnetic flux on this side must exactly cancel the outward magnetic flux from other parts of the surface. We see that the magnetic field in the vicinity of a north magnetic pole is really fundamentally different from the electric field of a positive point charge: this is part of what the difference between Gauss's law for the magnetic field and Gauss's law for the electric field is trying to tell us.

Gauss's magnetic law distinguishes physically possible and impossible field patterns

Gauss's law for the magnetic field essentially describes a constraint that any real magnetic field must satisfy (no matter how it was created), distinguishing field patterns that are physically possible from field patterns that are not. Example E10.7 illustrates how this works.

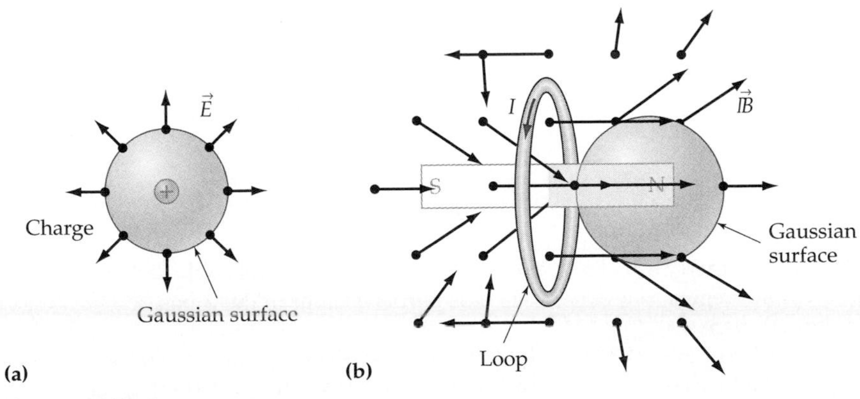

Figure E10.9
(a) A positive point charge creates an electric field that produces a definitely positive flux through a gaussian surface surrounding that charge. (b) A north pole is the closest magnetic analogy to a positive charge, and a loop conducting a current has an effective magnetic north pole on one side. But if we surround this loop's effective north pole with a gaussian surface, the positive flux flowing out the surface's right side is balanced by negative flux flowing in from the left.

Example E10.7 Impossible Magnetic Field Patterns

Problem The following drawings show field arrow diagrams for hypothetical magnetic fields. Assume that the field vectors are independent of position in the direction perpendicular to the page. Which of the fields depicted are forbidden by Gauss's law for the magnetic field?

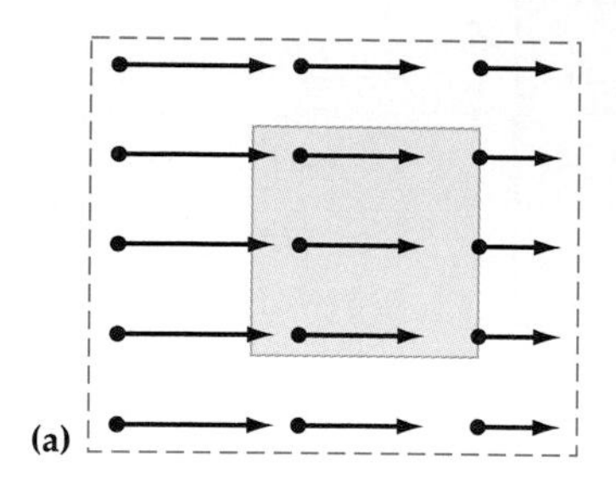
(a)

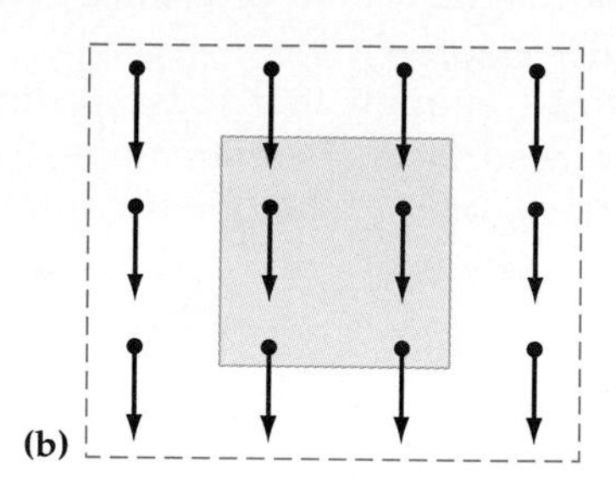
(b)

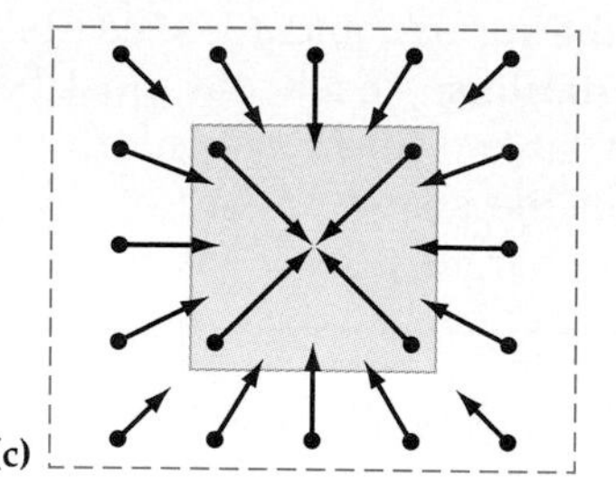
(c)

Solution Only the middle field shown is physically possible. To see this, imagine placing a cubical gaussian surface in the middle of each diagram (the colored squares in the diagrams represent cross-sectional views of these surfaces).

In (a), the negative flux due to field vectors entering the cube's left side is only partially canceled by the smaller positive flux due to field vectors leaving the cube's right side. All the other faces are parallel to the field and so contribute zero flux. Therefore the net flux through this cube is negative, contradicting Gauss's law for the magnetic field.

In (c), magnetic field vectors point inward across the four cube faces perpendicular to the plane of the drawing. The field vectors are parallel to the two faces above and below the plane of the drawing, so there is no flux across these faces. Therefore the net flux through this cube is negative, again contradicting Gauss's law for the magnetic field.

In (b), the field vectors entering the top face of the cube create negative flux through that face, but field vectors of the same magnitude leave the bottom face, creating an exactly opposite positive flux. The field vectors are parallel to the other faces, meaning that flux through these faces is zero. Therefore, the net flux through this cube is zero, which is *consistent* with Gauss's law for the magnetic field. Therefore of the three possible field patterns, only this one is physically realistic.

TWO-MINUTE PROBLEMS

E10T.1 Imagine that we know from some kind of symmetry argument that the electric field in a certain region everywhere points in the $+y$ direction and has a magnitude that depends only on y. For which of the following closed gaussian surfaces would it be easiest to calculate the net flux of an electric field fitting this description?

A. A cubical surface oriented so that two opposite faces are perpendicular to the y axis
B. A spherical surface centered on the origin
C. A cylindrical canlike surface oriented with its central axis perpendicular to the y axis
D. A pyramidlike surface whose bottom plane is perpendicular to the y axis.

E10T.2 Imagine that we place a positively charged particle inside an arbitrarily shaped closed gaussian surface, as shown in figure E10.10a, and we calculate the net flux through the surface and the magnitude of the electric field mag($\vec{E}$) at point P. We then bring in an identical particle, park it just *outside* the gaussian surface, as shown in figure E10.10b, and again calculate the net flux through the surface and mag($\vec{E}$) at point P.

(a) How does mag($\vec{E}$) at point P change?
(b) How does the net flux through the surface change?

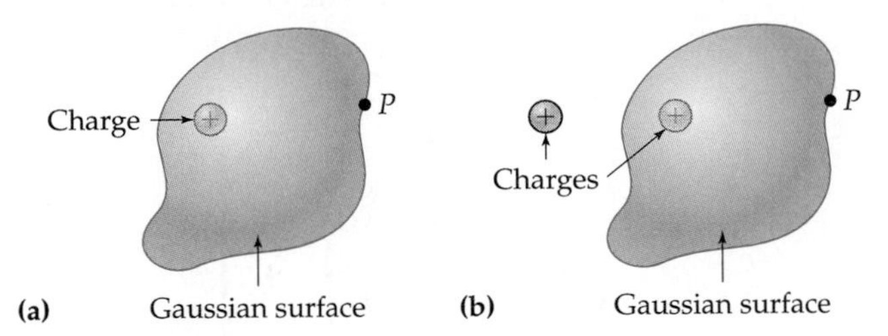

Figure E10.10
What is the change in the electric flux through the gaussian surface as we go from the situation shown in part (a) to the situation shown in part (b)? (See problem E10T.2.)

A. The quantity increases.
B. The quantity decreases.
C. The quantity remains the same.
D. The quantity changes, but whether it increases or decreases depends on details not given.
E. The quantity may or may not change, depending on details not given.

E10T.3 This drawing shows a cross-sectional view of three canlike gaussian surfaces embedded in an infinite slab of uniformly distributed charge. The left face in each case sits on the slab's central plane.

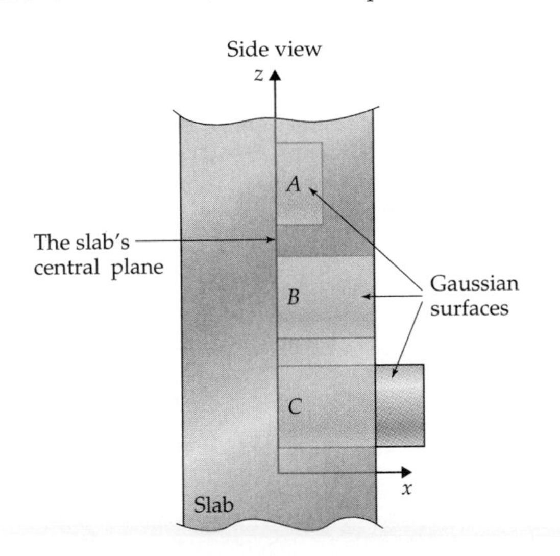

(a) Which of these surfaces encloses the largest amount of charge?
(b) Through which of these surfaces is the net electric flux the greatest?
(c) For which of these surfaces is the electric field at the surface's top cap the greatest?
A. The quantity is largest for surface A.
B. The quantity is largest for surface B.
C. The quantity is largest for surface C.
D. The quantity is the same for both surfaces B and C.
E. The quantity is equal for all the surfaces.
F. Nothing can be determined about this quantity without more information.

E10T.4 Figure E10.11 shows field arrow diagrams for some hypothetical electric fields. Assume that the field is independent of position in the direction perpendicular to the drawing. What can we say about the charge within the region enclosed by the dashed line? (*Hint:* The colored curve in each drawing represents the cross section of a possible gaussian surface that you might use to help you answer the question.)
A. There is no charge *anywhere* in the region.
B. There is positive charge *somewhere* in the region.
C. There is negative charge *somewhere* in the region.
D. There is charge in the region, but its sign depends on details not given.
E. There is positive charge *everywhere* in the region.
F. There is negative charge *everywhere* in the region.
T. We can say nothing definitive about the charge in the region.

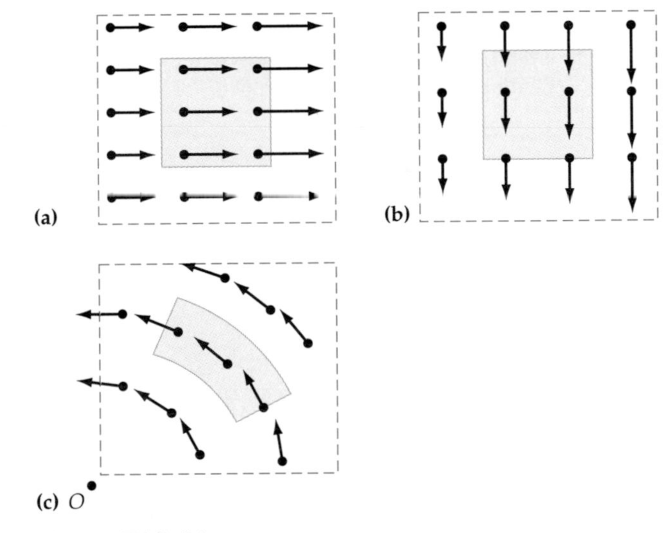

Figure E10.11
Field arrow diagrams for three possible electric fields. In (c), the field vectors are tangent to circles drawn around point O and have constant magnitude.

E10T.5 Consider a uniformly charged disk and a uniformly charged wire. These two charge distributions create very different electric fields. Even so, the flux of the disk's electric field through *any* closed surface enclosing a volume entirely in the empty space outside the disk is equal to the flux of the wire's electric field through the same surface enclosing the same volume in the empty space outside the wire, true or false (T or F)?

E10T.6 An electric field in a certain region inside an object has constant nonzero magnitude and direction. From this information alone, we can conclude

A. That there must be positive charge distributed throughout that region

B. That there must be negative charge distributed throughout the region

C. That there must be zero net charge at all points within that region

D. That such an electric field is physically impossible

E. Nothing useful about the presence or absence of charge in that region

E10T.7 A friend claims that in a certain region of space, a certain magnetic field being measured in an experiment points radially outward from some central point with a radial component of $\mathbb{B}_r \propto r^{-1}$. This is a physically possible magnetic field, T or F?

E10T.8 The drawing shows a field diagram for a hypothetical magnetic field in a certain region of space. Assume that the field is independent of position along the direction perpendicular to the diagram. This is a physically possible magnetic field, T or F?

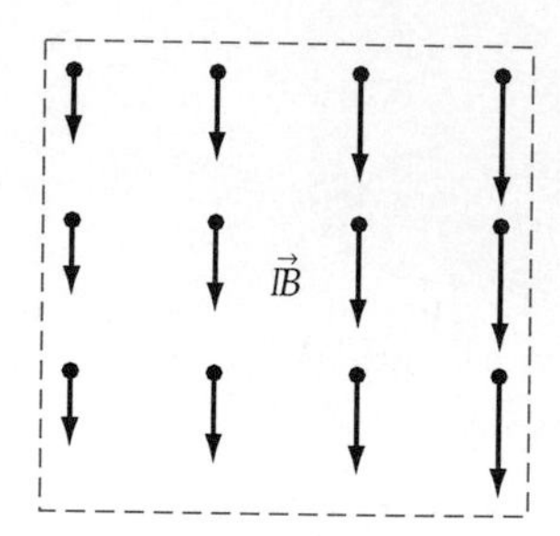

HOMEWORK PROBLEMS

Basic Skills

E10B.1 Figure E10.12 shows two electric dipoles, each consisting of point charges q and $-q$ separated by a distance d. What is the net electric flux through each of the five closed surfaces shown in cross section in the diagram?

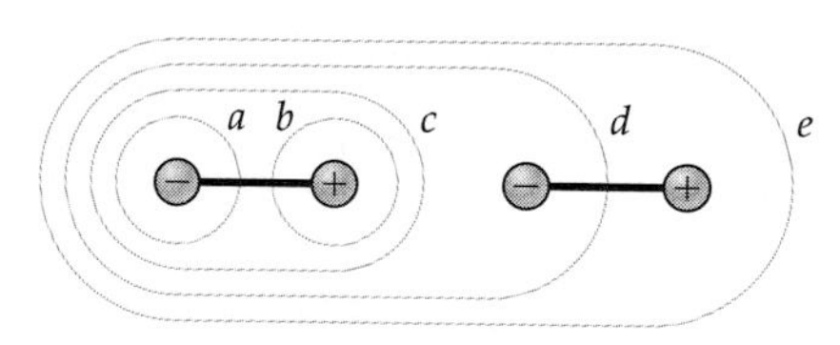

Figure E10.12
What are the electric fluxes through the gaussian surfaces *a*, *b*, *c*, *d*, and *e*? (See problem E10B.1.)

E10B.2 Figure E10.13 shows two small permanent magnets. What is the net magnetic flux through each of the five closed surfaces shown in cross section in the diagram?

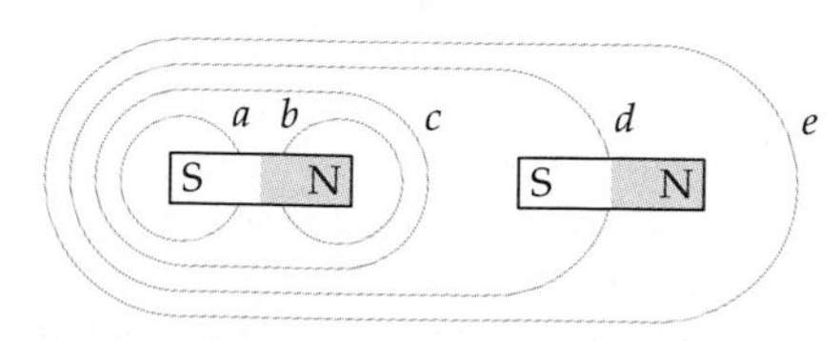

Figure E10.13
What are the electric fluxes through the gaussian surfaces *a*, *b*, *c*, *d*, and *e*? (See problem E10B.2.)

E10B.3 Imagine that a long, straight metal wire has an excess charge of only 1 nC on every meter of wire. Calculate the electric field strength a distance of 5 cm from the wire.

E10B.4 The electric field 10 cm away from a very long, straight wire has a magnitude of 180 N/C and points straight toward the wire. What is the excess charge per unit length on the wire?

E10B.5 Figure E10.14 shows a cross-sectional view of a nested pair of thin, infinite cylindrical pipes. The positively charged inner pipe has a radius of R, and the negatively outer pipe has a radius of $2R$. Both have the same charge per unit length. Draw a graph

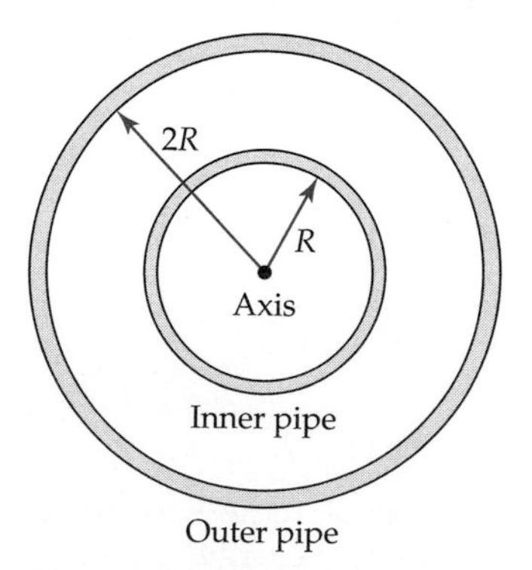

Figure E10.14
A cross-sectional view of two concentric charged pipes. What is the electric field as a function of distance *r* from the pipes' central axis? (See problem E10B.5.)

of the electric field as a function of r for $r = 0$ to $r = 3R$.

E10B.6 Figure E10.15 shows a cross-sectional view of two parallel, positively charged slabs. Each has the same width W and charge density ρ_0, and they are separated by a distance of $2W$. If we define the origin as shown, draw a graph of the x component of the electric field created by this arrangement as a function of x from $x = -3W$ to $x = +3W$.

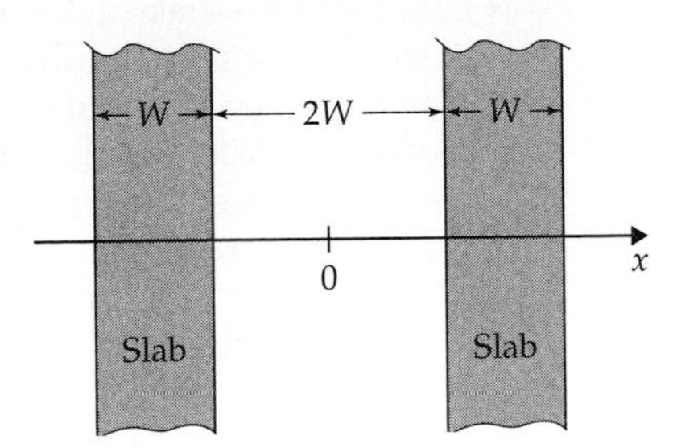

Figure E10.15
A cross-sectional view of two parallel infinite slabs. What is the electric field as a function of x? (See problem E10B.6.)

E10B.7 A plastic plate 1.0 m square carries a uniformly distributed positive total surface charge of 100 nC. Use the infinite slab approximation to determine the component of the electric field perpendicular to the plate at a point fairly near the plate's center and 2.0 cm from the nearest point on the plate.

E10B.8 Imagine that we want to create a magnetic field whose vectors all point in the $+z$ direction, but whose magnitude increases as z increases. Argue that such a field is physically impossible.

E10B.9 Imagine that we want to create a magnetic field that points radially outward from a central point and whose magnitude increases in proportion to r, where r is the distance from the central point. Argue that such a magnetic field is physically impossible.

Synthetic

E10S.1 Consider a spherical ball of radius R with a charge Q uniformly distributed throughout its interior.
(a) Use symmetry and Gauss's law to find the electric field (magnitude and direction) *inside* the ball in terms of k, Q, R, and the distance r that one is from the center.
(b) Draw a graph showing how the electric field magnitude depends on r from $r = 0$ to $r = 3R$.

E10S.2 Consider an infinite cylindrical rod of radius R that has charge with a uniform density ρ_0 throughout its interior.
(a) Use Gauss's law to find the electric field (magnitude and direction) *inside* the rod in terms of k, ρ_0, R, and the distance r that one is from the rod's axis.
(b) Draw a graph showing how the electric field magnitude depends on r from $r = 0$ to $r = 5R$.

E10S.3 Consider the electric field at points just outside the surface of an arbitrary conducting object in static equilibrium.
(a) In chapter E4, we argued that the field vectors at points just outside a charged conductor must point perpendicular to the surface. Explain in your own words why this must be true.
(b) On an arbitrarily shaped conductor, the surface charge density may vary from place to place on the conductor. Consider a patch of area small enough that the charge per unit area σ is essentially constant over the patch; and consider a small, pillboxlike gaussian surface that encloses such a patch, as shown in figure E10.16. Use Gauss's law to argue that the magnitude of the electric field at an arbitrary point P just outside the conductor is given by

$$E = 4\pi k\sigma = \frac{\sigma}{\varepsilon_0} \qquad \text{(E10.19)}$$

where σ is the charge per unit area on the patch of conductor nearest to P. Explain your reasoning very carefully.

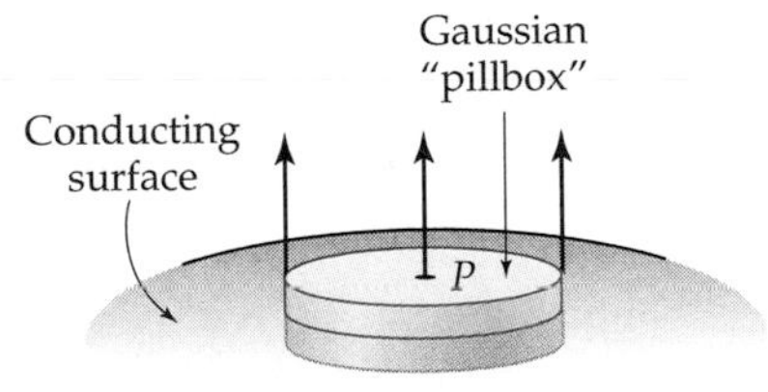

Figure E10.16
We can use the thin gaussian pillbox shown here embedded in the surface of a conducting object to determine the magnitude of the electric field near the surface of a charged conductor. (See problem E10S.3.)

E10S.4 Imagine that the charge density inside a spherical ball of radius R is not constant, but is rather given by

$$\rho(r) = \rho_0\left(1 - \frac{r}{R}\right) \qquad \text{(E10.20)}$$

(Note that this charge density is equal to ρ_0 at the ball's center, but decreases to zero at the ball's surface.) The charge distribution involved here is still spherically symmetric. Find the electric field at points both inside and outside the ball.

E10S.5 Imagine that in a different universe the radial component of the electric field of a point charge were given by $E_r = kQ/r^3$ instead of $E_r = kQ/r^2$ (assume that the field still points directly toward or away from the charge, though). Show that Gauss's law would *not* be true in such a universe.

E10S.6 Coulomb's law $\vec{F}_e = (kq_1q_2/r^2)\hat{r}$ is very similar to Newton's law of gravitation $\vec{F}_g = -(Gm_1m_2/r^2)\hat{r}$. Since (for static charges anyway) we can derive Gauss's law from Coulomb's law, it follows that there should be a Gauss's law for gravitation as well.

(a) By comparing the definition of the electric field vector $\vec{E}$ with that of the gravitational field vector $\vec{g}$, argue that the gravitational counterpart to the formula for the electric field of a point particle must be

$$\vec{g} = -\frac{Gm}{r^2}\hat{r} \tag{E10.21}$$

(b) Considering the ingredients of Gauss's law for the electric field and their gravitational counterparts, argue that Gauss's law for the gravitational field must be

$$\oint \vec{g} \cdot d\vec{A} = -4\pi G m_{\text{enc}} \tag{E10.22}$$

(c) Show that this formula works for a spherical gaussian surface of arbitrary radius r surrounding a particle of mass m at its center.

E10S.7 It would be very cool if we could use some suitable distribution of charges around a region of empty space to create an electric field in that region whose field vectors all point directly toward some particular point. If we could create such a field, a positively charged object could be stably suspended in space at that point (because if it drifted away from the point, the field would push it back). Use Gauss's law to show that a field whose vectors all point toward a point in empty space is, alas, physically impossible.

E10S.8 Section E10.1 states that one can argue that Gauss's law follows mathematically (for static charge distributions) from Coulomb's law. Here is such an argument.

(a) Imagine that a spherical gaussian surface of radius r surrounds a point particle with charge Q. By Coulomb's law, we know that the electric field at all points on the sphere's surface has a magnitude of

$$E = \frac{kQ}{r^2} \tag{E10.23}$$

Show the total flux through this *spherical* gaussian surface is equal to $4\pi kQ$, which is indeed $4\pi k$ times the charge enclosed.

(b) Now let's focus on a single tile on this surface. Imagine that you look at this tile while standing on the surface of the charge Q as if it were a planet. The tile would occupy a certain angular area in the sky. Let's say for the sake of argument (the actual numbers are not relevant) that from this vantage point the tile occupies a region of the sky that is $d\phi = 0.01$ rad wide by $d\alpha = 0.01$ rad high (this is very roughly the same angular area that the full moon occupies in the earth's sky). What fraction of the total flux going through the whole gaussian surface goes through this particular tile?

(c) Now imagine that we take this particular tile and move it closer to or farther away from charge Q, at the same time stretching it or shrinking it so that its angular size when viewed from the charge's surface never changes. This amounts to sliding the tile inward or outward so that its four vertices remain connected to four radial lines that go from the charge's center outward (the four radial lines and the tile together form a very tall and skinny pyramid, as shown in figure E10.17a). Argue that even though the tile gets larger as we move it away or smaller as we move it closer, Coulomb's law implies that the electric field gets weaker or stronger in the same proportion so that $E\,dA$ remains fixed for this tile.

(d) Now imagine tilting the tile by an angle θ from its original direction while keeping its vertices connected to the four radial lines, as shown in figure E10.17b. This ensures that the tile continues to span the same angular area when viewed from the charge. Show that in the limit that $d\phi$ and $d\alpha$ are so small that the radial lines are almost parallel, tipping and stretching the tile do not change the flux $d\Phi_e = \vec{E} \cdot d\vec{A}$ through that tile. (*Hint:* Show that when you tip it, the tile's area increases by a factor of $1/\cos\theta$.)

Comment So, parts (c) and (d) imply that moving a given tile closer to or farther from the charge Q or tilting it at an angle does not change the

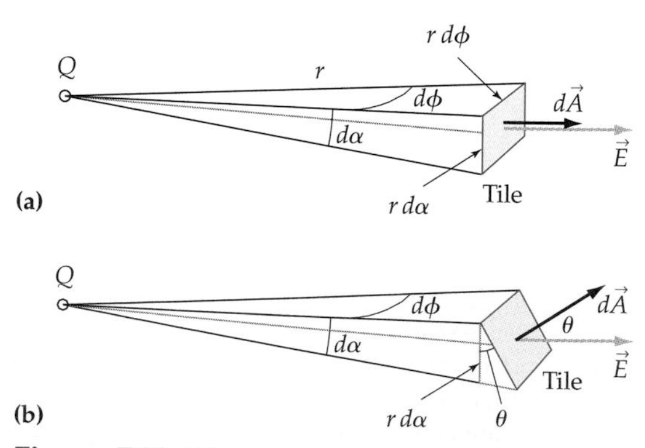

Figure E10.17
The electric flux across any tile that spans the same angular area $d\alpha\, d\phi$ in the "sky" of an observer on the point charge Q is the same, even though a more distant tile or a tilted tile has a larger area. (See problem E10S.8.)

flux through it (as long as we keep the angular area of the tile fixed). This applies to any arbitrary tile. I argue that by taking our initial spherical gaussian surface and moving tiles in and out and tilting them, I can create a closed surface of *any* arbitrary shape surrounding the charge. Since Gauss's law applied to the original spherical surface, it must therefore also apply to a surface of arbitrary shape around a point charge.

(e) Now consider a point charge *outside* our arbitrary surface. Again, imagine drawing four radial lines away from the charge, and let the points where these lines intersect with our surface mark out the four corners of tiles on the surface. Argue that these lines will mark out *two* tiles, one on the nearer side of the surface and one on the farther. Use the results of parts (c) and (d) to argue that the flux through the nearer tile is *canceled* by the flux through the farther.

Comment This means that a point charge *outside* the surface does not affect the net flux through the surface.

(f) The distributive property of the dot product implies that $(\vec{E}_1 + \vec{E}_2) \cdot d\vec{A} = \vec{E}_1 \cdot d\vec{A} + \vec{E}_2 \cdot d\vec{A}$. Argue that this and the superposition principle together imply that if Gauss's law applies for an arbitrary surface surrounding a single point charge, it must also apply to an arbitrary surface surrounding N arbitrarily placed point charges, that is, to *any* charge distribution inside the surface. Argue also that any distribution of external charges will not affect the net flux through the surface and so can be ignored.

ES10.9 In section E10.6, I claimed that the net magnetic flux through an arbitrarily shaped gaussian surface was zero, whether the surface contains moving charges or not. Here is an argument for this assertion.

(a) Consider a single charged particle moving at a constant velocity. We saw in section E10.6 that the magnetic flux through any sphere whose center lies on the particle's path is zero because the magnetic field is tangent to all points on the surface. Imagine now that we perturb the surface by extruding a small vertical tab from the surface, as shown in figure E10.18a. Assume that the height and width of the tab are small enough that the magnetic field is nearly constant over its face. Argue that the net magnetic flux through the perturbed surface is still zero.

(b) Now imagine tilting one of the faces of this tab at an angle θ away from the direction of the field, as shown in figure E10.18b. Argue that the net magnetic flux through the perturbed surface is *still* zero. (*Hint:* Show that when you tip it, the face's area increases by a factor of $1/\cos\theta$.)

Comment I claim that we can warp a sphere into any shape that we like by incrementally extruding such tabs, tilting either one or both faces, and then repeating the process until we get the shape that we want. Parts (a) and (b) therefore amount to a proof that the net magnetic flux through *any* surface created by the magnetic field of a moving charged particle is zero. Note that the proof really only involves the fact that the magnetic field of a moving charged particle always points tangent to a circle centered on the charge's direction of motion: unlike Gauss's law for the electric field, it does *not* depend on the fact that the magnetic field depends on the inverse square of the distance or any other features of the field.

(c) The distributive property of the dot product implies that $(\vec{\mathbb{B}}_1 + \vec{\mathbb{B}}_2) \cdot d\vec{A} = \vec{\mathbb{B}}_1 \cdot d\vec{A} + \vec{\mathbb{B}}_2 \cdot d\vec{A}$. Argue that this and the superposition principle together imply that if Gauss's law for the magnetic field applies to an arbitrary surface in the magnetic field created by a single point charge, it must also apply to an arbitrary surface in the magnetic field created by N arbitrarily moving charged particles.

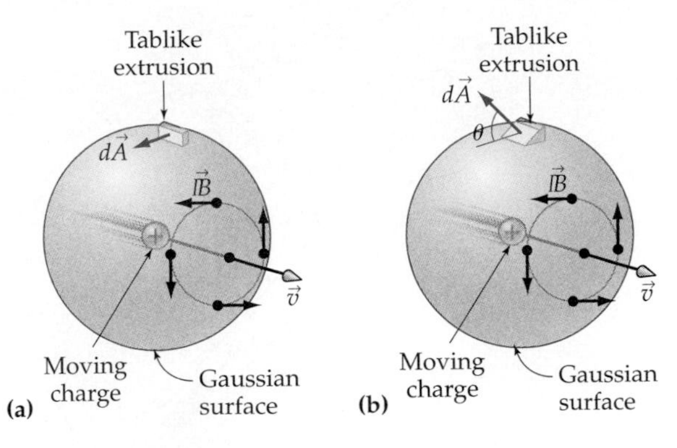

Figure E10.18
The flux through a small extruded tab on the surface of a gaussian sphere surrounding a moving charged particle is zero, no matter how the tab's faces are angled. (See problem E10S.9.)

Rich-Context

E10R.1 Imagine that you have some coaxial cable that consists of a central core of wire of radius $R_1 = 0.40$ mm surrounded by an insulating layer, which in turn is surrounded by pipelike conducting sheath with an inner radius of $R_2 = 1.2$ mm. It is often very important in computer or cable television applications to know the capacitance of the cables used. What is the capacitance per unit length of this particular cable? (*Hints:* You might find it helpful to review section E4.5 about calculating capacitance and section E3.2 about computing potential differences. The trick here is to calculate the potential difference between the inner and outer conductors: you will need to do a simple integral to do this.)

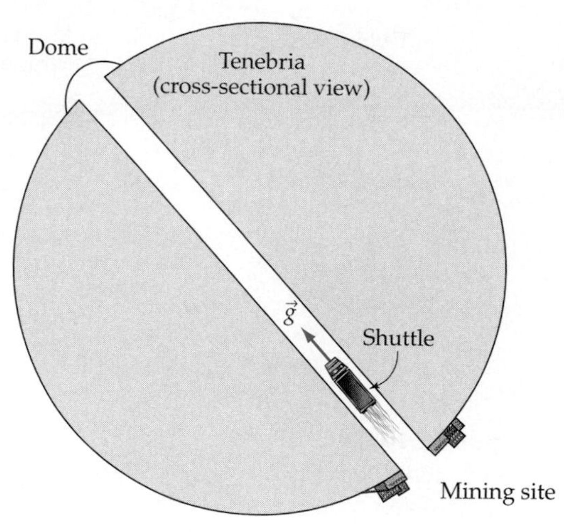

Figure E10.19
A cross-sectional view of the asteroid Tenebria, showing the tunnel used by the gravity-powered shuttle. (See problem E10R.2.)

E10R.2 The space colony on the asteroid Tenebria has built a shuttle that travels in a straight tunnel drilled from the main colony dome through the asteroid's center to the main mining site on the exact opposite side of the spherical asteroid (see figure E10.19). The asteroid has a diameter of 16 km and consists entirely of rock with an approximately constant density of 3200 kg/m^3. The shuttle requires no fuel: rather, it freely falls in the asteroid's own interior gravitational field throughout its trip from the dome to the mining site and back.

(a) Use Gauss's law for the gravitational field (see problem E10S.6) and the mathematics discussed in the chapter on harmonic oscillation (chapter N11) in unit N to determine how long it takes the shuttle to travel from the dome to the mining site.

(b) Compare this to the time required for a spaceship orbiting the asteroid just above its surface to travel between the two sites.

Advanced

E10A.1 A simple quantum-mechanical model of a hydrogen atom in its ground state visualizes a hydrogen atom as consisting of a proton of essentially infinitesimal radius surrounded by an "electron cloud," whose charge density varies with radius as

$$\rho(r) = \frac{-q_p}{\pi a_0^3} e^{-2r/a_0} \tag{E10.24}$$

where q_p is the (positive) charge of the proton and a_0 is the *Bohr radius* ($a_0 = 0.053$ nm).

(a) Find an expression for the magnitude of the electric field at any point inside a hydrogen atom. (You should find the integral you need in any good table of integrals.)

(b) Plot a graph of your function from essentially $r = 0$ to $r = 3a_0$. Label your vertical axis in terms of the field strength $E_0 \equiv \text{mag}(\vec{E})$ at $r = a_0$.

ANSWERS TO EXERCISES

E10X.1 Imagine that we *do* have a clump of charge inside a conductor. If I surround that clump with a gaussian surface, the surface will enclose a net charge of a certain sign. But the electric field at every point in the conductor must be zero, so the flux across the surface must be zero. This contradicts Gauss's law. The only way to avoid a logical contradiction is to say that we can't have a clump of net charge anywhere in a conductor.

E10X.2 Equation E10.9 tells us that

$$\vec{E} = \left(\frac{2k\lambda}{r}\right)\hat{r} \tag{E10.25}$$

If the charge on the wire is positive, then the charge per unit length λ is also positive, and the quantity in parentheses is positive, meaning that $\vec{E}$ points in the $+\hat{r}$ direction, which is radially away from the wire. On the other hand, if the charge on the wire is negative, then the charge per unit length λ is also negative, the quantity in parentheses is negative, and $\vec{E}$ points in the $-\hat{r}$ direction, which is radially *toward* the wire.

E10X.3 *Use Symmetry:* If an electric field exists inside the pipe's hollow interior, symmetry implies that in this region as well as the region outside the pipe the electric field must point either directly toward or directly away from the pipe's axis. *Choose a Gaussian Surface and Calculate the Flux:* Therefore, the flux across a can-shaped gaussian surface of radius r smaller than the pipe's radius will be $E_r(2\pi rL)$, just as in equation E10.7. *Apply Gauss's Law:* But in this case, the gaussian surface encloses no charge (since the volume it encloses is entirely within the pipe's hollow interior), so we must have $E_r(2\pi rL) = 0 \Rightarrow \vec{E} = 0$, since the radial component of $\vec{E}$ is the only component of $\vec{E}$ that could possibly be nonzero in this situation. *Evaluation:* This is what we were supposed to show!

E10X.4 Consider the two canlike gaussian surfaces shown in figure E10.20. One surface has circular faces of area A at $x = 0$ and $x = a$ (where a is a positive real number), while the other has circular faces of the same area at $x = 0$ and $x = -a$. Let the field's x component at $x = a$ be $E_x(a)$, and let the same at $x = -a$ be $E_x(-a)$. The net flux through each of these cans is simply equal to the flux through its outermost face, which is $[E_x(a)]A$ for the right can and

$$\int \vec{E} \cdot d\vec{A} = \int E_x \, dA_x = E_x(-a) \int (-dA)$$
$$= -[E_x(-a)]A \qquad \text{(E10.26)}$$

for the left can (since $dA_x = -dA$ for tile vectors on the left can's outer face). Since both enclose the same amount and sign of charge, the net fluxes through the cans must be equal in magnitude and sign, implying that

$$[E_x(a)]A = -[E_x(-a)]A$$
$$\Rightarrow \quad E_x(a) = -E_x(-a) \qquad \text{(E10.27)}$$

Q.E.D. (One could also show this by rotating the slab 180° around the y axis. Such a rotation moves the field vector at an arbitrary point P with coordinates $[x, y, z]$ on the slab's right side to a point S with coordinates $[-x, -y, z]$ on the slab's left side. Since the vector at P points entirely in the $\pm x$ direction, such a rotation will also reverse the direction of the vector as it is carried from P to S. Since this rotation does not affect the slab, it should not affect the field, so the vector originally at P must be equal in magnitude and opposite in direction to the vector originally at S if the vectors are to match: $\vec{E}(x, y, z) \equiv \vec{E}_P = -\vec{E}_S \equiv -\vec{E}(-x, -y, z)$. Since all field vectors of the infinite slab having the same x coordinate must be the same, it follows that $\vec{E}(x) = -\vec{E}(-x)$ for any choice of y and z coordinate.

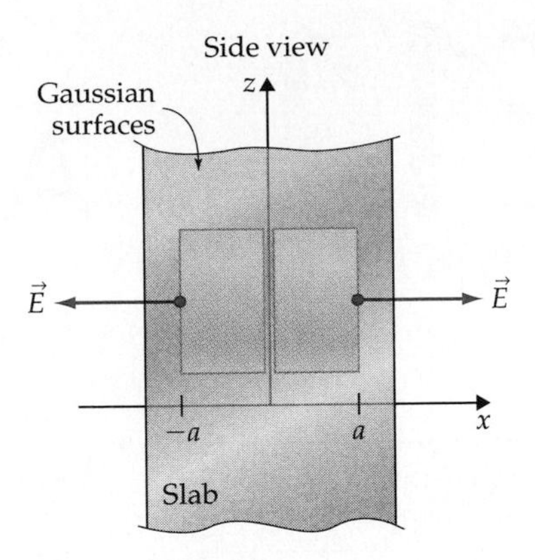

Figure E10.20
Gaussian surfaces that we can use to answer exercise E10X.4.

E11 Ampere's Law

▷ Electric Field Fundamentals
▷ Controlling Currents
▷ Magnetic Field Fundamentals
▽ Calculating Static Fields
Symmetry and Flux
Gauss's Law
Ampere's Law
▷ Dynamic Fields

Chapter Overview

Introduction

In chapter E10, we learned how to calculate static electric fields by using symmetry principles and Gauss's law. In this chapter we will learn how to do the same thing for static magnetic fields. The results will be useful to us in chapters E12 and E13.

Section E11.1: Ampere's Law

We can divide any arbitrary imaginary closed loop in space into infinitesimal segments each described by a **segment vector** $d\vec{S}$ whose magnitude is the segment's length and whose direction points tangent to the loop. We can choose the segment vectors to point either all clockwise or all counterclockwise around the loop. Given this, the magnetic analog to Gauss's law is **Ampere's law:**

$$\oint \vec{\mathbb{B}} \cdot d\vec{S} = \frac{4\pi k}{c} i_{\text{enc}} \quad \text{or} \quad \oint \vec{B} \cdot d\vec{S} = \mu_0 i_{\text{enc}} \tag{E11.2}$$

Purpose: This equation describes the link between a static magnetic field and the current distribution that creates it.

Symbols: $\oint$ is a shorthand notation in this case for "sum over all infinitesimal segments on a closed loop," $\vec{B}$ or $\vec{\mathbb{B}} = c\vec{B}$ is the magnetic field vector at a given loop segment, $d\vec{S}$ is the segment's segment vector, k is the Coulomb constant, $\mu_0 = 4\pi k/c^2$, c is the speed of light, and i_{enc} is the net current enclosed by the loop: $i_{\text{enc}} \equiv \Phi_J \equiv \int \vec{J} \cdot d\vec{A}$ (where the direction of the $d\vec{S}$ vectors defines the loop's circulation direction).

Limitations: This equation only applies to *static* magnetic fields. Each segment must be small enough that it is nearly straight and that $\vec{\mathbb{B}}$ is nearly uniform along its length. The loop can have any *closed* shape.

We call $\oint \vec{\mathbb{B}} \cdot d\vec{S}$ the magnetic **circulation** around the loop.

Section E11.2: Using Ampere's Law

This four-step process helps when using Ampere's law to calculate magnetic fields:

1. *Use symmetry* to learn as much as possible about the field's direction at all points and about what its magnitude does *not* depend on.
2. *Choose an amperian loop* shaped so that it is easy to calculate the magnetic circulation around the loop.
3. *Calculate the circulation* around the loop.
4. *Apply Ampere's law* to finish solving the problem.

Sections E11.3 through E11.5

These sections show how we can use the four-step process to determine the magnetic fields of axial current distributions, infinite planar slabs, and infinite solenoids, respectively. The results are summarized in table E11.1.

Section E11.6: Ampere's Law for the Electric Field

All *static* electric fields obey the following simple analog to Ampere's law:

$$\oint \vec{E} \cdot d\vec{S} = 0 \qquad \text{(E11.17)}$$

Purpose: This equation puts constraints on the kinds of static electric fields that can exist.

Symbols: $\oint$ in this context is a shorthand notation for "sum over all infinitesimal segments on a closed loop," $\vec{E}$ is the electric field evaluated at an infinitesimal segment on that loop, and $d\vec{S}$ is that segment's segment vector.

Limitations: This equation applies only to *static* electric fields (we will study this issue further in chapter E13).

This equation is equivalent to saying that (1) a static electric field cannot cause a current to flow around a loop and (2) such a field can be described by a well-defined potential function. This law constrains the kinds of electric fields that can exist.

Table E11.1 The magnetic fields of symmetric objects

Distribution	Axial	Infinite Solenoid	Infinite Slab
Description of distribution	An infinite cylindrical object carrying a steady total current I along the object's axis whose density depends at most on the distance r from the axis	An infinite cylindrical coil whose steady current flows circularly around the coil in a plane perpendicular to the coil's axis and whose current density depends at most on the distance r from the axis	A planar slab that has finite width W in one direction but is infinite in the other two perpendicular directions and that carries a uniform current density $\vec{J}_0$ in parallel to its surface
Diagram of the magnetic field	$\vec{J}$, r, $\vec{\mathbb{B}}$ End view	$\vec{\mathbb{B}} = 0$, $\vec{\mathbb{B}}$, I, $\vec{\mathbb{B}} = 0$	z, $\vec{\mathbb{B}}$, $\vec{J}_0$, x Side view
Equation for the magnetic field (k is the Coulomb constant, c is the speed of light)	$\mathbb{B} = \dfrac{2kI}{cr}$ where I is the total current enclosed by r	$\mathbb{B} = \dfrac{4\pi kI}{c}\left(\dfrac{N}{L}\right)$ inside where N/L is turns per unit length I is the current carried per turn	$\mathbb{B} = \begin{cases} (2\pi kJ_0/c)W & \text{outside} \\ (4\pi kJ_0/c)r & \text{inside} \end{cases}$ where r is the distance from the slab's central plane
Useful as an approximation for a finite object when:	$r \ll$ the distance to where the distribution is no longer straight	the coil's radius is much smaller than its length, and we evaluate the field far from the coil's ends	$r \ll$ the distance to the ends of the slab (as long as we are only interested in the component parallel to the slab)
The direction of the field	The field direction obeys the wire rule	The field direction is consistent with the loop rule	$\vec{\mathbb{B}}$ points parallel to the slab, $\perp$ to the current, and in the direction specified by the wire rule

E11.1 Ampere's Law

Introduction to Ampere's law

In chapter E10, we explored Gauss's laws for the electric and magnetic fields. Gauss's law for the magnetic field, however, does not link the magnetic field to current in the way that Gauss's law for the electric field links the electric field to the presence of charge. Therefore we cannot use Gauss's law for the magnetic field to calculate the magnetic field created by a current distribution. Our goal in this chapter is to explore a new law that we *can* use to calculate the magnetic field created by a given symmetric current distribution. This law is the real magnetic counterpart to Gauss's law for the electric field.

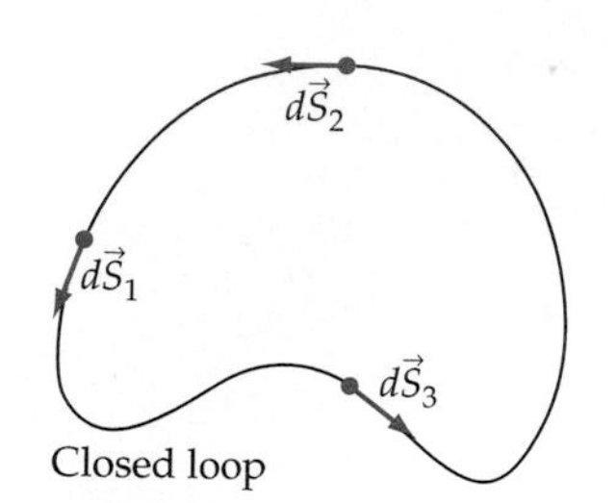

Figure E11.1
A closed loop in space and some of its segment vectors.

Imagine an arbitrary closed loop in space, as shown in figure E11.1. We can think of this loop as being a sequential set of infinitesimal segments, each of which is essentially straight. We can mathematically represent each segment by a **segment vector** $d\vec{S}$ whose magnitude is the segment's infinitesimal length dS and whose direction points along the curve. We can choose to have the segment vectors point either clockwise or counterclockwise around the loop, as long as we make the same choice for all the vectors: this defines the circulation direction for the loop.

Now imagine that this loop is in a region where there is a nonzero magnetic field. If we calculate the dot product $\vec{B} \cdot d\vec{S}$ of the magnetic field $\vec{B}$ evaluated at a given segment with that segment's segment vector $d\vec{S}$, we get an infinitesimal number that is positive if $\vec{B}$ at least leans in the direction of $d\vec{S}$, negative if $\vec{B}$ leans in the direction opposite to $d\vec{S}$, and zero if $\vec{B}$ and $d\vec{S}$ are perpendicular. We represent the sum of these small numbers around the loop by the symbol

The *circulation* of a magnetic field around a loop

$$\oint \vec{B} \cdot d\vec{S} \equiv \text{sum of } \vec{B} \cdot d\vec{S} \text{ for all segments around a closed loop} \quad \text{(E11.1)}$$

Do not let the integral symbol worry you: it just means "sum the following infinitesimal quantities around a closed loop" in this case. The circle in the integral sign means that the curve in this case is a *closed* loop, and the fact that we are doing the sum over terms involving segment vectors $d\vec{S}$ instead of tile vectors $d\vec{A}$ tells us that we are doing the sum over a loop instead of a surface. The symbol just expresses all this in a compact way.

We call this sum the **circulation** of the magnetic field around the closed loop. We can calculate the circulation of the electric field around the loop in a similar way. The circulation of a field around a loop is similar in many ways to the flux of a field through a closed surface, and we use similar notation to describe these quantities. Figure E11.2 summarizes the features of the notation so that you can see what the notation means and how you can distinguish flux from circulation.

Why are we interested in the circulation of a magnetic field? **Ampere's law** declares that in a situation where we have a *static* magnetic field, the circulation of the magnetic field $\vec{B}$ around a loop is proportional to the current

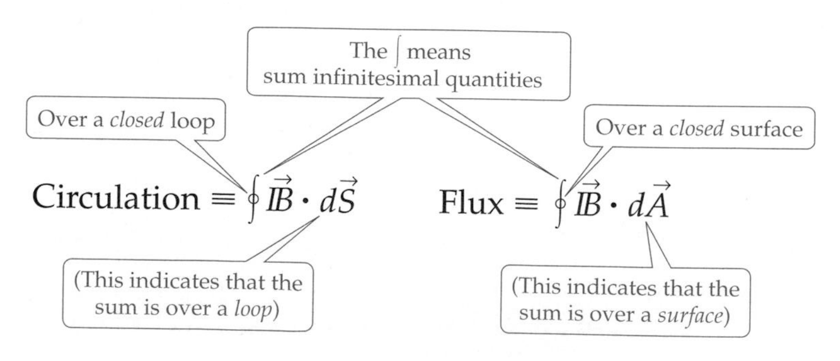

Figure E11.2
How to read the features of the notation we are using for circulation and flux through a closed surface.

enclosed by that loop:

Ampere's law

$$\oint \vec{\mathbb{B}} \cdot d\vec{S} = \frac{4\pi k}{c} i_{enc} \quad \text{or} \quad \oint \vec{B} \cdot d\vec{S} = \mu_0 i_{enc} \tag{E11.2}$$

Purpose: This equation describes the link between a static magnetic field and the current distribution that creates it.

Symbols: $\oint$ is a shorthand notation in this case for "sum over all infinitesimal segments on a closed loop," $\vec{B}$ or $\vec{\mathbb{B}} = c\vec{B}$ is the magnetic field vector at a given loop segment, $d\vec{S}$ is the segment's segment vector, k is the Coulomb constant, $\mu_0 = 4\pi k/c^2$, c is the speed of light, and i_{enc} is the net current enclosed by the loop (see the definition below).

Limitations: This equation only applies to *static* magnetic fields. Each segment must be small enough that it is approximately straight and that $\vec{\mathbb{B}}$ is nearly uniform along its length. The loop must be closed (but can otherwise have an arbitrary shape).

The technical definition of the quantity i_{enc}

The quantity i_{enc} is a signed quantity whose technical definition is as follows. Imagine dividing up the surface bounded by the loop into infinitesimal tiles. Since the current density $\vec{J}$ is defined to be the current per unit area, and since only the component of $\vec{J}$ perpendicular to a tiny tile on the surface bounded by the loop carries charge through that tile, $\vec{J} \cdot d\vec{A}$ is the current flowing through a given tile. The total current "enclosed by the loop" is simply the total current that flows through the surface bounded by the loop

$$i_{enc} \equiv \Phi_J \equiv \int \vec{J} \cdot d\vec{A} \tag{E11.3}$$

In other words, i_{enc} is simply the *flux* of the current density $\vec{J}$ through the surface enclosed by the loop.

In most cases that we will consider, it will be obvious how much current flows through the loop, and we will not have to do any actual sums. However, knowing this definition is important because it also defines the *sign* of i_{enc}. The flux of a field through a bounded surface is positive if the field at least leans in the same direction as the tile vectors for that surface, which in turn point in the direction indicated by the loop rule applied to the loop. In this case, the segment vectors define the circulation direction for the loop, so if the current flows in the direction indicated by your right thumb when your right fingers curl in the direction that the $d\vec{S}$ vectors go around the loop, the value of i_{enc} is positive; if it flows in the direction opposite to your thumb, i_{enc} is negative (see figure E11.3).

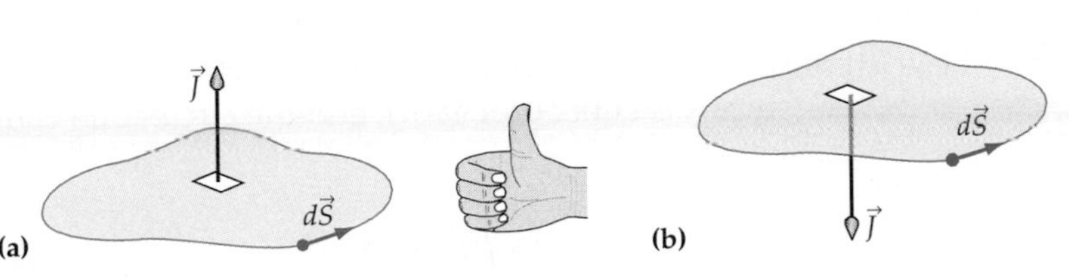

Figure E11.3
(a) The quantity $\vec{J} \cdot d\vec{A}$ will be positive through a tile on the loop if $\vec{J}$ points in the direction indicated by the loop rule when your right fingers curl in the direction of the loop's circulation. This means that the current flowing through the tile will contribute positively to i_{enc}. (b) Current flowing in the other direction contributes negatively to i_{enc}.

Ampere's law is the sought-after magnetic counterpart to Gauss's law. Combined with a symmetry argument, it allows us to determine the magnetic field of a suitably symmetric static current distribution without much calculation. It can be derived from the Biot-Savart law (equation E8.10), but is much easier to use for calculating the magnetic fields of symmetric current distributions.

Ampere's law works only for static magnetic fields

It is very important to realize, though, that Ampere's law, *unlike* Gauss's law, only works for *static* fields. We will learn how to correct Ampere's law to make it work for time-dependent fields in chapter E13.

E11.2 Using Ampere's Law

When using Ampere's law to calculate the magnetic field of a static current distribution, one uses a four-step process essentially identical to the steps we used when applying Gauss's law. The steps, when adapted for Ampere's law, read as follows:

A four-step process for solving problems involving Ampere's law

1. *Use symmetry.* You should first use symmetry and/or other arguments to determine as much as possible about the field's direction and magnitude at all points. Ampere's law will be easiest to apply when you know the field's precise direction at all points and you know at least what the field's magnitude does *not* depend on.
2. *Choose an amperian loop.* You should then pick a useful imaginary closed loop (which we will call an **amperian loop**) shaped so that it is easy to calculate the net circulation of the field around the loop.
3. *Calculate the circulation.* Then calculate the circulation around the loop. If you have chosen your loop well, this should be easy.
4. *Apply Ampere's law* to finish determining the field. In this step, you should compute the current i_{enc} enclosed by the loop defined in step 2 and use Ampere's law to link this current to the circulation computed in step 3 and thus to the magnetic field.

How to choose a useful amperian loop

As with Gauss's law, the crucial part of this process is the second step; choosing your amperian loop well is the key to making your solution easy. As with gaussian surfaces, the appropriate amperian loop is not always obvious, but there are some basic guidelines that you can follow. (I will express them assuming we are calculating magnetic circulation, but the same rules apply to calculating electric circulation.)

Calculating the flux through a particular section of the loop is *easiest* if either $\vec{\mathbb{B}} = 0$ or $\vec{\mathbb{B}}$ is perpendicular to the segment vectors at all points on that section, because either way $\vec{\mathbb{B}} \cdot d\vec{S} = 0$ for all segments in the section, meaning that that section contributes nothing to the net circulation. It is also very easy to calculate the circulation contributed by a loop section chosen so that $\vec{\mathbb{B}}$ is either *parallel* or *antiparallel* to the segment vectors $d\vec{S}$ in that section *and* $\mathbb{B} = \text{mag}(\vec{\mathbb{B}})$ is constant at all points along the section. Then $\vec{\mathbb{B}} \cdot d\vec{S} = \mathbb{B}\, dS \cos\theta = \pm\mathbb{B}\, dS$ (since θ is either 0°or 180°), and the circulation contributed by the section is simply

$$\int \vec{\mathbb{B}} \cdot d\vec{S} = \pm\mathbb{B} \int dS = \pm\mathbb{B}\,L \tag{E11.4}$$

where L is the total length of the segment. If your symmetry argument tells you that $\mathbb{B}$ can at most depend on some variable, call it x, then $\mathbb{B}$ will be constant along any curve of constant x.

So the trick is to construct a complete closed loop out of sections such that either (1) $\vec{\mathbb{B}}$ is zero at all points along the section, or (2) $\vec{\mathbb{B}}$ is perpendicular to all segment vectors along the section, or (3) $\vec{\mathbb{B}}$ is parallel or antiparallel to all segment vectors along the section.

Example E11.1

Problem Imagine our goal is to create a magnetic field in a given region of empty space (by using a suitable array of coils or wires surrounding that region) so $\vec{\mathbb{B}}$ points in the $+x$ direction, but increases in magnitude with increasing y according to the formula $\mathbb{B}_x = ay$, where a is some constant, as shown in figure E11.4. Using Ampere's law, prove that this is impossible.

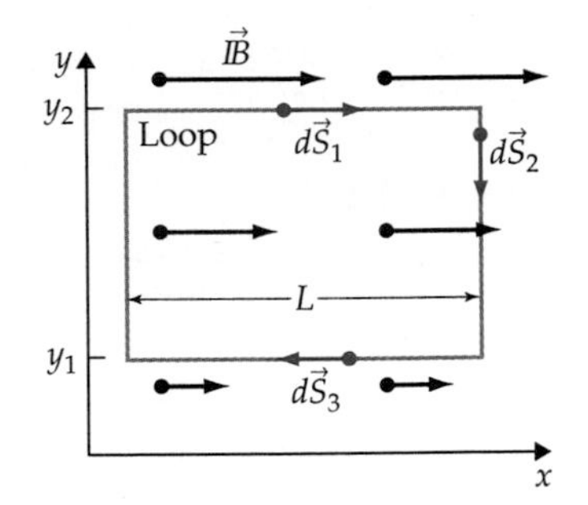

Figure E11.4
An appropriate amperian loop for the situation discussed in example E11.1. The black vectors show the magnetic field, and the colored arrows show selected segment vectors for segments on the loop. I have chosen L, y_1, and y_2 arbitrarily.

Use Symmetry In this case, the problem statement specifies the direction of the magnetic field and the fact that it does *not* depend on x or z.

Choose an Amperian Loop Figure E11.4 shows a suitable amperian loop for this problem. Each of the four legs of the closed loop involves segment vectors that are parallel, antiparallel, or perpendicular to the field, and $\mathbb{B}$ is constant along the legs where $\vec{\mathbb{B}} \cdot d\vec{s} \neq 0$.

Calculate the Circulation In the upper leg of the loop, the segment vectors are parallel to the field, and $\vec{\mathbb{B}}$ has the same magnitude $\mathbb{B} = ay_2$ at all segments along this leg, so $\int \vec{\mathbb{B}} \cdot d\vec{s} = +\mathbb{B}L = +ay_2L$ for this leg of the loop. Segment vectors along the bottom part of the loop are antiparallel to the field, and $\vec{\mathbb{B}}$ has the same magnitude $\mathbb{B} = ay_1$ at all segments along this leg, so $\int \vec{\mathbb{B}} \cdot d\vec{s} = -\mathbb{B}L = -ay_1L$ for this leg. The other two legs are everywhere perpendicular to $\vec{\mathbb{B}}$ so $\int \vec{\mathbb{B}} \cdot d\vec{s} = 0$ for both these legs. Therefore, the sum of $\vec{\mathbb{B}} \cdot d\vec{s}$ over the entire closed loop in this case is $\oint \vec{\mathbb{B}} \cdot d\vec{s} = ay_2L - ay_1L$.

Apply Ampere's Law Ampere's law in this case tells us that

$$\frac{4\pi k}{c} I_{\text{enc}} = \oint \vec{\mathbb{B}} \cdot d\vec{s} = aL(y_2 - y_1) \neq 0 \tag{E11.5}$$

Since there can be no current flowing in "empty space," such a magnetic field is impossible in a region of empty space, no matter what kind of current distributions we arrange around it.

Exercise E11X.1

Prove that it *is* possible for a magnetic field to be uniform in magnitude and direction in a region of empty space. (We would need to set up suitable current distributions *around* the empty space to create such a field in it.)

E11.3 The Magnetic Field of an Axial Current Distribution

Consider now an axial current distribution, which consists of an infinite cylindrical object that is unchanged by either rotation about its central axis or sliding along its central axis and that carries a steady current parallel to that axis. Infinite cylindrical wires, straight pipes, or straight coaxial cables could qualify as axial distributions. What does Ampere's law tell us about the external magnetic field of such a distribution?

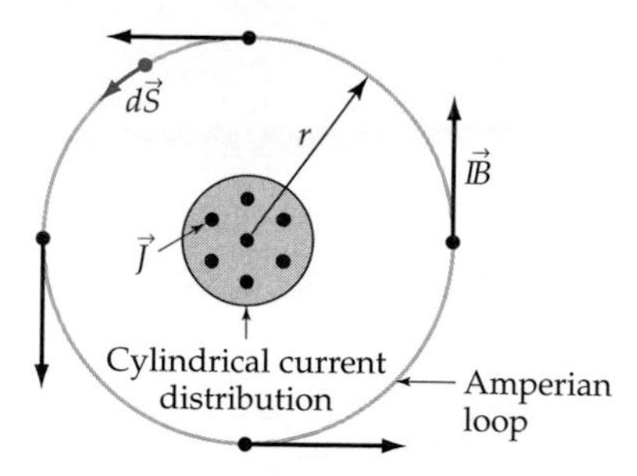

Figure E11.5
Magnetic field vectors and an appropriate amperian loop for an infinite cylindrical current distribution. This is a cross-sectional view of the distribution; the dots indicate that the current is flowing toward us. Symmetry implies that at all points on this loop, the magnetic field vectors (black arrows) created by this current must be tangent to the loop.

Use Symmetry

In example E9.4, we showed that the magnetic field created by an axial current distribution must point circularly around the distribution's central axis (that is, tangent to circles whose planes are perpendicular to that axis) and that its magnitude must depend at most on the distance r that one is from that central axis (see figure E11.5).

Choose an Amperian Loop

Since the magnetic field can at most depend on r, a convenient amperian loop would be a circle of radius r around the central axis (see the colored loop in figure E11.5) that lies in a plane perpendicular to that axis and goes through the point P where we would like to evaluate the field. The magnetic field strength $\mathbb{B} \equiv \text{mag}(\vec{\mathbb{B}})$ is the same at all points along this loop, and if we choose the segment vectors on our loop to point counterclockwise around the loop, then $\vec{\mathbb{B}}$ is either parallel or antiparallel to the segment vector at every point.

Calculate the Circulation

The net magnetic circulation around the loop is therefore $\oint \vec{\mathbb{B}} \cdot d\vec{S} = \pm\mathbb{B} \oint dS = \pm\mathbb{B}(2\pi r)$, where the positive result applies if the field goes counterclockwise around the loop and the negative result if the field goes clockwise.

Apply Ampere's Law

If the point P is outside the charge distribution, then our amperian loop completely encloses whatever total current I the distribution carries. Ampere's law in this case then tells us that

$$\oint \vec{\mathbb{B}} \cdot d\vec{S} = \frac{4\pi k}{c} i_{\text{enc}} \quad \Rightarrow \quad \pm\mathbb{B}(2\pi r) = +\frac{4\pi k}{c} I \qquad \text{(E11.6)}$$

We know that i_{enc} is positive here since in figure E11.5, current $\vec{I}$ is flowing in the direction indicated by your right thumb if you curl your fingers in the direction of $d\vec{S}$. To make the signs agree on both sides of this expression, we must choose the plus sign in front of $\mathbb{B}$, meaning that $\vec{\mathbb{B}}$ must go counterclockwise around the circle.

Exercise E11X.2

Show that this is consistent with the wire rule.

Solving for $\mathbb{B}$, we find that the magnetic field strength at points along this loop (and thus at the point P) is

The magnetic field outside an infinite cylindrical current distribution

$$\mathbb{B} = \frac{2kI}{cr} \quad \text{or} \quad B = \frac{\mu_0 I}{2\pi r} \qquad \text{(E11.7)}$$

Purpose: This equation describes the magnitude of magnetic field $\vec{B}$ or $\vec{\mathbb{B}} = c\vec{B}$ *outside* an infinite cylindrical current distribution carrying a total current I parallel to its axis.

Symbols: r is the distance between the axis and the point where the field is evaluated, k is the Coulomb constant, $\mu_0 = 4\pi k/c^2$, and c is the speed of light.

Limitations: The current must be steady. The current distribution must be unchanged by sliding along its axis (which means it has to be infinite) as well as rotations about its axis (which means $\vec{J}$ can at most depend on r).

Note: $\vec{\mathbb{B}}$ points circularly around the distribution in the plane perpendicular to the distribution's axis, in the direction indicated by the wire rule.

In the case where the distribution is not infinite, this equation is a useful approximation whenever the point P is much closer to the distribution's central axis than to the nearest point where it deviates from being an axially symmetric distribution.

Exercise E11X.3

The magnetic field inside an infinite pipe is zero

Use Ampere's law to prove that if the cylindrical current distribution is a circular pipe with a hollow interior, the magnetic field is *zero* in the hollow interior. (*Hint:* Remember that symmetry implies that a nonzero magnetic field *inside* the distribution would have to point circularly around the axis just as it does outside the distribution.)

E11.4 The Magnetic Field of an Infinite Planar Slab

Description of an infinite planar slab

Now consider an infinite planar slab that, for the sake of argument, is oriented perpendicular to the x axis with its central plane (the plane halfway between its surfaces) on the yz plane. The slab extends infinitely in the y and z directions, but has a finite width W along the x direction. Assume that this slab carries a uniform current density $\vec{J}_0$ in the $-y$ direction. What does the magnetic field created by this slab look like, both inside and outside the slab?

Use Symmetry

We saw in example E9.6 that the magnetic field of such a slab points in the $\pm z$ direction, has a magnitude that at most depends on x, and has a magnitude of zero on the slab's central plane.

Choose an Amperian Loop

This suggests that an appropriate amperian loop for this situation would be a rectangular loop with one vertical leg lying on the slab's central plane (see figure E11.6a). The segment vectors along these legs are all parallel, antiparallel, or perpendicular to the field vectors, and the field has a constant magnitude along the vertical legs.

Calculate the Circulation

For the sake of argument, let's orient the segment vectors so that they go counterclockwise around the loop. Since the magnetic field is zero on the central plane, the left-hand vertical leg contributes nothing to the magnetic circulation around the loop. The magnetic field vector $\vec{\mathbb{B}}$ is everywhere perpendicular to the segment vectors along the horizontal legs of the loop, so these legs also contribute nothing to the sum. On the right vertical leg, $d\vec{S}$ points in the $+z$ direction, so $\vec{\mathbb{B}} \cdot d\vec{S} \equiv \mathbb{B}_x\, dS_x + \mathbb{B}_y\, dS_y + \mathbb{B}_z\, dS_z = \mathbb{B}_z\, dS$ for each segment along that leg. Moreover, since $\vec{\mathbb{B}}$ can at most depend on x,

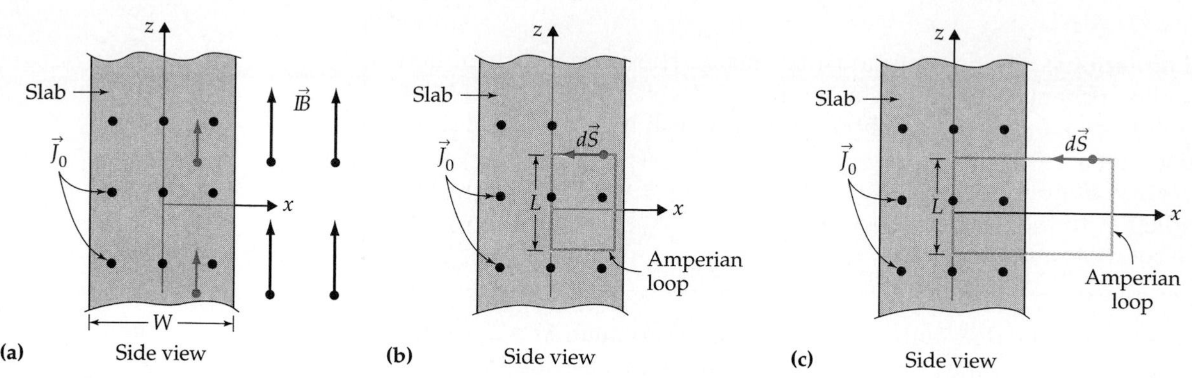

Figure E11.6
(a) Symmetry implies that the magnetic field vectors created by an infinite slab must point parallel to the slab but perpendicular to the current and must have a magnitude that depends at most on x. (I have chosen to make the magnetic field vectors all point upward, but I could just as easily have chosen them to point downward.) (b) An appropriate amperian loop for computing the magnetic field at a point inside the slab. (c) An appropriate amperian loop for computing the magnetic field at a point outside the slab.

$\mathbb{B}_z$ must have the same value for all segments on the leg. The contribution this leg makes to the total circulation is therefore

$$\int \vec{\mathbb{B}} \cdot d\vec{S} = \mathbb{B}_z \int dS = \mathbb{B}_z L \qquad \text{(E11.8)}$$

where L is the length of the right vertical leg.

Apply Ampere's Law

Consider first the case where the loop's right leg is inside the slab (that is, its x coordinate is such that $0 \leq x \leq \frac{1}{2}W$: see figure E11.6b). The current density is the current per unit area, so the loop encloses a current equal to area times the magnitude of the current density: $|i_{\text{enc}}| = xLJ_0$. Since this current *is* flowing in the direction indicated by one's right thumb when one's right fingers curl in the direction of $d\vec{S}$ around the loop, $i_{\text{enc}} = +xLJ_0$. So Ampere's law in this case implies that

$$\oint \vec{\mathbb{B}} \cdot d\vec{S} = \frac{4\pi k}{c} i_{\text{enc}} \quad \Rightarrow \quad \mathbb{B}_z L = \frac{4\pi k}{c} xLJ_0$$

$$\Rightarrow \quad \mathbb{B}_z = \left(\frac{4\pi k J_0}{c}\right) x \qquad \text{(E11.9}a\text{)}$$

So $\vec{\mathbb{B}}$ points in the *positive* z direction on the side of the slab where $x > 0$. Note also that $\mathbb{B}_z$ increases linearly with increasing x inside the slab.

Now imagine the loop's right leg to be *outside* the slab (see figure E11.6c). The current enclosed is now $i_{\text{enc}} = +(\frac{1}{2}W)LJ_0$, since the width of the current-carrying region is only $\frac{1}{2}W$ instead of extending all the way out to the x coordinate of the loop's right leg. Ampere's law in this case implies that

$$\oint \vec{\mathbb{B}} \cdot d\vec{S} = \frac{4\pi k}{c} i_{\text{enc}} \quad \Rightarrow \quad \mathbb{B}_z L = \frac{2\pi k}{c} WLJ_0$$

$$\Rightarrow \quad \mathbb{B}_z = \frac{2\pi k J_0 W}{c} \qquad \text{(E11.9}b\text{)}$$

which is a constant. Outside the slab, therefore, the magnetic field is a *uniform* field pointing in the $+z$ direction.

Exercise E11X.4

Use Ampere's law to show that the magnetic field on the left side of the slab points in the $-z$ direction but has the same magnitude as the field on the right at a point the same distance from the central plane. Show also that the field goes "around" the slab in the direction indicated by the wire rule.

The choice of coordinates is arbitrary, so generalizing equations E11.9*a* and E11.9*b* and the results of exercise E11X.4 for arbitrary slab orientations, we can say that

The magnetic field inside and outside an infinite planar slab

$$\mathbb{B} = \begin{cases} \dfrac{2\pi k J_0}{c} W \\ \dfrac{4\pi k J_0}{c} r \end{cases} \quad \text{or} \quad B = \begin{cases} \frac{1}{2}\mu_0 J_0 W & \text{at points outside the slab} \\ \mu_0 J_0 r & \text{at points inside the slab} \end{cases} \tag{E11.10}$$

Purpose: This equation describes the magnitude of the magnetic field $\vec{B}$ or $\vec{\mathbb{B}} = c\vec{B}$ inside and outside an infinite planar slab perpendicular to the x direction that carries a uniform current density $\vec{J}_0$.
Symbols: r is the distance between the slab's central plane and the point where the field is evaluated, W is the slab's width, k is the Coulomb constant, $\mu_0 = 4\pi k/c^2$, and c is the speed of light.
Limitations: The slab must be infinite and the current steady.
Note: The direction of the magnetic field is parallel to the slab's surface and perpendicular to the current in the direction indicated by the wire rule.

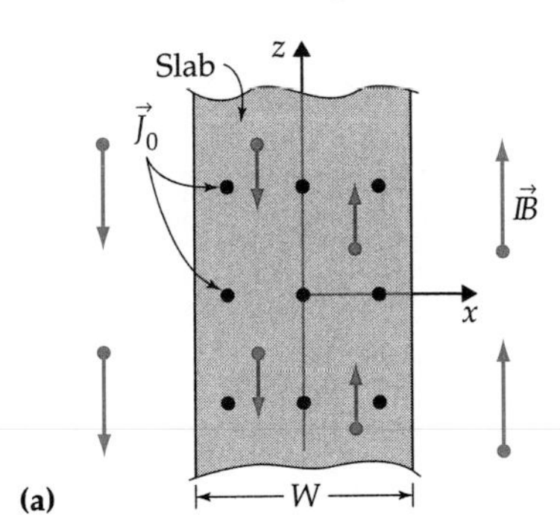

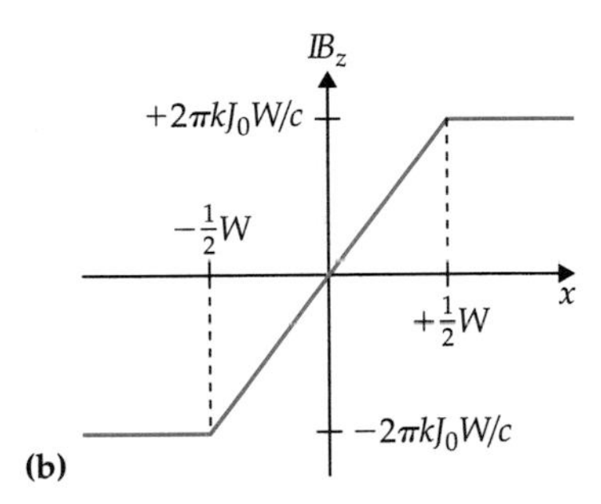

Figure E11.7
(a) A cross-sectional view of our infinite slab. The dots indicate that current density is flowing toward us. (b) A graph showing $\mathbb{B}_z$ as a function of y for points inside and outside the slab.

Figure E11.7 shows a graph of $\mathbb{B}_z$ as a function of x for our particular coordinate system.

Equation E11.10 is a pretty good approximation for the component of the magnetic field parallel to the slab and perpendicular to the current at a point near a finite slab as long as the point in the field is much closer to $x = 0$ than to the edges of the slab (or any other position where the symmetry assumptions are violated). However, we cannot assume that the other components of $\vec{\mathbb{B}}$ are negligible unless we are also very close to the slab's center.

E11.5 The Magnetic Field of an Infinite Solenoid

Description of an infinite solenoid

Now consider a **solenoid** consisting of N closely packed turns of wire wound in a coil around a given length L of an infinitely long, hollow, cylindrical, pipelike form of radius R. For the sake of argument, let us define the x axis to coincide with the solenoid's central axis. What does the magnetic field of such a solenoid look like, both inside and outside the solenoid?

Use Symmetry

As discussed in example E9.5, the symmetry of the solenoid implies that its magnetic field must everywhere point in the $\pm x$ direction (that is, that $\vec{\mathbb{B}} = \mathbb{B}_x\hat{x}$) and that its magnitude at a given point P can depend *at most* on the distance r that the point is from the solenoid's central axis.

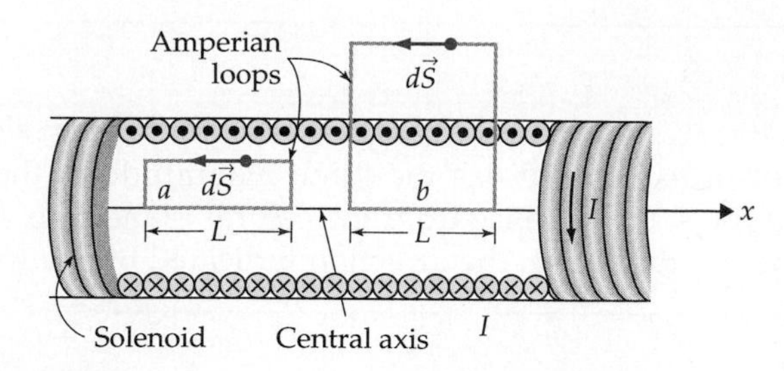

Figure E11.8
A cross-sectional side view of part of an infinite solenoid, showing two amperian loops we might use to calculate the solenoid's magnetic field. The $d\vec{S}$ vector shown for a segment on each loop indicates that we are choosing to orient these vectors in the counterclockwise direction around the loop in both cases. We have already learned from symmetry arguments that the magnetic field must point in the $\pm x$ direction.

Choose an Amperian Loop

Since the magnetic field everywhere is in the $\pm x$ direction, it is simple to calculate the magnetic circulation for square loops like those shown in figure E11.8. The segment vectors on both vertical legs of this loop are perpendicular to the field, so $\vec{\mathbb{B}} \cdot d\vec{S} = 0$ for each segment in these legs, meaning that these legs contribute nothing to the total sum around the loop. The segment vectors in the horizontal legs are parallel or antiparallel to the field, and the magnitude of the field is constant along these legs.

Calculate the Circulation

If we take the segment vectors to point counterclockwise around the loop, then $d\vec{S} = [+dS, 0, 0]$ for each segment along the bottom leg and $d\vec{S} = [-dS, 0, 0]$ for each segment along the top leg. Therefore $\vec{\mathbb{B}} \cdot d\vec{S} \equiv \mathbb{B}_x\, dS_x + \mathbb{B}_y\, dS_y + \mathbb{B}_z\, dS_z = +\mathbb{B}_x\, dS$ along the bottom leg and $\vec{\mathbb{B}} \cdot d\vec{S} = -\mathbb{B}_x\, dS$ along the top leg. Since $\mathbb{B}_x$ does not depend on x, its value can be pulled out in front of the sum for each leg. Therefore the net circulation is

$$\oint \vec{\mathbb{B}} \cdot d\vec{S} = -\mathbb{B}_x \int_{\text{top}} dS + 0 + \mathbb{B}_{x0} \int_{\text{bot}} dS + 0 = L(\mathbb{B}_{x0} - \mathbb{B}_x) \qquad \text{(E11.11)}$$

where $\mathbb{B}_x$ is the field's x component evaluated at the position of the top leg a distance r from the solenoid's central axis, $\mathbb{B}_{x0}$ is the same evaluated at $r = 0$ along the central axis, and L is the length of both legs.

Apply Ampere's Law

Consider first a loop whose top leg is a distance $r < R$ from the solenoid's central axis, so that the top and bottom legs are both inside the solenoid's hollow interior (such as loop a in figure E11.8). In this case, the loop encloses no current, so Ampere's law implies that

$$\oint \vec{\mathbb{B}} \cdot d\vec{S} = 0 \quad \Rightarrow \quad L(\mathbb{B}_{x0} - \mathbb{B}_x) = 0 \quad \Rightarrow \quad \mathbb{B}_x = \mathbb{B}_{x0} \qquad \text{(E11.12)}$$

This means that the magnetic field in the interior of an infinite solenoid is independent of r. Since the solenoid's symmetry implies that the field strength can *at most* depend on r, this means that *the magnetic field of an infinite solenoid is completely uniform throughout its interior.*

Consider now a loop whose top leg is a distance $r > R$ from the loop's axis, so that the top leg lies outside the solenoid (such as loop b in figure E11.8). If there are N turns of wire in a length L of the solenoid and each wire carries a current I, then the total current going through the loop is NI, no matter how far the top leg is from the axis. Since this current flows in the direction indicated by my right thumb when my fingers curl around the loop in the direction of the segment vectors $d\vec{S}$, in this case i_{enc} is positive. Ampere's law, therefore, tells us that for the second loop

$$\oint \vec{\mathbb{B}} \cdot d\vec{S} = \frac{4\pi k}{c} i_{enc} \quad \Rightarrow \quad L(\mathbb{B}_{x0} - \mathbb{B}_x) = \frac{4\pi k}{c} NI$$

$$\Rightarrow \quad \mathbb{B}_{x0} - \mathbb{B}_x = \frac{4\pi k I}{c}\frac{N}{L} \tag{E11.13}$$

Note that $\mathbb{B}_{x0} - \mathbb{B}_x$ is independent of r for $r > R$, so the magnetic field *outside* an infinite solenoid is uniform as well!

However, in this case, Ampere's law by itself only allows us to calculate the *difference* between $\mathbb{B}_{x0}$ and $\mathbb{B}_x$.† We find experimentally, however, that when we make a set of finite solenoids with the same radius but with increasing length, the measured magnetic field outside the solenoids approaches zero as the solenoid's length becomes large. If we assume that the limit in the case of infinite length is *exactly* zero, then equation E11.13 implies that the x component of the field *inside* the solenoid is

$$\mathbb{B}_{x0} = \frac{4\pi k I}{c}\frac{N}{L} \tag{E11.14}$$

Note that this is positive, so $\vec{\mathbb{B}}$ points in the $+x$ direction in this case.

Exercise E11X.5

Show that this magnetic field direction is consistent with the loop rule.

So, to summarize, the magnetic field created by an infinite cylindrical solenoid is

$$\mathbb{B} = \frac{4\pi k I}{c}\left(\frac{N}{L}\right) \quad \text{or} \quad B = \mu_0 I\left(\frac{N}{L}\right) \quad \text{inside the solenoid} \tag{E11.15a}$$

$$\mathbb{B} = 0 \quad \text{or} \quad B = 0 \quad \text{outside the solenoid} \tag{E11.15b}$$

Purpose: This describes the magnitude of the magnetic field $\vec{B}$ or $\vec{\mathbb{B}} = c\vec{B}$ of an infinite solenoid whose central axis is parallel to the x direction and which carries a current I in each of its turns.

Symbols: N/L is the number of turns per unit length on the solenoid, k is the Coulomb constant, $\mu_0 = 4\pi k/c^2$, and c is the speed of light.

Limitations: The solenoid must be infinite and the current steady.

Note: The direction of the field is parallel to the solenoid's central axis in the direction indicated by the loop rule.

The magnetic field inside and outside an infinite solenoid

†This is actually a *general* feature of Ampere's law. Note, for example, that in the infinite slab case, Ampere's law *really* only specifies the difference between the field strength at $x = 0$ with that at some other x. This feature is required so that Ampere's law works correctly even when the current distribution of interest is immersed in some external magnetic field (such an external field will cancel out in the difference). But this means that we can compute the absolute magnitude of the field created by a current distribution *only* if we know (by symmetry, for example) that the field created *specifically by that distribution* is zero along some line or plane: we can then use this as a reference, as we did in the slab case, to calculate the field strength elsewhere.

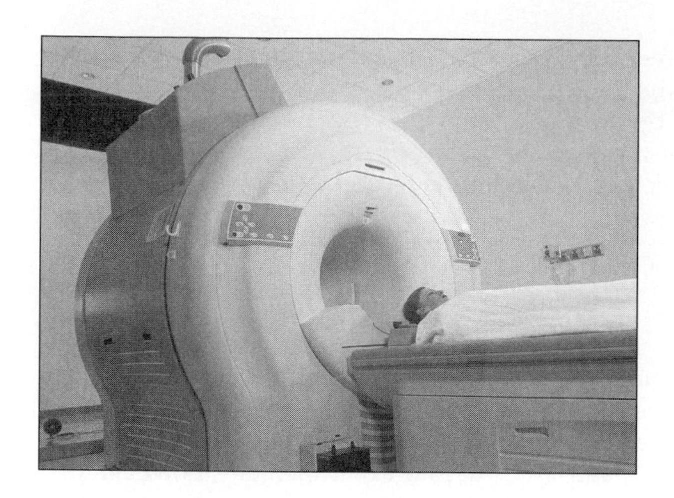

Figure E11.9
A patient being inserted inside the solenoid of an MRI machine.

These results are useful approximations for a finite solenoid

The field described by equations E11.15 turns out to be a good approximation to the magnetic field of a *finite* solenoid whose radius is much smaller than its length. A long solenoid provides the most practical way to create a strong and nearly uniform magnetic field in a fairly large volume of space. The solenoids used in medical magnetic resonance imaging (MRI) machines are long and narrow (and therefore claustrophobia-provoking) partly because good MRI images depend on having a strong and fairly uniform field over the imaging region (see figure E11.9).

Exercise E11X.6

A typical MRI solenoid creates a magnetic field of about 1.5 T (450 MN/C). If you want to create such a field using wires that can carry a maximum of 50 A of current, about how many turns per meter should your coil have?

E11.6 Ampere's Law for the Electric Field

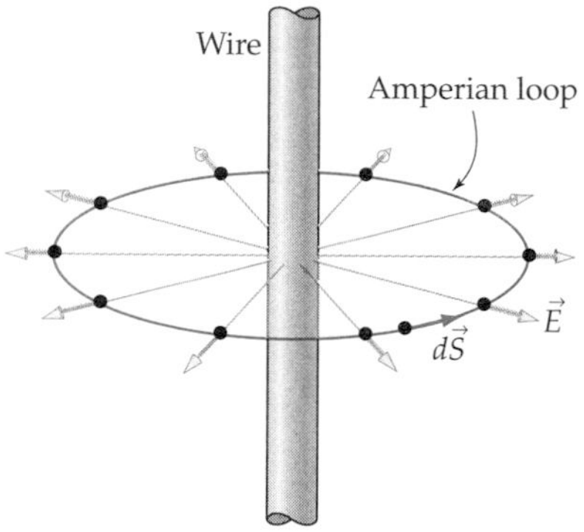

Figure E11.10
The electric field of an infinite charged wire points radially away from the wire and thus perpendicular to a circular amperian loop drawn around the wire.

Consider the electric field of an infinite, uniformly charged cylindrical wire. In chapter E10, we found that such a wire produces an electric field that points radially away from the wire's central axis. Imagine that we draw a circular amperian loop with a radius r centered on the wire in a plane perpendicular to its central axis (see figure E11.10). The wire's electric field $\vec{E}$ is everywhere perpendicular to this loop, so $\vec{E} \cdot d\vec{S} = 0$ for every segment on the loop. Therefore, even though the wire "carries" charge through this loop in the same way that the cylindrical current distribution carries current through the amperian loop in figure E11.5, if there is an equivalent to Ampere's law for the electric field at all, it must be

$$\oint \vec{E} \cdot d\vec{S} = 0 \qquad \text{(even when there is charge inside loop)} \qquad \text{(E11.16)}$$

A general argument based on energy conservation

It does not prove much that this law happens to work for a special loop in the special case of an infinite charged wire. But there is a more fundamental reason that this law must be true in general for *all* static (i.e., unchanging) electric fields. If $\vec{E} \cdot d\vec{S}$ is positive for a given segment of conducting wire, then the electric field has a nonzero component in the direction of $d\vec{S}$ and thus will push conventional current in that direction. If the electric circulation $\oint \vec{E} \cdot d\vec{S}$ is positive around a closed loop, the electric field will push current spontaneously around the loop in the direction indicated by the segment vectors. A negative circulation would mean that the field would push current around the loop in the opposite direction.

But we *never* observe a static electric field to push a current around a loop. Indeed, a static electric field that spontaneously pushes current around a closed loop would violate conservation of energy, because energy is dissipated in the wire as the current flows but nothing else is changing, so there is nowhere for this energy to come from. Therefore, the electric circulation around any closed loop must be zero for a static electric field.

One can also show, if one can describe an electric field well-defined potential energy function $\phi(x, y, z)$, that the field must have zero net circulation around an arbitrary loop (see problem E11S.9). Since we can describe any static electric field using a potential function, all such fields must have zero net circulation.

So to summarize, static electric fields must satisfy the following equation, which we call **Ampere's law for the electric field:**

Ampere's law for the electric field

$$\oint \vec{E} \cdot d\vec{S} = 0 \qquad \text{(E11.17)}$$

Purpose: This equation puts constraints on the kinds of static electric fields that can exist.

Symbols: $\oint$ in this context is a shorthand notation for "sum over all infinitesimal segments on a closed loop," $\vec{E}$ is the electric field evaluated at an infinitesimal segment on that loop, and $d\vec{S}$ is that segment's segment vector.

Limitations: This equation applies only to *static* electric fields (we will study this issue further in chapter E13).

Note: This equation is equivalent to saying that it is possible to describe an electric field with a well-defined potential function $\phi(x, y, z)$.

Like Gauss's law for the magnetic field, Ampere's law for the electric field puts constraints on the kinds of static electric fields that can exist, as example E11.2 illustrates.

Example E11.2 Identifying Realistic Electric Fields

Problem Figure E11.11 shows three field arrow diagrams for three hypothetical static electric fields. (Assume that the field is independent of position along the direction perpendicular to the diagram.) Which of these fields are physically possible? (*Hint:* The colored curve on each diagram is a suggested amperian loop for the field.)

Solution If we look at the suggested amperian loop for the first field, we can see that the segment vectors on the horizontal legs are all perpendicular to the field, so the horizontal legs contribute nothing to the circulation. The segment vectors on the left leg are antiparallel to the field, while the segment vectors on the right leg are parallel to the field. Therefore, the left leg makes a large negative contribution to the field while the right leg makes a smaller positive contribution. Therefore the net circulation is negative, which Ampere's law for the electric field tells us is impossible. (Remember this electric field pattern: we will encounter it again in chapter E13.)

The second graph illustrates that a radial electric field that depends only on r will always satisfy Ampere's law for the electric field. The field is perpendicular to the curved legs of the loop shown, so they contribute nothing to the circulation. The field strength varies along the straight legs, but is identical along those legs while the segment vectors point in opposite directions, so the contributions these legs make to the circulation will cancel. Therefore, the net circulation is zero, and this field is possible.

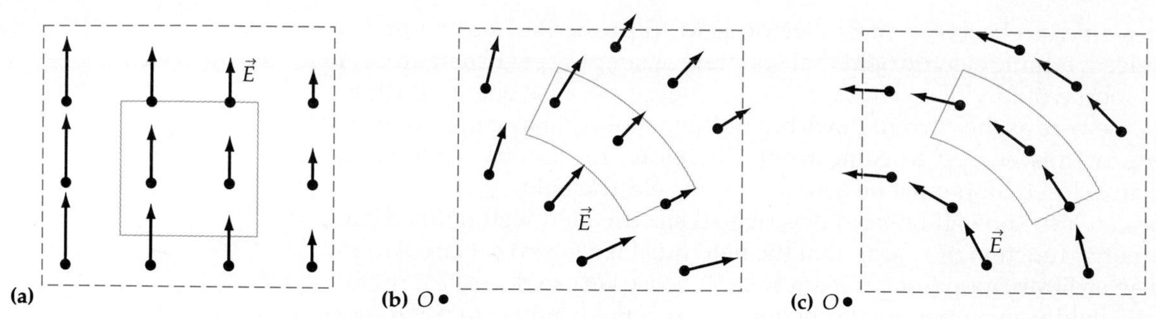

Figure E11.11
These diagrams show a field arrow diagram of the electric field within a certain area of space (bounded by the colored dashed lines). In all cases, the field is independent of position along the axis perpendicular to the drawing. (b) The field vectors point radially away from O and have a magnitude that depends only on the distance r from a vertical axis going through O. (c) The vectors point tangent to circles around O, but all have the same magnitude.

This field in the third graph is tangent to circles around a given point O and has a constant magnitude everywhere. In this case, the field is perpendicular to the segment vectors along the straight legs of the suggested loop, so these legs contribute nothing to the circulation. The field is parallel to the segment vectors on the curved leg to the upper right and antiparallel to the segment vectors on the curved leg to the lower left. The magnitude of the *field* is the same on both legs, but the upper right leg is longer; so while the lower left leg makes a negative contribution to the circulation, the upper right leg makes a larger positive contribution, leading to a net positive circulation, in violation of Ampere's law for the electric field.

Exercise E11X.7

Consider a circular field like the third field shown in figure E11.11. How would such a field have to depend on the radial distance r from the point O to satisfy Ampere's law for the electric field?

TWO-MINUTE PROBLEMS

E11T.1 Imagine a magnetic field that points radially outward from a point O with a magnitude that depends on the distance r that one is from O. If such a magnetic field could exist, for which of the amperian loops shown at the right would it be *easiest* to calculate the magnetic field circulation?

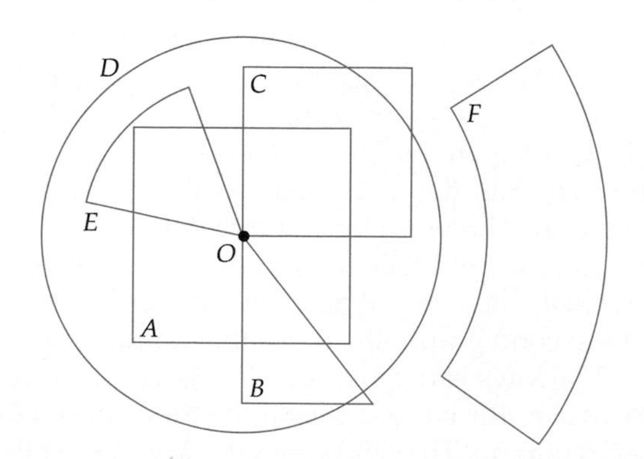

E11T.2 The radially outward magnetic field described in problem E11T.1 cannot exist (even in a region inside a current distribution) because

A. It violates Gauss's law for electric fields
B. It violates Gauss's law for magnetic fields
C. It violates Ampere's law
D. It cannot be a static field
E. (Specify another reason)

E11T.3 Consider an amperian loop in the plane perpendicular to our line of sight where the segment vectors $d\vec{S}$ point in the clockwise direction when viewed from our vantage point. This loop is embedded in an object that conducts a current. The quantity i_{enc} is positive if the current flowing through the surface enclosed by the loop goes
A. Toward our left
B. Toward our right
C. Away from us
D. Toward us
E. Clockwise around the loop
F. In some other direction (specify)

E11T.4 Imagine that we surround a current-carrying wire with an arbitrarily shaped closed amperian loop, as shown in figure E11.12a, and we calculate the net magnetic circulation around the loop and the magnitude $\mathbb{B}$ of the magnetic field at point P. We then bring another wire and park it just *outside* the amperian loop, as shown in figure E11.12b, and again calculate the net circulation around the loop and $\mathbb{B}$ at point P.
(a) How does $\mathbb{B}$ at point P change?
(b) How does the net circulation around the loop change?
A. The quantity increases.
B. The quantity decreases.
C. The quantity remains the same.
D. The quantity changes, but whether it increases or decreases depends on details not given.
E. The quantity may nor may not change, depending on details not given.

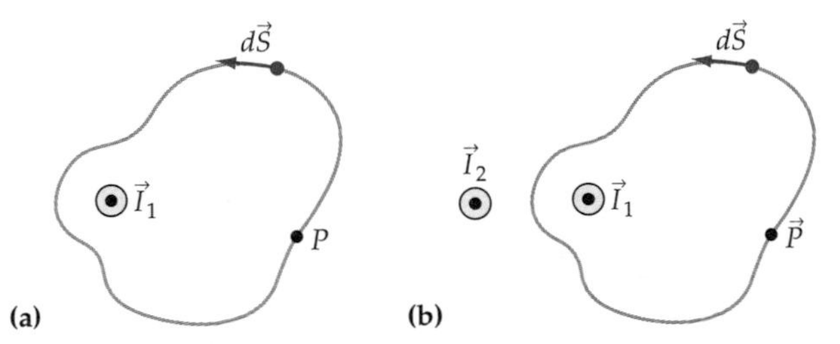

Figure E11.12
How do the magnetic field at P and the circulation through the amperian loop change as we move from the situation shown in (a) to the situation shown in (b)? (See problem E11T.4.)

E11T.5 Consider a finite planar slab perpendicular to the y axis that carries a current in the $+z$ direction. The y axis goes through the slab's geometric center. The magnetic field created by this slab at points along the positive side of the y axis points most nearly in the
A. $+x$ direction
B. $-x$ direction
C. $+y$ direction
D. $-y$ direction
E. $+z$ direction
F. $-z$ direction

E11T.6 Consider two coaxial solenoids, each with the same value of N/L (turns per meter) and each carrying the same current I, but in opposite directions. The outer solenoid has a radius of $2R$, and the inner solenoid has a radius of R. (a) What is the magnitude $\mathbb{B}$ of the magnetic field in the space between the two solenoid coils? (b) What is it inside the inner solenoid?
A. $8\pi kNI/cL$
B. $4\pi kNI/cL$
C. $2\pi kNI/cL$
D. $\pi kNI/cL$
E. 0
F. Some other value (specify)

E11T.7 Consider the magnetic field shown. This magnetic field always points tangent to circles going around point O, but (at least within the region enclosed by the dashed line) the magnetic field has a constant magnitude. Assume that the magnetic field is independent of position in the direction perpendicular to the drawing. This magnetic field is inside a conducting object that can *in principle* carry a current that, if it exists, would be perpendicular to the plane of the drawing at all points within the region in question. Judging from the characteristics of the magnetic field, the current is
A. Flowing toward us
B. Flowing away from us
C. Not zero, but we cannot determine its direction
D. Possibly zero
E. Definitely zero
F. Not at all specified by the known field characteristics

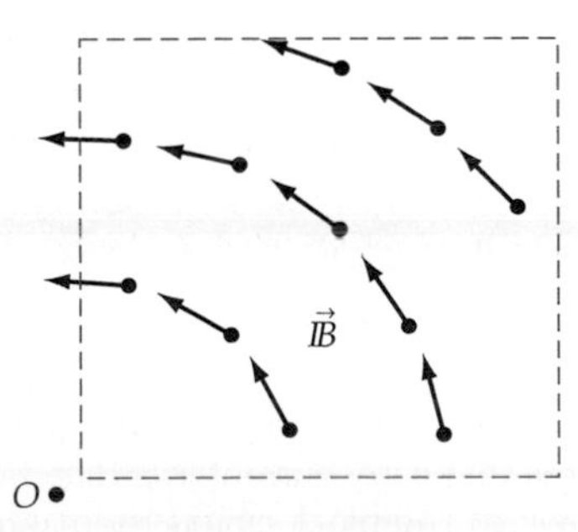

E11T.8 Consider the magnetic field shown. This magnetic field points in the $-y$ direction and decreases as x increases. Assume that the magnetic field is independent of position in the direction perpendicular to the drawing. This magnetic field is inside a conducting object that can *in principle* carry a current that, if it exists, would be perpendicular to the plane of the drawing at all points within the region in question. Judging from the characteristics of the magnetic field, the current is

A. Flowing toward us
B. Flowing away from us
C. Not zero, but we cannot determine its direction
D. Possibly zero
E. Definitely zero
F. Not at all specified by the known field characteristics

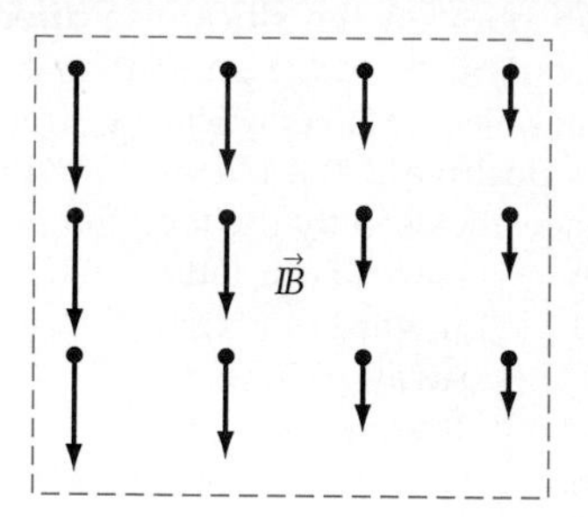

HOMEWORK PROBLEMS

Basic Skills

E11B.1 Imagine that the magnetic field in a certain region of space points in the $+z$ direction and has a magnitude $\mathbb{B} = ax^2$, where a is some constant. Consider a square amperian loop with sides of length L in the xz plane with one corner at the origin. Assume that the segment vectors $d\vec{S}$ go counterclockwise about this loop when we view it with the y axis facing us. Evaluate the magnetic circulation around this loop in terms of a and L.

E11B.2 Imagine that the magnetic field vectors in a certain region of space are always tangent to a circle drawn around the x axis that is perpendicular to that axis. The field has a magnitude of $\mathbb{B} = ar^2$, where r is the distance from the x axis and a is some constant. Consider a circular amperian loop of radius R centered on and perpendicular to the x axis. Evaluate the circulation for this loop in terms of a and R, and explain why there are two possible values for the result.

E11B.3 Figure E11.13 shows four wires, each carrying current perpendicular to the plane of the figure. The currents all have the same magnitude I. What is the net magnetic circulation around each of the five closed loops shown in the diagram?

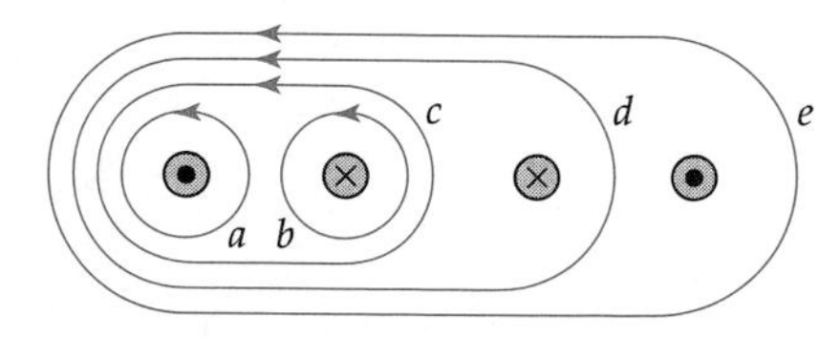

Figure E11.13
This figure shows various amperian loops surrounding a set of four wires that carry current in a direction perpendicular to the drawing. The black circles represent the wires (seen in cross section). If a circle has a dot in it, the current is flowing toward the viewer; if it has a cross, the current is flowing away from the viewer. What is the magnetic circulation around each loop? (See problem E11B.3.)

E11B.4 Figure E11.14 shows four charges that have the same magnitude of charge q. What is the net electrical circulation around each of the five closed loops shown in the diagram?

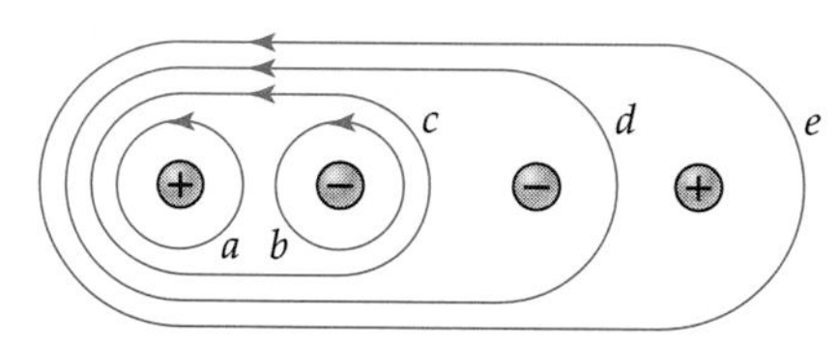

Figure E11.14
This figure shows various amperian loops surrounding a set of four charged particles. What is the net electric circulation around each loop? (See problem E11B.4.)

E11B.5 A long, thin wire carries a current of 3.0 A.
(a) Use the infinite wire approximation to calculate the magnitude of $\vec{\mathbb{B}}$ at a point 1.0 cm away from the wire
(b) How does your result compare to the earth's magnetic field?

E11B.6 A wire going between your car's battery and its starter motor may carry 200 A of current while the starter is cranking. Assuming that we can use the infinite wire approximation in this situation, what is the approximate magnetic field strength 2.0 cm from a wire carrying such a current?

E11B.7 A solenoid consists of 30,000 turns of wire wrapped around a cylinder 0.6 m in diameter and 3.0 m long.
(a) Is the infinite solenoid approximation going to be a reasonably good approximation here for points close to the center of the solenoid's interior? Explain.

(b) Using that approximation, calculate how much current the wire will have to carry to create a magnetic field with a strength of $B = 1.0$ T ($\mathbb{B} = 300$ MN/C) inside the solenoid far from either end.

E11B.8 A solenoid whose radius is 2.0 cm has 2000 turns of wire along its 50-cm length.
(a) Is the infinite solenoid approximation going to be a reasonably good approximation for points close to the center of the solenoid's interior? Explain.
(b) Using that approximation, what is the magnitude of $\vec{\mathbb{B}}$ in the hollow interior of the solenoid (far from either end) if the wire carries a current of 1.0 A?

E11B.9 Consider a flat, rectangular aluminum plate that is 1.0 m wide and 3.0 m long and that carries a total current of 500 A in a direction parallel to its length.
(a) Do you think that the infinite slab approximation is going to be a good approximation for calculating the magnetic field at points a few centimeters from the plate's center? Explain.
(b) What is the approximate magnetic field strength $\mathbb{B}$ at a point about 1.0 cm above the center of the plate?
(c) How does this compare to the earth's magnetic field strength?

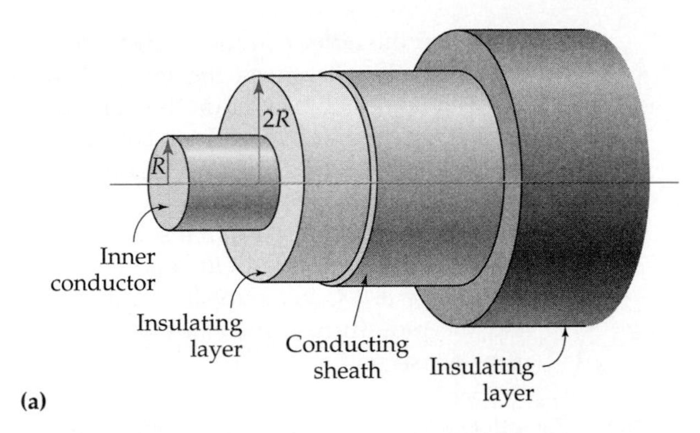

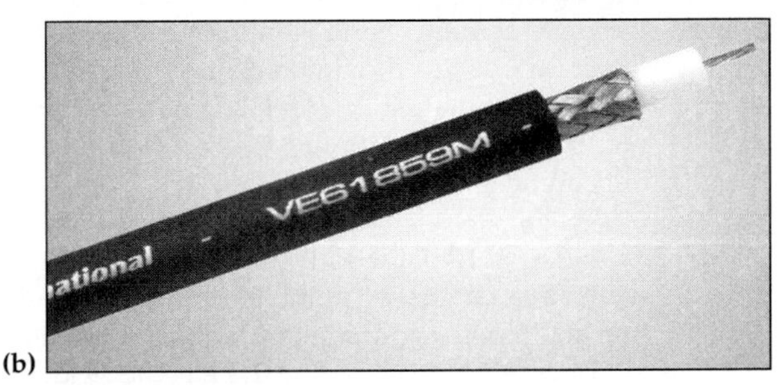

Figure E11.15
(a) A drawing showing the inner layers of a hypothetical coaxial cable. The cable has been cut and stripped so that you can see inside, but you should consider the cable to be infinite and straight. (b) This photograph shows that a real coaxial cable has a very similar structure.

Synthetic

E11S.1 The electron gun in a certain CRT emits 2×10^{14} electrons per second that travel at a speed of $0.03c$. What is the magnitude of the magnetic field 5.0 cm from the center of a beam that is many meters long?

E11S.2 Consider an infinite cylindrical wire of radius R whose central axis coincides with the x axis. The wire carries a current in the $+x$ direction that has a uniform current density $\vec{J}_0$. Use the four-step process to find the magnitude and direction of the magnetic field *inside* this wire.

E11S.3 Consider an infinite, straight coaxial cable consisting of a cylindrical wire, surrounded by a layer of insulation, which is surrounded by a conducting sheath (like a pipe), which is surrounded by another insulating layer, as shown in figure E11.15. Current flows one way in the wire, and an equal current flows the opposite way in the sheath. Use the four-step process presented in this chapter to argue that the magnetic field anywhere *outside* this cable is zero.

E11S.4 Consider an infinite, straight coaxial cable consisting of a cylindrical wire, surrounded by a layer of insulation, which is surrounded by a conducting sheath (like a pipe), which is surrounded by another insulating layer, as shown in figure E11.15. Assume that the radius of the inner conductor is R, the inner radius of the outer sheath is $2R$, and the sheath has a negligible thickness. Assume that the inner wire carries a current of I toward the viewer, and the sheath carries the same magnitude of current in the opposite direction. Use the four-step process presented in this chapter to calculate the magnetic field inside the inner wire, between the wire and the sheath, and outside the cable; and present your results by drawing a quantitatively accurate graph of $\mathbb{B}(r)$ for $0 \le r \le 3R$.

E11S.5 (Adapted from Purcell, *Electricity and Magnetism*, 2d ed., McGraw-Hill, New York, 1987). Consider an aluminum cylinder whose length is $L = 3.0$ m and whose radius is $R_1 = 6.0$ cm surrounded by a coaxial aluminum pipe of the same length whose inner radius is $R_2 = 8.0$ cm. These cylinders are connected to a power supply that gives the inner cylinder a charge Q and the outer pipe a charge $-Q$ such that the potential difference between them is 1500 V.
(a) At points both inside and outside the cylinder and pipe, describe the magnetic field created when the inner cylinder is rotated at 30 turns per second around its central axis, and compute the field's magnitude where it is nonzero. (*Hints:* Look at table E4.1 for an expression for the capacitance of a coaxial cable like this: you can use this to calculate the charge Q. The excess charge on the inner conductor will end up as a thin layer on its outer surface. The excess

charge on the outer conductor, because it is attracted to the charge on the inner conductor, will end up as a thin layer on the outer conductor's *inner* surface. When these conductors are rotated, these thin layers will become currents that flow circularly around the central axis much as the current does in a solenoid, right?)

(b) Describe the magnetic field (and compute its magnitude if it is nonzero) if *both* the cylinder and the pipe turn in the same direction at 30 turns per second.

E11S.6 Consider a hypothetical static electric field that within a certain region points entirely in the $+x$ direction but has a magnitude that depends on y. Is such a static field possible? If not, explain why not. If so, are there any restrictions on how E can depend on y?

E11S.7 Consider a hypothetical static electric field that within a certain region points entirely in the $+z$ direction and has a magnitude that at most depends on z.

(a) Does Ampere's law for the electric field forbid this field or otherwise put any restrictions on it? Explain.

(b) Does Gauss's law for the electric field forbid this field or otherwise put restrictions on it? Explain.

E11S.8 A *toriodal* coil (sometimes simply called a *toroid*) is a coil of wire wrapped around a donut-shaped form (or *torus*). One can also think of a toroidal coil as being a solenoid that has been bent around into a circle so that its two ends coincide. Toroidal coils are used as inductors and small transformers in electronic circuits and are used to confine the hot plasma in certain kinds of thermonuclear fusion reactors. Our goal in this problem will be to predict the magnetic field inside and outside such a toroid.

Consider the toroidal coil shown in cross section in figure E11.16. Let us define the origin of our coordinate system to be at the exact geometric center of the coil, and the coil's central plane (the plane that slices it in half like a bagel) is the xy ($z = 0$) plane. Let the radius of this coil's "donut hole" be R_1: we will call this the coil's *inner radius*. Let the coil's outer radius be R_2. The coil consists of N turns of wire that each carries current I. In each turn, current flows upward as it crosses the central $z = 0$ plane at $r = R_2$ and downward at $r = R_1$. We will assume that the thickness of the layer of wire is small enough compared with R_1 and R_2 to be negligible.

Symmetry in this case (see problem E9S.6) implies that the magnetic field created by such a toroidal coil, both inside and outside of the coil, (1) points tangent to a circle around the z axis lying in a plane parallel to the coil's central plane and (2) depends at most on r and z.

(a) Consider an amperian loop of radius r centered on the origin and lying in a plane parallel to the coil's central plane. Assume that the segment vectors $d\vec{S}$ for this loop point counterclockwise as we view the loop with the z axis facing us. Argue that in this case the magnetic circulation around this loop is $\pm 2\pi r \mathbb{B}$ (where $\mathbb{B}$ is the constant magnitude of the magnetic field on this loop) and that the sign is positive if $\vec{\mathbb{B}}$ points counterclockwise around the loop and negative if it points clockwise.

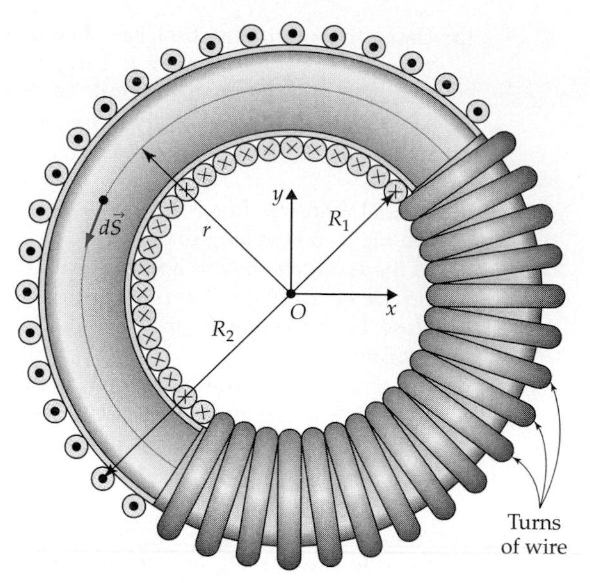

Figure E11.16
This is a top view of a toroidal coil whose central plane coincides with the $z = 0$ plane (the z axis points directly out of the plane of the drawing). The upper right part of the coil has been sliced along the $z = 0$ plane and the top half removed, so that you can see inside the coil's hollow interior. The circles with dots in them represent individual turns of wire that carry current toward us; the circles with crosses in them are turns that carry current away from us. The colored circle is a possible amperian loop of radius r such that $R_1 < r < R_2$.

(b) Consider first an amperian loop whose radius is $r < R_1$. Argue that Ampere's law implies that $\vec{\mathbb{B}} = 0$ everywhere on the $z = 0$ plane inside the donut hole.

(c) Now consider an amperian loop in the $z = 0$ plane whose radius r is such that $R_1 < r < R_2$ (such a loop is *inside* the coil). Argue that the current enclosed by the loop i_{enc} is negative in this case. What does this imply about the *direction* of the magnetic field in this region?

(d) Apply Ampere's law to the loop described in part (c) to show that the magnitude of the magnetic field on the $z = 0$ plane at a radius r such that $R_1 < r < R_2$ must be

$$\mathbb{B} = \frac{2kNI}{cr} \tag{E11.18}$$

(e) Now consider an amperian loop of radius r such that $r > R_2$ (such a loop is entirely *outside* the coil). Argue that Ampere's law applied to such a loop implies that $\vec{\mathbb{B}} = 0$ at all points on the $z = 0$ plane outside the coil.

(f) Now consider points *not* on the $z = 0$ plane. Argue that equation E11.18 still describes the magnetic field at *any* point actually inside the coil (that is, within the region completely surrounded by loops of wire), but that $\vec{\mathbb{B}} = 0$ at *all* points in the donut hole and otherwise outside the coil.

E11S.9 Imagine that the electric potential at point P has some value ϕ_P. Imagine that you then start from P and travel around a closed loop back to P, keeping track of changes in the potential for each step around the loop. In order for the potential at P to be well defined, you should find that the final potential you calculate at the end of the loop is the same as it was at the beginning. Show that this will be true only if Ampere's law for the magnetic field is true. (*Hint:* This problem is pretty easy, but you might need to review chapter E3 to remind yourself about how one might calculate the change in potential during a given step around the loop.)

Rich-Context

E11R.1 A normal electric power cord consists of two parallel conductors about 3 mm apart. Estimate the magnetic field strength about 1 ft from a linear 6-ft power cord connected to a 100-W lamp. (*Hints:* You may find the binomial approximation useful. Do the fields of the individual wires reinforce or oppose each other?)

E11R.2 You want to construct a solenoid 20 cm long that has an interior field strength $\mathbb{B}$ of about 1.5 MN/C. The coil has to be wound as a single layer of wire around a form whose diameter is 3.0 cm. You have two spools of wire handy. #18 wire has a diameter of about 1.02 mm and can carry a current of 6.0 A before overheating. #26 wire has a diameter of 0.41 mm and can carry up to 1.0 A. Which kind of wire should you use and why? What voltage do you want to put across the coil's ends?

E11R.3 Consider the MRI solenoid discussed in exercise E11X.6. Assume that the coil has an inside diameter of 60 cm and a length of 3.0 m. Copper wires able to carry 50 A of current (depending on the method used for cooling and how much power you are willing to dissipate) might need to be on the order of 3 mm across. Roughly estimate the power that such a magnet will dissipate if it uses copper wire of this diameter, compare it to the average electrical power used by a household of four (about 1000 W), and discuss why most modern MRI machines use superconducting magnets, which dissipate no power but require expensive cooling to near absolute zero. (Note that to get the required number of turns per meter, you are going to have to wind multiple layers of turns: be sure to take this into account.)

E11R.4 (See problem E11S.8.) Design a toroidal coil satisfying the following specifications. The toroid's magnetic field $\vec{\mathbb{B}}$ at points inside the coil but on its central plane must be within ± 10 percent of 100 MN/C, and this region must be at least 4.0 cm across. The turns of copper wire must be wound in a single layer around the toroidal form, and the wire's diameter cannot exceed 1.0 mm. Describe the dimensions of the toroid and the diameter of wire that we should use to produce a toroid satisfying these requirements that dissipates the least power, and specify the emf of the power supply that we need to provide the specified field.

Advanced

E11A.1 Much as Gauss's law is more fundamental than Coulomb's law, Ampere's law is more fundamental than the Biot-Savart law. However, just as we can derive Gauss's law for static charges from Coulomb's law, we can derive Ampere's law from the Biot-Savart law. The general proof is very difficult, but this problem outlines how we might do this proof for any current distribution consisting of infinite straight wires.

Figure E11.17 shows an arbitrary loop A surrounding a single infinite straight wire and a second loop B that lies entirely outside the wire. Let us set up our coordinate system so that the wire conducts current I along the z coordinate axis in the $+z$ direction, perpendicular to the plane of figure E11.17. We will *not* necessarily assume, however, that the two loops lie in the plane of the drawing (though this is hard to indicate in the drawing).

(a) Argue that at any point P, the vector $\hat{\theta} \equiv \hat{z} \times \hat{r}$, where $\hat{z}$ is a directional in the $+z$ direction and $\hat{r}$ is a radial directional indicating the direction directly away from the closest point on the z axis, is a directional that has unit magnitude and that points in a direction counterclockwise around a circle that contains P and is perpendicular to the z axis.

(b) Argue that equation E8.15 (which we derived using the Biot-Savart law) and what we know about the direction of the magnetic field of an infinite wire imply that the wire's magnetic field at a point P that is a distance r from the z axis is given by

$$\vec{\mathbb{B}} = \frac{2kI}{cr}\hat{\theta} \tag{E11.19}$$

(c) Since the $\hat{r}$, $\hat{z}$, and $\hat{\theta}$ directionals indicate three mutually perpendicular directions, we can write *any* segment vector in either loop as a sum of components in these directions

$$d\vec{S} = dr\,\hat{r} + r\,d\theta\,\hat{\theta} + dz\,\hat{z} \tag{E11.20}$$

where $d\theta$ is the angle around the z axis spanned by the segment vector (this is partially illustrated in figure E11.17). Argue, therefore, that

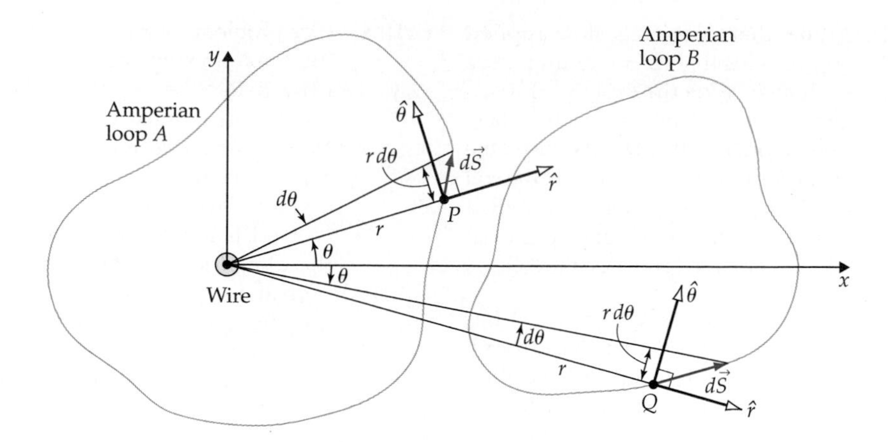

Figure E11.17
This drawing shows a wire and two amperian loops, one that surrounds the wire and one that does not. The wire carries a current that flows toward the viewer perpendicular to the plane of the picture along the z axis. The amperian loops do not necessarily lie in the plane of the picture. The drawing also shows how we define the quantities r, θ, and $d\theta$ and the directionals $\hat{r}$ and $\hat{\theta}$ for segment vectors at two arbitrary points P and Q. (Note that in this case the value of θ for point Q is negative, because it is being measured *clockwise* from the x axis.)

for any segment on either loop,

$$\vec{\mathbb{B}} \cdot d\vec{S} = \frac{2kI}{c}\,d\theta \tag{E11.21}$$

(d) Argue, therefore, that for loops A and B, we have

$$\oint \vec{\mathbb{B}} \cdot d\vec{S} = \frac{4\pi k}{c} I \quad \text{and} \quad \oint \vec{\mathbb{B}} \cdot d\vec{S} = 0$$

respectively. (*Hint:* Note that as the segment vectors go around loop B, θ starts from some initial value, goes to some maximum value, goes down to some minimum value, and then goes back to the initial value.)

(e) Note that if $\vec{\mathbb{B}} = \vec{\mathbb{B}}_1 + \vec{\mathbb{B}}_2 + \cdots$, then the distributive property of the dot product implies that $\vec{\mathbb{B}} \cdot d\vec{S} = \vec{\mathbb{B}}_1 \cdot d\vec{S} + \vec{\mathbb{B}}_2 \cdot d\vec{S} + \cdots$. Use this to argue that Ampere's law applies to any current distribution consisting of infinite wires in various orientations.

ANSWERS TO EXERCISES

E11X.1 If we choose our x axis to coincide with the direction of the uniform field, then we can address this problem by using the same general approach and amperian loop as we did in example E11.1. This time, though, the magnetic field is *uniform*, so $\vec{\mathbb{B}}$ has the same magnitude at all points along the top and bottom horizontal legs; thus the sums of $\vec{\mathbb{B}} \cdot d\vec{S}$ along the top and bottom horizontal legs will exactly cancel each other (because the segment vectors $d\vec{S}$ on the bottom legs are opposite to those on the top legs). The sum of $\vec{\mathbb{B}} \cdot d\vec{S}$ along the vertical legs is still zero because $\vec{\mathbb{B}}$ and $d\vec{S}$ are perpendicular. Therefore the circulation is zero for the whole loop, which by Ampere's law means that the loop encloses zero current, which is appropriate for empty space. Therefore, this is a possible field for empty space.

E11X.2 The wire rule says that if you stick your right thumb in the direction of the current carried by a wire, your curled right fingers indicate the direction that the magnetic field will go around the wire. In figure E11.5, the current is coming toward the viewer, so if you point your right thumb toward yourself, you should see your fingers wrapping counterclockwise around your thumb. This is consistent with the direction implied by Ampere's law, so Ampere's law is consistent with the wire rule.

E11X.3 Imagine a circular amperian loop like that shown in figure E11.5, except with a radius r small enough that the loop is entirely within the hollow space inside the pipe. The symmetry argument made in example E9.4 applies to *all* points inside or outside an infinite cylindrical current distribution, so it applies to the empty space inside the cylindrical pipe. If there is any magnetic field at all inside the pipe, then the magnetic circulation around this loop is $\pm(2\pi r)\mathbb{B}$, just as it would be for a loop outside the distribution. But our loop in this case encloses only empty space, so $i_{\text{enc}} = 0$. Ampere's law thus implies that $\pm(2\pi r)\mathbb{B} = 0$ for all r, which implies that $\vec{\mathbb{B}} = 0$ everywhere inside the pipe. (*Note:* Technically, Ampere's law does *not* require $\vec{\mathbb{B}}$ to be zero at $r = 0$, but symmetry does. Here is a quick argument. The mirror rule implies that $\vec{\mathbb{B}}$, if nonzero, must be perpendicular to any plane that contains the pipe's central axis, as discussed in example

E9.4. At any point P off the axis, there is only a single plane that contains both the axis and P, so the direction of the magnetic field is well defined. But there are an infinite number of possible mirror planes that contain both P and the axis if P lies *on* the axis. Since the magnetic field at P cannot be perpendicular to all these planes, it must be zero.)

E11X.4 Imagine two rectangular loops of vertical height L, both having one vertical leg at $x = 0$ and another at $x = -a$ and $x = a$, respectively (where a is a positive quantity), as shown in figure E11.18a. Define the circulation direction of the segment vectors in each loop to be counterclockwise. The legs at $x = 0$ contribute nothing to the circulation (because $\vec{\mathbb{B}} = 0$ along these legs), and the horizontal legs contribute nothing to the circulation (because $\vec{\mathbb{B}}$ is perpendicular to $d\vec{S}$ along these legs). Since the segment vectors $d\vec{S}$ point in the $+z$ direction in the right leg of the right loop, this leg contributes $+\mathbb{B}_z L$ to the circulation around the right loop. But since the segment vectors $d\vec{S}$ point in the $-z$ direction in the left leg of the left loop, this leg contributes $-\mathbb{B}_z L$ to the circulation in that loop. Now no matter what the value of a is, both loops surround the same amount area in the slab and so enclose the same amount of *positive* current, as you can check with your right hand. Therefore, Ampere's law requires that

$$-\mathbb{B}_z(-a)L = \frac{4\pi k}{c} i_{\text{enc, left}} = \frac{4\pi k}{c} i_{\text{enc, right}} = +\mathbb{B}(a)L$$

$$\Rightarrow \quad -\mathbb{B}_z(-a) = +\mathbb{B}(a) \tag{E11.22}$$

where $\mathbb{B}_z(-a)$ is the z component of the field on the left leg of the left loop and $\mathbb{B}(a)$ is the same on the right leg of the right loop. Therefore any field vector a certain distance to the left of the central plane has the same magnitude but opposite direction when compared to the field vector that is the same distance to the right of the central plane. Since we found that $\mathbb{B}_z$ was positive for $x > 0$, it must be negative for $x < 0$, as shown in figure E11.18b. Note that you can indeed use your right hand and the wire rule to determine the direction that the magnetic field goes "around" the slab.

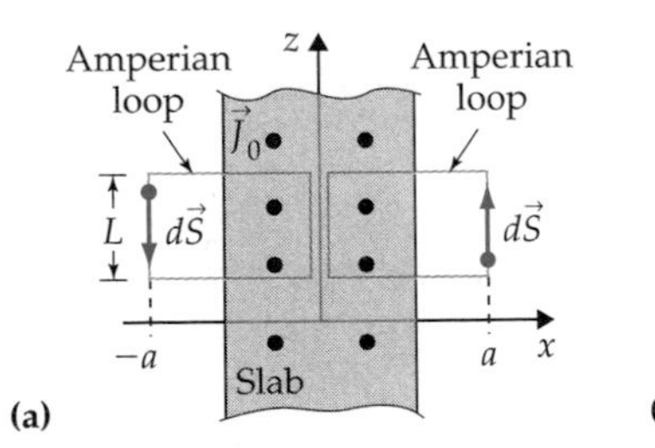

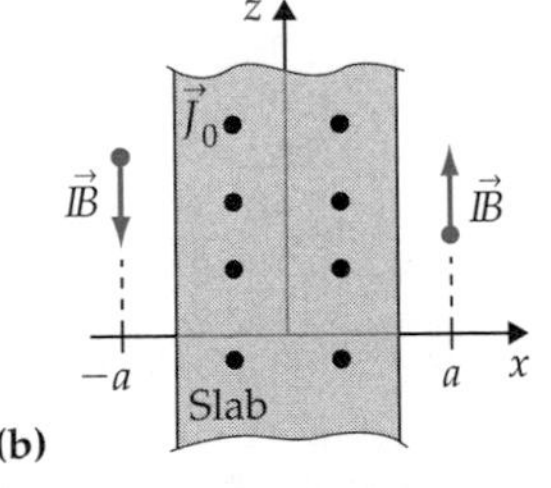

Figure E11.18
(a) This drawing shows a pair of amperian loops that one can use to compare the magnitude and direction of the magnetic field at $x = -a$ to that at $x = +a$. The dots indicate that the slab is carrying current density $\vec{J}_0$ toward the viewer. (b) If we apply Ampere's law to each loop and note that each encloses the same amount of current, we see that the magnetic field at $x = -a$ must be equal in magnitude but opposite in direction to the magnetic field at $x = a$.

E11X.5 Figure E11.19 illustrates how to hold your right hand.

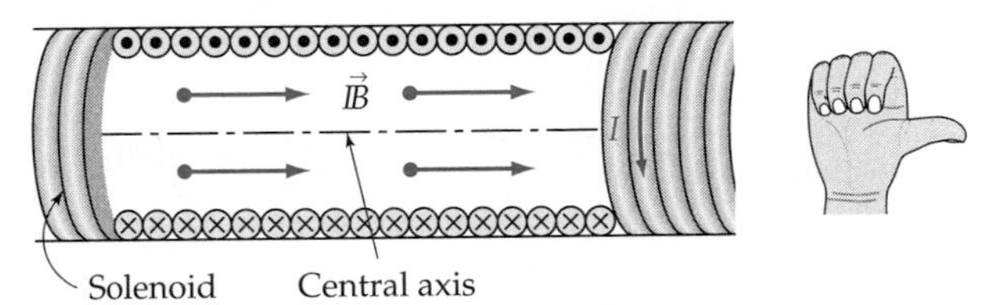

Figure E11.19
This diagram shows that the loop rule agrees with the direction of the magnetic field inside a solenoid determined from Ampere's law.

E11X.6 According to equation E11.15*a*,

$$\frac{N}{L} = \frac{\mathbb{B}c}{4\pi kI} = \frac{(450 \times 10^6\ \cancel{\text{N}}/\cancel{\text{C}})(3 \times 10^8\ \cancel{\text{m}}/\text{s})}{4\pi(8.99 \times 10^9\ \cancel{\text{N}} \cdot \text{m}^{\cancel{2}}/\text{C}^{\cancel{2}})(50\ \text{A})}$$

$$= 24{,}000 \frac{\cancel{\text{C}}/\cancel{\text{s}}}{\text{m} \cdot \cancel{\text{A}}} \left(\frac{1\ \cancel{\text{A}}}{1\ \cancel{\text{C}}/\cancel{\text{s}}}\right) = \frac{24{,}000}{\text{m}} \tag{E11.23}$$

E11X.7 The electric field is either parallel or antiparallel to the segment vectors along one of the curved legs in figure E11.12c. If we assume that the magnitude of the electric field depends on r alone, then the electric circulation contributed by each such leg is

$$\int \vec{E} \cdot d\vec{S} = \pm E \int dS = \pm EL \tag{E11.24}$$

where L is the leg's length and E is the electric field magnitude on the leg. Since the straight legs of the loop contribute nothing, the net circulation over the loop will be zero (as required by Ampere's law for the electric field) if the contributions from the curved legs cancel. Since $\vec{E}$ is parallel to the $d\vec{S}$ vectors along one such leg but antiparallel to the vectors in the other leg, these contributions will cancel if the absolute value of each is the same. Since L for a given curved leg is proportional to r, the contributions will have the same magnitude if $E \propto 1/r$. So the magnitude of the electric field should fall off as $1/r$ if the field shown in figure E11.12c is to be a possible static field in the region shown. Even so, such a field cannot go all the way around the point O, because then a loop around that point would yield a nonzero circulation (no matter how the field depends on r).

E12 The Electromagnetic Field

▷ Electric Field Fundamentals
▷ Controlling Currents
▷ Magnetic Field Fundamentals
▷ Calculating Static Fields
▽ Dynamic Fields
The Electromagnetic Field
Maxwell's Equations
Induction
Introduction to Waves
Electromagnetic Waves

Chapter Overview

Introduction

The final subdivision in this unit explores how to deal with time-dependent electric and magnetic fields. This chapter launches the subdivision by explaining why electric and magnetic fields are really different manifestations of a unified whole called the *electromagnetic field*. Chapter E13 draws on the results discussed here to develop a set of electromagnetic field equations (called *Maxwell's equations*) that are able to handle time-dependent fields. Chapters E14 to E16 explore applications of these equations.

Section E12.1: Why $\vec{E}$ and $\vec{B}$ Must Be Related

A moving charge creates a magnetic field whose magnitude is proportional to the charge's speed. Since observers in different reference frames observe a given particle to move with different velocities, the strength of the magnetic field the particle creates will be frame-dependent (and indeed vanishes in the frame where the particle is at rest!).

A simple puzzle exposes a similar problem with the electric field. A charged particle moving parallel to an electrically neutral current-carrying wire experiences a magnetic force either toward or away from the wire. In a frame where the particle is at rest, though, the particle cannot respond to a magnetic field, and yet still must feel a force (since *all* observers agree that it accelerates toward the wire). Since only an electric field can exert a force on a charge at rest, there must be an electric field in the particle's rest frame that does not exist in the wire's rest frame. A careful analysis shows that this field arises because the wire in fact has a nonzero electric charge in the moving particle's frame due to Lorentz contraction effects.

The ultimate solution to such puzzles is to recognize that the interaction between charged particles is mediated by an **electromagnetic field** that manifests itself as differing mixtures of electric and magnetic fields in different reference frames. This electromagnetic field exerts a force on a charged particle that in any frame we can calculate by using the **Lorentz force law:**

$$\vec{F}_{\text{em}} = q\left[\vec{E} + \frac{\vec{v}}{c} \times \vec{\mathbb{B}}\right] = q[\vec{E} + \vec{v} \times \vec{B}] \qquad \text{(E12.3)}$$

Purpose: This equation describes the electromagnetic force $\vec{F}_{\text{em}}$ exerted by an electromagnetic field on a particle with charge q in an inertial frame where it moves with velocity $\vec{v}$.

Symbols: $\vec{E}$ and $\vec{B}$ are the electric and magnetic manifestations of the electromagnetic field in that frame, respectively, and $\vec{\mathbb{B}} = c\vec{B}$.

Limitations: This equation strictly applies only to charged *particles*.

Note: This equation is relativistically correct (valid in *any* inertial frame) if we define $\vec{F}_{\text{em}} \equiv d\vec{p}/dt$, where t is the coordinate time and $\vec{p}$ is the particle's *relativistic* momentum $\equiv m\vec{v}/[1 - (v/c)^2]^{1/2}$ in that frame.

Note that if this force is zero (or nonzero) in any inertial frame, it must be zero (or nonzero) in *all* such frames, since all inertial observers agree about whether a particle is accelerating or not.

Special relativity thus implies that electric and magnetic fields are inextricably related. Indeed, if we only knew of the existence of the electric field, we could infer the existence of the magnetic field, given the principle of relativity!

Section E12.2: How the Fields Transform

If electric and magnetic fields are simply frame-dependent manifestations of a real electromagnetic field, it should be possible to compute these field vectors at a point in an inertial frame S' given *only* the field vectors at the same point in another frame S and the relative velocity of the frame. The transformation equations turn out to be

$$\begin{bmatrix} E'_x \\ E'_y \\ E'_z \end{bmatrix} = \begin{bmatrix} E_x \\ \gamma(E_y - \beta \mathbb{B}_z) \\ \gamma(E_z + \beta \mathbb{B}_y) \end{bmatrix} \quad \text{or} \quad \begin{bmatrix} E'_x \\ E'_y \\ E'_z \end{bmatrix} = \begin{bmatrix} E_x \\ \gamma(E_y - \beta c B_z) \\ \gamma(E_z + \beta c B_y) \end{bmatrix} \qquad \text{(E12.4}a\text{)}$$

$$\begin{bmatrix} \mathbb{B}'_x \\ \mathbb{B}'_y \\ \mathbb{B}'_z \end{bmatrix} = \begin{bmatrix} \mathbb{B}_x \\ \gamma(\mathbb{B}_y + \beta E_z) \\ \gamma(\mathbb{B}_z - \beta E_y) \end{bmatrix} \quad \text{or} \quad \begin{bmatrix} B'_x \\ B'_y \\ B'_z \end{bmatrix} = \begin{bmatrix} B_x \\ \gamma(B_y + \beta E_z/c) \\ \gamma(B_z - \beta E_y/c) \end{bmatrix} \qquad \text{(E12.4b)}$$

Purpose: These equations describe how, given $\vec{E}$ and $\vec{\mathbb{B}}$ (or $\vec{B}$) in one inertial reference frame, we can calculate $\vec{E}'$ and $\vec{\mathbb{B}}'$ (or $\vec{B}'$) in another inertial frame moving at a speed v in the $+x$ direction with respect to the first.

Symbols: $[E_x, E_y, E_z]$ are the components of $\vec{E}$ and so on; $\beta \equiv v/c$, where c is the speed of light; and $\gamma \equiv (1-\beta^2)^{-1/2}$.

Limitations: Both frames must be inertial. The frames' axes must point in the same directions with the x axes along the direction of relative motion.

The examples in this section check these equations by considering several charge distributions for which we can use the results of chapters E10 and E11 to calculate the electric and magnetic fields *directly* in both frames.

Section E12.3: A Bar Moving in a Magnetic Field

In section E8.1, we saw that if a conducting bar moves in a magnetic field, magnetic forces acting on the bar's charge carriers push them toward one end of the bar, leaving the other end oppositely charged. But in the bar's frame, the carriers are initially at rest and therefore cannot experience magnetic forces. How do we explain the charge separation (which every observer sees) in this frame? Equation E12.4 implies that an *electric* field in the bar's frame pushes the charges around.

Section E12.4: Motion in a Velocity Selector

The crossed electric and magnetic fields in a *velocity selector* exert zero electromagnetic force on a charged particle if and only if it passes through them with a constant velocity such that $v/c = E/\mathbb{B}$. In a frame moving with the particle, the force still has to be zero, since *all* observers agree about whether a particle accelerates or not. Since only the electric aspect of an electromagnetic field exerts a force on a charge at rest, the electric field must be zero in the particle's frame. We find that equation E12.4 does indeed predict this.

E12.1 Why $\vec{E}$ and $\vec{B}$ Must Be Related

In chapter E2, I argued that we need a *field* theory of electricity and magnetism because only a field theory can be made consistent with the requirements of special relativity. In chapter E13, we will finally construct equations for an electromagnetic field theory that fulfill this promise. As a first step toward this goal, though, we must understand more fully how electric and magnetic fields appear to observers in different reference frames.

The electric and magnetic fields are frame-dependent

According to equations E2.9 and E8.4, the electric and magnetic fields created at point P by a particle with charge q moving with a constant velocity $\vec{v}$ past point C are given (if $v \ll c$) by

$$\vec{E} = \frac{kq}{r_{PC}^2}\hat{r}_{PC} \quad \text{and} \quad \vec{\mathbb{B}} = \frac{kq}{r_{PC}^2}\left(\frac{\vec{v}}{c} \times \hat{r}_{PC}\right) \tag{E12.1}$$

where $\vec{r}_{PC}$ is the position of point P relative to point C. While these action-at-a-distance equations are not fully consistent with special relativity and fail when v becomes a significant fraction of c, they make it very clear (within their range of validity) that observers in different reference frames will *disagree* about the relative strength of the electric and magnetic fields in a given situation.

First example: the fields created by a moving charged particle

For example, in the frame where the charged particle has a constant velocity $\vec{v}$, equation E12.1 clearly implies that the particle creates a nonzero magnetic field at point P. But in an inertial frame traveling *with* that particle, the particle is at rest, so the same equation predicts that the particle creates *no* magnetic field in this frame! Here, the very *existence* of a magnetic field in a given situation depends on one's choice of inertial frame!

Can this be so? Is a magnetic field such an imaginary thing that it can exist in one frame and not exist in another? There has to be *something* real about the magnetic field, though, because it produces real, observable effects, doesn't it?

Observers must agree on experimental *results* but not on the explanation

The special theory of relativity is founded on the **principle of relativity,** which makes the following claim:

The laws of physics are the same in all inertial reference frames

This principle requires that *all* observers in all reference frames agree on the physical *outcome* of an experiment or physical process. For example, observers in all inertial frames must agree on whether a given particle is experiencing a force, since the *effects* of such a force on, say, the particle's trajectory are clear in all inertial reference frames. However, the principle of relativity does *not* require that all observers agree on an *explanation* for that force (nor even on the numerical value of such a force). This will turn out to be the key to solving the puzzle about the frame dependence of the magnetic field.

Second example: a charge moving near a current-carrying wire

To see how this works, consider the situation shown in figure E12.1a. Here we have an electron moving parallel to a long wire whose free electrons flow in the same direction. To make the subsequent discussion a bit simpler, let's assume that the electron outside the wire is also moving at the same *speed* as the electrons inside the wire. In the frame shown in the diagram, the wire is at rest (so we will call this the *wire frame*). Let's *assume* that in its rest frame, the wire is electrically neutral (that is, that the free electron

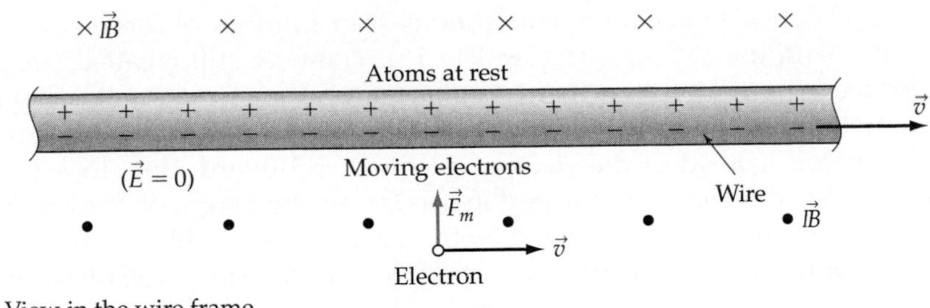

(a) View in the wire frame

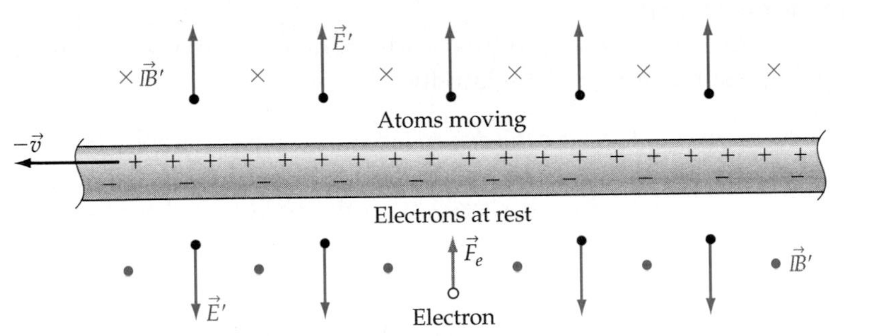

(b) View in the electron frame

Figure E12.1
(a) An electron moving near a wire through which electrons flow with the same velocity. In this frame, the external electron feels a magnetic force (gray arrow) toward the wires due to the current flowing in that wire. (The positive and negative charges are uniformly distributed throughout the wire, but have been separated to make the picture clearer.) (b) In an inertial frame moving with the electrons, the wire still creates a magnetic field, but the outside electron is at rest and so cannot feel a magnetic force. But in this frame, Lorentz contraction effects give the wire a net positive charge, so observers in this frame interpret the force on the electron as an electrostatic force. (This example is based on the argument presented in section 5.9 in Purcell, *Electricity and Magnetism*, 2d ed., McGraw-Hill, New York, 1987.)

density is the same as the positive ion density of the lattice atoms).[†] If this is true, the wire creates no electric field in the wire frame. It does, however, create a magnetic field having the direction shown in the diagram, and this field exerts on the external moving electron a magnetic force directed toward the wire.

Exercise E12X.1

Verify that the direction of the magnetic field produced by the wire and the magnetic force on the electron is correct in figure E12.1a.

[†]For real current-carrying wires, the situation is actually more complicated than this simple model would suggest, because there are surface charges on the wire and the electrons in the wire also respond in a complicated way to the magnetic field that they themselves create. However, the point of my argument is not necessarily to provide a good model of a current-carrying wire but rather to show how electric and magnetic fields must be interrelated to be logically consistent with the principle of relativity. Making the assumption that the wire is neutral in its own frame simplifies the argument.

Now consider the same situation as seen from an observer in a frame moving with the electrons (figure E12.1b). The wire still creates a magnetic field in this frame, because there is still a conventional current flowing to the left, though in this frame the current is due to the wire's positive atoms flowing leftward instead of the electrons flowing rightward. But since the electron outside the wire is also at rest in this frame, the magnetic force acting on it must be *zero:* $\vec{F}_m = q(\vec{v}/c) \times \vec{B} = 0$. Yet an observer in this frame must agree that the outside electron is pushed toward the wire, as *all* observers can see that it accelerates in that direction. How does the observer in the electrons' frame explain this?

Understanding the answer to this question requires that we know only the following result from special relativity:

> If we measure an object's dimensions in an inertial reference frame where it moves with velocity $\vec{v}$, we will find that its length L in the direction of its motion is *shorter* than the corresponding length L_R we would measure in the object's rest frame, as given by

$$L = L_R\sqrt{1 - \frac{v^2}{c^2}} \tag{E12.2}$$

This phenomenon is called *Lorentz contraction* and is discussed fully in chapter R7. For our purposes now, though, it is enough to know that this effect exists.

To return to our problem, this means that in the *wire's* frame, the distances between the electrons flowing in the wire (considered collectively as a moving object) were Lorentz-contracted due to their motion. In the electrons' frame, their separations are *uncontracted,* and thus their charge density is a bit *lower* than it was in the wire's frame. On the other hand, the distances between the wire's positive atoms are Lorentz-contracted in the *electrons'* frame (since *they* are moving leftward in that frame), and therefore they have a *higher* density in that frame than in the wire frame. Since we are assuming the wire was electrically neutral in the wire frame, this means it must have a net positive charge in the electrons' frame! This creates an *electric* field in that frame that attracts the electron toward the wire, explaining in this particular frame the acceleration that all observers see.

We see that observers in different reference frames *disagree* about how much electric field and how much magnetic field is actually present, and they disagree about whether the force the electron experiences is a magnetic force or an electrostatic force. However, they all agree that the charges in the wire create *some* kind of field that pushes the electron toward the wire.

The concept of an *electromagnetic field*

The solution to our puzzle is therefore to understand that the charges in the wire create an **electromagnetic field** whose reality and effects are frame-independent. The degree to which we interpret this electromagnetic field as being a purely electric field, a purely magnetic field, or some combination of both, however, depends on our frame of reference. The electromagnetic field of a point particle, for example, is purely electric in the frame in which it is at rest, while it consists of some mixture of electric and magnetic fields in a frame where the particle is moving. The wire in figure E12.1, on the other hand, creates a purely magnetic field in its own rest frame, but a mixture in other frames.

$\vec{E}$ + relativity implies $\vec{B}$

Note that the situation shown in figure E12.1 implies that magnetism is a necessary relativistic *consequence* of electrostatics. Imagine that we knew Coulomb's law and relativity but did not know anything about magnetism.

We could then start from figure E12.1b and argue that since the external electron experiences an electrostatic force toward the wire in its own frame, it must experience a force toward the wire (perpendicular to the electron's motion) even in the frame where the wire is electrically neutral. We could therefore *predict* the existence of magnetic effects just from Coulomb's law and relativity! Electric and magnetic fields are like opposite sides of the same coin: relativity implies that neither can exist without the other. *Both* are simply manifestations of a unified electromagnetic field, which we divide into electric and magnetic parts differently in different reference frames.

Electricity and magnetism are *unified*

We can calculate the *electromagnetic force* on a charged particle, no matter how it is moving or what particular mix of electric and magnetic fields exists in a given inertial frame from the formula

$$\vec{F}_{\text{em}} = q\left[\vec{E} + \frac{\vec{v}}{c} \times \vec{\mathbb{B}}\right] = q[\vec{E} + \vec{v} \times \vec{B}] \quad \text{(E12.3)}$$

The **Lorentz force law**

Purpose: This equation describes the electromagnetic force $\vec{F}_{\text{em}}$ exerted by an electromagnetic field on a particle with charge q in an inertial frame where it moves with velocity $\vec{v}$.

Symbols: $\vec{E}$ and $\vec{B}$ are the electric and magnetic manifestations of the electromagnetic field in that frame, respectively, and $\vec{\mathbb{B}} = c\vec{B}$.

Limitations: This equation strictly applies only to charged *particles*.

Note: This equation is relativistically correct (valid in *any* inertial frame) if we define $\vec{F}_{\text{em}} \equiv d\vec{p}/dt$, where t is the coordinate time and $\vec{p}$ is the particle's *relativistic* momentum $\equiv m\vec{v}/[1-(v/c)^2]^{1/2}$ in that frame.

We call this equation the **Lorentz force law.** No matter how the electromagnetic (EM) field manifests itself as electric and/or magnetic fields in a given reference frame, this force law correctly expresses the force exerted on a charged particle.

Note that this law essentially *defines* how an electromagnetic field manifests itself as electric and magnetic fields in a given frame. The *electric field* $\vec{E}$ in a given frame is that part of the electromagnetic field that exerts a force on a charge *at rest* (in that frame!). The *magnetic field* is the part that exerts a force proportional to a charge's *speed* (in that frame!). Since the distinction between the electric and magnetic fields hinges on whether the charge is moving or not, and since whether a charge is moving depends on one's choice of reference frame, the way that we divide the electromagnetic field into electric and magnetic parts in a given frame *necessarily* depends on our choice of frame!

How the electric and magnetic parts of an EM field are defined

It is an interesting side note to this discussion that the same kinds of arguments apply to the gravitational interaction as well. Einstein's general theory of relativity requires that a rapidly moving mass produce a "gravitomagnetic field" in addition to its normal gravitational field for pretty much the same reasons that a moving charge must produce a magnetic field in addition to an electric field. We don't notice gravitomagnetic effects in daily life because they are so weak, but astronomers recently announced tentative evidence of a gravitomagnetic effect called *frame dragging* in observations of matter falling into what is believed to be a rapidly spinning neutron star. Other astronomers have even more recently reported observing the consequences of gravitomagnetism in a binary star system containing a pulsar. A satellite

Gravitomagnetism

called the *Gravity Probe B*, currently scheduled for launch in early 2003, will soon measure the excruciatingly tiny effects of the rotating earth's gravitomagnetic field on perhaps the most perfect gyroscope ever made (which will behave essentially as a gravitomagnetic compass). If this experiment works, it will provide the first direct measurement of a gravitomagnetic field.

E12.2 How the Fields Transform

We should be able to find a transformation law for the six EM field components

In section E12.1, I argued that an electromagnetic field at a given point in space and time manifests itself in one frame as a pair of field vectors $\vec{E}$ and $\vec{\mathbb{B}}$, but in a different frame by a different pair of vectors $\vec{E}'$ and $\vec{\mathbb{B}}'$. Assuming that $\vec{E}$, $\vec{\mathbb{B}}$ and $\vec{E}'$, $\vec{\mathbb{B}}'$ are really just different "camera angles" on the same real electromagnetic field at the point in space and time in question, then given $\vec{E}$ and $\vec{\mathbb{B}}$ in an inertial reference frame S and the velocity $\vec{v}$ of some other inertial frame S' relative to S, we should be able to *determine* $\vec{E}'$ and $\vec{\mathbb{B}}'$ in S' without having to know anything about the charge and/or distribution that creates these fields.

(It is sometimes hard to keep track of which quantities are measured in which frame. The prime notation I have been using is traditional, but it is sometimes hard to see. Since it is essential in this chapter and in chapter E13 that we clearly see which quantities correspond to which frames, I will supplement the prime notation by also setting all the quantities that are measured in the primed reference frames in *color*, as displayed in the last paragraph.)

The transformation equations for the electromagnetic field

If the two frames' axes are aligned and we choose the x direction so that frame S' moves in the $+x$ direction relative to frame S, it turns out that the following simple equations describe how the fields transform in *all* situations.

$$\begin{bmatrix} E'_x \\ E'_y \\ E'_z \end{bmatrix} = \begin{bmatrix} E_x \\ \gamma(E_y - \beta \mathbb{B}_z) \\ \gamma(E_z + \beta \mathbb{B}_y) \end{bmatrix} \quad \text{or} \quad \begin{bmatrix} E'_x \\ E'_y \\ E'_z \end{bmatrix} = \begin{bmatrix} E_x \\ \gamma(E_y - \beta c B_z) \\ \gamma(E_z + \beta c B_y) \end{bmatrix} \quad \text{(E12.4}a\text{)}$$

$$\begin{bmatrix} \mathbb{B}'_x \\ \mathbb{B}'_y \\ \mathbb{B}'_z \end{bmatrix} = \begin{bmatrix} \mathbb{B}_x \\ \gamma(\mathbb{B}_y + \beta E_z) \\ \gamma(\mathbb{B}_z - \beta E_y) \end{bmatrix} \quad \text{or} \quad \begin{bmatrix} B'_x \\ B'_y \\ B'_z \end{bmatrix} = \begin{bmatrix} B_x \\ \gamma(B_y + \beta E_z/c) \\ \gamma(B_z - \beta E_y/c) \end{bmatrix} \quad \text{(E12.4}b\text{)}$$

Purpose: These equations describe how, given $\vec{E}$ and $\vec{\mathbb{B}}$ (or $\vec{B}$) in one inertial reference frame, we can calculate $\vec{E}'$ and $\vec{\mathbb{B}}'$ (or $\vec{B}'$) in another inertial frame moving at a speed v in the $+x$ direction with respect to the first.

Symbols: $[E_x, E_y, E_z]$ are the components of $\vec{E}$ and so on; $\beta \equiv v/c$, where c is the speed of light; and $\gamma \equiv (1-\beta^2)^{-1/2}$.

Limitations: Both frames must be inertial. The frames' axes must point in the same directions with the x axes along the direction of relative motion.

Exercise E12X.2

Argue that the versions of equations E12.4a and E12.4b that involve $\vec{B}$ and $\vec{B}'$ are consistent with those involving $\vec{\mathbb{B}}$ and $\vec{\mathbb{B}}'$.

Actually *deriving* these transformation equations by using the mathematics we currently have available is complicated and is not particularly illuminating. Rather than derive these equations, in this section we will look at a number of specific cases where we can calculate the electric and magnetic fields of the charge/current distribution in *both* frames, using the methods of chapters E10 and E11 (*without* using equation E12.4), and then check that equation E12.4 yield the same results for the fields in S' given the field of the charge distribution in frame S. The solutions to all the examples in this chapter will therefore involve completing these steps:

1. *Calculate the fields in frame S* (using results from chapters E10 and E11 and charge and/or current densities *measured in frame S*).
2. *Calculate the fields in frame S′* (using results from chapters E10 and E11 and charge and/or current densities *measured in frame S′*).
3. *Check for consistency,* that is, check that the fields calculated for frame S, when plugged into equation E12.4, do in fact yield the fields calculated for frame S'.

The results from chapters E10 and E11 are valid in the context of static fields. One of the great advantages of the infinite distributions we considered in those chapters is that an infinite charge or current distribution can be moving and yet still be unchanging (and thus still create a static field). The principle of relativity also ensures that the same equations apply without modification in *any* inertial frame, as long as we use in those equations quantities measured *in that frame.* We will take full advantage of these features in this chapter.

Example E12.1 Nested Cylindrical Charge Distributions

Problem As a first case study, let us examine a more precisely defined version of the case discussed qualitatively in section E12.1. Imagine that we have an infinite cylindrical charged sheath that surrounds an infinite wire-like charged core. Imagine that in the S frame, the sheath is at rest and has a positive charge per unit length of λ_S, and the core is moving in the $+x$ direction with speed v and has a charge per unit length (in the S frame) that is equal in magnitude but opposite in sign to that of the sheath: $\lambda_C = -\lambda_S$. The situation is illustrated in figure E12.2a. We will also look at this system in a frame S' that is moving with the core, as illustrated in figure E12.2b. What are the electric and magnetic fields in both frames? Are they consistent with equation E12.4?

Calculate the Fields in the S Frame The combination is a cylindrical charge distribution of the type discussed in section E10.4. The *net* charge per unit length $\lambda = \lambda_S + \lambda_C$ on the sheath/wire combination is therefore zero; according to equation E10.9, the electric field outside the combination is zero.

On the other hand, the moving core represents a cylindrical current distribution of the type discussed in section E11.4. To calculate the magnetic field created by this moving charge, we need to know the current that the moving charge represents. This current is by definition the charge per unit time carried by the core past a given point. Since the core moves a distance $v\,dt$ during a time dt, the moving core carries a total magnitude of charge of $|\lambda_C|(v\,dt)$ past a given point during that time, so the magnitude of the current

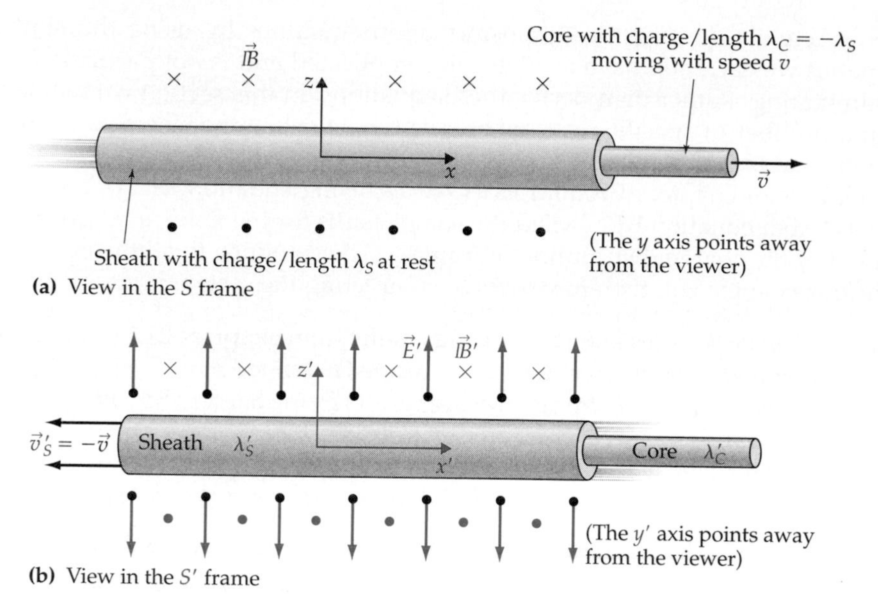

Figure E12.2
An infinite cylindrical charged sheath surrounds an infinite wirelike charged core. (The diagram shows a kind of a cutaway view: both sheath and core are infinitely long along the $\pm x$ directions.) (a) In the S frame, the cylinder's net charge per unit length is zero. However, the moving core creates a magnetic field that is oriented as shown outside the sheath. (b) In the S' frame, Lorentz contraction effects imply that the sheath and core have *different* charges per unit length λ'_S and λ'_C that no longer cancel, so there is an electric field in this frame.

it carries is

$$I = \frac{|\lambda_C| v \, dt}{dt} = |\lambda_C| v \tag{E12.5}$$

Since λ_C is negative, the current actually flows in the $-x$ direction. This means that the magnetic field outside the combination will be a circular field going around the distribution as shown in figure E12.2a. According to equation E11.7, the magnitude of this field at a distance r from the core's axis is

$$\mathbb{B} = \frac{2kI}{cr} = \frac{2k|\lambda_C| v}{cr} = \frac{2k|\lambda_C|\beta}{r} = \frac{2k\lambda_S\beta}{r} \tag{E12.6}$$

since $\lambda_S = |\lambda_C|$ and I have defined $\beta \equiv v/c$.

Calculate the Fields in the S' Frame Now let's look at the magnetic field of the object in an S' frame that moves in the $+x$ direction with speed v along with the core. In this frame, the core will be at rest, and the sheath will move backward with speed $v'_S = v = \beta c$. Since the sheath will be Lorentz-contracted by a factor $[1 - (v/c)^2]^{1/2} = (1 - \beta^2)^{1/2}$, the linear separation of its charges will decrease by this factor, implying that its charge density will be *larger* in this frame than it was in the S frame: $\lambda'_S = \lambda_S/(1 - \beta^2)^{1/2}$. In this frame, it is the moving positive charge on the *sheath* that creates the magnetic field. The wire rule tells us that this current creates a circular magnetic field around the sheath that has the same direction as the magnetic field in the S frame. However, equation E11.7 implies that the magnitude of the magnetic field in the S' frame will be

$$\mathbb{B}' = \frac{2kI'}{cr'} = \frac{2k\lambda'_S v'}{cr'} = \frac{2kv}{cr}\frac{\lambda_S}{\sqrt{1-\beta^2}} = \gamma\frac{2k\lambda_S\beta}{r} = \gamma\,\mathbb{B} \qquad \text{(E12.7)}$$

since $\gamma \equiv (1-\beta^2)^{-1/2}$ and λ_S is positive by hypothesis. (Note that the distance r to any particular given point, since it is measured perpendicular to the line of relative motion, is not affected by Lorentz contraction and is the same in both frames. Note also that universal constants, such as k and c, have the same value in *all* inertial frames.)

Since the core *was* Lorentz-contracted in the S frame but is *un*contracted in this frame, the core's charge is more stretched out in the S' frame than in the S frame and thus the charge density is *lower* by the Lorentz contraction factor: $\lambda'_C = \lambda_C(1-\beta^2)^{1/2} = -\lambda_S(1-\beta^2)^{1/2}$, since $\lambda_C = -\lambda_S$. The cylindrical distribution's net charge density in this frame is therefore

$$\lambda' = \lambda'_S + \lambda'_C = \frac{\lambda_S}{\sqrt{1-\beta^2}} - \lambda_S\sqrt{1-\beta^2} = \frac{\lambda_S[1-(1-\beta^2)]}{\sqrt{1-\beta^2}} = \gamma\beta^2\lambda_S \quad \text{(E12.8)}$$

This is a positive charge density, so there will be an outward electric field in this frame, as shown in figure E12.2b. According to equation E10.9, this field will have a magnitude of

$$E' = \frac{2k\lambda'}{r'} = \frac{2k}{r}\gamma\beta^2\lambda_S = \gamma\beta\frac{2k\lambda_S\beta}{r} = \gamma\beta\,\mathbb{B} \qquad \text{(E12.9)}$$

Note that as long as the velocities are constant, all these fields are static fields.

Check for Consistency This is what the results of chapters E10 and E11 say the electric and magnetic fields should be in both frames. What do equations E12.4 say?

So that we can determine specific field components, let us consider the case of a point lying in the xz plane on the $+z$ side outside the distribution. According to figure E12.2a, the magnetic field at such a point has components $[0, +\mathbb{B}, 0]$ in the S frame, and there is no electric field in that frame. According to figure E12.2b and equations E12.7 and E12.9, the electric and magnetic fields in the S' frame are $[0, 0, +E'] = [0, 0, +\gamma\beta\mathbb{B}]$ and $[0, +\mathbb{B}', 0] = [0, +\gamma\mathbb{B}, 0]$.

According to equations E12.4, the field vectors in the S' frame at such a point should be

$$\begin{bmatrix} E'_x \\ E'_y \\ E'_z \end{bmatrix} = \begin{bmatrix} E_x \\ \gamma(E_y - \beta\mathbb{B}_z) \\ \gamma(E_z + \beta\mathbb{B}_y) \end{bmatrix} = \begin{bmatrix} 0 \\ \gamma(0-0) \\ \gamma(0+\beta\mathbb{B}) \end{bmatrix} = \begin{bmatrix} 0 \\ 0 \\ \gamma\beta\mathbb{B} \end{bmatrix} \qquad \text{(E12.10}a\text{)}$$

$$\begin{bmatrix} \mathbb{B}'_x \\ \mathbb{B}'_y \\ \mathbb{B}'_z \end{bmatrix} = \begin{bmatrix} \mathbb{B}_x \\ \gamma(\mathbb{B}_y + \beta E_z) \\ \gamma(\mathbb{B}_z - \beta E_y) \end{bmatrix} = \begin{bmatrix} 0 \\ \gamma(\mathbb{B}+0) \\ \gamma(0-0) \end{bmatrix} = \begin{bmatrix} 0 \\ \gamma\mathbb{B} \\ 0 \end{bmatrix} \qquad \text{(E12.10}b\text{)}$$

Equations E12.10 therefore yield the *same* results as a detailed analysis of the charge and current densities in the S' frame (and with a lot less work!).

Exercise E12X.3

Check that equations E12.4 also correctly predict the field magnitudes and directions in the S' frame at a point in the xy plane.

Example E12.2 A Moving Infinite Slab

Problem Consider now the case of an infinite slab whose central plane coincides with the xy plane of the S frame (see figure E12.3). Assume that in that frame, the slab's thickness is W and that it is filled with uniform positive charge density ρ_0. Determine the electric and magnetic fields at an arbitrary point outside the slab as measured in S and also in a frame S' that moves with velocity $\vec{v}$ in the $+x$ direction relative to S; and show that equations E12.4 are consistent with your results.

Calculate the Fields in the S Frame We found in section E10.5 that the electric field vectors created by such a slab in its rest frame point entirely perpendicular to the slab, so in this case, $\vec{E} = [0, 0, E_z]$ everywhere. According to equation E10.15, E_z outside the slab has the constant value

$$E_z = \pm 2\pi k \sigma_0 \tag{E12.11}$$

where $\sigma_0 \equiv \rho_0 W$ is the charge per unit area on the slab. Since the charge is stated to be positive, the vectors point away from the slab, so E_z is positive at points where z is positive and is negative where z is negative (see figure E12.3a). There is no moving charge in this frame, so there is no magnetic field.

Calculate the Fields in the S′ Frame An observer in the S' frame will see the slab moving in the $-x'$ direction with speed $v'_S = v = \beta c$. The slab is Lorentz-contracted in that frame along the x direction, so since the distance

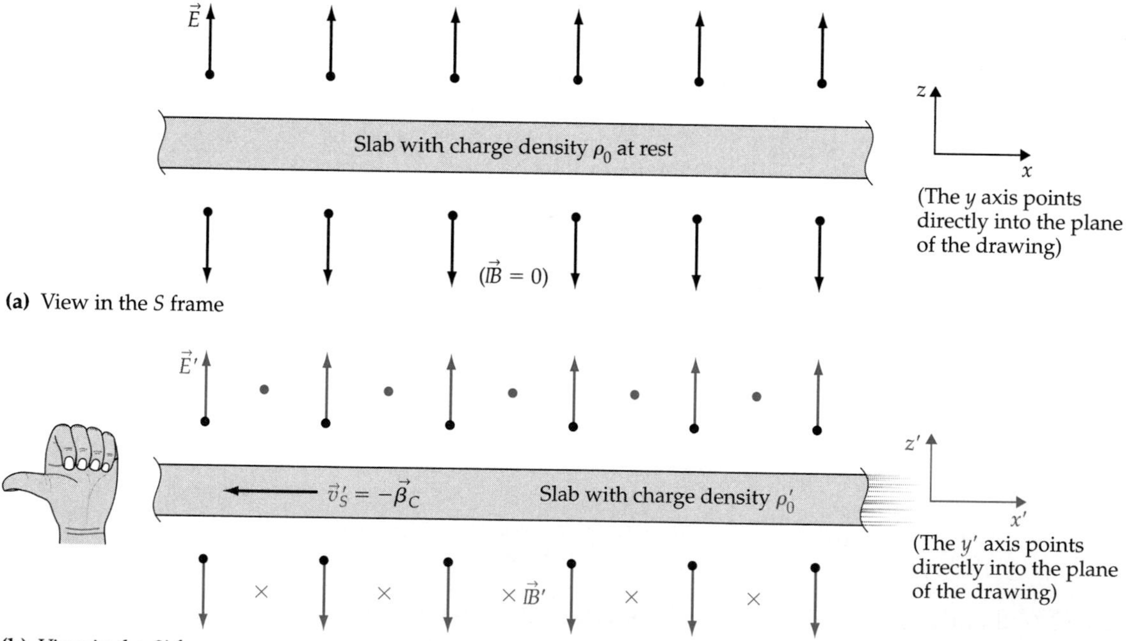

Figure E12.3
This diagram shows an infinite planar slab lying parallel to the $z = 0$ plane. (The diagram shows a cross-sectional view: the slab extends infinitely in the $\pm x$ and $\pm y$ directions.) (a) In the S frame, the slab produces an outward electric field perpendicular to its central plane. (b) In an S' frame that moves in the $+x$ direction with respect to the S frame, the slab produces a similar electric field but also a magnetic field in the $\pm y'$ direction due to the motion of its charge in the $-x'$ direction. The hand shows how the wire rule specifies the directions of these vectors.

between charges decreases by the Lorentz contraction factor $(1-\beta^2)^{1/2}$, the charge density in the S' frame must be *larger* than that in the S frame by the same factor: $\rho_0' = \rho_0/(1-\beta^2)^{1/2} = \gamma\rho_0$. The width of the slab is perpendicular to the line of motion, though, so it is unaffected by Lorentz contraction: $W = W'$.

Even though the slab is moving, it still represents an unchanging charge distribution, so Gauss's law still applies, and the results discussed in section E10.5 still apply. Therefore, equation E10.15 implies that E_z' outside the slab in S' is related to the charge per unit area σ_0' in that same frame by

$$E_z' = \pm 2\pi k\sigma_0' \qquad \text{(E12.12)}$$

where again the sign of E_z' is the same as the sign of z'. Since $\sigma_0' \equiv \rho_0' W'$, $W' = W$, and $\rho_0' = \gamma\rho_0$, we can express this in terms of quantities measured in the S frame as follows:

$$E_z' = \pm 2\pi k\rho_0' W' = \pm 2\pi k(\gamma\rho_0)W = \pm 2\pi k\gamma\sigma_0 = \gamma E_z \qquad \text{(E12.13)}$$

Because the slab is moving, it also creates a magnetic field. In section E11.4, we saw that if an infinite slab carries a uniform current density in a direction parallel to its surface, its external magnetic field has a constant magnitude and goes around the slab in the direction perpendicular to the current as indicated by the wire rule. The particular case we have in the S' frame here fits this description: we have a slab whose motion converts its uniform charge density to a uniform current density that flows in the $-x'$ direction and whose magnitude is

$$J_0' = \rho_0' v_S' = \rho_0' v \qquad \text{(E12.14)}$$

and which flows parallel to the slab's surface. Applying the wire rule and adapting equation E11.10 to our situation, we find that the slab's external magnetic field in this case should be

$$\mathbb{B}_y' = \pm\frac{2\pi k J_0' W'}{c} \qquad \text{(E12.15)}$$

where $\mathbb{B}_y'$ has the same sign as z. This field is shown in figure E12.3b. We can express this in terms of S frame quantities using equation E12.14, $W' = W$, and $\rho_0' = \gamma\rho_0$:

$$\begin{aligned}\mathbb{B}_y' &= \pm\frac{2\pi k J_0' W'}{c} = \pm\frac{2\pi k\rho_0' v W'}{c} = \pm 2\pi k(\gamma\rho_0)\beta W \\ &= \pm\gamma\beta(2\pi k\rho_0 W) = \gamma\beta E_z \end{aligned} \qquad \text{(E12.16)}$$

Check for Consistency Equations E12.4 then imply in this situation that the relationship between the fields in the S and the S' frames should be

$$\vec{E}' = \begin{bmatrix} E_x' \\ E_y' \\ E_z' \end{bmatrix} = \begin{bmatrix} E_x \\ \gamma(E_y - \beta\mathbb{B}_z) \\ \gamma(E_z + \beta\mathbb{B}_y) \end{bmatrix} = \begin{bmatrix} 0 \\ \gamma(0-0) \\ \gamma(E_z - 0) \end{bmatrix} = \begin{bmatrix} 0 \\ 0 \\ \gamma E_z \end{bmatrix} \qquad \text{(E12.17a)}$$

$$\vec{\mathbb{B}}' = \begin{bmatrix} \mathbb{B}_x' \\ \mathbb{B}_y' \\ \mathbb{B}_z' \end{bmatrix} = \begin{bmatrix} \mathbb{B}_x \\ \gamma(\mathbb{B}_y + \beta E_z) \\ \gamma(\mathbb{B}_z - \beta E_y) \end{bmatrix} = \begin{bmatrix} 0 \\ \gamma(0 + \beta E_z) \\ \gamma(0-0) \end{bmatrix} = \begin{bmatrix} 0 \\ \gamma\beta E_z \\ 0 \end{bmatrix} \qquad \text{(E12.17b)}$$

These again provide results completely consistent with the results stated in equations E12.13 and E12.16 with *much* less work.

E12.3 A Bar Moving in a Magnetic Field

Example: looking at a bar moving in a magnetic field in two different frames

As another application of equations E12.4, consider the case of a conducting bar moving in a magnetic field. In chapter E8, we saw that if such a bar moves with a velocity $\vec{v}$ perpendicular to a magnetic field $\vec{\mathbb{B}}$ (see figure E12.4a), magnetic forces are exerted on the free electrons in the bar that drive them toward one of the bar's ends, leaving the other end positively charged. According to equation E8.1 (or equation E12.3 for that matter), since the electrons in the bar are moving (along with the bar) with velocity $\vec{v}$ perpendicular to $\vec{\mathbb{B}}$, the magnetic force on any given electron in the bar has a magnitude of

$$F_m = \frac{ev\mathbb{B}}{c} \tag{E12.18}$$

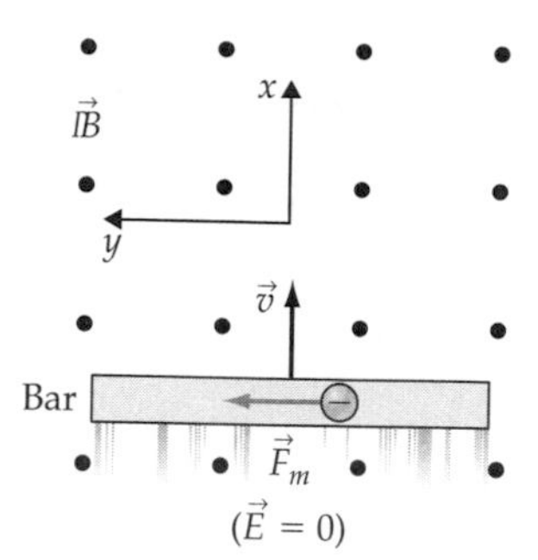

(a) View in the S frame

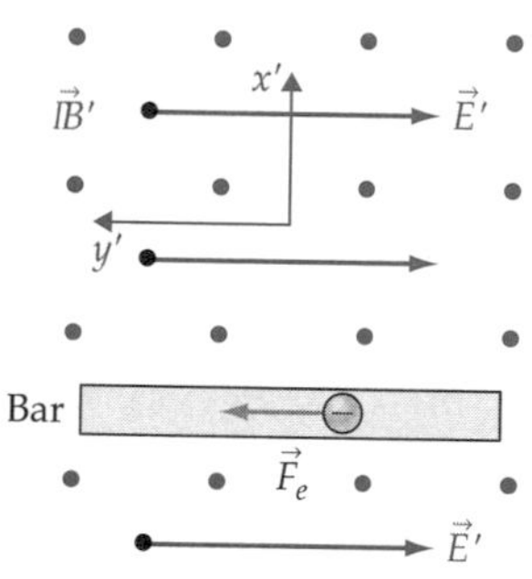

(b) View in the S' frame

Figure E12.4
(a) Electrons in a bar moving in a magnetic field experience a magnetic force that drives them toward the bar's left end. (b) In the bar's rest frame, there is an electric field $\vec{E}'$ that drives the electrons to the left. (Both diagrams show the initial situation, before many electrons have moved.)

This force will drive electrons toward the left in figure E12.4a until the electric field $\vec{E}$ created by the charge concentration creates an opposing force strong enough that there is no longer any net force on the electrons, that is, until

$$eE = F_e = F_m = \frac{ev\mathbb{B}}{c} \quad \Rightarrow \quad E = \frac{v}{c}\mathbb{B} \tag{E12.19}$$

However, now let's look at this situation in the reference frame of the bar. In this frame, the bar is not moving, so its electrons cannot respond to the magnetic field. Yet *all* observers must agree that the opposite ends of the bar do become charged. What drives the electrons through the bar in this case?

Equation E12.4 provides the answer. The rest frame of the bar is moving with a speed of v in the $+x$ direction with respect to our original reference frame. Initially, there was no electric field in our original reference frame, but equation E12.4 implies that in the bar's frame there is an initial electric field (see figure E12.4b) whose components are

$$\begin{bmatrix} E'_x \\ E'_y \\ E'_z \end{bmatrix} = \begin{bmatrix} E_x \\ \gamma(E_y - \beta\mathbb{B}_z) \\ \gamma(E_z + \beta\mathbb{B}_y) \end{bmatrix} = \begin{bmatrix} 0 \\ \gamma(0 - \beta\mathbb{B}) \\ \gamma(0 + \beta 0) \end{bmatrix} = \begin{bmatrix} 0 \\ -\gamma(v/c)\mathbb{B} \\ 0 \end{bmatrix} \tag{E12.20}$$

since $\beta \equiv v/c$. Therefore in this frame, the electrons initially experience an *electrostatic* force driving them in the $+y$ direction whose magnitude is

$$F'_e = qE' = \gamma e\frac{v}{c}\mathbb{B} \tag{E12.21}$$

This is the force that causes the electrons to move.[†] According to our analysis in the original frame, they will continue to move until they create an electric field whose y component in the original frame is $E = (v/c)\mathbb{B} = \beta\mathbb{B}$. If this is so, the final electric field $\vec{E}'$ inside the bar in the bar's frame will then be

$$\begin{bmatrix} E'_x \\ E'_y \\ E'_z \end{bmatrix} = \begin{bmatrix} E_x \\ \gamma(E_y - \beta\mathbb{B}_z) \\ \gamma(E_z + \beta\mathbb{B}_y) \end{bmatrix} = \begin{bmatrix} 0 \\ \gamma(E - \beta\vec{\mathbb{B}}) \\ \gamma(0 + \beta 0) \end{bmatrix} = \begin{bmatrix} 0 \\ \gamma(\beta\mathbb{B} - \beta\mathbb{B}) \\ 0 \end{bmatrix} = 0 \tag{E12.22}$$

Different observers offer different explanations for the bar's polarization

so we would conclude that the electrons no longer move in this frame as well.

We see that an analysis of the situation in either reference frame leads to the same *conclusion:* electrons will pile up on the bar's left side until they

[†]While all observers must agree on whether an electromagnetic force *exists* or not, they do not have to agree on what the *magnitude* of that force is. Since the force exerted by an interaction is defined to be the rate at which the interaction transfers momentum, and since observers in different frames disagree about the durations of time intervals and how much momentum an object has, they will also disagree about the magnitude of a force. We won't concern ourselves with the details of this in this course, but equations E12.18 and E12.21 do correctly express the same physical force as measured in two different reference frames (see problem E12A.1).

reach the concentration that creates an electric field of magnitude $E = (v/c)\mathbb{B}$ inside the bar in the original reference frame. However, the explanations offered for *why* the electrons do this are strikingly different. Observers in the frame where the bar is moving say that the electrons move in response to magnetic forces acting on them, while observers in the bar's frame say that the conducting bar is simply polarized by the electric field in that frame. This again illustrates the idea that the *physics* of a situation is frame-independent, but any explanations based on the electric and magnetic parts of an electromagnetic field will be frame-dependent.

E12.4 Motion in a Velocity Selector

Example: looking at an ion in a velocity selector in two different frames

Figure E12.5 shows a schematic diagram of a device known as a **velocity selector.** The device consists of a box with holes in the middle of two faces. Charge and current distributions outside the box create approximately uniform electric and magnetic fields $\vec{E}$ and $\vec{\mathbb{B}}$, respectively, that are perpendicular to each other and to the line connecting the holes (which we will take to define the x axis). Imagine that we take a positively charged ion with charge q and send it through the hole in the left face with a velocity $\vec{v}$ in the $+x$ direction. Once the ion is in the box, the electric field exerts an upward force on it, while (as you can check with your right hand) the magnetic field exerts a downward force on it. According to the Lorentz force law, the net electromagnetic force on the ion is

$$F_{em} = |F_{em,z}| = q\left|E_z + \left[\frac{\vec{v}}{c} \times \vec{\mathbb{B}}\right]_z\right| = q\left|E - \frac{v}{c}\mathbb{B}\right| \tag{E12.23}$$

You can see that the electromagnetic force on the ion will be zero irrespective of its charge if and only if $v/c = E/\mathbb{B}$ (or equivalently, $v = E/B$). If the ion happens to have this speed, it will follow a straight-line trajectory through the box and go out the hole on the far side: otherwise, it will be vertically deflected by the fields and hit the far wall. Thus ions that escape through the far hole will be those that have the specific speed v/c defined by the ratio of the electric and magnetic fields. Such a device can be very useful for creating a beam of particles with a well-defined speed.

Now, observers in *all* reference frames will agree that a successful ion experiences no net force in this box, as they can all see that it undergoes no acceleration. In particular, an observer in a frame traveling with the ion (call this the primed frame S') must find that the fields in that frame exert zero electromagnetic force on the ion. In the frame where the ion is at rest, the only field that *can* exert a force on the ion is the electric field, so in this frame, saying that $F'_{em} = 0$ means that we must have $E' = 0$.

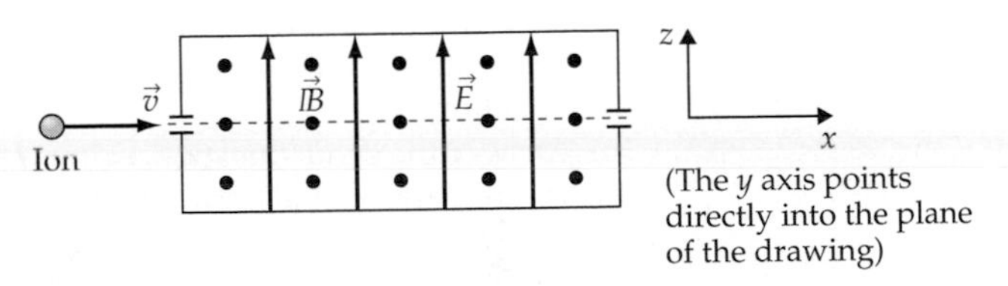

Figure E12.5
A schematic diagram of a *velocity selector.* In the white region, the electric field $\vec{E}$ is uniform and points in the $+z$ direction while the magnetic field $\vec{\mathbb{B}}$ is uniform and points in the $-y$ direction. The ion will travel in a straight line through the region and make it through the far gap if and only if it has a velocity in the x direction such that $v/c = E/\mathbb{B}$.

Is this consistent with the predictions of equations E12.4? In the laboratory frame S, we have $\mathbb{B}_y = -\mathbb{B}$, and $E_z = +(v/c)\mathbb{B}$, while $E_x = E_y = 0$ and $\mathbb{B}_x = \mathbb{B}_z = 0$. Note also that the velocity of the primed (ion) frame relative to the laboratory frame is the same as the ion's velocity $\vec{v}$, so $\beta = v/c$. Equation E12.4*a* thus implies that

$$\begin{bmatrix} E'_x \\ E'_y \\ E'_z \end{bmatrix} = \begin{bmatrix} E_x \\ \gamma(E_y - \beta\mathbb{B}_z) \\ \gamma(E_z + \beta\mathbb{B}_y) \end{bmatrix} = \begin{bmatrix} 0 \\ \gamma(0-0) \\ \gamma(\beta\mathbb{B} - \beta\mathbb{B}) \end{bmatrix} = \begin{bmatrix} 0 \\ 0 \\ 0 \end{bmatrix} \tag{E12.24}$$

since $E_z = +(v/c)\mathbb{B} = \beta\mathbb{B}$. We see that equation E12.4 indeed predicts that the electric field, and thus the force on the ion, is indeed zero in the ion's frame.

Exercise E12X.4

Is the magnetic field also zero in the ion's frame?

Exercise E12X.5

Imagine that we rotate our coordinates so that $\vec{E}$ and $\vec{\mathbb{B}}$ point in the $+y$ and $+z$ directions, respectively. Assume that $E = (v/c)\mathbb{B}$ in the laboratory frame. Show that the net electromagnetic force on the ion is zero in both laboratory and ion frames in this case also.

Concluding comments

The point of this section and section E12.3 is that the transformation rule for the electromagnetic field given by equations E12.4 provides helpful and quantitatively correct explanations, in multiple reference frames, for various kinds of physical phenomena. In chapter E13, we will use these transformation equations to determine how we have to correct Gauss's and Ampere's laws to handle dynamic fields.

TWO-MINUTE PROBLEMS

E12T.1 Two charged particles move past each other in the laboratory frame, as shown.
(a) Does particle A exert a magnetic force on particle B in this frame? If so, in what direction?
(b) Answer the same question as in (a) in the rest frames of particle A and particle B.

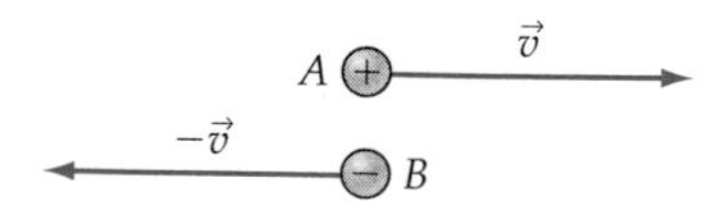

A. Particle B experiences a magnetic force toward A.
B. Particle B experiences a magnetic force away from A.
C. Particle B experiences a magnetic force toward the right in the drawing.
D. Particle B experiences a magnetic force toward the left in the drawing.
E. Particle B experiences a magnetic force out of the plane of the drawing.
F. Particle B experiences a magnetic force into the plane of the drawing.
T. Particle B experiences no magnetic force in this frame.

E12T.2 Two charged particles move with the same velocities in the laboratory frame, as shown.
(a) In the reference frame where charge B is at rest, does particle A exert a magnetic force on particle B in this frame? If so, in what direction?
(b) Answer the same question as in (a) in the frame where particle A and particle B are at rest.

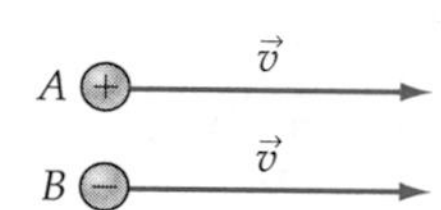

A. Particle B experiences a magnetic force toward A.
B. Particle B experiences a magnetic force away from A.

C. Particle B experiences a magnetic force toward the right in the drawing.
D. Particle B experiences a magnetic force toward the left in the drawing.
E. Particle B experiences a magnetic force out of the plane of the drawing.
F. Particle B experiences a magnetic force into the plane of the drawing.
T. Particle B experiences no magnetic force in this frame.

E12T.3 Imagine that in one inertial reference frame, there is a uniform electric field and no magnetic field. It is possible to find another inertial reference frame where there is a magnetic field but no electric field, true (T) or false (F)?

E12T.4 If the electric and magnetic fields are both zero in any given frame, they are zero in all frames, T or F?

The remaining two-minute problems all refer to the following situation. Imagine that we have a sheath and core like those shown in figure E12.2. Imagine that *in their own rest frames*, these objects have uniform charge-per-unit lengths that are equal in magnitude but opposite in sign. In the laboratory frame, the sheath and core move in opposite directions with the same speed v.

E12T.5 An observer in the laboratory frame would say that a charged particle at rest in the laboratory frame experiences
A. A magnetic force
B. An electric force
C. Both electric and magnetic forces, and $\vec{F}_e + \vec{F}_m \neq 0$
D. Both electric and magnetic forces, but $\vec{F}_e + \vec{F}_m = 0$
E. Neither an electric nor a magnetic force

E12T.6 An observer in the core's frame sees a charged particle at rest in the laboratory frame to move with v in the $-x$ direction. The core frame observer would say that this particle experiences . . . (choose from the answers for problem E12T.5).

E12T.7 According to an observer in the core frame, a charged particle at rest in the core frame experiences . . . (choose from the answers for problem E12T.5).

E12T.8 An observer in the laboratory frame sees a charged particle at rest in the core's frame to move with v in the $+x$ direction. The laboratory frame observer would say that this particle experiences . . . (choose from the answers for problem E12T.5).

HOMEWORK PROBLEMS

Basic Skills

E12B.1 What is the total electromagnetic force on a particle with charge $q = 10$ nC moving with a speed $v = \frac{3}{5}c$ in the $+x$ direction in a region where the electric and magnetic fields are $\vec{E} = [0, 0, 3000 \text{ N/C}]$ and $\mathbb{B} = [0, 5000 \text{ N/C}, 0]$?

E12B.2 In a certain reference frame, there is a uniform electric field $\vec{E} = [0, 0, E]$, where $E = 100$ N/C, and no magnetic field. What are the magnitude and direction of the electric field $\vec{E}'$ in a frame moving with speed $v = \frac{3}{5}c$ in the $+x$ direction? Is there a magnetic field $\mathbb{B}'$ in this frame? If so, what are its magnitude (in both newtons per coulomb and teslas) and direction?

E12B.3 In a certain reference frame, there is a uniform magnetic field $\mathbb{B} = [0, \mathbb{B}, 0]$, where $\mathbb{B} = 100$ N/C, and no electric field. What are the magnitude and direction of the magnetic field $\vec{E}'$ in a frame moving with speed $v = \frac{3}{5}c$ in the $+x$ direction? Is there a magnetic field in this frame? If so, what are its magnitude (in newtons per coulomb and teslas) and direction?

E12B.4 In the situation described in problem E12B.1, calculate the electric field $\vec{E}'$ and magnetic field $\mathbb{B}'$ in the frame of the moving particle. What is the electromagnetic force on the particle in this frame?

E12B.5 Imagine that in the laboratory frame we have two infinite and oppositely charged slabs that both lie parallel to the $z = 0$ plane and are moving in opposite directions along the x axis with speed v relative to the laboratory frame, as shown in figure E12.6. (*more*)

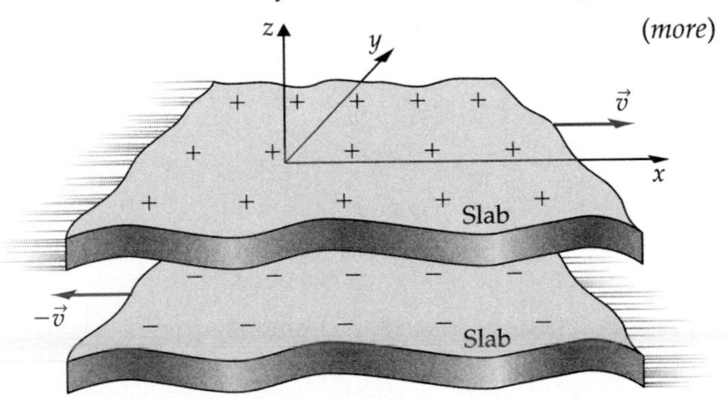

Figure E12.6
This diagram shows two infinite slabs, both parallel to the $z = 0$ plane, that in the laboratory frame S are moving in opposite directions with the same speed. The slabs have the same thickness but charge densities with opposite signs. What can we say about the electric and magnetic fields in the laboratory frame and in frames moving with the slabs? (See problem E12B.5.)

Assume that the positively charged slab has a charge density ρ_R measured in its own rest frame, that the negative slab has a charge density of $-\rho_R$ in its own rest frame (where ρ_R has the same numerical magnitude in both cases), and that the slabs have the same width. *Qualitatively* answer the following questions, using what you know about the fields of slabs and Lorentz contraction. (You can *check* your work with equation E12.4 if you so desire.)

(a) Is there a nonzero electric field in the region outside the plates in the laboratory frame? Explain.
(b) Is there a nonzero magnetic field in the region outside the plates in that frame? Explain.
(c) Is there a nonzero electric field outside the plates in a frame moving with the positive plate? Explain.

E12B.6 Imagine that in the laboratory frame we have two identical infinite, uniformly charged slabs that both lie parallel to the $z = 0$ plane and are moving in opposite directions along the x axis with speed v relative to the laboratory frame, as shown in figure E12.7. Assume that both slabs have the same positive charge density as measured in their own rest frames. *Qualitatively* answer the following questions, using what you know about the fields of slabs and Lorentz contraction. (You can *check* your work with equation E12.4 if you so desire.)

(a) Is there a nonzero electric field in the region between the plates in the laboratory frame? Explain.
(b) Is there a nonzero magnetic field between the plates in that frame? Explain.
(c) Is there a nonzero electric field outside the plates in a frame moving with the positive plate? Explain.

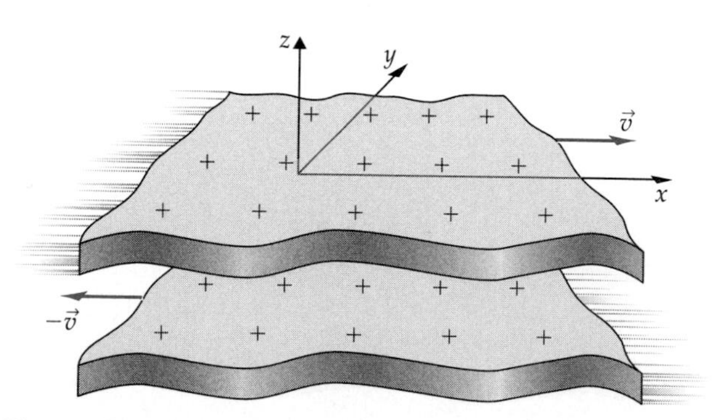

Figure E12.7
This diagram shows two infinite slabs, both parallel to the $z = 0$ plane, that in the laboratory frame S are moving in opposite directions with the same speed. The slabs have the same thickness and charge densities having the same sign. What can we say about the electric and magnetic fields in the laboratory frame and in frames moving with the slabs? (See problem E12B.6.)

Synthetic

E12S.1 Imagine that in the laboratory reference frame a charge q moving at a speed $v = \beta c$ passes a point P a distance r from the line along which q is moving. Coulomb's law $E = kq/r^2$ exactly describes the electric field of charge q only in that charge's rest frame. By transforming from that frame to the laboratory frame, show that at the instant the charge passes closest to point P, the electric field of the charge at P is $\gamma \equiv (1 - \beta^2)^{-1/2}$ times larger than we might expect from $E = kq/r^2$. (Be sure to discuss whether the value of r is the same in both frames.)

E12S.2 Consider the infinite slab described in example E12.2.

(a) Use the results of sections E10.5 and E11.4 to determine the electric and magnetic fields *inside* the slab in the S frame.
(b) Use the results of the same sections to determine electric and magnetic fields *inside* the slab in the S' frame, expressing them ultimately in terms of the measured field strengths in the S frame.
(c) Verify that equations E12.4 yield the same result.

E12S.3 Consider an infinite charged slab with a uniform charge density ρ_0 that is perpendicular to the x axis. In the S frame, the slab is at rest; in the S' frame, which moves in the $+x$ direction with speed $v = \beta c$, the plate moves with speed βc in the $-x$ direction (see figure E12.8).

(a) Using the results of chapters E10 and E11, directly compute the slab's external electric and magnetic fields in the S frame.

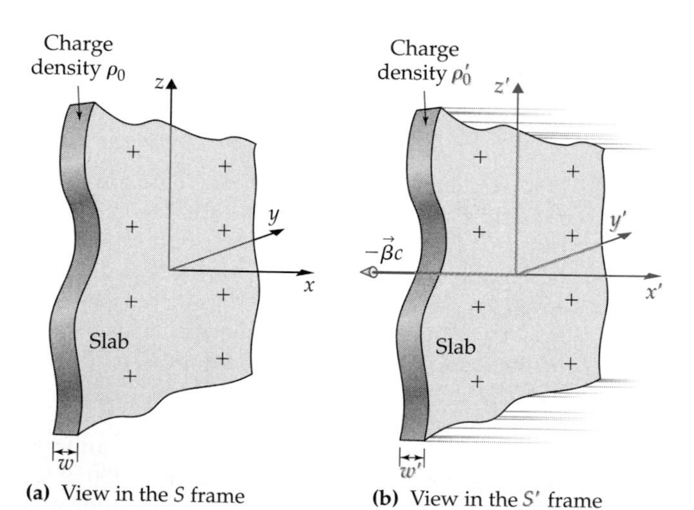

Figure E12.8
(a) An infinite slab with width W oriented perpendicular to the x axis. The slab is at rest and has a uniform charge density ρ_0. (b) In a frame S' moving with speed $\vec{v} = \vec{\beta}c$ in the $+x$ direction relative to the S frame, the slab will be seen to move backward with velocity $-\vec{\beta}c$. What are its electric and magnetic fields in this frame? (See problem E12S.3.)

(b) In the S' frame, the slab is moving in such a way that the electric field is time-dependent. In chapter E13, however, we will see that Gauss's law is correct even for time-dependent fields. This means that the results of chapter E10 for infinite slabs apply even to a slab moving like this one. Use this information to calculate the slab's external electric field in the S' frame.

(c) Use a symmetry argument to prove that the slab's external magnetic field in the S' frame must be zero, in spite of the moving charge.

(d) Use equations E12.4 to calculate $\vec{E}'$ and $\vec{\mathbb{B}}'$ in the S' frame from $\vec{E}$ and $\vec{\mathbb{B}}$ in the S frame, and show that you get the same results that you got in the direct calculation you did in parts (a) through (c).

E12S.4 Consider two infinite charged plates, both parallel to the xy plane a distance d apart and at rest in the S frame. The upper plate has a uniform positive charge per unit area σ_0, and the lower plate has a uniform charge per unit area of $-\sigma_0$. In the S' frame, which moves in the $+x$ direction with speed $v = \beta c$, the plates move with speed $v'_p = v$ in the $-x'$ direction.

(a) Use the results of chapters E10 and E11 to determine the electric and magnetic fields both between the plates and outside the plates in both frames. (You should find that the electric and magnetic fields outside the plates are zero in both frames.)

(b) Use equations E12.4 to calculate $\vec{E}'$ and $\vec{\mathbb{B}}'$ in the S' frame from $\vec{E}$ and $\vec{\mathbb{B}}$ in the S frame, and show that you get the same results that you got in the direct calculation you did in part (a).

E12S.5 Consider a point P whose y and z coordinates are $y = r\cos\theta$ and $z = r\sin\theta$ and whose x coordinate might be anything. (Note that such a point is a distance r from the x axis but does not necessarily lie on either the xy or xz plane.) (a) Using equations E12.7 and E12.9 and the geometry of the situation, find the components of the electric and magnetic fields produced by the sheath/core distribution in both the S and S' frames. (b) Use equations E12.4 to calculate $\vec{E}'$ and $\vec{\mathbb{B}}'$ in the S' frame from $\vec{E}$ and $\vec{\mathbb{B}}$ in the S frame, and show that you get the same results that you got in the direct calculation you did in part (a). (This shows that equations E12.4 yield the correct results for *all* points exterior points in the sheath/core situation.)

E12S.6 Use equations E12.4 to prove that if there is no magnetic field at a given point P in an inertial reference frame S, then in any frame S' moving with velocity $\vec{v}$ relative to S the magnetic field at P will be

$$\vec{\mathbb{B}}' = -\frac{\vec{v}}{c} \times \vec{E}' \tag{E12.25}$$

(*Hint:* I really do mean there to be a prime on the $\vec{E}'$!)

E12S.7 Use equations E12.4 to show that $\vec{E}\cdot\vec{\mathbb{B}} = \vec{E}'\cdot\vec{\mathbb{B}}'$ no matter what the fields might be and no matter what β is. This means that $\vec{E}\cdot\vec{\mathbb{B}}$ is a quantity that has the same numerical value in all reference frames. (*Hint:* Write out the dot product in component form.)

E12S.8 Use equations E12.4 to show that $E^2 - \mathbb{B}^2 = (E')^2 - (\mathbb{B}')^2$ no matter what the fields might be and no matter what β is. This means that $E^2 - B^2$ is a quantity that has the same numerical value in all reference frames. (*Hint:* $E^2 \equiv [\text{mag}(\vec{E})]^2 \equiv E_x^2 + E_y^2 + E_z^2$.)

E12S.9 Consider the velocity selector problem discussed in section E12.4. The electromagnetic force on the charged particle must be zero in *every* reference frame, not just the two considered in the section. Let us prove that this is so. Consider a frame S' that moves in the $+x$ direction with respect to the laboratory frame at a speed βc that is *not* equal to the particle's speed v with respect to the laboratory frame S. Assume that in the S frame, $\vec{E}$ points in the $+z$ direction, $\vec{\mathbb{B}}$ points in the $-y$ direction, and $E = (v/c)\mathbb{B}$.

(a) Use equations E12.4 to show that the electric field $\vec{E}'$ in the primed frame still points entirely in the $+z$ direction and has z component of

$$E'_z = \gamma B(v - \beta c) \tag{E12.26}$$

(b) Use equation E12.4 to show that the magnetic field $\vec{\mathbb{B}}'$ in the primed frame still points in the $-y$ direction and has y component of

$$\mathbb{B}'_y = -\gamma\mathbb{B}\left(1 - \beta\frac{v}{c}\right) \tag{E12.27}$$

(c) Assuming that the particle moves in the $+x$ direction in the laboratory frame, its x-velocity in the primed frame is

$$v'_x = \frac{v - \beta c}{1 - \beta(v/c)} \tag{E12.28}$$

(See chapter R8.) Using this equation, show that for all β such that $0 < \beta < 1$

$$E'_z = -v'_x B'_y \tag{E12.29}$$

(d) Use this to show that $\vec{F}'_{em} = 0$ in the primed frame, no matter what β is. (Since the sign of v'_x depends on β, make sure that your argument works no matter what the sign of v'_x turns out to be.)

Rich-Context

E12R.1 The electric field of a charged particle in its own rest frame is $\vec{E} = (kq/r_{PC}^2)\hat{r}_{PC}$.

(more)

(a) Use equations E12.4 to calculate the magnetic field in a frame where the charge is moving with a constant speed $\vec{v}$, and show that the result is

$$\mathbb{B} = \frac{kq}{r_{PC}^2}\left(\frac{\vec{v}}{c} \times \hat{r}_{PC}\right) \tag{E12.30}$$

in some appropriate limit. What is the limit?

(b) Reflect on the implied limitations of this equation, which we first saw in chapter E8.

E12R.2 The electric field vectors of a positively charged particle in its rest frame all point directly away from that particle. Prove that this statement is *still* true in a frame where the particle moves with a constant velocity $\vec{v}$, in spite of the way that equation E12.4*a* messes up the electric field components. (*Hint:* Let's say that the point where we want to evaluate the field is P. You will want to consider the angle that a line between the charge's position at an instant P makes with respect to the x axis in each frame. Why are these angles *not* the same?)

Advanced

E12A.1 (This problem requires detailed knowledge of special relativity.) In special relativity, we define the force $\vec{F}$ on an object in a given frame that an interaction exerts on an object to be the rate in that frame at which the interaction transfers relativistic momentum $\vec{p}$ to that object:

$$\vec{F} \equiv \frac{d\vec{p}}{dt} \qquad \text{where } \vec{p} \equiv \frac{m\vec{v}}{\sqrt{1-v^2}} \tag{E12.31}$$

where m is the invariant mass of the object and $\vec{v}$ is its velocity in the frame in question. According to the Lorentz transformation equations (see chapter R6),

$$dt' = \gamma(dt - \beta\, dx) \tag{E12.32a}$$

$$dx' = \gamma(-\beta\, dt + dx) \tag{E12.32b}$$

$$dy' = dy \qquad dz' = dz \tag{E12.32c}$$

when the primed frame moves at speed β in the $+x$ direction relative to the unprimed frame [$\gamma \equiv (1-\beta^2)^{-1/2}$]. According to chapter R9, the Lorentz transformation equations for the spatial components of the relativistic momentum are

$$p'_x = \gamma(\beta p_t - p_x) \tag{E12.33a}$$

$$p'_y = p_y \qquad p'_z = p_z \tag{E12.33b}$$

(a) Use these equations to show that the transformation law for the force acting on a particle instantaneously at rest in the primed frame is

$$F'_x = F_x \qquad F'_y = \gamma F_y \qquad F'_z = \gamma F_z \tag{E12.34}$$

(*Hint:* Note that for the particle in the unprimed frame $dx/dt = \beta$ in this case. Why?)

(b) Argue that this result applies to the situation discussed in section E12.3 and explains the difference between equations E12.18 and E12.21.

E12A.2 (This problem requires detailed knowledge of special relativity.) We can argue that the transformation law for the component of the magnetic field along the axis of relative motion between two frames must be $\mathbb{B}'_x = \mathbb{B}_x$ as follows. Consider an infinite solenoid at rest in the S' frame aligned with the x axis in both frames. According to the results of section E11.5, the magnetic field inside the solenoid in this frame is

$$\mathbb{B}'_x = \frac{4\pi k}{c} I' \left(\frac{N}{L}\right)' \qquad \mathbb{B}'_y = \mathbb{B}'_z = 0 \tag{E12.35}$$

where I' is the current and $(N/L)'$ is the number of turns per unit length measured in the solenoid's rest frame. Now look at the solenoid in the S frame, where the solenoid is observed to move in the $+x$ direction with speed $v = \beta c$. An infinite solenoid is unchanged by this motion, so its fields are all still static and Ampere's law applies.

(a) The current no longer moves purely in the direction perpendicular to the axis of rotation, but has a component in the x direction. We can break up the current into two parts, one parallel to the solenoid's axis and the other perpendicular to the solenoid's axis. Argue that the component of the current parallel to the axis contributes *nothing* to the magnetic field *inside* the moving solenoid. (*Hint:* See exercise E11X.3.)

(b) According to the results of section E11.5, the component of the current perpendicular to the axis should create a magnetic field in the solenoid whose components are

$$\mathbb{B}_x = \frac{4\pi k}{c} I \left(\frac{N}{L}\right) \tag{E12.36}$$

Argue that N/L is larger than $(N/L)'$ because of Lorentz contraction, but I is smaller than I' by the exact same factor because the coordinate time required for a charge to go once around the solenoid in the unprimed frame is longer than in the unprimed frame. (*Hint:* Let the event of a charge carrier leaving a given point on the solenoid be event A; let the event that the carrier returns to essentially the same point *on the solenoid* after circling the solenoid once be event B. Calculate the coordinate time between these events in the S frame from that in the solenoid frame, using either the metric equation or the Lorentz transformations.)

ANSWERS TO EXERCISES

E12X.1 This is a matter of applying appropriate right-hand rules. In both parts of figure E12.1, the conventional current is flowing toward the left. Your fingers will curl in the direction of the magnetic field if you point your thumb in the direction of the conventional current. The magnetic field thus points upward in the region below the wire and downward in the region above, as shown. The electron is negatively charged, so the vector $q\vec{v}/c$ points to the left in figure E12.1a. Since $\vec{F}_m = q(\vec{v}/c) \times \vec{\mathbb{B}}$, your right thumb will indicate the direction of $\vec{F}_m$ if you point your right index finger in the direction of $q\vec{v}/c$ and your second finger in the direction of $\vec{\mathbb{B}}$. When you do this, you should find that $\vec{F}_m$ points toward the wire, as shown.

E12X.2 Remember that $\vec{\mathbb{B}} = c\vec{B}$. If you plug this directly into the left equation of E12.4*a*, you get the equation on the right. If you plug this into the equation on the left side of E12.4*b*, you get the right side if you also divide through by c.

E12X.3 At points in the xy plane outside the distribution, the magnetic field vectors in S point in the $-z$ direction if $y > 0$ and in the $+z$ direction if $y < 0$. We expect the same to be true in S'. There is no electric field in S, but we expect a radially outward electric field in S' that at a point in the $x'y'$ plane points in the $+y'$ direction if $y' > 0$ and in the $-y'$ direction if $y' < 0$. So consider a point on the $y > 0$ side (the case for $y < 0$ will be analogous with all signs reversed). Equations E12.37 show that the fields in S' predicted by equations E12.4 at such a point will have the expected directions and the magnitudes:

$$\begin{bmatrix} E'_x \\ E'_y \\ E'_z \end{bmatrix} = \begin{bmatrix} E_x \\ \gamma[E_y - \beta\mathbb{B}_z] \\ \gamma[E_z + \beta\mathbb{B}_y] \end{bmatrix} = \begin{bmatrix} 0 \\ \gamma[0 - \beta(-\mathbb{B})] \\ \gamma(0+0) \end{bmatrix} = \begin{bmatrix} 0 \\ \gamma\beta\mathbb{B} \\ 0 \end{bmatrix} \quad \text{(E12.37}a\text{)}$$

$$\begin{bmatrix} \mathbb{B}'_x \\ \mathbb{B}'_y \\ \mathbb{B}'_z \end{bmatrix} = \begin{bmatrix} \mathbb{B}_x \\ \gamma(\mathbb{B}_y + \beta E_z) \\ \gamma(\mathbb{B}_z - \beta E_y) \end{bmatrix} = \begin{bmatrix} 0 \\ \gamma(0+0) \\ \gamma(-\mathbb{B} - 0) \end{bmatrix} = \begin{bmatrix} 0 \\ 0 \\ -\gamma\mathbb{B} \end{bmatrix} \quad \text{(E12.37}b\text{)}$$

E12X.4 According to equation E12.4*b*, the magnetic field in the ion's frame is

$$\begin{bmatrix} \mathbb{B}'_x \\ \mathbb{B}'_y \\ \mathbb{B}'_z \end{bmatrix} = \begin{bmatrix} \mathbb{B}_x \\ \gamma(\mathbb{B}_y + \beta E_z) \\ \gamma(\mathbb{B}_z - \beta E_y) \end{bmatrix} = \begin{bmatrix} 0 \\ \gamma[-\mathbb{B} + \beta(v/c)\mathbb{B}] \\ \gamma(0-0) \end{bmatrix} = \begin{bmatrix} 0 \\ -\gamma\mathbb{B}(1-\beta^2) \\ 0 \end{bmatrix} = \begin{bmatrix} 0 \\ -\mathbb{B}\sqrt{1-\beta^2} \\ 0 \end{bmatrix} \quad \text{(E12.38)}$$

since $v/c = \beta$ and $\gamma \equiv (1-\beta^2)^{-1/2}$. So the magnetic field is *not* zero in this frame, but it cannot exert any force on a charged particle at rest. (If *both* the electric and magnetic fields were zero in this or any frame, then equations E12.4 would imply that there would be no electric or magnetic field in *any* frame.)

E12X.5 If $\mathbb{B}_z = +\mathbb{B}$ and $E_y = +(v/c)\mathbb{B}$ and all other components of $\vec{E}$ and $\vec{\mathbb{B}}$ are zero, then

$$F_{em} = F_{em,y} = q\left(E_y + \left[\frac{\vec{v}}{c} \times \vec{\mathbb{B}}\right]_y\right) = q(v\mathbb{B} - v\mathbb{B}) = 0 \quad \text{(E12.39)}$$

In the particle's rest frame, $\vec{F}'_{em} = q(\vec{E}' + 0 \times \vec{\mathbb{B}}') = q\vec{E}'$ and

$$\begin{bmatrix} E'_x \\ E'_y \\ E'_z \end{bmatrix} = \begin{bmatrix} E_x \\ \gamma(E_y - \beta\mathbb{B}_z) \\ \gamma(E_z + \beta\mathbb{B}_y) \end{bmatrix} = \begin{bmatrix} 0 \\ \gamma(\beta\mathbb{B} - \beta\mathbb{B}) \\ \gamma(0 + \beta 0) \end{bmatrix} = \begin{bmatrix} 0 \\ 0 \\ 0 \end{bmatrix} \quad \text{(E12.40)}$$

E13 Maxwell's Equations

▷ Electric Field Fundamentals
▷ Controlling Currents
▷ Magnetic Field Fundamentals
▷ Calculating Static Fields
▽ Dynamic Fields
The Electromagnetic Field
Maxwell's Equations
Induction
Introduction to Waves
Electromagnetic Waves

Chapter Overview

Introduction

In this chapter, we will extend the field equations developed in chapters E10 and E11 to handle time-dependent electric and magnetic fields. The resulting set of equations provides a complete description of electromagnetic phenomena. We will explore applications of these equations in chapters E14 through E16.

Section E13.1: Introduction to Dynamic Fields

In chapters E10 and E11, we explored four powerful equations (Gauss's laws for the electric and magnetic fields and Ampere's laws for the magnetic and electric fields) for calculating fields in *static* situations. The generalizations of these equations that apply in *dynamic* situations (where the charge and/or current distributions vary with time) may involve additional time derivative terms that are zero in static situations. Our task in this chapter is to determine the needed terms.

Section E13.2: Correcting Ampere's Law

Equations E12.4 imply that in a frame where a uniformly charged infinite slab moves in a direction perpendicular to its plane, the slab has *no* magnetic field. If we set up an arbitrary loop perpendicular to the direction of motion, then as the slab passes the loop, some of its charge flows *through* the loop, and yet there is no magnetic circulation around the loop, contradicting our current form of Ampere's law.

However, Ampere's law *is* satisfied in this situation if we add a term involving the time derivative of the electric flux through the amperian loop used to calculate the magnetic circulation, so that the law becomes

$$\oint \vec{\mathbb{B}} \cdot d\vec{S} - \frac{1}{c}\frac{d\Phi_E}{dt} = \frac{4\pi k}{c} i_{\text{enc}} \qquad \text{(E13.10}b\text{)}$$

This is the **Ampere-Maxwell law.** Even though we have derived it for a special case, experiments show that it works in all circumstances.

Section E13.3: Faraday's Law

In chapter E8, we saw that magnetic forces on moving charge carriers in a conducting loop moving through a nonuniform magnetic field loop drive a current around this loop. This current must still flow in the frame where the loop is at rest, but since only an electric field can exert forces on charges at rest, an electric field must exist in this frame that is capable of driving charges around a loop. However, this contradicts Ampere's law for the electric field, which states that such a field is *impossible.*

Again, we can make this law work at all places and times in a similar situation involving a moving current-carrying infinite slab if we add a term that involves a time derivative of the magnetic flux through the amperian loop that we use to calculate the electric circulation, so that the law becomes

$$\oint \vec{E} \cdot d\vec{S} + \frac{1}{c}\frac{d\Phi_{\mathbb{B}}}{dt} = 0 \qquad \text{(E13.21}b\text{)}$$

We call this **Faraday's law,** and it again seems to be generally valid.

Section E13.4: Gauss's Laws Need No Correction

We also find that Gauss's laws for the electric and magnetic fields are satisfied without modification in the situations discussed in sections E13.2 and E13.3, even in the presence of changing electric and magnetic fields.

Section E13.5: Maxwell's Equations

Gauss's laws and the corrected forms of Ampere's laws together comprise **Maxwell's equations** (in integral form):

$$\oint \vec{E} \cdot d\vec{A} = 4\pi k q_{\text{enc}} \quad \text{or} \quad \oint \vec{E} \cdot d\vec{A} = \frac{q_{\text{enc}}}{\varepsilon_0} \tag{E13.26a}$$

$$\oint \vec{\mathbb{B}} \cdot d\vec{S} - \frac{1}{c}\frac{d\Phi_E}{dt} = 4\pi k \frac{i_{\text{enc}}}{c} \quad \text{or} \quad \oint \vec{B} \cdot d\vec{S} = \mu_0 i_{\text{enc}} + \varepsilon_0 \mu_0 \frac{d\Phi_E}{dt} \tag{E13.26b}$$

$$\oint \vec{\mathbb{B}} \cdot d\vec{A} = 0 \quad \text{or} \quad \oint \vec{B} \cdot d\vec{A} = 0 \tag{E13.26c}$$

$$\oint \vec{E} \cdot d\vec{S} + \frac{1}{c}\frac{d\Phi_{\mathbb{B}}}{dt} = 0 \quad \text{or} \quad \oint \vec{E} \cdot d\vec{S} = -\frac{d\Phi_B}{dt} \tag{E13.26d}$$

Purpose: These equations provide a complete and relativistically valid description of the behavior of dynamic electromagnetic fields.

Symbols: Φ_E, $\Phi_{\mathbb{B}}$, and Φ_B are the fluxes of $\vec{E}$, $\vec{\mathbb{B}}$, and $\vec{B}$, respectively, through the amperian loops used to calculate the circulations in equations where these fluxes appear. See the definitions of Gauss's laws and Ampere's laws for a discussion of other symbols.

Limitations: These equations are correct for all electromagnetic fields whenever their quantum nature is unimportant.

Note: The versions on the right express the equations in a more historical format and in terms of the constants $\varepsilon_0 = 1/4\pi k$ and $\mu_0 = 4\pi k/c^2$.

These equations represent one of the greatest intellectual achievements in physics.

Section E13.6: Local Field Equations (optional)

We can reexpress Maxwell's equations as follows

$$\text{div}(\vec{E}) \equiv 4\pi k\rho \tag{E13.30a}$$

$$\text{curl}(\vec{\mathbb{B}}) - \frac{1}{c}\frac{d\vec{E}}{dt} = 4\pi k \frac{\vec{J}}{c} \tag{E13.30b}$$

$$\text{div}(\vec{\mathbb{B}}) = 0 \tag{E13.30c}$$

$$\text{curl}(\vec{E}) + \frac{1}{c}\frac{d\vec{\mathbb{B}}}{dt} = 0 \tag{E13.30d}$$

where the **divergence** and **curl** of a field at a given point P are defined to be

$$\text{div}(\vec{E}) \equiv \lim_{dV \to 0} \frac{\oint \vec{E} \cdot d\vec{A}}{dV} \quad \text{for a gaussian surface of volume } dV \text{ surrounding } P \tag{E13.29}$$

$$\text{curl}(\vec{\mathbb{B}})_x \equiv \lim_{dA \to 0} \frac{\oint \vec{\mathbb{B}} \cdot d\vec{S}}{dA} \quad \text{around a loop of area } dA \text{ around } P \text{ whose tile vectors all point in } +x \text{ direction} \tag{E13.31}$$

(with similar definitions for the y and z components of the curl).

Since in any given situation we can express the divergence and curl in terms of derivatives of field quantities, we call equations E13.30 the **differential forms of Maxwell's equations:** they are mathematically equivalent to the integral forms.

Since the divergence and curl have well-defined values at a mathematical point P and ignore the fields created by any charges outside the infinitesimal neighborhood of P, equations E13.30 consitute **local field equations** that link a field characteristic only to the *local* presence of charge and current. Only equations that we can cast in such a form can be compatible with relativity.

E13.1 Introduction to Dynamic Fields

A summary of the electromagnetic field equations for static fields

In chapters E10 and E11, we discussed the following field equations:

$$\oint \vec{E} \cdot d\vec{A} = 4\pi k q_{\text{enc}} \quad \text{(E13.1}a\text{)} \qquad \oint \vec{\mathbb{B}} \cdot d\vec{A} = 0 \quad \text{(E13.1}b\text{)}$$

$$\oint \vec{\mathbb{B}} \cdot d\vec{S} = 4\pi k \frac{i_{\text{enc}}}{c} \quad \text{(E13.1}c\text{)} \qquad \oint \vec{E} \cdot d\vec{S} = 0 \quad \text{(E13.1}d\text{)}$$

Equations E13.1*a* and E13.1*c* link static electric and magnetic fields with their sources (charge and current, respectively) and provide a means of calculating the fields created by known sources. Equations E13.1*b* and E13.1*d* impose physical restrictions on the structure of electric and magnetic fields that are independent of how those fields are created. Together, these four equations provide a correct, complete, and time-tested description of electromagnetic fields that (as we saw in chapter E12) is also consistent with the principle of relativity.

Our goal is to update these equations to handle time-dependent fields

However, so far we have only examined the behavior of these equations in contexts where the charge and/or current distributions are *static* (time-independent). More general versions of these equations that apply in *dynamic* (time-dependent) situations might contain additional terms (such as time derivatives of field quantities) that are zero when the fields are static.

Our task in this chapter is to examine some simple time-dependent situations and look for evidence that any of equations E13.1 might be inadequate. We will find that Ampere's laws for the magnetic and electric fields (equations E13.1*c* and E13.1*d*) both fail in cases where the fields depend on time, but that adding a simple time-derivative term to each fixes the problems. In contrast, we will (in the same kinds of situations) see no evidence that Gauss's laws have similar problems, suggesting that these laws are correct as they stand.

E13.2 Correcting Ampere's Law

An infinite charged slab at rest in the S frame

The following situation displays the problem with Ampere's law for the magnetic field. Consider a uniformly charged, infinite slab oriented perpendicular to the x axis, as shown in figure E13.1a. Assume that in frame S where

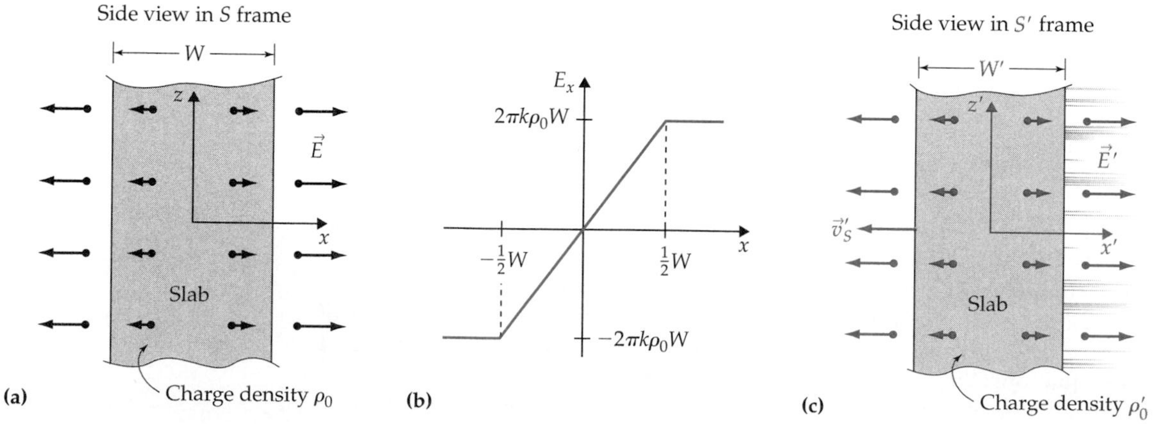

Figure E13.1
(a) A cross-sectional side view of a uniformly charged infinite slab and its electric field in the S frame, where the slab is at rest. (The y axis points away from us perpendicular to the plane of the drawing.) (b) A graph of E_x versus position along the x axis in the S frame. (c) The same slab and its electric field as seen in the S' frame, which moves in the $+x$ direction with speed v relative to the S frame. In this frame the slab is moving backward with speed $v'_S = v$ and is somewhat Lorentz-contracted.

the slab is at rest, it has width W in the x direction and a uniform positive charge density ρ_0 throughout its interior. Also assume that we have defined our coordinates so that the $x = 0$ plane (that is, the yz plane) coincides with the slab's central plane. Because this situation is entirely static, we can use the results of chapter E10 for a stationary slab, which imply that the electric field created at an arbitrary point by this slab is

$$E_x = \begin{cases} +2\pi k\rho_0 W & \\ 4\pi k\rho_0 x & \text{if} \\ -2\pi k\rho_0 W & \end{cases} \begin{array}{ll} x > \frac{1}{2}W & \text{(outside on right)} \\ -\frac{1}{2}W \le x \le \frac{1}{2}W & \text{(inside slab)} \\ x < -\frac{1}{2}W & \text{(outside on left)} \end{array} \tag{E13.2a}$$

$$E_y = E_z = 0 \tag{E13.2b}$$

where x is the x coordinate of the point where we are evaluating the field. Note that the field is uniform outside the slab, but E_x increases linearly with x inside the slab (see figure E13.1b). Because the charge in the slab is at rest, it has no *magnetic* field in this frame.

The slab viewed in the S' frame

Now let us look at this slab in a frame S' that moves at a speed $v \equiv \beta c$ in the $+x$ direction. Observers in this frame will see the slab moving in the $-x'$ direction with speed $v'_s = v$, as shown in figure E13.1c. It turns out that even though the charge in the slab is moving in the S' frame, the slab creates *zero* magnetic field in that frame. This is a surprising result, but we can see that in fact this result is *required* by symmetry. Figure E13.2a shows that the mirror image of the slab seen in a mirror parallel to the $x'z'$ plane is identical to the part of the slab behind the mirror. Therefore, the mirror rule implies that if this slab has a magnetic field at all, it must point in the $\pm y'$ direction. However, we can also rotate the slab around any axis parallel to the x axis without changing the slab. Figure E13.2b shows that if we rotate the slab around such an axis going through P, any nonzero field vector at P that originally points in the $\pm y'$ direction will change direction. But this rotation does not change the slab, so it should not change the field. We can avoid the contradiction only if $\vec{\mathbb{B}}' = 0$ at P. Since P was an arbitrary point, $\vec{\mathbb{B}}' = 0$ *everywhere.*

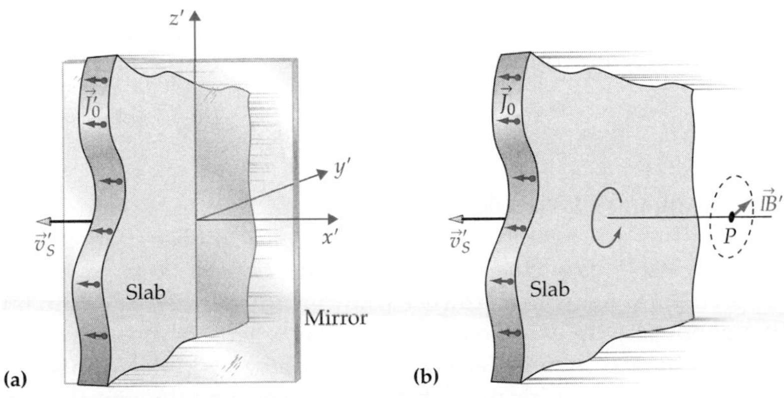

Figure E13.2
(a) The image of the moving slab in a mirror parallel to the $x'z'$ plane looks the same (even considering the current direction) as the part of the slab behind the mirror. Therefore, any magnetic field created by the slab must point in the $\pm y'$ direction. (The colored arrows show the current density associated with the slab's moving charge.) (b) The slab is unchanged by a rotation around an axis parallel to the x' axis. However, such a rotation changes the magnetic field vector at P unless it is zero.

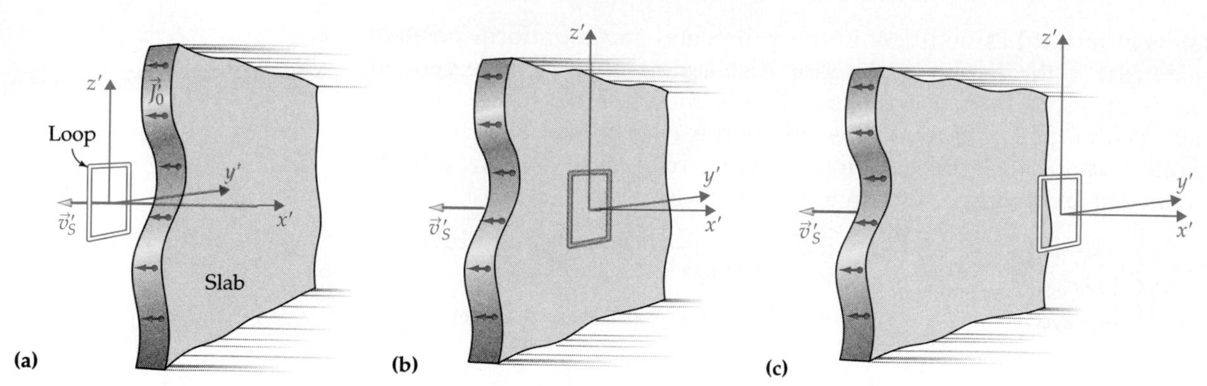

Figure E13.3
Images at successive times of the slab moving past an amperian loop that is at rest in the S' frame. (In the middle image, the loop is inside the slab.) The colored arrows show the current density associated with the slab's moving charge.

Equations E12.4*a* and E12.4*b*, which are quite generally valid, corrabo-rate this result:

$$\begin{bmatrix} E'_x \\ E'_y \\ E'_z \end{bmatrix} = \begin{bmatrix} E_x \\ \gamma(E_y - \beta \mathbb{B}_z) \\ \gamma(E_z + \beta \mathbb{B}_y) \end{bmatrix} = \begin{bmatrix} E_x \\ \gamma(0 - \beta 0) \\ \gamma(0 + \beta 0) \end{bmatrix} = \begin{bmatrix} E_x \\ 0 \\ 0 \end{bmatrix} \quad \text{(E13.3a)}$$

$$\begin{bmatrix} \mathbb{B}'_x \\ \mathbb{B}'_y \\ \mathbb{B}'_z \end{bmatrix} = \begin{bmatrix} \mathbb{B}_x \\ \gamma(\mathbb{B}_y + \beta E_z) \\ \gamma(\mathbb{B}_z - \beta E_y) \end{bmatrix} = \begin{bmatrix} 0 \\ \gamma(0 + \beta 0) \\ \gamma(0 - \beta 0) \end{bmatrix} = \begin{bmatrix} 0 \\ 0 \\ 0 \end{bmatrix} \quad \text{(E13.3b)}$$

Equation E13.3*a* also gives us the electric field in the S' frame, which we will find useful shortly.

Ampere's law is violated in the S' frame in this case!

However, having $\vec{\mathbb{B}} = 0$ inside the slab violates Ampere's law! Figure E13.3 shows the slab moving past a small square loop at rest in the S' frame. During the time interval that the loop is actually inside the slab as it passes (see Figure E13.3b), there is charge moving through the plane of the loop, so the loop encloses a *nonzero* current. However, there is zero magnetic field everywhere in the slab, so the magnetic circulation around the loop must be *zero*. Therefore, during that time interval

$$0 = \oint \vec{\mathbb{B}}' \cdot d\vec{S}' \quad \text{and} \quad i'_{\text{enc}} \neq 0 \quad \Rightarrow \quad \oint \vec{\mathbb{B}}' \cdot d\vec{S}' \neq \frac{4\pi k}{c} i'_{\text{enc}} \quad \text{(E13.4)}$$

contrary to Ampere's law. Since the principle of relativity requires that all good laws of physics work in all inertial reference frames, this means that Ampere's law itself (as we have it currently) *must be incorrect*.

Can we fix this law with a term involving a time derivative of $\vec{E}$?

The problem is really that Ampere's law is *incomplete*. We gain an important clue about how to complete the equation when we recognize that the problem only arises when the electric field is *changing* in the loop's vicinity. As long as the loop is outside the slab, there is no current flowing through the loop and thus no problem with Ampere's law, and in this region, the electric field in the loop's vicinity is constant in time (even as the slab moves) because the slab's external electric field is uniform. It is only when the loop is *inside* the passing slab that Ampere's law fails, but this is precisely when the electric field in the loop's vicinity is changing with time. Therefore, the problem may be that Ampere's law is missing a term that involves a time derivative of the electric field.

What would the electric field's time derivative be in the S' frame? The time $\Delta t'$ that it takes the slab to pass the loop in this frame is simply the slab's width W' divided by its speed $v'_S = v$:

$$\Delta t' = \frac{W'}{v'_S} \tag{E13.5}$$

Since $E'_x = E_x$, the electric field in the vicinity of the loop goes from $-2\pi k\rho_0 W$ (its value to the left of the slab) to $+2\pi k\rho_0 W$ (its value to the right of the slab) while the loop is within the slab. Because E'_x changes *linearly* during this time interval, the time derivative of the electric field at the loop's location as the slab passes is

$$\frac{dE'_x}{dt'} = \frac{\Delta E'_x}{\Delta t'} = \frac{4\pi k\rho_0 W}{W'/v'_S} \tag{E13.6}$$

This gives us the time derivative of E'_x evaluated in the primed frame, but the right side of equation E13.6 contains quantities measured in both frames. Let us see if we can simplify this so that it only involves quantities measured in the primed frame. The slab's width W' in the S' frame is Lorentz-contracted compared to its width W in the S frame (where the slab is at rest). This compresses the slab's charge in the x direction in the S' frame relative to the S frame, so the charge density in S' is larger than that in S by the Lorentz contraction factor. Therefore, Lorentz contraction in this case implies that

$$W' = W\sqrt{1-\beta^2} \quad \text{and} \quad \rho'_0 = \frac{\rho_0}{\sqrt{1-\beta^2}} \tag{E13.7}$$

If we plug these results into equation E13.6, we find that (since $4\pi k$ has the same value in both frames) the latter simplifies to

$$\frac{dE'_x}{dt'} = \frac{4\pi k\rho_0 \cancel{W} v'_S}{\cancel{W}\sqrt{1-\beta^2}} = 4\pi k\rho'_0 v'_S = -4\pi k J'_{0x} \tag{E13.8a}$$

since $\vec{J}'_0 = \rho'_0\vec{v}'_S$ is the uniform current density of the moving charge inside the slab in this frame, and the velocity of the charge in this case is v'_S in the $-x'$ direction. Since $E'_y = 0$, $E'_z = 0$, $J'_y = 0$, and $J'_z = 0$, this means more generally that

$$\frac{d\vec{E}'}{dt'} = -4\pi k\vec{J}'_0 \tag{E13.8b}$$

Connecting the time derivative to the current enclosed

We can now connect the current density $\vec{J}'_0$ to the current i'_{enc} flowing through the loop as follows. The current enclosed by the loop is defined to be the flux of current density through the loop, so if we take the dot product of both sides of equation E13.8*b* with a tile vector $d\vec{A}'$ on the loop and sum over all the tiles on the loop's bounded surface, we find that

$$-4\pi k i'_{enc} = -4\pi k\int \vec{J}'_0 \cdot d\vec{A}' = \int \frac{d\vec{E}'}{dt'}\cdot d\vec{A}' = \frac{d}{dt'}\int \vec{E}'\cdot d\vec{A}' \equiv \frac{d\Phi'_E}{dt'} \tag{E13.9}$$

Exercise E13X.1

See if you can explain why it is legal to move the time derivative outside the integral in the second-to-last step. (*Hint:* A dot product obeys the same rules in calculus as an ordinary product. Think of the integral as simply being a sum.)

A corrected version of Ampere's law (the *Ampere-Maxwell law*)

Equation E13.9 means that the following corrected version of Ampere's law is satisfied in the S' frame:

$$\oint \vec{\mathbb{B}}' \cdot d\vec{S}' - \frac{1}{c}\frac{d\Phi'_E}{dt'} = \frac{4\pi k}{c} i'_{\text{enc}} \tag{E13.10a}$$

because while the slab is passing the loop, this equation reduces to

$$0 - \left(-\frac{4\pi k}{c} i'_{\text{enc}}\right) = \frac{4\pi k}{c} i'_{\text{enc}} \quad ! \tag{E13.11}$$

It also works before and after the slab passes, because then the magnetic circulation is zero, Φ'_E is constant, and $i'_{\text{enc}} = 0$, implying that equation E13.10 reads $0 + 0 = 0$.

Exercise E13X.2

The principle of relativity requires that the equation E13.10*a* *without* the primes

$$\oint \vec{\mathbb{B}} \cdot d\vec{S} - \frac{1}{c}\frac{d\Phi_E}{dt} = \frac{4\pi k}{c} i_{\text{enc}} \tag{E13.10b}$$

should be satisfied in the S frame as well. Show that this is true.

Equation E13.10, therefore, is the simplest equation that works at all times in both frames in this particular case. Of course, showing that the equation works in this special case does not prove that it works in *all* cases. However, we have used this simple case to find the *minimum* change that we have to make to Ampere's law to bring it into compliance with special relativity, and if the universe is kind to us, this will be what is required to make it work in other cases as well. The universe is indeed kind; experiments show that in all known situations involving dynamic macroscopic electric and magnetic fields, this equation, which we call the **Ampere-Maxwell law,** accurately describes those fields.

E13.3 Faraday's Law

A moving loop example that displays a problem with Ampere's law for the electric field

The following situation displays an analogous problem with Ampere's law for the electric field. Consider a conducting loop that (in the laboratory frame S) moves at a constant velocity through a static but nonuniform magnetic field, as shown in figure E13.4a. We saw in section E8.1 that as the loop moves, it carries its charge carriers with it, so these moving charges experience magnetic forces that (because the forces are different in the different legs of the loop) end up pushing a current around the loop. But now consider this situation in the reference frame S' where the loop is at rest. In this frame, the charge carriers in the loop are also at rest, and therefore they cannot respond to whatever magnetic field is present in that frame. Only an *electric* field can exert forces on charges at rest and thus drive the current. Indeed, equation E12.4*a* implies that the electric field in the S' frame in this situation must look as shown in figure E13.4b. Since this electric field pushes harder on charges in the front leg than on those in the back leg, it will indeed drive electric charges around the loop. This provides an explanation in the loop frame for the experimental fact that current does flow in this situation.

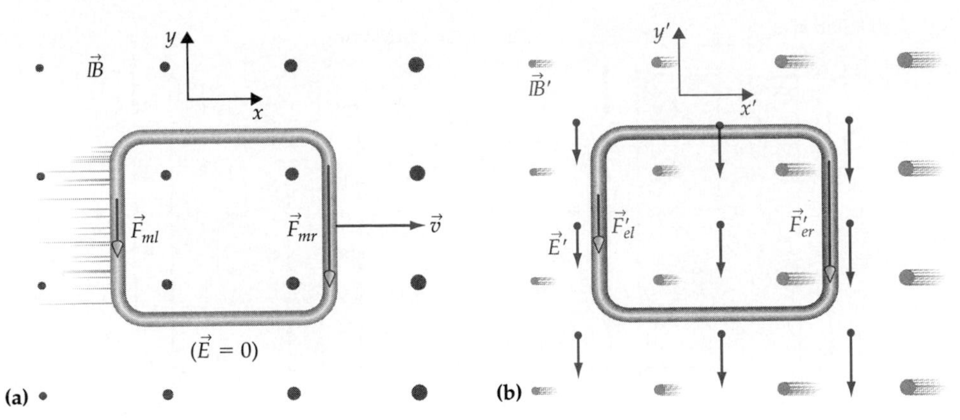

Figure E13.4
(a) A square loop moving through a magnetic field whose strength increases toward the right. Since the loop's charge carriers are moving with the loop, they experience magnetic forces that produce a net conventional current flowing clockwise. (The arrows show the direction of the magnetic force on positive charge carriers.) (b) In a frame where the loop is at rest, the loop's charge carriers feel no magnetic forces. Observers in that frame will see an *electric* field driving the current. But such a field violates Ampere's law for the electric field!

Exercise E13X.3

Show that equation E12.4*a* implies that the electric field in this case will indeed look like the field shown in figure E13.4b.

The problem is that, as we discussed in section E11.6, Ampere's law for the electric field implies that a static electric field *cannot* drive a current around a loop! Indeed, the electric field shown in figure E13.4b is one of electric field patterns that in example E11.2 we argued were specifically *excluded* by that law!

Exercise E13X.4

Ampere's law for the electric field evaluated in the S' frame would read

$$\oint \vec{E}' \cdot d\vec{S}' = 0 \qquad \text{(E13.12)}$$

Argue that this law is indeed violated by the electric field shown in figure E13.4b.

Just as we had to correct Ampere's law for the magnetic field in dynamic situations, so we have to correct Ampere's law for the electric field in such cases. While the moving loop example we have just discussed vividly and very physically illustrates the *problem* with this law, it is somewhat easier to find the right correction term by considering an example similar to the one discussed in section E13.2.

An infinite slab example that also displays the problem

Consider an infinite slab that in the S frame is at rest, is oriented perpendicular to the x direction, and carries a uniformly distributed current in the $-y$ direction, as shown in figure E13.5a. We will assume that in this

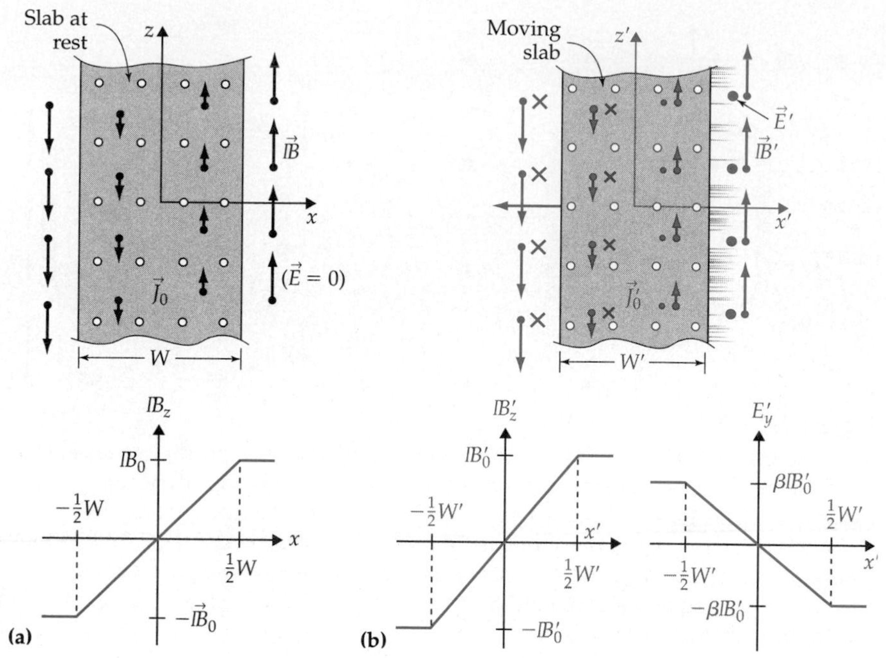

Figure E13.5
(a) A cross-sectional side view of an infinite slab carrying a uniform current density in the $-y$ direction (white dots) in the S frame, where the slab is at rest. (The y axis points away from us perpendicular to the plane of the drawing.) The graph shows how the z component of the magnetic field varies with x in this frame. (b) The same slab viewed in a frame S' that moves with speed v in the $+x$ direction. In this frame, the slab moves with speed $v_S' = v$ in the $-x'$ direction, and the slab has an electric field (dots and crosses) in the $\pm y'$ direction in addition to the magnetic field. The graphs below the drawing show how the nonzero components of the electric and magnetic fields vary with position x' in this frame.

frame, the slab has no net electric charge and therefore does not create an electric field. According to the results of section E11.4 (which are fully correct in this static situation), this slab does create the following magnetic field

$$\mathbb{B}_z = \begin{cases} +2\pi k\left(\dfrac{J_0}{c}\right)W & x > \frac{1}{2}W \quad \text{(outside to right)} \\ 4\pi k\left(\dfrac{J_0}{c}\right)x \quad \text{if} & -\frac{1}{2}W \le x \le \frac{1}{2}W \quad \text{(inside slab)} \\ -2\pi k\left(\dfrac{J_0}{c}\right)W & x < -\frac{1}{2}W \quad \text{(outside to left)} \end{cases} \tag{E13.13a}$$

$$\mathbb{B}_x = \mathbb{B}_y = 0 \tag{E13.13b}$$

where J_0 is the magnitude of the current density in the slab. Figure E13.5a shows a graph of how this magnetic field depends on x inside and outside the slab.

Now let us examine the same situation in a frame S' that is moving in the $+x$ direction with a speed $v = \beta c$ relative to the S frame. In this frame, the slab is moving backward with a constant speed of $v_S' = v$. Again, we can use equations E12.4 to calculate the electric and magnetic fields in this frame:

$$\begin{bmatrix} \mathbb{B}'_x \\ \mathbb{B}'_y \\ \mathbb{B}'_z \end{bmatrix} = \begin{bmatrix} \mathbb{B}_x \\ \gamma(\mathbb{B}_y + \beta E_z) \\ \gamma(\mathbb{B}_z - \beta E_y) \end{bmatrix} = \begin{bmatrix} 0 \\ \gamma(0 + \beta 0) \\ \gamma(\mathbb{B}_z - \beta 0) \end{bmatrix} = \begin{bmatrix} 0 \\ 0 \\ \gamma \mathbb{B}_z \end{bmatrix} \tag{E13.14a}$$

$$\begin{bmatrix} E'_x \\ E'_y \\ E'_z \end{bmatrix} = \begin{bmatrix} E_x \\ \gamma(E_y - \beta \mathbb{B}_z) \\ \gamma(E_z + \beta \mathbb{B}_y) \end{bmatrix} = \begin{bmatrix} 0 \\ \gamma(0 - \beta \mathbb{B}_z) \\ \gamma(0 + \beta 0) \end{bmatrix} = \begin{bmatrix} 0 \\ -\gamma\beta\mathbb{B}_z \\ 0 \end{bmatrix} = \begin{bmatrix} 0 \\ -\beta\mathbb{B}'_z \\ 0 \end{bmatrix} \tag{E13.14b}$$

We see that the magnetic field is multiplied by γ but is otherwise the same as in the S frame. However, there is in this frame an electric field $\vec{E}'$ that points in the $\pm y'$ direction and whose y' component is proportional to $\mathbb{B}'_z$. Figure E13.5b shows graphs of how $\mathbb{B}'_z$ and E'_y vary with position along the x' axis in this frame ($\mathbb{B}'_0$ is the magnitude of the magnetic field outside the slab in the S' frame).

Now consider a hypothetical rectangular amperian loop that lies at rest in the $z' = 0$ plane of the S' frame. If we define this loop's segment vectors $d\vec{S}'$ to go counterclockwise around the loop, then the segment vectors have components $[0, dS', 0]$ along the loop's front leg (the one at the largest value of x'), $[0, -dS', 0]$ along its rear leg, and $[\pm dS', 0, 0]$ along the other two legs parallel to the x' axis. Because $E'_x = E'_z = 0$, the component definition of the dot product and equation E13.14b imply that

$$\vec{E}' \cdot d\vec{S}' = E'_x\,dS'_x + E'_y\,dS'_y + E'_z\,dS'_z = 0 + E'_y\,dS'_y + 0 = -\beta\mathbb{B}'_z\,dS'_y \tag{E13.15}$$

This product will be zero for the two legs parallel to the x' axis. Note that the value of $\mathbb{B}'_z$ is constant along each of the other two legs, so if we define $\mathbb{B}'_{zf}$ and $\mathbb{B}'_{zr}$ to be the values of $\mathbb{B}'_z$ at points on the front and rear legs respectively, then the net electric circulation around this loop is

$$\oint \vec{E}' \cdot d\vec{S}' = \int_{\text{front}} \vec{E}' \cdot d\vec{S}' + \int_{\text{top}} \vec{E}' \cdot d\vec{S}' + \int_{\text{rear}} \vec{E}' \cdot d\vec{S}' + \int_{\text{bottom}} \vec{E}' \cdot d\vec{S}'$$
$$= -\beta\mathbb{B}'_{zf}\int_{\text{front}} dS' + 0 + \beta\mathbb{B}'_{zr}\int_{\text{rear}} dS' + 0 = -\beta(\mathbb{B}'_{zf} - \mathbb{B}'_{zr})\,\Delta y' \tag{E13.16}$$

where $\Delta y'$ is the length of the front and rear legs.

Note that the electric circulation is nonzero (violating Ampere's law for the electric field) only when the magnetic field has different values at the loop's front and rear legs. Because the magnetic field is uniform outside the slab, Ampere's law for the electric field holds both before and after the slab passes the loop. It is only *while* the slab is passing the loop that the magnetic field evaluated at the two legs is different and Ampere's law is violated. Not coincidentally, this is exactly when the magnetic field in the stationary loop's vicinity is *varying with time*. This suggests that we need to add a term involving the time derivative of the magnetic field to Ampere's law for the electric field if it is to handle dynamic cases correctly.

Figure E13.6 shows how the magnetic field at a given point in the S' frame varies with time as the slab passes (the graph assumes that $t' = 0$ is the instant that the slab's central plane passes the origin of the S' frame). The slab is moving at speed $v'_S = v = \beta c$, so it takes the slab a time equal to its width W' divided by its speed for the slab to pass a given point. Since $\mathbb{B}'_z$ goes linearly from $-\mathbb{B}'_0$ to $+\mathbb{B}'_0$ during this time, the time derivative of $\mathbb{B}'_z$ at any point in the slab while the slab is passing is

$$\frac{d\mathbb{B}'_z}{dt'} = \frac{2\mathbb{B}'_0}{W'/v} = \frac{2\mathbb{B}'_0 v}{W'} \tag{E13.17}$$

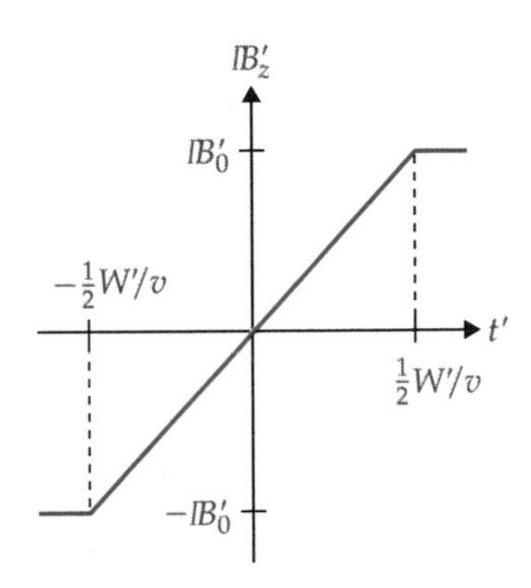

Figure E13.6
A graph of how the z' component of the magnetic field varies with time t' at a fixed point in the S' frame (the slab's central plane passes the point at time $t' = 0$). The magnetic field is constant when the point is outside the slab, but varies linearly with time when the point is inside the passing slab.

Can we fix the problem with a term involving the time derivative of $\vec{\mathbb{B}}$?

Note that the magnetic field also varies linearly with *position* inside the slab (see figure E13.5b). This means that at any time when the loop is completely inside the slab,

$$\frac{\mathbb{B}'_{zf} - \mathbb{B}'_{zr}}{\Delta x'} = \frac{d\mathbb{B}'_z}{dx'} = \frac{2\mathbb{B}'_0}{W'} \tag{E13.18}$$

since $\mathbb{B}'_z$ changes linearly from $-\mathbb{B}'_0$ to $+\mathbb{B}'_0$ as we go a distance of W' from one side of the slab to the other. Putting equations E13.16, E13.17, and E13.18 together, we find that

$$\oint \vec{E}' \cdot d\vec{S}' = -\frac{1}{c}\frac{d\mathbb{B}'_z}{dt'}\Delta x'\,\Delta y' \tag{E13.19}$$

Exercise E13X.5

Fill in the missing steps leading to equation E13.19.

Now, since any tile vector on the surface of this loop would have components $[0, 0, dA']$ and since $d\mathbb{B}'_z/dt'$ is the same at all points inside the slab as it passes, we can write

$$-\frac{1}{c}\frac{d\mathbb{B}'_z}{dt'}\Delta x'\,\Delta y' = -\frac{1}{c}\frac{d\mathbb{B}'_z}{dt'}\int dA' = -\frac{1}{c}\int\frac{d\mathbb{B}'_z}{dt'}\,dA'$$

$$= -\frac{1}{c}\int\frac{d}{dt'}(\vec{\mathbb{B}}'\cdot d\vec{A}') = -\frac{1}{c}\frac{d}{dt'}\int\vec{\mathbb{B}}'\cdot d\vec{A}' = -\frac{1}{c}\frac{d\Phi'_{\mathbb{B}}}{dt'} \tag{E13.20}$$

where $\Phi'_{\mathbb{B}}$ is the magnetic flux through the loop in the S' frame.

Exercise E13X.6

Explain why each step in equation E13.20 is valid in this situation.

A corrected version of Ampere's law for the electric field (*Faraday's law*)

Therefore, if we correct Ampere's law for the electric field to read

$$\oint \vec{E}' \cdot d\vec{S}' + \frac{1}{c}\frac{d\Phi'_{\mathbb{B}}}{dt'} = 0 \tag{E13.21a}$$

then it will be satisfied in the S' frame while the slab is passing the loop (because the two terms on the left will cancel out) as well as before and after the slab has passed (since the magnetic field is uniform and time-independent in this case, the electric circulation is zero by equation E13.16 and the derivative of the magnetic flux through any surface is zero). The same equation without the primes,

$$\oint \vec{E} \cdot d\vec{S} + \frac{1}{c}\frac{d\Phi_{\mathbb{B}}}{dt} = 0 \tag{E13.21b}$$

also works in the S frame (there is no electric field, and the magnetic field is static, so both terms on the left are zero). This law also reduces to Ampere's law for the electric field in any of the static conditions we considered in chapter E11, because the time derivative of the magnetic flux will be zero in such cases.

This equation is, therefore, the simplest equation that works at all times in both frames in this particular case and yet reduces to the correct equation in static circumstances. Again, showing that the equation works in this special case does not prove that it works in *all* cases (though the similarity to equation E13.10 should give us hope). The universe is again kind, though: experiments show that equation E13.20*b*, which we call **Faraday's law,** works in all inertial frames in all known situations involving dynamic macroscopic electric and magnetic fields. This very important law implies that a changing magnetic flux through a conducting loop causes current to flow around the loop, an idea we will explore more fully in chapter E14.

E13.4 Gauss's Laws Need No Correction

The case of the moving charged slab

Let us check to see whether either Gauss's law for the electric field or Gauss's law for the magnetic field shows similar signs of trouble in the frame where the fields are dynamic. Consider first the example of the moving charged slab considered in section E13.2. Imagine a can-shaped gaussian surface whose circular end faces have area A' that is at rest in the S' frame and oriented so that its central axis is parallel to the x' direction. Imagine that we evaluate the electric flux across this gaussian surface at the instant in the S' frame when the slab's central plane passes the surface's left face; and assume for the sake of argument that the surface's right face lies somewhat outside the slab at that instant, as shown in figure E13.7. The slab's electric field is zero on the surface's left face, and it is perpendicular to all tiles on the curved part of the surface, so the flux through these surface elements is zero. The electric field evaluated at points on the right face is parallel to all tile vectors $d\vec{A}'$ on that face and has a constant magnitude of $E' = E = 2\pi k \rho_0 W$. Therefore the net electric flux through this surface is

$$\oint \vec{E}' \cdot d\vec{A}' = \int_{\text{r. face}} E'\,dA' + 0 = E' \int_{\text{r. face}} dA' = E'A' = 2\pi k \rho_0 W A' \qquad \text{(E13.22)}$$

How is this related to the charge enclosed by the surface in that frame? The charge enclosed by this surface is the volume $\frac{1}{2}W'A'$ of the charge enclosed by the can in this frame times the density of charge ρ_0' in that frame:

$$q'_{\text{enc}} = \tfrac{1}{2}\rho_0' W'A' \qquad \text{(E13.23)}$$

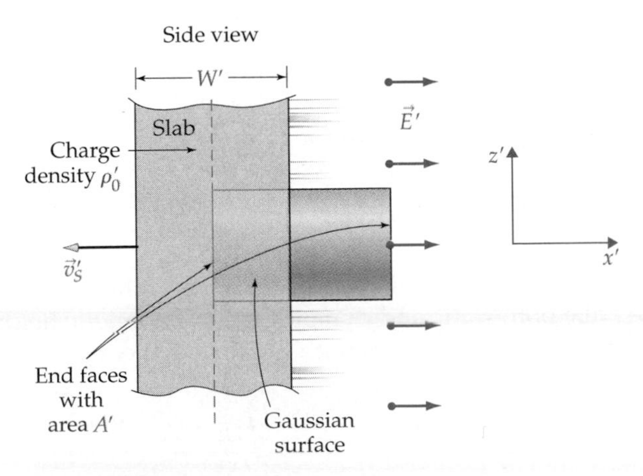

Figure E13.7
A cross-sectional side view of the moving slab and a can-shaped gaussian surface that is at rest in the S' frame. At the instant shown, the surface is embedded in the slab, and the slab's central plane is just passing the circular left face of the gaussian surface.

Note however, that equations E13.7 imply that

$$q'_{\text{enc}} = \frac{1}{2}\frac{\rho_0}{\sqrt{1-\beta^2}}W\sqrt{1-\beta^2}\,A' = \frac{1}{2}\rho_0 W A' \qquad \text{(E13.24)}$$

Substituting this into equation E13.12, we find that

$$\oint \vec{E}' \cdot d\vec{A}' = 2\pi k \rho_0 W A' = 4\pi k\left(\tfrac{1}{2}\rho_0 W A'\right) = 4\pi k q'_{\text{enc}} \qquad \text{(E13.25)}$$

in agreement with Gauss's law.

Exercise E13X.7

Use the same gaussian surface to argue that Gauss's law for the magnetic field holds in the S' frame in this case as well. (*Hint:* This is pretty easy.)

Therefore, we see no evidence in this particular situation that either of Gauss's laws suffers the same kind of problem that Ampere's law did, even though the electric field throughout most of the volume enclosed by our gaussian surface is strongly time-dependent.

The case of the current-carrying slab

Now let us check that Gauss's laws work in the case of the moving current-carrying slab considered in section E13.3. Consider a gaussian surface shaped like a rectangular box that is very thin in the x direction compared to the distance over which $\vec{E}'$ and $\vec{\mathbb{B}}'$ vary (see figure E13.8). All the faces of this surface are parallel to the electric field shown in Figure E13.5b except for the two thin faces perpendicular to the y' axis, so only these faces contribute anything to the net flux. Since the electric field that exists in the S' frame does not depend on y', the electric flux entering one of these two faces will be equal to that leaving the other face, so the net flux through the surface is zero whether the slab is passing or not. Since the box also encloses zero net charge in either case, Gauss's law for the electric field is satisfied in this frame.

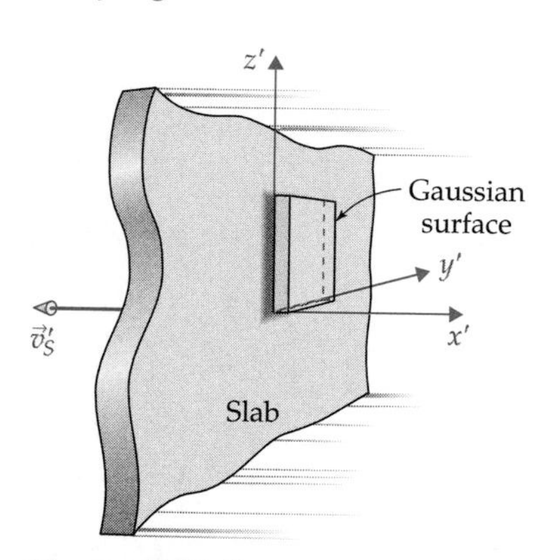

Figure E13.8
A gaussian surface at rest in the S' frame that we can use to test Gauss's law in that frame. The surface is a rectangular box that is very thin in the x direction (so that the electric field has an essentially constant value over its faces). The slab has just passed the surface.

Exercise E13X.8

Argue that Gauss's law for the magnetic field is also satisfied for the gaussian surface shown in figure E13.8, whether or not the slab is passing.

Again, we find no evidence to suggest that either one of Gauss's laws need correction in this case, even though in this case *both* the electric and magnetic fields vary with time in frame S'.

E13.5 Maxwell's Equations

Maxwell's equations (in integral form)

So we have seen that (after removing the primes) the following set of equations seems to describe both static and dynamic electromagnetic fields in all reference frames in a manner consistent with the special theory of relativity:

$$\oint \vec{E} \cdot d\vec{A} = 4\pi k q_{\text{enc}} \quad \text{or} \quad \oint \vec{E} \cdot d\vec{A} = \frac{q_{\text{enc}}}{\varepsilon_0} \qquad \text{(E13.26}a\text{)}$$

$$\oint \vec{\mathbb{B}} \cdot d\vec{s} - \frac{1}{c}\frac{d\Phi_E}{dt} = 4\pi k \frac{i_{\text{enc}}}{c} \quad \text{or} \quad \oint \vec{\mathbb{B}} \cdot d\vec{s} = \mu_0 i_{\text{enc}} + \varepsilon_0\mu_0\frac{d\Phi_E}{dt} \qquad \text{(E13.26}b\text{)}$$

$$\oint \vec{\mathbb{B}} \cdot d\vec{A} = 0 \qquad \text{or} \qquad \oint \vec{B} \cdot d\vec{A} = 0 \qquad \text{(E13.26}c\text{)}$$

$$\oint \vec{E} \cdot d\vec{s} + \frac{1}{c}\frac{d\Phi_{\mathbb{B}}}{dt} = 0 \qquad \text{or} \qquad \oint \vec{E} \cdot d\vec{s} = -\frac{d\Phi_B}{dt} \qquad \text{(E13.26}d\text{)}$$

Purpose: These equations provide a complete and relativistically valid description of the behavior of dynamic electromagnetic fields.

Symbols: Φ_E, $\Phi_{\mathbb{B}}$, and Φ_B are the fluxes of $\vec{E}$, $\vec{\mathbb{B}}$, and $\vec{B}$, respectively, through the amperian loops used to calculate the circulation in equations where these fluxes appear. See the definitions of Gauss's laws and Ampere's laws for a discussion of other symbols.

Limitations: These equations are correct for all electromagnetic fields whenever their quantum nature is unimportant.

Note: The versions on the right express the equations in a more historical format and in terms of the constants $\varepsilon_0 = 1/4\pi k$ and $\mu_0 = 4\pi k/c^2$.

Physicists call these equations the **(integral forms of) Maxwell's equations.** (Note how the deep symmetry between the electric and magnetic fields is vividly apparent in the equations on the left.)

A short history of Maxwell's equations

These equations were first proposed by James Clerk Maxwell in 1864 (though not quite in this form). I should emphasize here that Maxwell did *not* historically arrive at these equations in the way that we have done in this chapter, as the theory of relativity had not yet been invented (and would not be for another 40 years). Maxwell began working on electromagnetic theory in an effort to put work done by Michael Faraday in the 1830s on a firmer mathematical foundation. (Faraday had brilliant physical intuition but had little mathematical training or inclination.) Toward this end, Maxwell developed an intricate mechanical model of the behavior of electromagnetic fields. In the process, he found that his model required the additional time dependent term in Ampere's law that we discovered in section E11.2, a term that not only had not been previously observed but also had little hope of *being* observed with equipment then available. Once this missing piece was added, though, Maxwell was able to express everything that was then known about electricity and magnetism in a mathematically coherent whole.

Especially when viewed in the context of disarray in electromagnetic theory at the time, Maxwell's compression of humanity's total understanding of electricity and magnetism into a coherent structure involving only a handful of powerful equations ranks as one of the greatest intellectual achievements in the history of physics, on a par with Newton's mechanics and Einstein's theories of relativity. Not only did relativity fail to unseat Maxwell's equations (as it did so many other ideas of physics), but also it was in a real sense *inspired* by them: since Einstein was convinced that equations so beautiful *must* be correct, he developed special relativity to explain how they could work in all reference frames (Einstein's original paper on relativity was titled *On the Electrodynamics of Moving Bodies*). One can also see Maxwell's influence on Einstein's formulation of gravitational field equations for his general theory of relativity. Even quantum theory has done little more than to underline the basic validity of Maxwell's equations.

However, the line of reasoning that historically led Maxwell to his equations is complex (this is part of why it was an astonishing achievement!). The argument presented in this chapter takes full advantage of 21st-century hindsight to make the ideas simpler and clearer.

E13.6 Local Field Equations (optional)

A potential relativistic problem with these equations

There is one remaining issue with regard to these equations that we should consider. In section E2.1, I argued that Coulomb's law must be incompatible with relativity because it is an action-at-a-distance law that links distant charges in such a way that moving a charge at one point would have an instantaneous effect on a charge at a distant point. We might criticize Maxwell's equations on exactly the same grounds: it certainly *looks* as if moving a charge at a point could lead to instantaneous changes in the field quantities at distant points on an enclosing surface or loop.

What is a *local field equation?*

The most straightforward way to see that Maxwell's equations avoid this kind of problem is to express them in the form of **local field equations.** A *local* field equation avoids making *any* statements about how charges at a point might affect the fields at a distant point: instead it links a field quantity at a point *only* to presence of charge or current at the *same* point. A local field equation will therefore have the form

$$\begin{pmatrix} \text{a quantity that describes} \\ \text{some } \textit{characteristic} \text{ of the} \\ \text{field at a point in space} \end{pmatrix} = \begin{pmatrix} \text{a quantity that describes} \\ \text{the presence of charge or} \\ \text{current } \textit{at the same point} \end{pmatrix} \tag{E13.27}$$

Note that we must construct the "field characteristic" quantity on the left so that (1) it is well defined at a single point in space and (2) it *completely ignores* any electromagnetic fields created by charges distant from that point, so as to focus entirely on what is happening locally.

Gauss's law as a local field equation

We can in fact express Maxwell's equations in such a form. As an example, consider Gauss's law for the electric field. Imagine shrinking our gaussian surface so that it encloses an essentially infinitesimal volume surrounding a point P. In this limit, the charge enclosed by infinitesimal volume dV will essentially be $\rho\, dV$, where ρ is the density of charge at point P. If we divide both sides of Gauss's law in this case by dV and take the limit that $dV \to 0$, we get

$$\lim_{dV\to 0} \frac{\oint \vec{E}\cdot d\vec{A}}{dV} = 4\pi k\rho \tag{E13.28}$$

Since the quantity on the right has a well-defined limiting value at point P, the field quantity on the left must also. Moreover, notice that this quantity *does* in fact ignore the electric field of any charge not enclosed by the infinitesimal volume surrounding P: as discussed in chapter E10, a charge outside the gaussian surface does *not* affect the net flux through that surface. Therefore, this quantity has all the right characteristics to appear on the left side of a local field equation. We call this quantity (which is positive if the field in the immediate vicinity of a point radiates away from that point and negative if the field converges toward the point) the **divergence** of the electric field at a point:

$$\text{div}(\vec{E}) \equiv \lim_{dV\to 0} \frac{\oint \vec{E}\cdot d\vec{A}}{dV} \tag{E13.29}$$

The differential (local) forms of Maxwell's equations

Therefore, a fully local version of Gauss's law for the electric field reads

$$\text{div}(\vec{E}) \equiv 4\pi k\rho \tag{E13.30a}$$

Similarly, we can express the other three of Maxwell's equations in fully local form:

$$\text{curl}(\vec{\mathbb{B}}) - \frac{1}{c}\frac{d\vec{E}}{dt} = 4\pi k\frac{\vec{J}}{c} \tag{E13.30b}$$

$$\text{div}(\vec{\mathbb{B}}) = 0 \qquad \text{(E13.30}c\text{)}$$

$$\text{curl}(\vec{E}) + \frac{1}{c}\frac{d\vec{\mathbb{B}}}{dt} = 0 \qquad \text{(E13.30}d\text{)}$$

where the **curl** of a field is a *vector* quantity defined so that its x component is

$$\text{curl}(\vec{\mathbb{B}})_x \equiv \lim_{dA\to 0} \frac{\oint \vec{\mathbb{B}} \cdot d\vec{S}}{dA} \quad \begin{array}{l}\text{around loop of area } dA \text{ whose tile}\\ \text{vectors all point in } +x \text{ direction}\end{array} \qquad \text{(E13.31)}$$

and its other components are defined analogously. As the name suggests, the x component of this field quantity describes the extent to which the field "curls" around the x axis at the point in question as a magnetic field "curls" around a wire. (If you visualize the field as flowing water, this quantity reflects how rapidly and in what direction the water would cause a paddle-wheel to turn around an axis parallel to the x direction when placed at the point in question).

We call equations E13.30 the **differential form of Maxwell's equations.** Because these equations *are* completely local, they sidestep the action-at-a-distance problem.† We should therefore consider them to be the most fundamental expression of Maxwell's equations (Maxwell himself originally proposed his equations in differential form). It turns out, however, that the integral form of Maxwell's equations (equations E13.26) are mathematically *equivalent* to these local equations. Therefore, even though it *looks* as if they might suffer from action-at-a-distance problems, they do not.

One of the advantages of equations E13.30 is that, with the help of some multivariable calculus, one can express the divergence and curl of a vector field in terms of partial derivatives of that field. Example E13.1 illustrates that in special cases, even an *ordinary* derivative can suffice.

Example E13.1 A Derivative Expression for the Curl

Problem Imagine that we know that an electric field $\vec{E}$ always points in the $+y$ direction but has a magnitude that varies with x alone. Find a derivative expression for the z component of curl($\vec{E}$) in this special case.

Model Consider a rectangular loop with sides Δx and Δy lying in a plane perpendicular to the z axis. If we define the loop's segment vectors to go counterclockwise around the loop, as shown in figure E13.9, then tile vectors on the loop's surface will point in the $+z$ direction, as required by the definition of this component of the curl. For the sake of argument, let's define the coordinates of the loop's left corner to be $[x, y, z]$.

The component definition of the dot product implies that in this case

$$\vec{E} \cdot d\vec{S} = E_x\,dS_x + E_y\,dS_y + E_z\,dS_z = 0 + E_y\,dS_y + 0 = E_y\,dS_y \qquad \text{(E13.32)}$$

Note that dS_y is only nonzero for segment vectors along the legs parallel to the y axis. Since E_y does not depend on y, we can write its value at all points along the left leg as $E_y(x)$ and its value at all points along the right leg as $E_y(x + \Delta x)$. The segment vectors have y component $dS_y = -dS$ on the left leg and $dS_y = +dS$ on the right leg, so the net electric circulation around this loop

†Of course, it *is* possible to write local field equations that are not physically correct: being able to cast a field equation in local form is thus a necessary but not sufficient criterion for consistency with relativity.

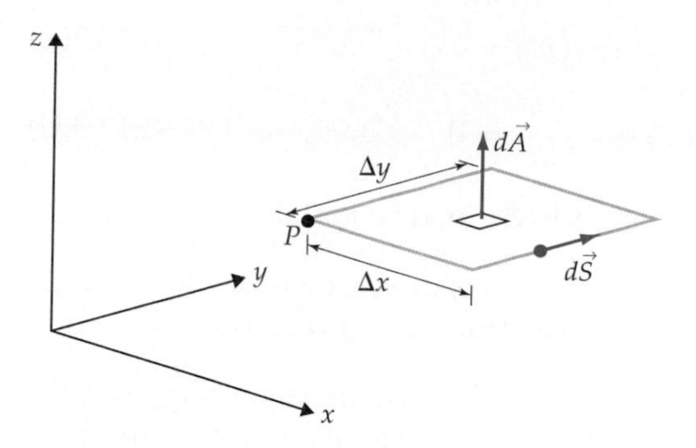

Figure E13.9
A hypothetical small amperian loop parallel to the *xy* plane. If we define the loop's segment vectors to go in the direction shown, a tile vector for a tile on the loop's surface will point in the $+z$ direction. The coordinates of point *P* are $[x, y, z]$.

is

$$\oint \vec{E} \cdot d\vec{r} = +E_y(x+\Delta x)\int dS - E_y(x)\int dS = [E_y(x+\Delta y) - E_y(x)]\,\Delta y \tag{E13.33}$$

Since the area of the loop is $\Delta x\,\Delta y$, we can evaluate the curl by dividing this by $\Delta x\,\Delta y$ and taking the limit.

Solution Therefore

$$\text{curl}(\vec{E})_z \equiv \lim_{\Delta x\,\Delta y \to 0} \frac{\oint \vec{E} \cdot d\vec{S}}{\Delta x\,\Delta y} = \lim_{\Delta x\,\Delta y \to 0} \frac{E_y(x+\Delta x) - E_y(x)}{\Delta x} \equiv \frac{dE_y}{dx} \tag{E13.34}$$

Evaluation We see that in this special case, this component of the curl has a very simple derivative expression.

Exercise E13X.9

Argue that the expression for $\text{curl}(\vec{E})_y$ in this case is even simpler.

In most of the cases that we have considered in this text, the integral forms are easier to understand and evaluate, but in many practical cases it is easier to solve differential equations for the fields than integral equations. In principle one can always use *either* form of Maxwell's equations to solve electromagnetic field problems.

TWO-MINUTE PROBLEMS

E13T.1 Consider a infinite, uniformly charged slab perpendicular to the x direction that in the laboratory frame moves in the $+x$ direction. Such a slab violates the *un*corrected version of Ampere's law in the laboratory frame because

A. The slab creates a magnetic field in that frame, but there is no current to create a magnetic field.

B. The moving slab creates a changing electric field at points inside the slab as it passes, but such a field contradicts Coulomb's law.

C. Ampere's law predicts that the current in the slab should everywhere be zero, but it is clearly *not* zero in the laboratory frame at points inside the slab.

D. The moving slab creates an electric field that violates Ampere's law for electric fields.

E13T.2 Imagine that at a certain instant of time, measurements in a certain region inside a conductor determine that the magnetic field is as shown. Assume that in the region within the dotted lines, the conductor is carrying a uniform current density $\vec{J}_0$ (black dots) directly toward the viewer, which we will take to be the $+z$ direction. What can we conclude about the electric field at all points in this region?

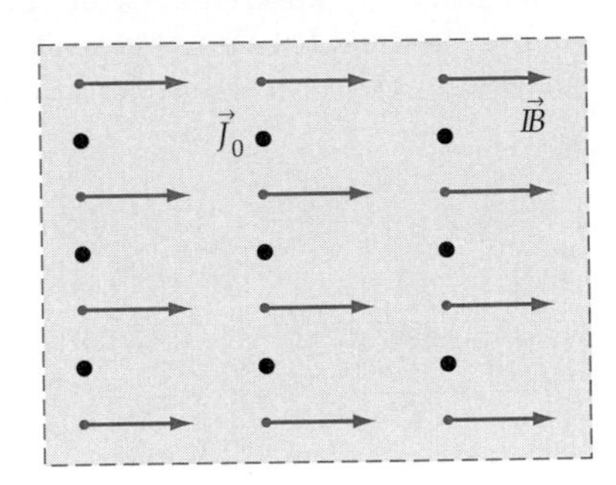

A. E_z must be zero.
B. E_z must be positive.
C. E_z must be negative.
D. E_z must be growing more positive.
E. E_z must be growing more negative.
F. We can conclude something about other electric field components (specify).
T. We do not have enough information to determine anything about the field.

E13T.3 Consider a parallel-plate capacitor, and imagine that we have a small amperian loop lying in a plane parallel to the capacitor plates in the region between the plates, as shown. There will be a nonzero magnetic circulation around this loop when

A. The capacitor is uncharged
B. The capacitor is being charged
C. The capacitor is being discharged
D. The capacitor is fully charged but that charge is not changing
E. In cases B and C
F. In all cases but A
T. In some other combination of cases (specify)

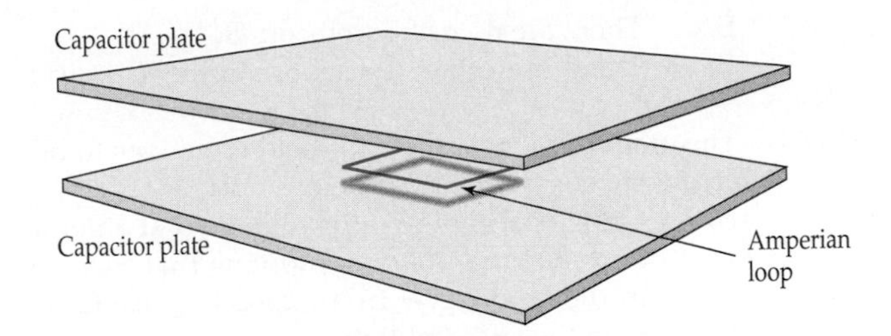

E13T.4 In each of the regions shown in figure E13.10, the electric field at an instant of time is as shown by the arrows. Assume that the z axis points directly toward us, perpendicular to the plane of the drawing. What can we conclude about the magnetic field in each of these regions? (The curves show possible amperian loops you might consider in your deliberations.)

A. B_z must be zero.
B. B_z must be positive.
C. B_z must be negative.
D. B_z must be growing more positive.
E. B_z must be growing more negative.
F. We can conclude something about other magnetic field components (specify).
T. We do not have enough information to determine anything about the magnetic field.

E13T.5 Anytime that we have a varying magnetic field in a given frame, we cannot describe the electric field in that frame using a potential function $\phi(x, y, z)$, true (T) or false (F)?

E13T.6 Imagine that in a certain region of space of a given reference frame, an electric field is uniform in magnitude and direction. What can we conclude about the magnetic field in that region and in that frame?

A. $\vec{B} = 0$
B. $\vec{B}$ is uniform in space.
C. $\vec{B}$ is constant in time.
D. $\vec{B}$ is both uniform in space and constant in time.

(more)

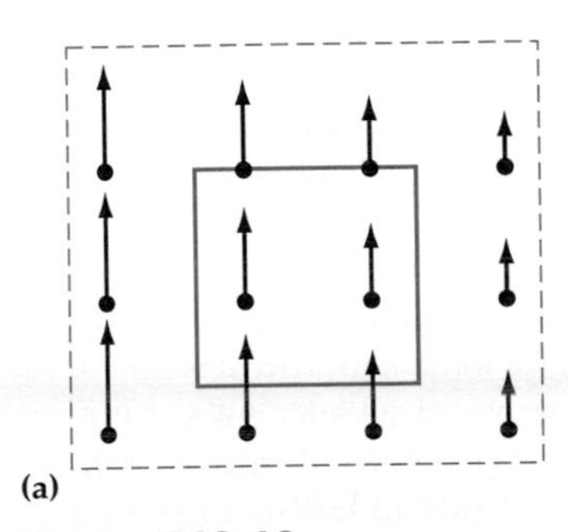

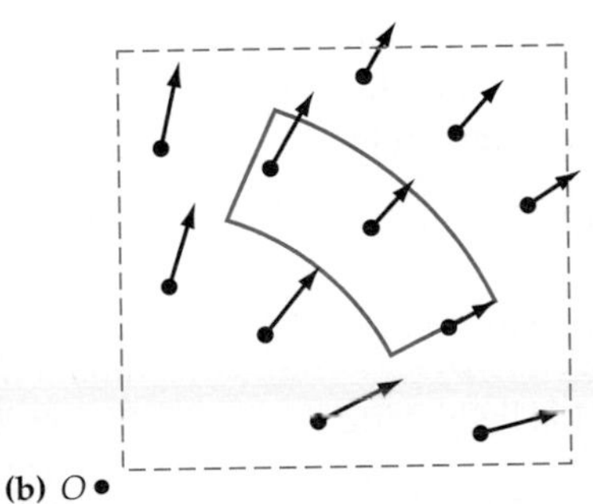

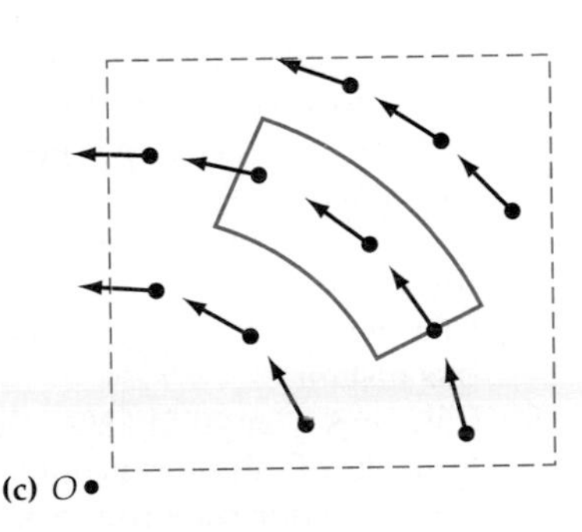

Figure E13.10
Each of these diagrams is a field arrow diagram of the electric field within a certain area of space (bounded by the colored dashed lines). In all cases, the field is independent of position along the axis perpendicular to the drawing. (b) The field vectors point radially away from O and have a magnitude that depends only on the distance r from a vertical axis going through O. (c) The vectors point tangent to circles around O, but all have the same magnitude. You may want to consider the colored amperian loops as you answer the questions. (See problems E13T.4 and E13B.6.)

E. There are no restrictions on $\vec{\mathbb{B}}$.
F. There are other restrictions on $\vec{\mathbb{B}}$ (specify).

E13T.7 The main point of using *local* field equations to describe the electromagnetic field is that
A. Only local charges affect the field at a point.
B. Such equations avoid implying that changes in the field travel from place to place faster than the speed of light.
C. Taking derivatives is easier than doing big sums.
D. Only local field equations can be made to work for time-dependent fields.

E13T.8 A local field equation equates the value of the electric or magnetic field at a point in space with the density of charge or current at that same point, T or F?

HOMEWORK PROBLEMS

Basic Skills

E13B.1 In the case of the moving slab discussed in section E13.2, assume that the slab has a charge density of $\rho_0' = 0.0010\ \text{C/m}^3$ and a thickness $W' = 1$ cm and is traveling at 10 m/s in the S' frame. Consider an imaginary square amperian loop 1 cm on a side that is at rest in that frame and is oriented perpendicular to the x' axis.
(a) What is the total current enclosed by this loop while the slab is passing by?
(b) Use the Ampere-Maxwell law to determine the rate at which E_x' changes with time.

E13B.2 Consider an amperian loop of area A that is small enough that the electric field $\vec{E}$ is essentially uniform over its surface. Argue that if the loop is perpendicular to the x direction and its segment vectors go counterclockwise when viewed from the positive side of the x axis, then $\Phi_E = E_x A$. (Similar statements apply for loops oriented perpendicular to the y and z axes.)

E13B.3 Consider an amperian loop that is small enough that the electric field $\vec{E}$ is essentially uniform over its surface. Assume that the loop is perpendicular to the x direction and its segment vectors go counterclockwise around the loop when viewed from the positive side of the x axis. Assume that $d\Phi_E/dt$ through this loop is positive at a certain instant of time.
(a) Can we say anything about the direction of $\vec{E}$ at this instant?
(b) Can we say that E_x is increasing with time at this instant?
(c) Can we say anything about what is happening to E_y and E_z at this instant?
Explain your response to each question. (*Hint:* See problem E13B.2.)

E13B.4 Consider an imaginary amperian loop in empty space. Assume that the net magnetic circulation around this loop at a certain instant of time is 200 N · m/C. What is the magnitude of the rate of change of the electric flux through this loop?

E13B.5 Consider a conducting ring whose plane is oriented perpendicular to our line of sight, and imagine that we define segment vectors for this ring that go counterclockwise around the ring. Imagine that the magnetic flux through the surface bounded by this ring is decreasing with time. According to Faraday's law, in what direction does conventional current flow around the ring? Explain your reasoning.

E13B.6 Consider the electric fields shown in figure E13.10. Assume that in each case we are looking at the field in the xy plane, with the z axis pointing toward us perpendicular to the drawing. In each case, use the shape of the electric field to infer the most that you can about any magnetic field that might or might not be present. (For example, you might say that only the z component can be nonzero and it must be positive; or that the x and z components of the magnetic field must be constant, but the y component must be increasing; or something like that.) Carefully explain your reasoning in each case.

Synthetic

E13S.1 One can write the Ampere-Maxwell law in the form

$$\oint \vec{\mathbb{B}} \cdot d\vec{S} = 4\pi k \frac{i_{\text{enc}}}{c} + \frac{1}{c}\frac{d\Phi_E}{dt} \qquad \text{(E13.35)}$$

This tells us that either a changing electric flux or a current can act as the source of a magnetic field.
(a) Consider a long, cylindrical wire whose diameter is 1 mm and which carries 1.0 A of current. (This is typical current in a household circuit.) What is the strength of the magnetic field that this current creates at the wire's surface? Is this a very strong magnetic field? (*Hint:* Consider a circular amperian loop that just barely encloses the entire wire.)
(b) Imagine now that instead of enclosing a wire, an amperian loop of the same radius is placed in empty space in an electric field oriented

perpendicular to the loop. Assume that a loop of this size is small enough that the electric field is essentially uniform over its face. What is the magnitude of the time derivative of the electric field in this case if it is to create the same magnetic field strength at points around the loop that the wire did in part (a)?

(c) In typical electric circuits available in Maxwell's time, electric fields might be as large as thousands of newtons per coulomb and might vary on time scales as small as 1 ms. Speculate as to why no one had experimentally noticed the electric flux term in the Ampere-Maxwell law before Maxwell.

E13S.2 Imagine that we have a parallel-plate capacitor consisting of two circular plates with a radius of 10 cm and a separation of 1 mm. Assume that we charge this capacitor to a potential difference of 100 V within 1 ms.

(a) Argue that the symmetry of this situation implies that any magnetic field that might exist between the two plates must point tangent to circles around the capacitor's central axis.

(b) Argue that the magnitude of the average rate at which the electric field between the plates changes as the capacitor charges is 1×10^8 (N/C)/s.

(c) Calculate the magnetic field strength while the capacitor is charging at a point that is between the plates and a distance of 9 cm from the capacitor's central axis. (*Hint:* Model the charging as occurring at a constant rate.)

E13S.3 Imagine that the magnetic field within a certain region of empty space is accurately modeled by the expression $\vec{\mathbb{B}} = [0, 0, Ay\cos\omega t]$, where A and ω are positive constants.

(a) What are the units of A? What are the units of ω?

(b) Find an expression for dE_x/dt at all points within the region in this case. (*Hints:* Consider an arbitrarily placed amperian loop oriented so that tile vectors on its bounded surface point in the $+x$ direction. Assume that the loop is small enough that E_x is approximately uniform over its surface. Argue that you get the same result for dE_x/dt no matter where you place the loop in this field.)

E13S.4 Imagine that the electric field within a certain region of space is accurately modeled by the expression $\vec{E} = [0, axe^{-bt}, 0]$, where a and b are positive constants.

(a) What are the units of a? What are the units of b?

(b) Find an expression for $d\mathbb{B}_z/dt$ at all points within the region in this case. (*Hints:* Consider an arbitrarily placed amperian loop oriented so that tile vectors on its bounded surface point in the $+z$ direction. Assume that the loop is small enough that $\mathbb{B}_z$ is approximately uniform over its surface. Argue that you get the same result for $d\mathbb{B}_z/dt$ no matter where you place the loop in this field.)

E13S.5 Consider a rectangular conducting loop lying in the xy plane that in the laboratory (S) frame is moving with speed $v = \beta c$ in the $+x$ direction in a static magnetic field that (in the xy plane at least) points purely in the z direction and depends only on x. There is no electric field in this frame. This is the same situation as is illustrated in figure E13.4 and discussed qualitatively in section E13.3. Our goal in this problem is to show *quantitatively* that we need the time derivative term in Faraday's law if that law is to correctly describe the electromagnetic fields observed in the frame S' in which the loop is at rest. (The solution to the problem will be very similar to the argument regarding the moving slab presented in section E13.3.)

(a) Use equations E12.4 to argue that the electric field in the S' frame is described by

$$E'_y = -\beta\mathbb{B}'_z \tag{E13.36}$$

(b) Assume that the loop has length L' (measured along the x' direction) and width W' (measured along the y' direction) in the S' frame, and let $\mathbb{B}'_{zr}$ and $\mathbb{B}'_{zl}$ be the values of $\mathbb{B}'_z$ at points along the loop's right and left legs, respectively, at a given instant in that frame. Show that at that instant, the electric circulation around the loop in that frame is

$$\oint \vec{E}' \cdot d\vec{S}' = -\beta(\mathbb{B}'_{zr} - \mathbb{B}'_{zl})W' \tag{E13.37}$$

Comment: Note that since $\mathbb{B}'_z$ depends on x at a given instant of time, this will not be zero, meaning that Ampere's law for the electric field does not work in the S' frame.

(c) Here is the tricky part. In the S frame, the magnetic field depends on position but is completely static. However, the magnetic field at a given point in the S' frame changes with time because the magnetic field is essentially moving backward relative to the S' frame (visualize the field vectors whizzing by backward!). This means that a fixed position in the S' frame will, as time passes, experience the magnetic fields at what corresponds to different *positions* in the S frame. Since the magnetic field does depend on position in frame S, the magnetic field at a fixed position in frame S' will depend on time. In fact, the magnetic field strength at the loop's right leg at a given time t' will *become* the magnetic field at the loop's left leg at a time $t' + \Delta t'$, where $\Delta t' = L'/v$ is the time required for the field to move backward the loop's length L' at the field's speed v. This means that the value of the magnetic field on the loop's left leg will change from $\mathbb{B}'_{zl}$ to $\mathbb{B}'_{zr}$ during that time. Argue that this means that (as long as the loop is small enough that the magnetic field in the S frame changes essentially linearly with position over the region spanned by the loop)

$$\frac{d\mathbb{B}'_z}{dt'} \approx \frac{\beta c(\mathbb{B}'_{zr} - \mathbb{B}'_{zl})}{L'} \tag{E13.38}$$

everywhere on the surface of the loop.

(d) Argue therefore that Faraday's law

$$\oint \vec{E}' \cdot d\vec{S}' + \frac{1}{c}\frac{d\Phi'_{\mathbb{B}}}{dt'} = 0 \tag{E13.39}$$

holds in the frame of the loop.

(e) Argue that Faraday's law is also satisfied (rather trivially) in the laboratory (S) frame.

E13S.6 Consider an infinite slab perpendicular to the x direction that is at rest, has width W, and has a uniform charge density ρ_0. In such a case, we know that the electric field inside the slab is given by $\vec{E} = [0, 0, 4\pi k\rho_0 x]$, and that $\vec{E}$ is constant and points in the x direction outside of the slab.

(a) Prove that in general whenever $E_y = E_z = 0$ and E_x depends at most on x, we have

$$\text{div}(\vec{E}) = \frac{dE_x}{dx} \tag{E13.40}$$

(b) Use the result of part (a) to show that the differential form of Gauss's law is satisfied at any point *inside* the slab.

(c) Use the result of part (a) to show that the differential form of Gauss's law is satisfied at any point *outside* the slab. (*Hint:* What is the charge density outside the slab?)

E13S.7 Consider an infinite slab perpendicular to the x direction that is at rest, has width W, and carries a uniform current density $\vec{J}_0$ in the $-y$ direction. In such a case, we know that the magnetic field inside the slab is described by $\vec{\mathbb{B}} = [0, 0, 4\pi k(J_0/c)x]$.

(a) Prove that in general whenever $\mathbb{B}_x = \mathbb{B}_y = 0$ and $\mathbb{B}_z$ depends at most on x, we have

$$\text{curl}(\vec{\mathbb{B}})_x = 0 \quad \text{curl}(\vec{\mathbb{B}})_y = -\frac{d\mathbb{B}_z}{dx} \quad \text{curl}(\vec{\mathbb{B}})_z = 0 \tag{E13.41}$$

(b) Use the result of part (a) to show that the differential form of Ampere's law is satisfied at any point *inside* the slab.

(c) Use the result of part (a) to show that the differential form of Ampere's law is satisfied at any point *outside* the slab. (*Hint:* What is the charge density outside the slab?)

E13S.8 Here is an example of how one can use the differential form of Gauss's law to compute an electric field. Consider an infinite slab perpendicular to the x direction that is at rest, has width W along the x direction, and has a uniform charge density ρ_0. Assume that we choose our origin so that the slab's central plane corresponds to the $x = 0$ plane. We know from symmetry in this case that the slab has an electric field that must point in the $\pm x$ direction and whose magnitude must depend at most on x. Symmetry also implies that the electric field is zero at $x = 0$.

(a) Prove that in general whenever $E_y = E_z = 0$ and E_x depends at most on x, we have

$$\text{div}(\vec{E}) = \frac{dE_x}{dx} \tag{E13.42}$$

(b) Using the result of part (a), integrate the differential form of Gauss's law from 0 to x to determine E_x at an arbitrary point whose x coordinate is inside the slab.

(c) Use the result of part (a) to argue that the electric field outside the slab must be constant.

(d) Determine the value of this constant by using the result of part (b) to determine the value of the electric field at the slab's surface.

Rich-Context

E13R.1 Imagine that you are working on an experiment immersed in a magnetic field that is essentially uniform, static, and vertical in a cubical region about 50 cm on a side between two horizontal plates. Your supervisor wants you to set things up so that the field remains vertical but falls to zero within a few millimeters of the region's boundary, as shown in figure E13.11. However, you must also have complete access to the region inside the magnetic field, meaning that for the four vertical walls of this region, the field has to fall to zero in essentially empty space.

(a) Your supervisor argues that this should be possible if one sets up an appropriate electric field

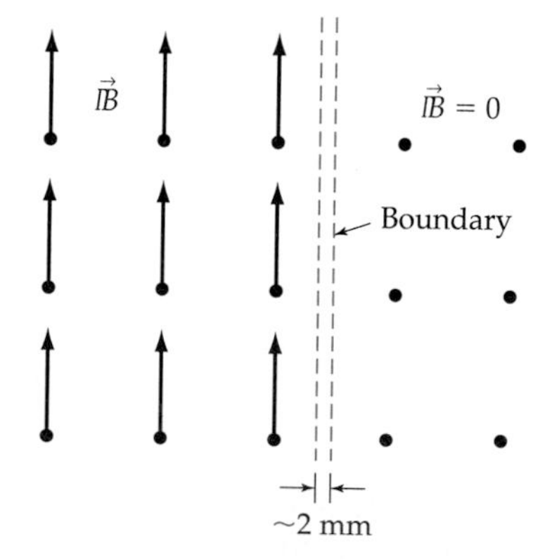

Figure E13.11
A schematic diagram of the magnetic field that your supervisor wants you to create (see problem E13R.1). The diagram shows magnetic field vectors in the neighborhood of one of the boundaries of the experimental region evaluated on a vertical plane that goes through the boundary in a direction perpendicular to that boundary.

in addition to the magnetic field. Describe the characteristics of the required electric field as completely as possible.

(b) Explain to your supervisor (respectfully but persuasively) that not only can this field not be sustained for long, but that it cannot exist at all in empty space if the magnetic field is static.

Advanced

E13A.1 Consider a particle with charge q moving at a constant speed v along the x axis in the $+x$ direction. As long as its speed is not relativistic, the particle's electric and magnetic fields at a point P at the instant that the charge is passing point C are reasonably accurately given by

$$\vec{E} = \frac{kq}{r_{PC}^2}\hat{r}_{PC} \quad \text{and} \quad \vec{\mathbb{B}} = \frac{kq}{r_{PC}^2}\left(\frac{\vec{v}}{c} \times \hat{r}_{PC}\right) \tag{E13.43}$$

where $\vec{r}_{PC}$ is the position of point P relative to point C and $\hat{r}_{PC} = \vec{r}_{PC}/r_{PC}$. For the sake of simplicity, consider the electric and magnetic fields at a point P a distance r from the origin along the z axis at the instant the particle passes the origin. The particle's magnetic field in the yz plane at that instant will point circularly around the x axis, as shown in figure E13.12.

(a) Argue that in the limit that $\Delta r \ll r$, the magnetic circulation around the loop centered on P shown in figure E13.12 is given by

$$\oint \vec{\mathbb{B}} \cdot d\vec{S} \approx -\frac{kq}{r^3}\left(\frac{v}{c}\right)A \tag{E13.44}$$

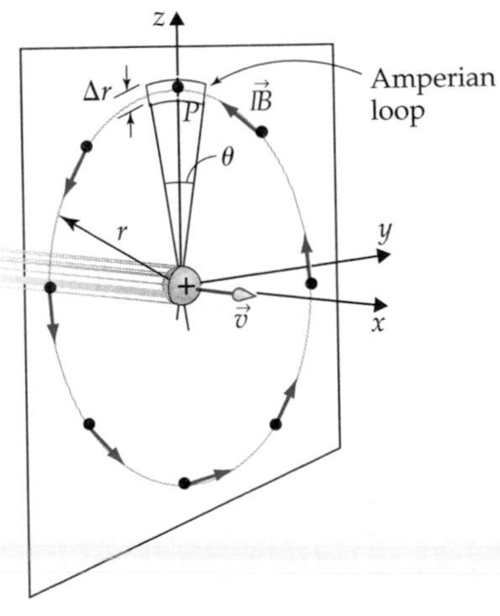

Figure E13.12
This diagram shows the magnetic field along a circle of radius r lying in the yz plane at the instant that a moving charged particle crosses the plane as it travels at speed v in the $+x$ direction. A small colored amperian loop that tracks the circle and has radial thickness Δr encloses the point P on the z axis. Note that by the definition of an angle in radians, the loop's top leg has length $(r + \Delta r/2)\theta$ and the bottom leg has length $(r - \Delta r/2)\theta$.

where $A = r\theta\,\Delta r$ is the loop's area in this limit. (*Hints:* Note that the top of the loop has length $r + \frac{1}{2}\Delta r$ while the bottom of the loop has length $r - \frac{1}{2}\Delta r$. Use the binomial approximation to simplify your expression for the circulation.)

(b) Show that the rate at which the electric field's x component at point P is changing at the instant that the particle passes the origin is

$$\frac{dE_x}{dt} = -\frac{kqv}{r^3} \tag{E13.45}$$

(*Hints:* Draw a side view of the situation, and use similar triangles to evaluate E_x when the particle is near the origin but not exactly at the origin.)

(c) Argue that Ampere's law is not satisfied in this situation for the loop in question, but that the Ampere-Maxwell law *is* satisfied. [*Hint:* In the limit that $\Delta r \ll r$, E_x will be nearly the same everywhere on the surface bounded by the loop. In the strict limit that $\Delta r \to 0$, this approximation and the approximations involved in part (a) become exact.]

E13A.2 (This problem requires some knowledge of partial derivatives.) Consider a gaussian surface consisting of a box with sides of length Δx, Δy, and Δz in the x, y, and z directions, respectively. Let the coordinates of the box's center be x, y, and z. Assume that the box is so small that the electric field averaged over all points on a given face is essentially the same as the value of the electric field at the face's center. (This approximation will become more and more exact as the size of the box decreases.) Now, the definition of the partial derivative of a multivariable function $f(x, y, z)$ with respect to x can be written

$$\frac{\partial f}{\partial x} \equiv \lim_{\Delta x \to 0} \frac{f(x + \frac{1}{2}\Delta x, y, z) - f(x - \frac{1}{2}\Delta x, y, z)}{\Delta x} \tag{E13.46}$$

Using the information presented, show that in an arbitrary (continuous) electric field,

$$\text{div}(\vec{E}) = \frac{\partial E_x}{\partial x} + \frac{\partial E_y}{\partial y} + \frac{\partial E_z}{\partial z} \tag{E13.47}$$

This is the general differential expression of the divergence.

E13A.3 (This problem requires some knowledge of partial derivatives.) Consider a rectangular amperian loop lying in a plane perpendicular to the z axis and centered on an arbitrary point whose coordinates are x, y, and z. Assume that we define this loop's segment vectors so that they go counterclockwise around the loop when viewed from the positive z axis. Assume also that the loop has sides of length Δx and Δy along the x and y directions, respectively, and that these quantities are so small that the value of the magnetic field averaged over all points along any given leg is essentially the same as the value of the magnetic field at the leg's center. (This approximation will become more and more exact as the size of the loop decreases.) Now, the definition of the partial derivative of a multivariable function with respect to x can be written as given in equation E13.46.

(a) Using the information presented, argue carefully that in an arbitrary (continuous) magnetic field,

$$\text{curl}(\vec{\mathbb{B}})_z = \frac{\partial \mathbb{B}_y}{\partial x} - \frac{\partial \mathbb{B}_x}{\partial y} \tag{E13.48}$$

(b) Find similar expressions for the other two components of the curl. (The resulting set of three expressions provides a general differential expression of the curl.)

E13A.3 (This problem requires some knowledge of partial derivatives.) According to problem E13A.1, the differential form of Gauss's law can be quite generally written in the form

$$\frac{\partial E_x}{\partial x} + \frac{\partial E_y}{\partial y} + \frac{\partial E_z}{\partial z} = 4\pi k \rho \tag{E13.49}$$

In this problem, we will show from first principles that this law correctly ignores the field of *any* point charge that is not located at the position where the derivatives are being evaluated. Consider evaluating the divergence of the field at a given point P due to a charge q at some other point C. To make the mathematics somewhat easier, let's take advantage of our freedom to define coordinates to set the origin of our reference frame at point C. Let the position of point P be $\vec{r} = [x, y, z]$.

(a) Argue that the electric field vector $\vec{E}(x, y, z)$ created at point P by the point charge q has components

$$\vec{E}(x, y, z) = \begin{bmatrix} kqx/r^3 \\ kqy/r^3 \\ kqz/r^3 \end{bmatrix} \tag{E13.50}$$

where $r \equiv (x^2 + y^2 + z^2)^{1/2}$.

(b) By substituting in $r \equiv (x^2 + y^2 + z^2)^{1/2}$ and using the chain rule, show that

$$\frac{\partial r}{\partial x} = \frac{x}{r} \tag{E13.51}$$

and argue that by analogy, $\partial r/\partial y = y/r$ and $\partial r/\partial z = z/r$.

(c) Use this result and the chain and product rules to prove

$$\frac{\partial}{\partial x}\left(\frac{x}{r^3}\right) = -\frac{3x^2}{r^5} + \frac{1}{r^3} \tag{E13.52}$$

Again, the derivatives $\partial(yr^{-3})/\partial y$ and $\partial(zr^{-3})/\partial z$ are completely analogous.

(d) Use the result of part (c) to show that the divergence at point P of the electric field created by a charge at another point Q is identically zero:

$$\text{div}(\vec{E}) \equiv \frac{\partial E_x}{\partial x} + \frac{\partial E_y}{\partial y} + \frac{\partial E_z}{\partial z} = 0 \tag{E13.53}$$

(e) Argue (by writing out the partial derivatives and applying the sum rule) that

$$\text{div}(\vec{E}_1 + \vec{E}_2) = \text{div}(\vec{E}_1) + \text{div}(\vec{E}_2) \tag{E13.54}$$

Comment: The result of part (d) implies that the divergence at point P simply does not register the electric field created by a charge at any *other* point C. (Note that our proof breaks down if $r = 0$.) The result of part (e) implies that this applies to the field of any sum of point charges, and thus to *any* distant charge distribution. Therefore, Gauss's law correctly ignores the field created by any distant charge distributions, as a good *local* field equation must.

ANSWERS TO EXERCISES

E13X.1 If the dot product obeys the same rules in calculus as an ordinary product, then the product rule implies that

$$\frac{d}{dt'}(\vec{E}' \cdot d\vec{A}') = \frac{d\vec{E}'}{dt'} \cdot d\vec{A}' + \vec{E}' \cdot \frac{d\vec{A}'}{dt'} \tag{E13.55}$$

But the loop is fixed in space, so its tile vectors are constant, implying that the last term in equation E13.55 is zero. Therefore

$$\int \frac{d\vec{E}'}{dt'} \cdot d\vec{A}' = \int \frac{d}{dt'}(\vec{E}' \cdot d\vec{A}') \tag{E13.56}$$

But the integral is just a sum, and the sum rule of calculus implies that the derivative of a sum is the same as the sum of the derivatives of its terms. Therefore,

$$\int \frac{d}{dt'}(\vec{E}' \cdot d\vec{A}') = \frac{d}{dt'} \int \vec{E}' \cdot d\vec{A}' \equiv \frac{d\Phi'_E}{dt'} \tag{E13.57}$$

E13X.2 In the S frame, the corrected version of Ampere's law reads

$$\oint \vec{\mathbb{B}} \cdot d\vec{S} - \frac{1}{c}\frac{d\Phi_E}{dt} = \frac{4\pi k}{c} i_{\text{enc}} \tag{E13.58}$$

In this frame, the slab is at rest, so there is no current. There is also no magnetic field, so the magnetic circulation is zero. Finally, there is an electric field, but it is static, so the time derivative of the electric flux is zero. Therefore, this equation simply reads $0 + 0 = 0$, which is (trivially) true.

E13X.3 We are told that the magnetic field in the S frame points in the $+z$ direction but increases in magnitude as x increases. According to equation E12.4a, the electric field in the S' frame (the loop's rest frame) is

$$\begin{bmatrix} E'_x \\ E'_y \\ E'_z \end{bmatrix} = \begin{bmatrix} E_x \\ \gamma(E_y - \beta \mathbb{B}_z) \\ \gamma(E_z + \beta \mathbb{B}_y) \end{bmatrix} = \begin{bmatrix} 0 \\ \gamma(0 - \beta \mathbb{B}_z) \\ \gamma(0 + \beta 0) \end{bmatrix} = \begin{bmatrix} 0 \\ -\gamma \beta \mathbb{B}_z \\ 0 \end{bmatrix} \tag{E13.59}$$

Therefore, the electric field in this frame will point in the $-y$ direction and (like $\mathbb{B}_z$) increase in magnitude as x increases, as shown in figure E13.4b.

E13X.4 Consider a square loop in the $x'y'$ plane in figure E13.4. The segment vectors on the legs that are parallel to the x axis are perpendicular to the electric field, so they contribute nothing to the circulation. The segment vectors on the loop's other two legs are antiparallel and parallel to the field, respectively, and the electric field is constant over these legs (since it only depends on x). If we define E'_{yr} to be the field's y' component at points along the right leg and E'_{yl} to be the same along the left leg, the net electric circulation around this loop is

$$\oint \vec{E}' \cdot d\vec{S}' = (E'_{yl} - E'_{yr})W' \tag{E13.60}$$

This is *not* zero, because $E'_y = -\gamma\beta\mathbb{B}_z$ depends on position, just as $\mathbb{B}_z$ does. Therefore, Ampere's law for the electric field is indeed violated in this frame.

E13X.5 According to equation E13.16,

$$\oint \vec{E}' \cdot d\vec{S}' = -\beta(\mathbb{B}'_{zf} - \mathbb{B}'_{zr})\,\Delta y' = -\frac{v}{c}\left(\frac{\mathbb{B}'_{zf} - \mathbb{B}'_{zr}}{\Delta x'}\right)\Delta x'\,\Delta y' \tag{E13.61}$$

where in the last step, I multiplied the top and bottom of the right side by $\Delta x'$ and substituted $\beta = v/c$. But according to equation E13.18, the quantity in parentheses has the value $2\mathbb{B}'_0/W'$, which according to equation E13.17 is the same as $(d\,\mathbb{B}'_z/dt')/v$. Substituting the latter for the quantity in parentheses and canceling the v, we get

$$\oint \vec{E}' \cdot d\vec{S}' = -\frac{1}{c}\frac{d\,\mathbb{B}'_z}{dt'}\Delta x'\,\Delta y' \tag{E13.62}$$

which is the same as equation E13.19.

E13X.6 Equation E13.20 reads

$$-\frac{1}{c}\frac{d\,\mathbb{B}'_z}{dt'}\Delta x'\,\Delta y' = -\frac{1}{c}\frac{d\,\mathbb{B}'_z}{dt'}\int dA' = -\frac{1}{c}\int \frac{d\,\mathbb{B}'_z}{dt'}dA' = -\frac{1}{c}\int \frac{d}{dt'}(\vec{\mathbb{B}}'\cdot d\vec{A}') = -\frac{1}{c}\frac{d}{dt'}\int \vec{\mathbb{B}}'\cdot d\vec{A}' = -\frac{1}{c}\frac{d\Phi'_{\mathbb{B}}}{dt'} \tag{E13.63}$$

The first step simply notes that $\Delta x'\,\Delta y'$ is the total area of the loop, so it will be equal to the sum of the areas of all tiles on the loop. The second step is legal because the derivative has the same value at all points on the loop (assuming that the loop is completely inside the moving slab at the time in question), so we can bring it into the sum. Since the tile vectors for all tiles on the loop point in the $+z$ direction, the component definition of the dot product implies that

$$\vec{\mathbb{B}}'\cdot d\vec{A}' = \mathbb{B}'_x\,dA'_x + \mathbb{B}'_y\,dA'_y + \mathbb{B}'_z\,dA'_z = 0 + 0 + \mathbb{B}'_z\,dA' \tag{E13.64}$$

This implies that

$$\frac{d}{dt'}(\vec{\mathbb{B}}'\cdot d\vec{A}') = \frac{d}{dt'}(\mathbb{B}'_z dA') = \frac{d\,\mathbb{B}'_z}{dt'}dA' \tag{E13.65}$$

because the tile vectors have a fixed area that does not depend on time. Therefore the third step is valid. The fourth step is valid because the integral is just a sum, and the sum rule for derivatives in calculus implies that the derivative of a sum is the same as the sum of the derivatives of all the terms in the sum. The last step is valid because the integral on the left side of the final equals sign is the definition of the flux of $\vec{\mathbb{B}}'$ through the bounded surface.

E13X.7 The magnetic field everywhere in the S' frame is zero, as we have discussed, so

$$\oint \vec{\mathbb{B}}'\cdot d\vec{A}' = \oint 0\cdot d\vec{A}' = 0 \tag{E13.66}$$

which does indeed mean that Gauss's law for the magnetic field is satisfied in this frame.

E13X.8 All the faces of this surface are parallel to the magnetic field except for the two thin faces perpendicular to the z' axis, so only the latter two faces contribute anything to the net flux. Since the magnetic field does not depend on z', the magnetic flux entering one of these two faces will be equal to that leaving the other face, so the net flux through the surface is zero whether the slab is passing or not, as required by Gauss's law for the magnetic field.

E13X.9 In this case, we should consider a loop lying in a plane perpendicular to the y direction, with sides Δx and Δz. But since the electric field points in the y direction by hypothesis, this means that the segment vectors on this loop will all be perpendicular to the electric field, implying that $\vec{E}\cdot d\vec{S} = 0$ for every segment vector on the loop. Therefore,

$$\text{curl}(\vec{E})_y \equiv \lim_{\Delta x\,\Delta z\to 0}\frac{\oint \vec{E}\cdot d\vec{S}}{\Delta x\,\Delta z} = \lim_{\Delta x\,\Delta z\to 0}\frac{0}{\Delta x\,\Delta z} = 0 \tag{E13.67}$$

E14 Induction

- ▷ Electric Field Fundamentals
- ▷ Controlling Currents
- ▷ Magnetic Field Fundamentals
- ▷ Calculating Static Fields
- ▽ Dynamic Fields
  - The Electromagnetic Field
  - Maxwell's Equations
  - **Induction**
  - Introduction to Waves
  - Electromagnetic Waves

Chapter Overview

Introduction

In this chapter, we will be exploring the fascinating implications and important technological applications of Faraday's law, one of the new equations we discovered in chapter E13. In the process, we will learn how to calculate the energy density of an electromagnetic field, a result we will find useful in chapter E16.

Section E14.1: Magnetic Flux and Induced EMF

Faraday's law implies that *a changing magnetic flux in the interior of a closed conducting loop will induce a current in that loop*, a phenomenon we call **induction.** The effective electromotive force (emf) driving that current through the loop is given by **Faraday's law of induction:**

$$\mathscr{E}_{\text{loop}} = -\frac{1}{c}\frac{d\Phi_{\mathbb{B}}}{dt} \quad \text{or} \quad \mathscr{E}_{\text{loop}} = -\frac{d\Phi_B}{dt} \tag{E14.2}$$

Purpose: This equation describes the effective emf $\mathscr{E}_{\text{loop}}$ induced in a conducting loop by the time-dependent magnetic flux $\Phi_{\mathbb{B}}$ going through the loop.
Symbols: $\Phi_{\mathbb{B}}$ is the magnetic flux through the loop; c is the speed of light.
Limitations: There are no known limitations at the macroscopic level (that is, as long as the quantum nature of the electromagnetic field is unimportant).
Note: The emf is defined so that $\mathscr{E}_{\text{loop}} \equiv \oint \vec{E} \cdot d\vec{S}$ in the frame of the loop, where $\vec{E}$ is the electric field evaluated at a loop segment represented by $d\vec{S}$. A positive value of $\mathscr{E}_{\text{loop}}$ means that conventional current will flow in the direction indicated by $d\vec{S}$.

Section E14.2: Lenz's Law

One *can* calculate the direction of an induced current by carefully following the loop rule conventions in equation E14.2, but it is generally easier to use it to calculate the *magnitude* of the induced emf and determine the *direction* by using **Lenz's law:**

> An induced current seeks to *oppose* the change that creates it.

For example, the current induced in a loop flows in the direction that creates a magnetic field that reinforces a decreasing external magnetic field or opposes an increasing field.

This law is a consequence of conservation of energy. As discussed in the section, if an induced current were to reinforce a change instead of opposing it, it would be possible to create energy from nothing.

Section E14.3: Self-induction

A time-varying current flowing through a coil creates a changing magnetic field, which in turn induces an emf *in the coil itself* that opposes the original variation in current. The strength of this effect is characterized by the coil's **inductance:**

$$|\mathscr{E}| = L\left|\frac{dI}{dt}\right| \qquad \text{(E14.5)}$$

Purpose: This equation defines the (self-) inductance L of a coil or loop.
Symbols: dI/dt is the time rate of change of the current I in the coil or loop, and $|\mathscr{E}|$ is the resulting induced emf in the coil or loop.
Limitations: Since this is a definition, there are no limitations.
Notes: The value of a coil's self-inductance L turns out to be a fixed characteristic of the shape, size, and number of turns in the coil.

Example E14.2 illustrates how we can calculate L for a long solenoid.

Section E14.4: "Discharging" an Inductor

Imagine that we use a battery to drive a current through a coil with inductance L connected in series with a resistor R. If we suddenly remove the battery from such an ***LR* circuit,** the induced emf seeks to oppose the change by continuing to push current through the coil and the resistor. A simple calculation shows that the current driven by the coil in this case decays exponentially with time.

$$I(t) = I_0 e^{-tR/L} \qquad \text{(E14.14)}$$

Purpose: This equation describes the current I as a function of time t in a coil of inductance L whose self-induced emf is pushing the current through both the coil and a resistor with resistance R connected in series.
Symbols: t is the elapsed time, and I_0 is the current at time $t = 0$.
Limitations: This assumes L is independent of I (which *is* generally true).

Section E14.5: The Energy in a Magnetic Field

A detailed analysis of the energy in an LR circuit shows that a magnetic field must store energy. The total energy per unit volume stored in an electromagnetic field is

$$u_{\text{EM}} = \frac{1}{8\pi k}(E^2 + \mathbb{B}^2) = \frac{\varepsilon_0 E^2}{2} + \frac{B^2}{2\mu_0} \qquad \text{(E14.20)}$$

Purpose: This equation expresses the energy density u_{EM} (energy per unit volume) of an electromagnetic field at a point where the electric and magnetic field magnitudes are E and $\mathbb{B} = cB$, respectively.
Symbols: k is the Coulomb constant, $\varepsilon_0 = 1/4\pi k$, and $\mu_0 = 4\pi k/c^2$, where c is the speed of light.
Limitations: There are no known limitations at the macroscopic level (that is, as long as the quantum nature of the electromagnetic field is unimportant).

Section E14.6: Transformers

A **transformer** consists of a pair of nested coils, a *primary* and a *secondary.* If the primary is connected to a source of **alternating** (sinusoidally varying) **current,** an alternating current will be induced in the secondary. If the secondary is attached to something that uses electricity, their interacting magnetic fields transfer energy from the primary to the secondary without any direct electrical connection between the coils! By varying the ratio of turns in the two coils, one can set the secondary's output emf to be any arbitrary multiple or fraction of the primary's emf. Transformers make our technological civilization possible by making it practical to ship electrical power over large distances economically at high emf's while still delivering the power at useful and safe low emf's.

E14.1 Magnetic Flux and Induced EMF

Faraday's law

Faraday's law

$$\oint \vec{E} \cdot d\vec{S} + \frac{1}{c}\frac{d\Phi_{IB}}{dt} = 0 \qquad \text{(E14.1)}$$

is the most interesting of the two Maxwell equations we discovered in chapter E13 because it has some fascinating implications and useful applications. (The correction to Ampere's law, by contrast, is difficult to even observe.) Our purpose in this chapter is to explore some of these implications and applications.

The meaning of $\oint \vec{E} \cdot d\vec{S}$ in this context

This equation has a very simple physical interpretation. Note that $\vec{E} \cdot d\vec{S} = (\vec{F}_e/q) \cdot d\vec{S}$ is the k-work delivered per unit charge by the electric field to a charge carrier moving from one end of a loop segment $d\vec{S}$ to the other. The sum of this for all segments in the loop is the total energy per unit charge, that is, the emf $\mathscr{E}_{\text{loop}}$ transferred to a charge carrier by the electromagnetic field as the charge carrier travels once around the loop. In particular, if the loop in question happens to correspond to a conducting loop, this equation implies that a changing magnetic flux *through* the loop will drive a current *around* that loop exactly as if the loop were connected to a battery with emf

Faraday's law of induction

$$\mathscr{E}_{\text{loop}} = -\frac{1}{c}\frac{d\Phi_{IB}}{dt} \quad \text{or} \quad \mathscr{E}_{\text{loop}} = -\frac{d\Phi_{B}}{dt} \qquad \text{(E14.2)}$$

Purpose: This equation describes the effective emf $\mathscr{E}_{\text{loop}}$ induced in a conducting loop by time-dependent magnetic flux Φ_{IB} going through the loop.

Symbols: Φ_{IB} is the magnetic flux through the loop, and c is the speed of light.

Limitations: There are no known limitations at the macroscopic level (that is, as long as the quantum nature of the electromagnetic field is unimportant).

Note: The emf is defined so that $\mathscr{E}_{\text{loop}} \equiv \oint \vec{E} \cdot d\vec{S}$ in the frame of the loop, where $\vec{E}$ is the electric field evaluated at a loop segment represented by $d\vec{S}$. A positive value of $\mathscr{E}_{\text{loop}}$ therefore means that conventional current will flow in the direction indicated by the segment vectors.

Physicists call this phenomenon **induction** and equation E14.2 **Faraday's law of induction.** Michael Faraday is generally credited as being the first to recognize and describe this law in qualitative terms in 1831. Figure E14.1 shows how a changing magnetic field can create a current.

Exercise E14X.1

Argue that $\Phi_B = c\Phi_{IB}$.

Example E14.1 illustrates how we can use this law.

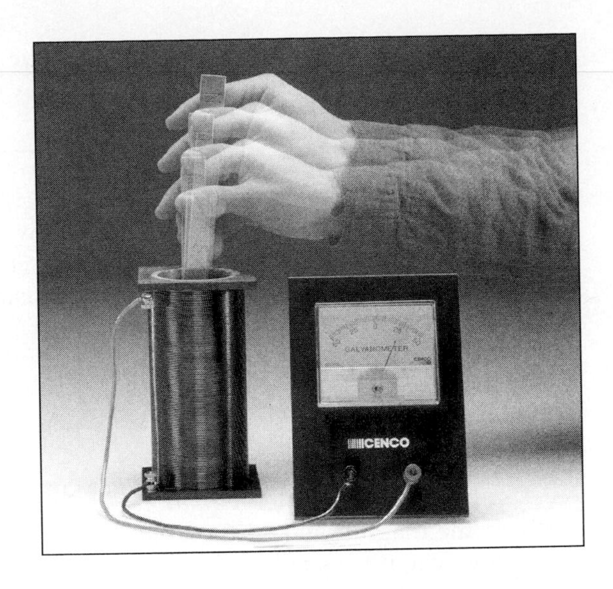

Figure E14.1
As we move the magnet into or out of this coil, the magnetic flux through the coil changes, inducing a current in the coil.

Example E14.1

Problem Imagine that we place a loop with a radius $r = 2.0$ cm inside a long solenoid with radius $R = 3.0$ cm. The magnetic field inside the solenoid points in the $+x$ direction and has a uniform magnitude $B = 1.0$ T. The loop is perpendicular to the solenoid's axis (the x axis here). The field inside suddenly drops to zero within 0.1 s. What is the approximate emf induced in the loop during this time?

Translation Figure E14.2 shows a cross-sectional drawing of the situation with the x axis pointing toward the viewer.

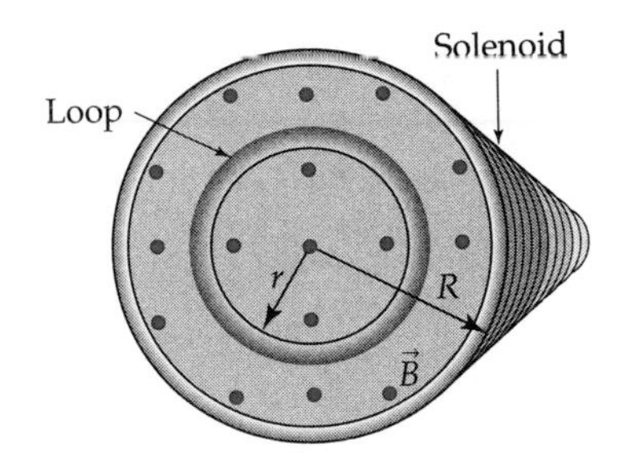

Figure E14.2
A cross-sectional cutaway view of the solenoid and loop, with the x axis facing us. The dots indicate that the solenoid's magnetic field is also pointing directly toward us.

Model Since the magnetic field inside a long solenoid is approximately uniform, this loop should be small enough that $\vec{B} \approx$ constant over its face. If we define the loop's $d\vec{S}$ vectors to point counterclockwise around the loop as shown in the diagram, the loop rule tells us that the enclosed surface's tile vectors point in the $+x$ direction. Since the magnetic field inside the solenoid is uniform and also points in the $+x$ direction, the flux through the loop is $\Phi_B = \int \vec{B} \cdot d\vec{A} = \int B\,dA \cos 0° = B \int dA = B(\pi r^2)$, because the loop's total area is πr^2. We are not given any details about how the magnetic field falls to zero, so let's assume that it falls to zero at a roughly *constant* rate (as a first approximation anyway):

$$\frac{d\Phi_B}{dt} \approx \frac{\Delta\Phi_B}{\Delta t} = \frac{\Phi_{B,\text{final}} - \Phi_{B,\text{initial}}}{\Delta t} = \frac{0 - \pi r^2 B}{\Delta t} = -\frac{\pi r^2 B}{\Delta t} \tag{E14.3}$$

Solution According to equation E14.2, the induced emf in the loop is thus

$$\begin{aligned}\mathscr{E}_{\text{loop}} &= -\frac{d\Phi_B}{dt} \approx +\frac{\pi r^2 B}{\Delta t} = \frac{\pi(0.02\text{ m})^2(1.0\text{ T})}{0.1\text{ s}} \\ &= 0.013\,\frac{\cancel{\text{T}}\cdot\cancel{\text{m}^2}}{\text{s}}\left(\frac{1\,\cancel{\text{N}}\cdot\cancel{\text{s}}\cdot\cancel{\text{C}^{-1}}\cdot\cancel{\text{m}^{-1}}}{1\text{ T}}\right)\left(\frac{1\,\cancel{\text{J}}}{1\,\cancel{\text{N}}\cdot\cancel{\text{m}}}\right)\left(\frac{1\text{ V}}{1\,\cancel{\text{J}}\cdot\cancel{\text{C}^{-1}}}\right) \\ &= 0.013\text{ V}\end{aligned} \tag{E14.4}$$

Since $\mathscr{E}_{\text{loop}} > 0$ and since we have defined the loop's $d\vec{S}$ vectors to point counterclockwise, we see that the induced current flows counterclockwise.

Evaluation Note that 1.0 T is a *huge* magnetic field, so the effect is pretty small!

Exercise E14X.2

In the situation described in exercise E14X.1, what will be the approximate magnitude of the induced current I during the change if the loop has a resistance of 0.10 Ω? (*Hint:* Use the definition of resistance.)

E14.2 Lenz's Law

We can find the direction of the induced current flow by using Faraday's law of induction . . .

As example E14.1 illustrates, equation E14.2 provides us with a way of finding both the magnitude and the flow direction of the induced current in a loop. However, it is usually easier to use the *absolute value* of equation E14.2 to determine the *magnitude* of an induced emf or current and to use another method to determine its direction. *Lenz's law* provides both the most straightforward way to determine this direction as well as providing insight into the physical implications of the signs in Faraday's law of induction.

This law (first proposed by Heinrich Lenz in 1834) asserts the following:

. . . but Lenz's law provides a memorable alternative

Lenz's law: An induced current seeks to *oppose* the change that creates it.

Let's see how this works in a specific situation. Imagine pushing the north end of a magnet toward a conducting ring, as shown in figure E14.3 (remember that magnetic field vectors point away from the north end of a magnet). Pushing the magnet toward the loop makes the average x component of the applied magnetic field inside the loop increase. Lenz's law means that the magnetic field produced by the current induced in the loop should *oppose* this change. In this case, this means that the loop should produce a magnetic field in the $-x$ direction to try to counteract the growing applied magnetic field in the $+x$ direction. Knowing the direction of the induced field, we can then determine the direction of the induced current by using the loop rule introduced in chapter E8: if you curl your right fingers in the direction of the current, your thumb points in the direction of the loop's north pole, which is also the direction of the induced magnetic field.

Lenz's law implies not only that the induced field opposes the change in the applied field, but also the loop actually exerts a magnetic force on the bar magnet that opposes its motion! Note that the induced current in the loop

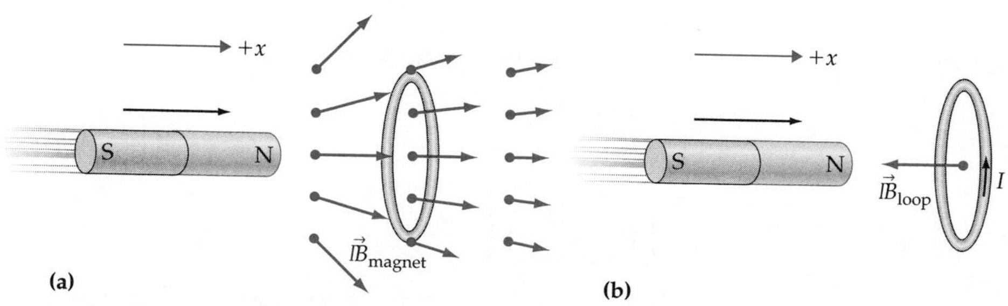

Figure E14.3
(a) If we push a bar magnet toward the ring, the average magnetic field within the ring increases in the $+x$ direction. (b) Lenz's law implies that the induced current will *oppose* this change by creating a magnetic field in the $-x$ direction.

above makes the loop equivalent to a bar magnet whose north magnetic pole faces the oncoming north pole of the bar magnet. The loop and bar magnet's north poles repel each other, meaning that you actually have to *push* on the magnet to keep it moving.

This is in fact a necessary consequence of conservation of energy. The changing magnetic field is creating an electric field that drives current through the loop. Electrons flowing in the loop pick up energy from the electric field and convert it to thermal energy through collisions with the metal atoms in the loop. Where does this energy come from? It has to come from the energy that we transfer to the bar magnet by pushing it against the opposing induced field of the loop. As we push the magnet through a distance, we transfer energy from our arm muscles to the system: this energy ends up as thermal energy in the loop.

Lenz's law expresses conservation of energy

The link between Lenz's law and conservation of energy is especially vivid when you consider what would happen if the induced current acted to *support* the change in the applied magnetic field. In our case, just nudging the bar magnet a bit toward the loop would set up an induced field that would tend to attract the magnet. This would cause the magnet to speed up toward the loop, which would create a stronger induced current in the loop, which would tug harder on the magnet, and so on. In such a case, the kinetic energy of both the magnet and the ring would increase spontaneously even while thermal energy is being dissipated by the growing current in the loop. Energy would therefore be created from (apparently) nowhere if this "inverted" version of Lenz's law were true.

It is important to recognize that Lenz's law does not *add* anything to Faraday's law of induction: it is simply a vivid way of expressing what is already implicit there. As pointed out at the beginning of this section, we can determine the direction of the induced current by remembering the various sign conventions and subtleties of Faraday's law of induction. But you will probably find it easier to use the absolute value of Faraday's law to find the *magnitude* of the induced emf and/or current and to use Lenz's law to find the direction.

Exercise E14X.3

Determine the sign of $d\Phi_{\mathbb{B}}/dt$ in the situation shown in figure E14.3, and use the sign conventions regarding Faraday's law to show that the direction of the induced current predicted by Faraday's law is the same as that predicted by Lenz's law.

Exercise E14X.4

Imagine that the bar magnet is pulled *away* from the ring. Does the current go in the same direction or the opposite direction? Does the loop attract or repel the magnet?

E14.3 Self-induction

One of the most fascinating implications of Faraday's law is that a loop or coil that conducts a changing current can induce an emf *in itself* that opposes changes in the applied current. To consider an extreme example, imagine a loop that carries a current I supplied by a battery. This creates a magnetic field that goes through the face of the loop, creating a nonzero magnetic flux

A current-carrying coil can induce an emf in itself!

through the loop. Imagine that we now disconnect the loop from the battery, so the current suddenly drops to zero. This causes the magnetic field and thus the flux through the loop to decrease very rapidly. But by Faraday's law, this changing flux will induce an emf in the loop, and by Lenz's law, this emf will seek to drive a current through the loop in the *same* direction that the battery did, so as to prop up the failing magnetic field. In situations where the current falls to zero almost instantly (as when a switch is opened), the induced emf can be so large that it causes current to arc briefly between the opening contacts of the switch; the loop will do *anything* to keep the current flowing at least for a short time.

The (self-) **inductance** L of any coil or loop is defined by the equation

Definition of a coil's (self-) inductance L

$$|\mathcal{E}| = L\left|\frac{dI}{dt}\right| \tag{E14.5}$$

Purpose: This equation defines the (self-) inductance L of a coil or loop.

Symbols: dI/dt is the time rate of change of the current I in the coil or loop, and $|\mathcal{E}|$ is the resulting induced emf in the coil or loop.

Limitations: This is a definition, so there are no limitations.

Notes: The value of a coil's self-inductance L turns out to be a fixed characteristic of the shape, size, and number of turns in the coil.

Equation E14.5 specifies only the magnitude of the induced emf. The direction of the emf is most easily determined by Lenz's law.

Exercise E14X.5

Inductance L is measured in the SI unit of **henrys** (H). What is the henry in terms of more basic units?

Exercise E14X.6

If the current in a 100-mH coil decreases from 1 A to zero in a time interval of about 1 ms, what is the approximate emf induced in the coil during the time the current decreases?

Example E14.2 The Inductance of a Long Solenoid

Problem Consider a long solenoid consisting of N turns of wire around a cylindrical form having a radius of r and length $\ell \gg r$. Use an infinite solenoid model to *estimate* the inductance of the real solenoid.

Model The magnetic field everywhere inside an *infinite* solenoid with N turns per length ℓ points parallel to its axis and has a uniform strength given by

$$\mathbb{B} = \frac{4\pi kI}{c}\left(\frac{N}{\ell}\right) \tag{E14.6}$$

(see equation E11.15*a*). Each turn of wire in this coil is essentially a loop whose enclosed surface is perpendicular to this field direction, so tile vectors $d\vec{A}$ on this surface can be chosen to be parallel to the field, implying that

$\vec{\mathbb{B}} \cdot d\vec{A} = +\mathbb{B}\,dA$. The magnitude of the magnetic field is also approximately uniform over the enclosed surface, so the magnetic flux going through a given turn in the coil is

$$\Phi_{\mathbb{B}} = \int_S \vec{\mathbb{B}} \cdot d\vec{A} = \mathbb{B} \int_S dA = \mathbb{B}(\pi r^2) = \frac{4\pi^2 r^2 N}{c\ell} I \qquad \text{(E14.7)}$$

since the turn spans an area of πr^2.

Solution Combining this with Faraday's law of induction, we find that the induced emf in one turn of the coil is given by

$$|\mathscr{E}_{\text{per turn}}| = \left| \frac{1}{c} \frac{d\Phi_{\mathbb{B}}}{dt} \right| = \frac{4\pi^2 k r^2 N}{c^2 \ell} \left| \frac{dI}{dt} \right| \qquad \text{(E14.8)}$$

because all the factors in equation E14.7 are independent of time except for the current I. Since the turns of the coil are essentially a series of loops in series, the total emf induced in the coil is the sum of the emf's induced in each turn, so

$$|\mathscr{E}_{\text{tot}}| = N|\mathscr{E}_{\text{per turn}}| = \frac{4\pi^2 k r^2 N^2}{c^2 \ell} \left| \frac{dI}{dt} \right| \qquad \text{(E14.9)}$$

Comparing this to equation E14.5, we see the coil's inductance is roughly

$$L = \frac{4\pi^2 k r^2 N^2}{c^2 \ell} = \frac{\mu_0 \pi r^2 N^2}{\ell} \quad \left(\text{since } \mu_0 = \frac{4\pi k}{c^2}\right) \qquad \text{(E14.10)}$$

Note that this only depends on fixed features of the coil, as I claimed before.

Evaluation Now, this is only an approximation, since the magnetic field inside a finite coil will get weaker as we get close to its ends, meaning that equation E14.7 overestimates the flux through a turn near the coil's ends, meaning that the inductance given by equation E14.10 is an overestimate of the coil's actual inductance. But the inductance of a long, skinny coil (whose length l is much greater than its radius R) will be pretty close to this value.

Exercise E14X.7

Use the result found above to estimate the inductance of a coil consisting of 500 turns of wire wound around a cylindrical form 1.0 cm in diameter and 10 cm long.

E14.4 "Discharging" an Inductor

Imagine that we connect a coil to a battery in parallel with a resistor, as shown in figure E14.4a. After a steady state is achieved, the coil will conduct a certain current I (that depends on its resistance) and will have some magnetic field in its interior. Now, imagine that we cut the connection with the battery but leave the resistor connected (see figure E14.4b). As the current through the coil begins to decrease, the magnetic field in the coil will begin to decrease, and the change in the magnetic field induces an emf in the coil that (by Lenz's law) opposes the change by continuing to push current through

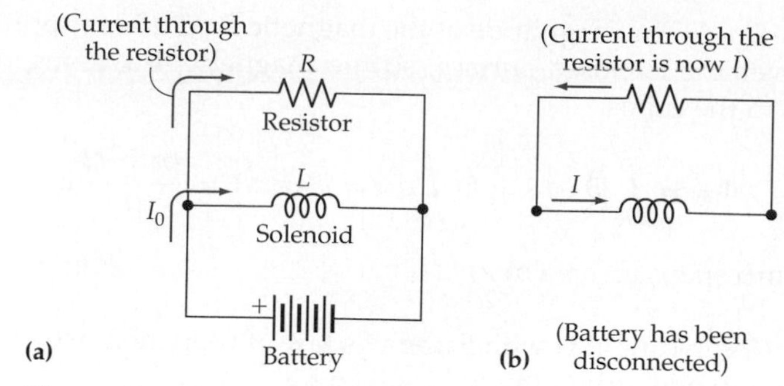

Figure E14.4
(a) A solenoid coil connected in parallel to a resistor and a battery. After the circuit settles down to a steady state, the solenoid conducts a steady current I_0. (b) If the battery is disconnected at $t = 0$, an emf will be induced in the solenoid that will continue to drive current through the resistor for a while.

the coil. According to equation E14.5, the magnitude of this emf is

$$|\mathscr{E}| = L\left|\frac{dI}{dt}\right| \tag{E14.11}$$

So in this situation, the coil's self-generated emf continues to drive current through both the coil and the resistor after the battery has been removed. How long can this last? Assume that our coil's resistance is very small compared to that of the resistor; then the total resistance in the circuit shown in figure E14.4b is essentially R. The current that flows in this circuit at any given time is thus $I = |\mathscr{E}|/R$. If we solve this for $|\mathscr{E}|$ and plug this into the above, we get

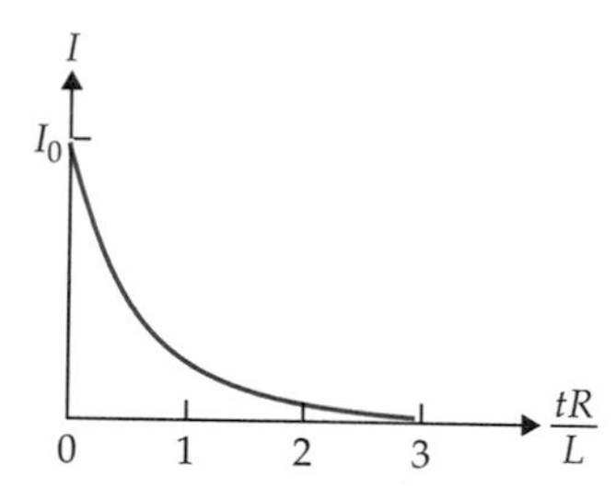

Figure E14.5
A graph of the current as a function of time when an inductor (such as our solenoid) pushes current through a resistor in an effort to maintain the current flowing in the inductor.

$$IR = L\left|\frac{dI}{dt}\right| = L\left(-\frac{dI}{dt}\right) \tag{E14.12}$$

Note that $|dI/dt| = -dI/dt$ in this case because I is decreasing, so dI/dt is negative and $-dI/dt$ is positive. We can rewrite equation E14.12 in the form

$$\frac{dI}{dt} = -\frac{R}{L}I \tag{E14.13}$$

This says that the current I is that function of time whose time derivative is a negative constant times itself. We encountered a similar equation in the context of a discharging capacitor (see equation E5.20). There we found that a decaying exponential was the solution. Similarly, you can verify that in this case, the solution to equation E14.13 is

The formula for the time-dependence of the current

$$I(t) = I_0 e^{-tR/L} \tag{E14.14}$$

Purpose: This equation describes the current I as a function of time t in a coil of inductance L whose self-induced emf is pushing the current through both the coil and a resistor with resistance R connected in series.

Symbols: t is the elapsed time, and I_0 is the current at time $t = 0$.

Limitations: This equation assumes L is independent of I (which *is* generally correct).

Note: A graph of this function is shown in figure E14.5.

Figure E14.6
This photograph shows a toroidal inductor in an electronic circuit.

Exercise E14X.8

Verify that $I(t) = I_0 e^{-tR/L}$ is indeed a solution to equation E14.13.

Exercise E14X.9

Check that R/L has units of inverse seconds.

Just as Lenz's law implies that an inductor opposes any decrease in the current flowing through it, it also opposes any increase in that current. This means that when we connect an inductor to a battery, it can take some time for the current through the inductor to reach its maximum value (see problem E14S.2). Inductors are used in electric circuits to smooth out current variations or as parts of timing circuits. Figure E14.6 shows a photograph of an inductor in an electronic circuit.

E14.5 The Energy in a Magnetic Field

This energy must come from the magnetic field!

So we see that a coil, like a capacitor, can act for a short time as if it were a battery, driving a current through a resistor. But driving a current through a resistor takes energy (which is ultimately converted to thermal energy in the resistor). Where does this energy come from?

It makes sense that this energy is coming from the internal energy of the magnetic field, which is decreasing in magnitude as the current decreases. Let's accept this as a hypothesis and see if we can figure out how much energy is stored per unit volume in a magnetic field.

First we need to determine the total energy that is dissipated by the resistor. At every instant during the decay, the *power* involved in the conversion of electrical energy to thermal energy in the resistor is given by

$$\frac{dU^{\text{th}}}{dt} \equiv P = |\mathcal{E}|I = I^2R = I_0^2(e^{-tR/L})^2R = I_0^2e^{-2tR/L}R \qquad \text{(E14.15)}$$

The total energy stored in the magnetic field of an inductor

You can find the total energy dissipated in the resistor by integrating this expression from $t = 0$ to $t = \infty$: if you do this, you should find that

$$U_{\text{tot}}^{\text{th}} = \tfrac{1}{2}LI_0^2 \qquad \text{(E14.16)}$$

This is, therefore, the energy stored in the magnetic field at time $t = 0$.

Exercise E14X.10

Verify equation E14.16.

Our hypothesis is that this energy came from the internal energy of the magnetic field that was in the solenoid at time $t = 0$. We can relate this to the magnetic field strength most easily if we assume that the coil is a solenoid whose length $\ell \gg$ radius r. In this case, the coil is well approximated by an infinite solenoid, so its magnetic field at time $t = 0$ has a uniform magnitude $\mathbb{B}_0$ throughout its solenoid's interior and is essentially zero outside. According to equation E11.15*a*, the relationship between $\mathbb{B}_0$ and the initial current I_0 is

$$\mathbb{B}_0 = \frac{4\pi kN}{c\ell}I_0 \quad \Rightarrow \quad I_0 = \frac{c\ell\mathbb{B}_0}{4\pi kN} \qquad \text{(E14.17)}$$

If you now plug this into equation E14.16 and substitute for the inductance L the formula given by equation E14.10, you can show that the energy in the solenoid's magnetic field at time $t = 0$ must have been

$$U_0^{\text{field}} = \frac{r^2\ell\mathbb{B}_0^2}{8k} \qquad \text{(E14.18)}$$

Exercise E14X.11

Verify equation E14.18.

The density of energy in the solenoid's magnetic field

But in an *infinite* solenoid, the field is uniform throughout its interior and zero outside. So in the limit that our solenoid becomes infinite, the energy density stored in the magnetic field is simply the total energy that we have just calculated divided by the volume $\pi r^2\ell$ of the solenoid's cylindrical interior

$$u_B = \frac{U_0^{\text{field}}}{\pi r^2\ell} = \frac{1}{\pi r^2\ell}\frac{r^2\ell\mathbb{B}_0^2}{8k} = \frac{\mathbb{B}_0^2}{8\pi k} \qquad \text{(E14.19)}$$

This equation is essentially identical to the equation $u_E = E^2/8\pi k$ for the energy density of an electric field (equation E3.29)! This may not come as a great surprise if you have really been convinced that the electric and magnetic fields are just two aspects of a greater whole, but it is a gratifying sign that this equation is probably correct. Though our "derivation" of this equation does not ensure it, this formula does apply even to a magnetic field that is *not* uniform in space.

Combining equations E14.19 and E3.29, we see that the total energy density in a general electromagnetic field is

A general expression for the energy per unit volume in an electromagnetic field

$$u_{EM} = \frac{1}{8\pi k}(E^2 + \mathbb{B}^2) = \frac{\varepsilon_0 E^2}{2} + \frac{B^2}{2\mu_0} \quad \text{(E14.20)}$$

Purpose: This equation expresses the energy density u_{EM} (energy per unit volume) of an electromagnetic field at a point where the electric and magnetic field magnitudes are E and $\mathbb{B} = cB$, respectively.

Symbols: k is the Coulomb constant, $\varepsilon_0 = 1/4\pi k$, and $\mu_0 = 4\pi k/c^2$, where c is the speed of light.

Limitations: There are no known limitations at the macroscopic level (that is, as long as the quantum nature of the electromagnetic field is unimportant).

This formula will be useful to us in chapter E16.

E14.6 Transformers

An idealized transformer

Consider now two independent, long, skinny solenoid coils of wire, one (which we will call the *primary* coil) consisting of N_1 turns of wire wrapped around a cylindrical form of radius R and length ℓ, and the other (the *secondary* coil) consisting of N_2 turns wrapped directly on top of the first. For the sake of simplicity, imagine that both coils have negligible resistance.

Imagine now that we connect the ends of the primary coil to a source of a sinusoidally varying emf

$$\mathscr{E}_s(t) = \mathscr{E}_0 \sin \omega t \quad \text{(E14.21)}$$

where $\mathscr{E}_s(t)$ is the emf imposed on the primary coil as a function of time and $\mathscr{E}_0$ is a constant specifying the *amplitude* of that oscillating emf. Let us assume for the present that we leave the two ends of the secondary coil unconnected to anything, so that no current can flow through that coil.

If the primary coil's resistance is small, the induced emf is the applied emf

The oscillating emf connected to the primary coil drives an oscillating current through that coil that creates an oscillating magnetic field inside *both* coils. This oscillating magnetic field, in turn, induces an oscillating *opposing* emf in both coils. In the case of the primary coil, this opposing emf acts to limit the current flowing in that coil. Indeed, if the primary coil's resistance is negligible, the opposing induced emf must almost exactly balance the emf imposed by the outside source so that energy is conserved in the coil (negligible resistance means that only a negligible amount of energy is being converted to thermal energy in the wire). Therefore, we must have

$$\mathscr{E}_0 \sin \omega t = \mathscr{E}_s(t) \approx \mathscr{E}_1(t) \quad \text{(E14.22)}$$

where $\mathscr{E}_1(t)$ is the emf induced in the primary coil.

Now, since the same magnetic field goes through each turn of each coil and the turns in each coil have essentially the same area, the magnetic flux Φ_B going through each turn in the primary *and* the secondary has the same value at all times. By Faraday's law of induction, this means that the emf induced in each turn of each coil must be the same:

$$\mathscr{E}_{turn} = -\frac{d\Phi_B}{dt} \quad \text{(E14.23)}$$

Since the total induced emf in the primary coil is $\mathcal{E}_1 = N_1\mathcal{E}_{\text{turn}}$ and the emf in the secondary coil is $\mathcal{E}_2 = N_2\mathcal{E}_{\text{turn}}$, it follows that the emf induced in the secondary coil is related to the emf induced (and imposed) on the primary as follows:

The ratio of induced emf's is equal to the ratio of the number of the coils' turns

$$\frac{\mathcal{E}_2}{N_2} = \mathcal{E}_{\text{turn}} = \frac{\mathcal{E}_1}{N_1} \quad \Rightarrow \quad \mathcal{E}_2(t) = \frac{N_2}{N_1}\mathcal{E}_1(t) \tag{E14.24}$$

We see, therefore, that if we connect a low-resistance primary coil to a source of sinusoidally varying emf, we induce a sinusoidally varying emf in the secondary coil whose magnitude is related to that of the primary-coil emf by the ratio of the number of turns in the two coils. What we have here is a simple example of a **transformer.** A transformer is any device that uses induction to transfer electrical energy from one coil to another while changing the emf at which that electrical energy is delivered.

Why transformers are important

Why is being able to change the emf advantageous? For safety reasons, we would like to deliver electrical power to people at a low enough emf that accidentally touching a bare wire will not certainly cause a person to receive an instantly fatal current. It also turns out to be economically advantageous to generate electrical power at a relatively low emf. On the other hand, shipping power at a low emf over large distances is very wasteful. This is so because the amount of electrical power delivered to a device when we apply an emf $\mathcal{E}$ to it is given by

$$P = |\mathcal{E}|I \tag{E14.25}$$

where I is the current flowing through the device. Therefore, the lower the emf, the greater the current that has to flow to supply a given power. Driving a large current through a wire causes more thermal energy to be wasted in the wire than would be the case if we could supply the same power using lower currents (see problem E14S.4). If we ship the power at a very high emf, only a very small current is needed to carry substantial amounts of energy, and thus less energy is lost to thermal energy in the wires. Interstate power lines therefore carry energy at an emf that may be as high as 500 kV.

Transformers that can shift emf levels thus make it economically feasible to transport electrical power over long distances and yet generate it and deliver it at a usefully low emf. The transformer is therefore one of the core inventions that makes our technological civilization possible.

Transformers depend on alternating current

Note that the transformer design described above assumes that the current in the primary coil is oscillating sinusoidally. The current flowing in a transformer indeed *must* be oscillating to create an oscillating magnetic field that in turn creates the oscillating induced emf; remember that only a *changing* magnetic field can induce an emf. We call a sinusoidally oscillating current an **alternating current** (**AC**). Electrical power is delivered to our homes as alternating current precisely because it is so much easier to use induction-based transformers to shift the emf of an alternating current than it is to shift the emf of a steady current (called a **direct current,** or **DC**).

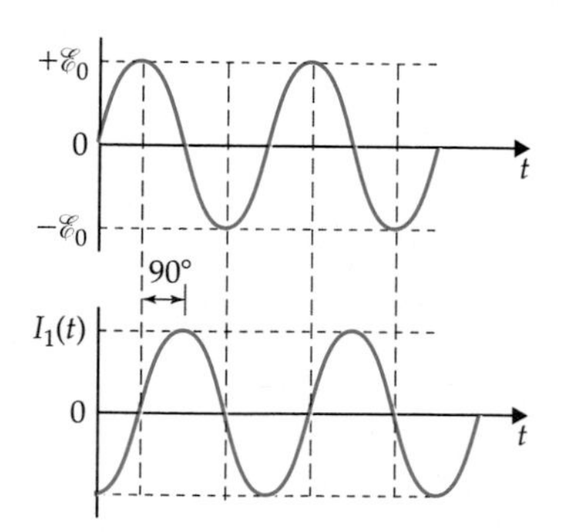

Figure E14.7
The current in the primary of a transformer lags behind the emf by $\frac{1}{4}$ cycle (when no current flows in the secondary).

There are some aspects of transformer design that are interesting to consider more carefully. According to the definition of inductance, the induced emf in the primary (which is equal to the applied emf) is

$$L_1\frac{dI_1}{dt} = \mathcal{E}_1(t) = \mathcal{E}_0 \sin\omega t \tag{E14.26}$$

If we divide both sides by L and integrate with respect to time, we get

$$I_1(t) = -\frac{\mathcal{E}_0}{L_1\omega}\cos\omega t = \frac{\mathcal{E}_0}{L_1\omega}\sin\left(\omega t - \frac{1}{2}\pi\right) \tag{E14.27}$$

since $-\cos\theta = \sin(\theta - \frac{1}{2}\pi)$. This tells us two very interesting things about the current flowing in the primary coil: (1) It lags the applied emf by a quarter-cycle of the sine wave, as shown in figure E14.7 (a full cycle of the wave would correspond to ωt changing by 2π), and (2) the current becomes smaller as the inductance of the coil increases. Because losses to thermal energy in the primary coil are going to be primarily determined by the current it carries, there is a significant advantage to making the inductance of the primary coil as large as possible. For this reason, transformer coils are usually wrapped not around long, empty cylindrical forms but rather around specially shaped iron cores that both greatly increase the inductance of the transformer coils and confine the magnetic fields created mostly to the interior of the iron (see figure E14.8).

High primary inductance keeps primary current (and thus thermal losses) small

The average power that the primary coil absorbs from its source will be given by the average value of

$$P_1(t) = \mathscr{E}_1(t)I_1(t) = -\frac{\mathscr{E}_0^2}{L\omega}\sin\omega t\cos\omega t = -\frac{\mathscr{E}_0^2}{2L\omega}\sin 2\omega t \qquad \text{(E14.28)}$$

The average value of $\sin 2\omega t$ over a cycle is *zero*, since it will spend as much time in positive territory as in negative territory. We see then that because the current lags the emf by a quarter cycle, the primary coil *on the average* does not draw any power (other than small thermal energy losses that we are neglecting) from the source of its emf.

Inductive feedback from the secondary regulates power drawn by the primary

However, this analysis assumes that the transformer *secondary* coil does not conduct any current. When we connect this coil to some kind of load, *it* will begin to conduct an alternating current. This alternating current creates an additional oscillating magnetic field inside both coils that in turn induces an additional emf in both coils. The details are complex to analyze, but the new oscillating magnetic field ultimately has the effect of causing the primary coil to draw additional current from the source of its emf and shifts the time relationship between the current and emf in the primary coil so that the primary actually draws power from the source. This power is then converted to power in the secondary circuit that is ultimately dissipated by whatever that coil is connected to.

The point is that inductive feedback effects ensure that the primary coil draws essentially exactly the power from the source of the primary coil's emf that is required to power the device connected to the secondary coil. A well-designed transformer experiences losses to thermal energy that are typically on the order of 1 percent of the power transformed.

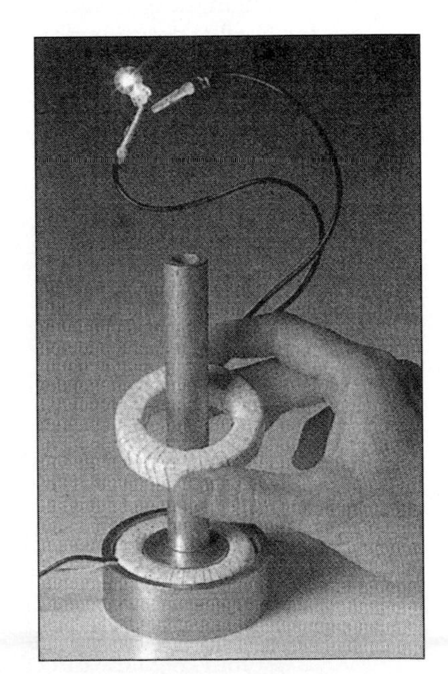

Alternating current flowing through the lower coil creates a sinusoidal magnetic field (mediated by the metal rod) that induces a current in the upper coil, which causes the bulb to glow. This illustrates how energy can be transported from one coil to the other across empty space.

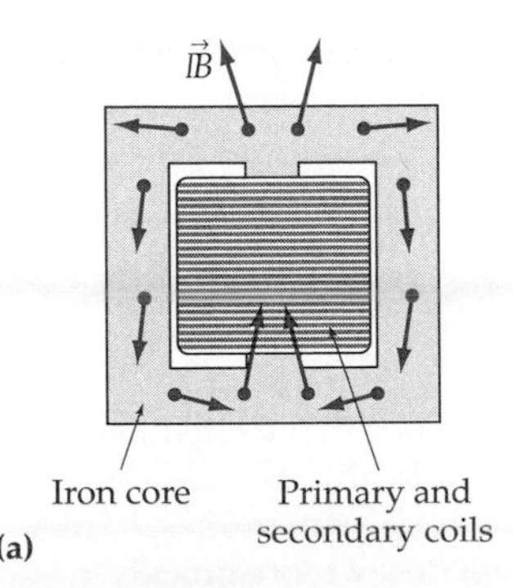

Figure E14.8
(a) A schematic diagram of a common transformer design. The magnetic field created by the coils is mostly confined to the interior of the specially shaped iron core. (b) An actual transformer that uses this design.

TWO-MINUTE PROBLEMS

E14T.1 Imagine that you place a loop with its face perpendicular to a uniform, static magnetic field. If the magnetic field points directly at you as you look at the loop, in what direction does the induced current flow around the loop?
- A. It flows clockwise.
- B. It flows counterclockwise.
- C. There is no induced current.
- D. Something else happens (specify).
- E. There is not enough information to determine the flow direction.

E14T.2 In the situation described in problem E14T.1, imagine that the current is *increasing* in magnitude. What is the direction of the induced current now?
- A. It flows clockwise.
- B. It flows counterclockwise.
- C. There is no induced current.
- D. Something else happens (specify).
- E. There is not enough information to determine the flow direction.

E14T.3 Loop A and a long, straight, current-carrying wire lie near each other on a tabletop in the top view shown in figure E14.9. Loop B is perpendicular to the wire and concentric with it. Assume that the current in the wire suddenly decreases.
(a) In what direction does the induced current flow in loop A?
(b) In what direction does the induced current flow in loop B?

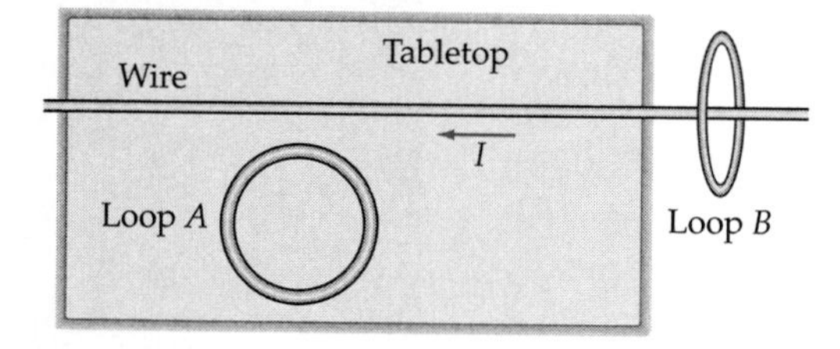

Figure E14.9
The situation discussed in problem E14T.3.

- A. It flows clockwise.
- B. It flows counterclockwise.
- C. There is no induced current.
- D. Something else happens (specify).
- E. There is not enough information to determine the flow direction.

E14T.4 Two identical concentric loops are arranged as shown in figure E14.10. One loop has a steady current flowing through it (provided by the power supply shown). When the power is turned off, in what direction does the induced current flow in loop 2?
- A. It flows clockwise.
- B. It flows counterclockwise.
- C. There is no induced current.

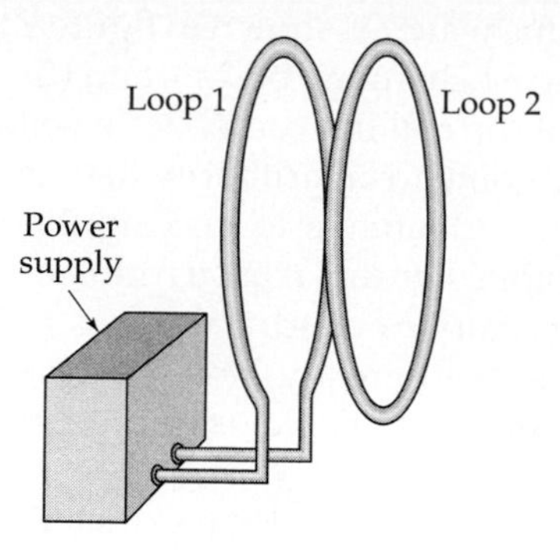

Figure E14.10
The two loops discussed in problems E14T.4 and E14T.5.

- D. Something else happens (specify).
- E. The answer depends on what direction the current was flowing in loop 1.

E14T.5 In the situation described in problem E14T.4 and shown in figure E14.10, will the two loops briefly attract or repel each other when the current is cut off?
- A. They will attract each other.
- B. They will repel each other.
- C. They will neither attract nor repel each other.
- D. The answer depends on the direction current was flowing in loop 1.

E14T.6 Imagine that we wrap a solenoid coil around a hollow cardboard cylinder. Now imagine that we insert an iron core into the solenoid. What will happen to the coil's inductance when we insert the iron core? (Explain.)
- A. The induction increases.
- B. The induction decreases.
- C. The induction is unchanged.
- D. We do not have enough information to determine how the induction changes.

E14T.7 A coil with a self-inductance of 1 H is connected in parallel to a 10-Ω lightbulb and a 5-V battery. Very roughly how long would it keep the lightbulb lighted if the battery were disconnected from the circuit? (Choose the closest answer.)
- A. 100 s
- B. 20 s
- C. 2 s
- D. 0.2 s
- E. 0.02 s

E14T.8 If we increase the resistance in the circuit shown in figure E14.4b, how would this change the rate of decay of the current through the coil?
- A. The current will decay more quickly.
- B. The current will decay more slowly.
- C. There will be no change in the rate of decay.
- D. There is not enough information to answer definitively.

HOMEWORK PROBLEMS

Basic Skills

E14B.1 The face of a loop with a radius of 10 cm is perpendicular to a magnetic field that decreases in magnitude from about 200 MN/C to zero in 0.5 s. What is the magnitude of the induced emf in the loop? If you look at the loop with the magnetic field coming toward you, in what direction (counterclockwise or clockwise) does the current flow around the loop?

E14B.2 The face of a loop with a radius of 3 cm is perpendicular to a magnetic field that increases in magnitude from zero to about $B = 0.10$ T ($\mathbb{B} = 30$ MN/C) in 1.0 ms. What is the magnitude of the induced emf in the loop? If you look at the loop with the magnetic field coming toward you, in what direction (counterclockwise or clockwise) does the current flow around the loop?

E14B.3 The magnetic field *inside* a long solenoid is approximately uniform, while the magnetic field outside the solenoid is approximately zero. Imagine that we have a solenoid that is 3.0 cm in diameter and has an internal field strength of $B = 0.17$ T ($\mathbb{B} = 50$MN/C). This solenoid goes through the center of a square loop 10 cm on a side, with the axis of the solenoid perpendicular to the loop. What is the magnetic flux going through the square loop?

E14B.4 Imagine that a horizontal conducting ring falls toward the south pole of a vertical bar magnet. As we look at the ring from above, which way does the induced current flow? Does the bar magnet exert a force on the falling ring? Explain your reasoning carefully.

E14B.5 Imagine that I have two coils of wire arranged as shown in figure E14.11. When I connect the battery to deliver power to coil A, which end (left or right) of the lightbulb connected to coil B becomes positive? What about when I disconnect the battery? What about when coil A conducts a steady current?

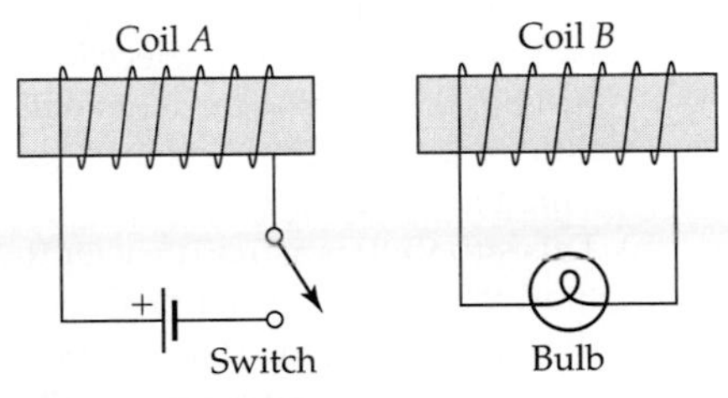

Figure E14.11
The coils discussed in problem E14B.5.

E14B.6 A coil has 1000 turns of wire wrapped around a cylindrical form 1.0 cm in diameter and 10 cm long. What is the approximate inductance L of such a coil?

E14B.7 A coil has an inductance of 0.5 H. It is conducting a current of 10 mA. If this current drops to zero within a time span of 0.02 s, what is the emf induced in the coil?

E14B.8 Imagine that we connect a 10-mH coil to a battery through a resistor, and it settles down to conducting 1.0 A. How much energy is now stored in its magnetic field?

E14B.9 The magnetic field between the poles of an extremely strong horseshoe magnet has a magnitude of $B = 1.0$ T ($\mathbb{B} = 300$ MN/C), which is fairly well confined to a region about 5 cm in diameter and 2 cm long. Estimate the total energy in this magnetic field.

Synthetic

E14S.1 Imagine that we have a coil consisting of N turns of wire shaped like a circular loop with radius R. We place this coil in a uniform magnetic field that points perpendicular to the face of the coil. Imagine that the field strength decreases exponentially in time according to the formula

$$\mathbb{B}(t) = \mathbb{B}_0 e^{-at} \tag{E14.29}$$

where a is a constant. Find an expression for the induced emf in the loop as a function of time as a function of N, R, $\mathbb{B}_0$, and a (and whatever other constants you might need).

E14S.2 Self-inductance in a coil means that it also takes time to get a current started in a coil, because (by Lenz's law) as the current in the coil increases, an emf will be created to oppose the change. If we connect a resistor with resistance R in series with a coil with negligible resistance and inductance L and a battery with voltage V_B, show that the current flowing through the coil as a function of time is

$$\mathbb{B}(t) = I(t) = \frac{V_B}{R}(1 - e^{-tR/L}) \tag{E14.30}$$

[*Hints:* The potential difference across the resistor plus the emf across the coil must be equal to that across the battery. You might also find it helpful to define a new variable $H(t) = V_B/R - I$: if you do this, note that $dH/dt = -dI/dt$.]

E14S.3 Imagine that you want to wind a solenoid that can keep a 10-Ω lightbulb lighted longer than 1 s after the circuit shown in figure E14.4b is disconnected from the battery (the bulb replaces the resistor in that circuit). If the form around which you wrap the coil has a radius of 5 cm, and you can wrap 2000 turns of superconducting (zero-resistance) wire around each linear meter of the form, how long will you have to make the form? Will this solenoid be approximately infinite (in the sense that its length is much greater than its radius)?

E14S.4 Imagine that we connect a given device (such as a motor or a lightbulb) to a power supply (ps) with an emf $\mathscr{E}_{ps}$ by two long wires, each with a given resistance R. Note that because of potential differences across the wires, the emf $\mathscr{E}_{dev}$ delivered to the device will be smaller than $\mathscr{E}_{ps}$. Argue that the power that is wasted in the wires in delivering a certain fixed power to the device is proportional to $1/\mathscr{E}^2_{dev}$. (This is why power companies like to ship power over long distances at the highest possible emf.)

E14S.5 Imagine that we have somehow set up a static magnetic field that in the xy plane points in the $+z$ direction and decreases linearly with increasing x according to

$$\mathbb{B}_z = \mathbb{B}_0\left(1 - \frac{x}{L}\right) \qquad \text{(E14.31)}$$

between $x = 0$ and $x = L$. Imagine that we also have a square loop with sides of length $w \ll L$ lying flat in the xy plane whose electrical resistance is R. The situation is shown in figure E14.12.

(a) Use Lenz's law and energy concepts to argue qualitatively that if we pull on this loop, we will have to exert some force F to cause the loop to move through this field in the $+x$ direction, even at a constant speed v. (*Hint:* Newton's laws will not particularly help you here. Think about energy flows. If you are pulling on the loop and moving it through a displacement, you are transferring energy to the loop, right? Where is this energy going and why?)

(b) Now let's make this discussion quantitative. Explain carefully why the rate at which energy is transferred to the loop from your hand is Fv (assuming the force directly opposes the displacement), and argue carefully that this must be equal to $\mathscr{E}^2_{loop}/R$.

(c) Show that in this case, as the loop moves at speed v in the $+x$ direction, the rate of change of flux through this loop is given by

$$\frac{d\Phi_{\mathbb{B}}}{dt} = \frac{\mathbb{B}_0 w^2}{L} v \qquad \text{(E14.32)}$$

(d) By combining the results from parts (b) and (c), show that the magnitude of the force must be

$$F = \frac{\mathbb{B}_0^2 w^4}{RLc^2} v \qquad \text{(E14.33)}$$

(e) Explain qualitatively (and separately) why it makes good physical sense that the force *should* increase as $\mathbb{B}_0$, w, and v increase.

(f) Explain qualitatively (and separately) why it makes good physical sense that the force should decrease as R or L increases.

(g) Now imagine moving the loop at a constant speed v in the $-x$ direction from $x = L$ toward $x = 0$. Will you have to push on the loop to do this, or will you have to pull back to prevent it from speeding up? Will the force you apply at a given speed be the same as when you pull it out?

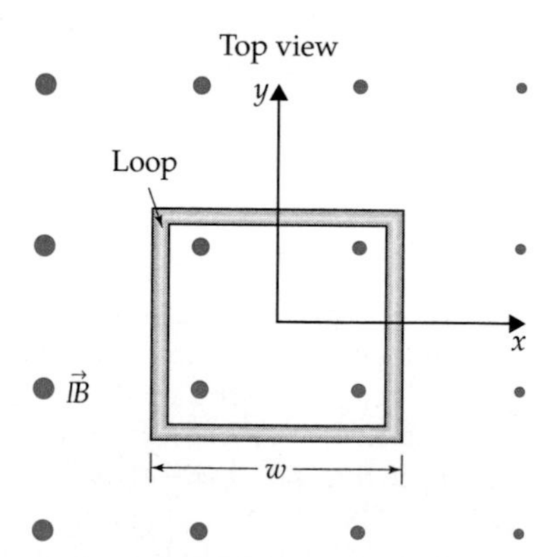

Figure E14.12
This diagram shows a square loop in a magnetic field whose magnitude decreases linearly as we go to the right. How much force is required to move this loop at speed v in the +x direction? (See problem E14S.5.)

E14S.6 (Adapted from an example in Lea and Burke, *Physics*, Brooks/Cole, Belmont, Calif; 1997.) Consider a horseshoe magnet having square poles that are separated by a very small gap. Imagine that we can model the magnetic field in the gap as having an essentially uniform magnitude $\mathbb{B}_0$ in a square region whose sides have a length of approximately L and essentially zero magnitude outside this region. Now imagine that we pull an aluminum plate through the gap with speed v, as shown in figure E14.13a. Assume that the aluminum plate is much larger than the rectangular region, has thickness w, and has a conductivity of σ_c.

(a) Look at this situation in the frame of the plate. By considering some fixed amperian loops in the plate, argue *qualitatively* that the passing magnet must induce electric currents in the plate and that these currents will *oppose* the motion of the passing magnet. (Such currents are called *eddy currents.*)

(b) This situation is actually easier to analyze quantitatively in the frame of the magnet. Assume that we define coordinates so that the plate moves in the $+x$ direction and the magnetic field points in the y direction. As the plate drags charge carriers through the magnetic field, the carriers experience magnetic forces that push them in the $+z$ direction through the region where $\mathbb{B} \neq 0$, as shown in figure E14.13b. Argue that the magnetic force exerted on a charge

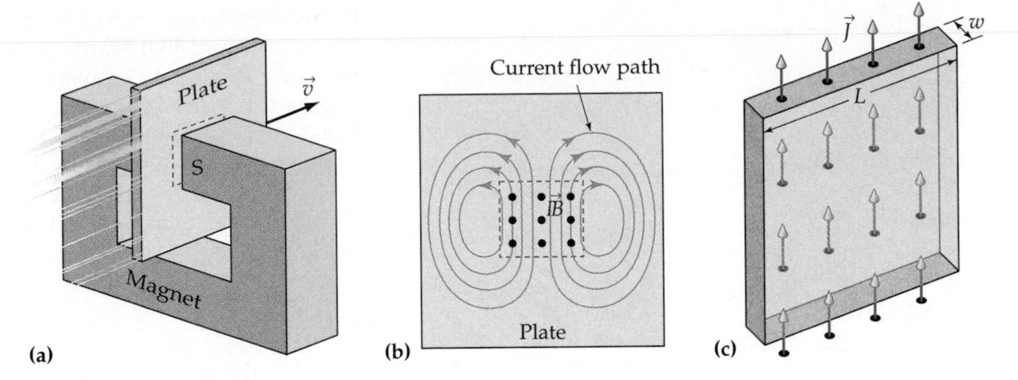

Figure E14.13
(a) An aluminum plate being pulled through the gap between the square poles of a horseshoe magnet. (b) As the plate moves to the right, the magnetic field in the square region enclosed by the dotted line exerts a magnetic force on charge carriers within that region that pushes conventional current upward in this region. This *eddy current* curves back around outside this region to complete the circuit. (c) This diagram shows just the part of the plate where the magnetic field is nonzero. The current density $\vec{J}$ will be essentially uniform and upward throughout the volume of this region. If the magnet's poles have sides of length L, this region will be a flat box with sides L and a thickness equal to the plate's thickness w. The current goes into the box's bottom face and emerges from its top face.

carrier with charge q in this region is given by

$$F_m = \frac{qv\mathbb{B}_0}{c} \quad \text{(E14.34)}$$

(c) Equation E4.11 says that when it is an electric field $\vec{E}$ that exerts forces on charge carriers in a conductor, the current density is $\vec{J} = \sigma_c \vec{E}$. Argue by analogy that in this case, the magnitude of the current density within the region in the aluminum where $\mathbb{B} \neq 0$ should be

$$J \approx \frac{\sigma_c v \mathbb{B}}{c} \quad \text{(E14.35)}$$

(d) This current flows in the bottom and out the top of the rectangular volume in the aluminum block where $\mathbb{B} \neq 0$, as shown in figure E14.13c (the current completes the circuit by flowing in the opposite direction outside of this region, as illustrated in figure E14.13b). Argue that the total current flowing through this region is therefore

$$I \approx \frac{\sigma_c v w L}{c} \mathbb{B} \quad \text{(E14.36)}$$

(This estimate is actually too large, as the surface charges that redirect the current outside of the region where $\mathbb{B} \neq 0$ somewhat oppose the current flowing inside the region. A detailed calculation shows that the actual current is smaller than this estimate by very roughly a factor of 2, so equation E14.36 yields only a rough approximation.)

(e) We can consider this rectangular volume of aluminum as a chunk of wire carrying a current in a magnetic field. Argue that if the current flows in the $+z$ direction and has the magnitude given by equation E14.36, the aluminum plate must experience a force in the $-x$ direction with magnitude

$$F_{\text{eddy}} \approx \frac{\sigma_c \mathbb{B}^2 V v}{c^2} = \sigma_c B^2 V v \quad \text{(E14.37)}$$

where V is the volume of metal immersed in the magnetic field. As noted previously, this is only an estimate, but it should be good to within a factor of 2 or so.

(f) Assume that the horseshoe magnet's pole faces are 4 cm on a side, the magnetic field strength in the region between the poles is 0.1 T, the speed of the plate is 10 cm/s, and its thickness is 0.5 cm. Estimate the magnetic force on the plate.

E14S.7 Imagine that a coil consisting of 75 turns of wire shaped as a circular loop with a radius of 12 cm is placed with its face perpendicular to the direction of an alternating (sinusoidal) magnetic field whose frequency is 60 Hz and whose maximum magnitude is roughly $B_{\text{max}} = 0.13$ T ($\mathbb{B}_{\text{max}} = 40$ MN/C). Will this coil be able (in principle) to illuminate a normal 120-V lightbulb? (*Note:* The emf of "120-V" household alternating current actually oscillates sinusoidally between ± 170 V. Technically 120 V is the square root of the average of the squared emf, the square ensuring that what you are averaging is always positive.)

Rich-Context

E14R.1 Consider the circuit shown in figure E14.14. Assume that the coil has a huge inductance (so that $LR \approx 1$ s) and that both the coil and the resistor

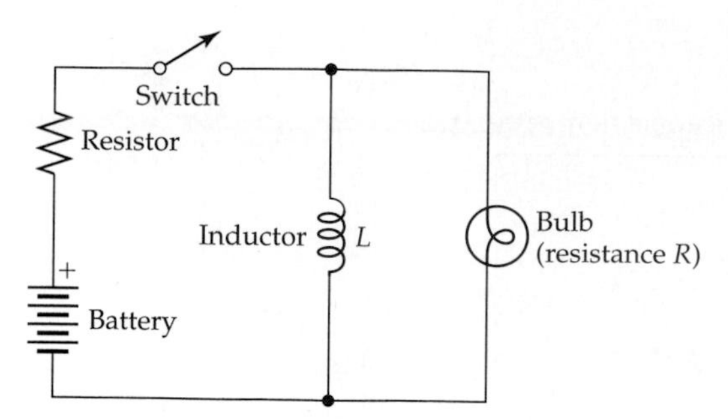

Figure E14.14
The circuit considered in problem E14R.1.

have a resistance about 20 times smaller than that of the bulb. Use Lenz's law and the idea of self-inductance to answer the following questions. *Be sure to explain your reasoning in detail in each case.*

(a) Predict what will happen to the bulb's brightness when the switch is closed. (*Hint:* Before the switch is closed, the coil has zero current flowing through it, and the coil is initially going to want to keep it that way when the switch is closed. How can the coil arrange to have essentially zero current flowing through it? If we define the coil's effective resistance to be the potential difference across it divided by the current flowing through it, what is the coil's effective resistance at the instant that the switch is closed? How does this effective resistance change with time? You might study problem E14S.2 for some ideas.)

(b) Predict what will happen to the bulb's brightness when the switch is opened.

(c) When the switch contacts are opened, one often sees bright sparks arcing across the gap between the opening contacts. Why?

E14R.2 A kilogram of gasoline contains about 46 MJ of useful energy. Anyone who can develop an energy storage device that can store electrical energy with anything approaching the same density as gasoline will be the world's richest person (and save the planet, too!). An inductor stores electrical energy. Estimate the magnitude of the magnetic field inside a long solenoid that would store energy at about the same density as gasoline. Is this a reasonable approach toward solving the world's energy problems? (*Hint:* Gasoline floats on water, so it must be somewhat less dense; but as anyone who has carried a gasoline can will tell you, it is not *much* less dense. Make a suitable estimate: the exact answer is not important.)

Advanced

E14A.1 Calculate the inductance of a toroidal coil having N turns whose inner radius (the radius of the "donut hole") is R_1 and whose outer radius is R_2. (See problem E11S.8 for a discussion of the magnetic field inside a toroidal coil. You will have to do an integral in this case to calculate the total magnetic flux through the toroid's interior.)

E14A.2 Design a transformer whose coils are wrapped around a cylindrical form 20 cm long and 3.0 cm in diameter that changes 120-V household alternating current in the primary coil to 6.0 V in the secondary coil. Assume that you use an iron core that increases the inductance of each coil by a factor of 20 above the value it would have if the cylindrical form were hollow. Assume also that you must use #26 copper wire (which has a diameter is 0.41 mm) for both coils. Your secondary coil must be able to provide 300 mA of current without the emf at its ends dropping below 5.7 V.

(a) How many turns must the primary coil have?

(b) About how much power is wasted in the primary even when the secondary is disconnected?

(c) How could the design be improved if we could use wire having a different diameter?

(*Hints:* Use the infinite solenoid approximation. You can also *usually* treat alternating currents and emf's in Ohm's law and power calculations as if they were ordinary direct currents: *assume* this. Be sure to state any other approximations that you use.)

ANSWERS TO EXERCISES

E14X.1 By definition, $\vec{\mathbb{B}} = c\vec{B}$. Therefore

$$\Phi_{\mathbb{B}} \equiv \int \vec{\mathbb{B}} \cdot d\vec{A} = \int c\vec{B} \cdot d\vec{A} = c \int \vec{B} \cdot d\vec{A} \equiv c\Phi_B \tag{E14.38}$$

E14X.2 $I = \mathcal{E}/R = 13\text{ mV}/0.10\ \Omega = 130\text{ mA}$.

E14X.3 *Model:* We have some freedom to decide which way the loop's segment vectors go around the loop. Let us choose to orient these vectors so that they point counterclockwise around the loop as it is shown in figure E14.3. *Solution:* If I curve my right fingers in the direction that these vectors go around the loop, my right thumb points in the $-x$ direction. This means that since the magnetic field points mostly in the $+x$ direction, $\vec{\mathbb{B}} \cdot d\vec{A}$ for all tiles and thus $\Phi_{\mathbb{B}}$ will be *negative*. Since the field strength is increasing, $\Phi_{\mathbb{B}}$ becomes increasingly negative with time, so $d\Phi_{\mathbb{B}}/dt$ is also negative. Equation E14.2 then says that $\mathcal{E}_{\text{loop}}$ is *positive*, so it will drive current counterclockwise around the loop in the same direction as the $d\vec{s}$ vectors. *Evaluation:* This is the same direction that we deduced using Lenz's law (see figure E14.3b).

E14X.4 Lenz's law implies that the induced current will again resist the change, this time by flowing in a clockwise direction to create a magnetic field in the $+z$ direction to support the failing magnetic field from the departing bar magnet. This means that the loop will attract the magnet.

E14X.5 Looking at equation E14.5, we see that the inductance must have units of emf over current per time, or $\text{V} \cdot \text{s/A}$. Since $1\ \Omega = 1\ \text{V/A}$, $1\ \text{H} = 1\ \text{V} \cdot \text{s/A} = 1\ \Omega \cdot \text{s}$.

E14X.6 Plugging numbers into equation E14.5 yields about 100 V.

E14X.7 Plugging numbers into equation E14.10, we get about 0.25 mH.

E14X.8 Taking the time derivative of the proposed solution $I(t)$, we get

$$\frac{dI}{dt} = \frac{d}{dt}(I_0 e^{-tR/L}) = I_0 e^{-tR/L}\left(-\frac{R}{L}\right)$$
$$= -\frac{R}{L}(I_0 e^{-tR/L}) = -\frac{R}{L}I \qquad \text{(E14.39)}$$

This is what equation E14.13 says we should get, so this is a solution to that equation.

E14X.9 According to the answer to exercise E14X.5, $1\ \text{H} = 1\ \Omega \cdot \text{s}$, so R/L has units of $\Omega/(\Omega \cdot \text{s}) = \text{s}^{-1}$. This is necessary, because the argument $-tR/L$ in the exponent has to be a unitless number.

E14X.10 Integrating, we get

$$U^{\text{th}}_{\text{tot}} = \int_0^\infty \frac{dU^{\text{th}}}{dt}\, dt = \int_0^\infty I_0^2 e^{-2tR/L} R dt$$
$$= I_0^2 R\left[-\frac{L}{2R}e^{-2tR/L}\right]_0^\infty = -\frac{1}{2}L I_0^2 (0-1)$$
$$= \frac{1}{2}L I_0^2 \qquad \text{(E14.40)}$$

E14X.11 Following the suggested steps, we get

$$U_0^{\text{field}} = U^{\text{th}}_{\text{tot}} = \frac{1}{2}L I_0^2 = \frac{1}{2}\frac{4\pi^2 k r^2 N^2}{c^2 \ell}\left(\frac{\mathbb{B}_0 c l}{4\pi k N}\right)^2$$
$$= \frac{4\pi^2 k r^2 N^2}{2c^2\ell}\frac{\mathbb{B}_0^2 c^2 \ell^2}{16\pi^2 k^2 N^2} = \frac{r^2 \ell}{8k}\mathbb{B}_0^2 \qquad \text{(E14.41)}$$

E15 Introduction to Waves

▷ Electric Field Fundamentals
▷ Controlling Currents
▷ Magnetic Field Fundamentals
▷ Calculating Static Fields
▽ Dynamic Fields
The Electromagnetic Field
Maxwell's Equations
Induction
Introduction to Waves
Electromagnetic Waves

Chapter Overview

Introduction

Maxwell's crowning achievement was his discovery that his equations allowed for the possibility of waves moving through an electromagnetic field. In chapter E16 we will explore the nature of these wavelike solutions of Maxwell's equations. When we talk about "electromagnetic waves," though, we are really constructing an analogy to mechanical waves (such as water waves) with which we have more experience. The purpose of *this* chapter is, therefore, to discuss how we can describe such waves physically and mathematically, so that we can better appreciate the analogy.

Section E15.1: What Is a Wave?

In general, a **wave** is *a disturbance that moves through a medium while the medium remains essentially at rest*. Examples include **water waves, sound waves, tension waves** on a vibrating string or spring, **seismic waves,** and "the wave" at a stadium.

In this chapter, we will focus primarily on **mechanical waves,** where the disturbance involves some kind of temporary displacement of the medium. Almost all such waves can be classified as being either **transverse** or **longitudinal,** which involve displacements of the medium that are *perpendicular* to or *parallel to* the wave motion, respectively. Such mechanical waves carry energy.

Section E15.2: A Sinusoidal Wave

Fourier's theorem states that a wave of any shape can be treated as a superposition of **sinusoidal waves.** Therefore, if we fully understand how sinusoidal waves behave in a given situation, we essentially understand how *any* wave would behave.

In this unit, we will consider only **one-dimensional waves,** waves whose disturbance function $f(t, x)$ depends on time and only one spatial coordinate. The equations describing a one-dimensional sinusoidal wave and its associated quantities are

$$f(t, x) = A \sin(kx - \omega t) \tag{E15.7}$$

$$\text{where} \quad k \equiv \frac{2\pi}{\lambda} \qquad \omega \equiv \frac{2\pi}{T} \qquad f = \frac{\omega}{2\pi} = \frac{1}{T} \tag{E15.8}$$

Purpose: These equations describe an idealized one-dimensional sinusoidal traveling wave that varies with time t and position x along the x axis.

Symbols: $f(t, x)$ quantifies the "disturbance" the wave represents at point x at time t, A is the wave's **amplitude,** k (not the Coulomb k!) its **wavenumber,** λ its **wavelength,** ω its **angular frequency,** T its **period,** and f its **frequency** (don't confuse this with the disturbance function).

Limitations: This is an idealization of a real wave.

The value of the disturbance oscillates from $+A$ to $-A$, where A is the wave's *amplitude*. The **wavenumber** k expresses (in radians per meter) how rapidly the wave oscillates with increasing position at a given instant. It is related to the **wavelength** λ, which specifies the distance between wave crests at a given instant. The **angular frequency** ω specifies (in radians per second) how rapidly the wave oscillates with increasing time at a given position. It is related to the **frequency** f of the wave (the number of complete oscillations per unit time at a given position) and the wave's **period** T (which is the time required for a complete oscillation at a given position).

Section E15.3: The Phase Velocity of a Wave

Another important feature of the sinusoidal wave $f(t,x) = A\sin(kx - \omega t)$ is that its shape moves in the $+x$ direction as time passes. Many of the waves we encounter in nature are **traveling waves** of this type. In this section, we see that a given crest of a sinusoidal wave moves in the $+x$ direction with a **phase speed** v of

$$v = \frac{\omega}{k} = \frac{\lambda}{T} = \lambda f \tag{E15.12}$$

Purpose: This equation describes how we can calculate a sinusoidal wave's phase speed v from information about its angular frequency ω, its wavenumber k, its wavelength λ, its period T, and/or its frequency f.

Limitations: This expression applies only to sinusoidal traveling waves.

(The wave's **phase velocity** is a vector that specifies the direction as well as the rate of the motion.)

Section E15.4: The Wave Equation

One of the most important equations in physics is the **wave equation:**

$$0 = \frac{1}{v^2}\frac{d^2f}{dt^2} - \frac{d^2f}{dx^2} \tag{E15.14}$$

Purpose: If this equation (where v is a constant independent of t and x) accurately describes the behavior of a disturbance $f(t,x)$ in a medium, that disturbance will travel through the medium as a traveling wave moving in the $\pm x$ direction with speed v.

Limitations: This equation applies only to cases where the disturbance depends only on one spatial coordinate x.

Note: This equation assumes that when we evaluate the derivative of $f(t,x)$ with respect to one of the variables t or x, we treat the other variable as if it were a constant.

A medium where disturbances obey this wave equation has a number of nice properties: (1) the medium supports sinusoidal traveling waves, (2) it obeys the principle of superposition, and (3) waves preserve their shape as they move.

The **wave rule** allows us to identify such media quickly:

> A medium obeys the wave equation if a sudden disturbance f_{i-1} at a given point in the otherwise undisturbed medium causes the disturbance value f_i at a point a distance Δx away to *accelerate in direct proportion to* $f_{i-1}/\Delta x^2$ (in the limit that we take $\Delta x \to 0$ while holding the medium's macroscopic properties constant). The constant of proportionality is the squared wave speed: $d^2f/dt^2 = v^2(f_{i-1}/\Delta x^2)$.

We will find this useful in chapter E16.

E15.1 What Is a Wave?

Drop a pebble in a still pond; the splash of the pebble creates a series of concentric ripples that move out from the disturbance at a sedate and constant pace. When these ripples arrive at the location of a small object floating in the pond (such as a leaf or small stick) some distance away, they cause the object to bob up and down. The fact that the object bobs up and down instead of being swept in the direction of the wave's motion indicates that the pond water that carries the wave does not substantially move along with the wave. The waves move *through* the medium of the water: while the water itself is disturbed by the passing wave and moves slightly in response to it, there is no net displacement of the water in the direction of the wave.

There are many kinds of waves in the natural world

A **wave** in general can be described as being *a disturbance that moves through a medium while the medium remains basically at rest*, at least compared to the velocity of the wave. Examples of such waves in nature are abundant: **water waves** (from tiny ripples to tsunamis), **sound waves** (from tiny whispers to explosion shock waves), **tension waves** on a vibrating string or spring, **seismic waves** that radiate through the earth's crust from an earthquake, and so on. Figure E15.1 shows some everyday waves.

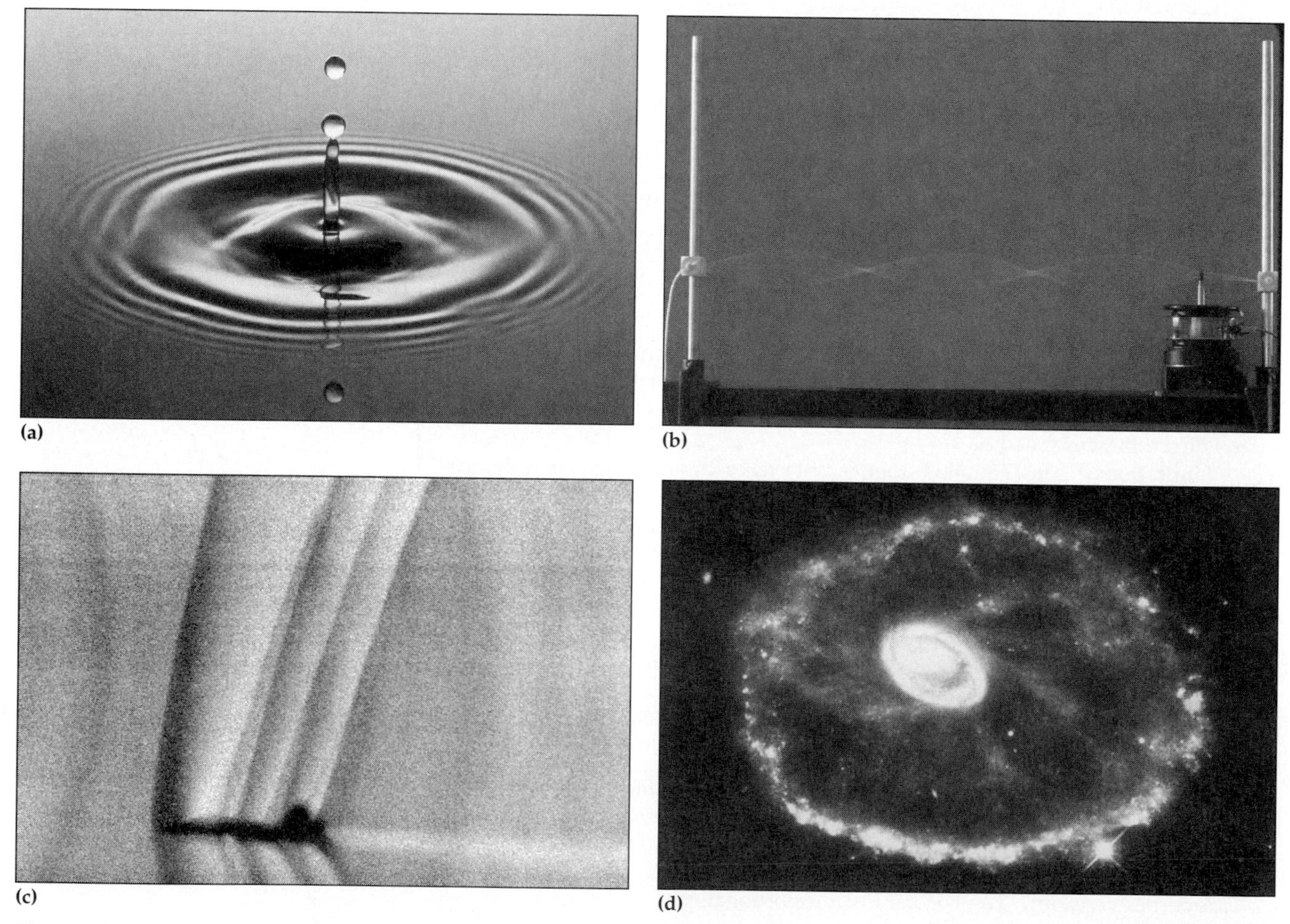

Figure E15.1
Various examples of waves. (a) Circular ripples on the surface of a body of water. (b) Waves on a vibrating string. (c) A Schlieren photograph of the shock waves in the air surrounding a supersonic jet. (d) A Hubble photograph of the Cartwheel galaxy. A head-on collision with another galaxy has caused a circular shock wave to move radially outward through the galaxy's gas. As the shock wave passes, it compresses the gas, triggering a burst of star formation just behind the wave's leading edge.

The list given above by no means exhausts the kinds of waves that occur in nature. A crowd doing "the wave" in a stadium provides a good example of a disturbance that moves through a medium (in this case, the human beings involved in the wave) without a net motion of the medium in the direction of the wave's motion. If you observe a traffic jam from a helicopter, you can sometimes see waves of disturbance radiate through obstructed traffic at speeds much higher than that at which any of the individual cars are moving. Recently, astrophysicists have discovered that star formation in galaxies often moves in waves away from some disturbance in the galaxy's structure (say as the result of a collision with another galaxy). The growth of cells in a petri dish can sometimes proceed in waves. The list goes on and on.

Indeed, waves occur so commonly in the physical world and in such a wide variety of contexts that a general study of wave behavior is an indispensable part of a scientist's education. Studying wave behavior in this course would be worthwhile in and of itself even if we weren't using it here as a stepping stone for the study of electromagnetic waves.

While waves of star formation or biological growth are definitely "disturbances in a medium," we will focus in the next few sections on **mechanical waves,** where the disturbance is some kind of temporary *physical displacement* of the medium. Almost all such waves can be classified as being either **transverse** or **longitudinal** waves. A transverse wave causes the medium to displace in a direction *perpendicular* to the direction of the wave motion. A ripple generated on a rope by a sideways flick of the wrist and "the wave" in a stadium are examples of transverse waves. A longitudinal wave causes the medium to move back and forth *parallel* to the direction in which the wave is moving. Sound waves (which are waves of compression and rarefication in air) and/or "car waves" in a traffic jam are examples of longitudinal waves. Transverse and longitudinal waves are illustrated in figure E15.2.

Water waves are somewhat peculiar in that as a wave passes, a given "piece" of water actually moves in a small vertical circle around its rest position (see figure E15.3). Those of you who have played in the surf at a beach know that in front of a wave crest, water moves backward toward the wave and upward as the crest approaches, but after it has passed, the water moves forward with the wave and downward. Thus water waves exhibit *both* longitudinal and transverse motions (though the net displacement of the water after the wave has passed is still zero). Most mechanical waves, though, are either clearly longitudinal or clearly transverse.

Waves carry energy

One of the most important features of mechanical waves is that they carry not only information that a disturbance has occurred but also *energy* away from the disturbance. For example, the water waves moving away

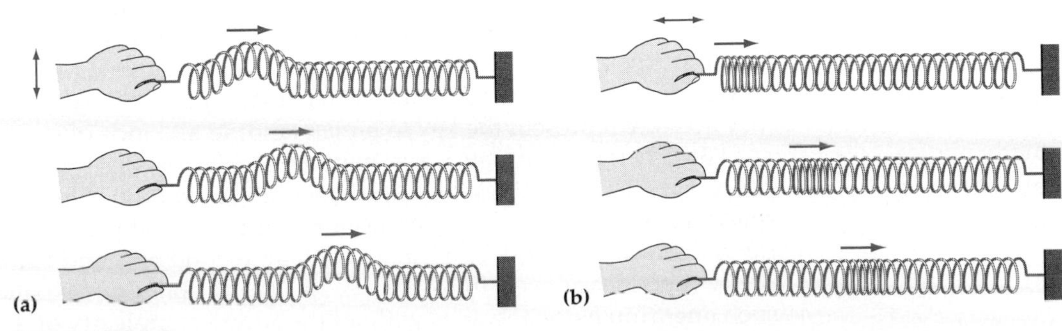

Figure E15.2
(a) A transverse wave moving along a stretched spring. As the wave passes, each element of the spring displaces perpendicular to the wave's motion. (b) A longitudinal wave on a stretched spring. As the wave passes, each element of the spring displaces parallel to the wave's motion.

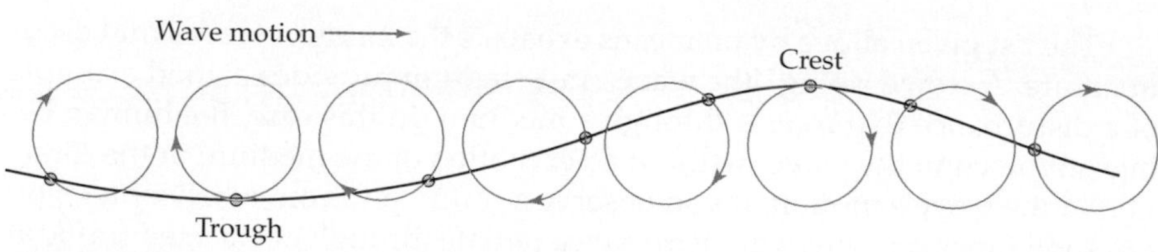

Figure E15.3
This diagram illustrates how particles on the surface of a body of water go around in nearly circular paths as a water wave passes.

from a splash can cause a distant floating bottle to bob up and down as the waves pass; the waves thus transfer energy from the splash and convert it to kinetic energy in the bobbing bottle.

Exercise E15X.1

An earthquake occurs when part of the earth's crust suddenly slips relative to its surroundings. Such an event radiates energy in the form of two different types of *seismic waves* in the crust of the earth: *P waves* cause the crust to oscillate back and forth toward and away from the earthquake epicenter, and *S waves* cause the crust to oscillate up and down. Which of these types is a transverse wave? Which is a longitudinal wave?

Exercise E15X.2

Describe some evidence that seismic waves carry energy.

E15.2 A Sinusoidal Wave

Why sinusoidal waves are worth studying

A **sinusoidal wave** is a special kind of wave that is especially easy to describe mathematically. Realistic waves are often *approximately* sinusoidal, so a sinusoidal wave represents a convenient simplified model of such waves. But sinusoidal waves are important for another reason. A mathematical theorem called **Fourier's theorem** states that any wave, no matter how complicated in shape or behavior, can be treated as a superposition of sinusoidal waves. This means that if we fully understand how *sinusoidal* waves behave in a given situation, we essentially understand how *any* wave would behave.

Fourier's theorem: any wave = sum of sine waves

Fourier's theorem is an extremely important and useful theorem which you will certainly encounter more than once if you proceed in the study of physics and/or engineering. Its proof, unfortunately, is somewhat beyond our means and would be tangential to our purposes in any case. It is sufficient for our purposes at present for you to understand that not only do sinusoidal waves represent a good approximation to many kinds of real waves, but also they actually represent the key to understanding all kinds of waves.

The general mathematical representation of a wave

We can represent *any* wave mathematically by describing a function $f(t, x, y, z)$ that quantifies the disturbance of the medium at every position in space at every instant of time. In this text, I am essentially going to ignore the y and z coordinates and focus on waves that depend on x and t alone: we can pretty much learn everything we need to know about wave behavior from such **one-dimensional waves** without the added complexity of dealing with the y and z coordinates.

A sinusoidal wave

A one-dimensional sinusoidal wave has the simple mathematical form

$$f(t, x) = A \sin(kx - \omega t) \qquad \text{(E15.1)}$$

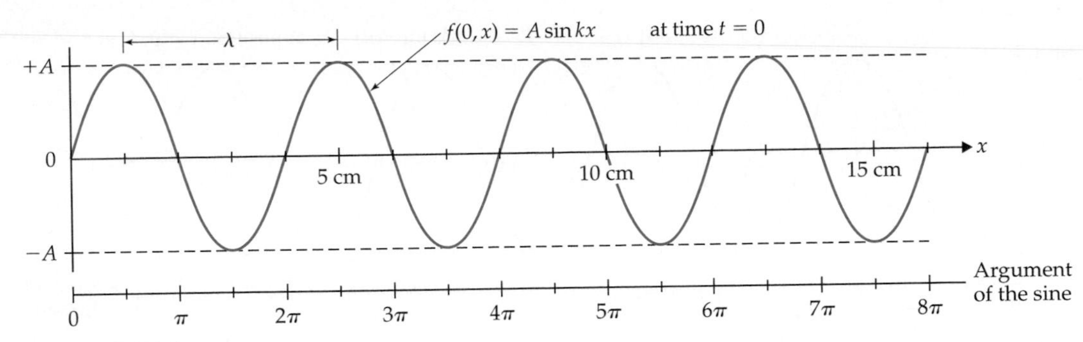

Figure E15.4
A graph of a sinusoidal wave as a function of x at time $t = 0$. In this case, $k = 2\pi/(5\text{ cm}) \approx 1.3\text{ cm}^{-1}$.

where $f(t, x)$ quantifies the disturbance of the medium at time t and position x, and A, k, and ω are constants. (Please do not confuse k in this context with the k that we have previously encountered as the Coulomb constant.) What does this sinusoidal wave look like?

A "snapshot" of the wave helps us define its amplitude, wavelength, and wavenumber

We can take a "snapshot" of this wave at time $t = 0$ by setting $t = 0$ in equation E15.1 and drawing a graph of how the disturbance $f(x)$ depends on x at this time. Such a graph is shown in figure E15.4. Notice how the wave looks like an undulating sequence of hills and valleys (called **crests** and **troughs**). We can see also that the wave disturbance value oscillates between $+A$ and $-A$. The quantity A, which is called the **amplitude** of the wave, thus characterizes the maximum strength of the disturbance.

The distance between two adjacent crests in such a graph is called the **wavelength** λ of the sinusoidal wave. This wavelength is related to the constant k as follows. The first crest of the wave to the right of $x = 0$ occurs where $kx_1 = \pi/2$, as you can see from figure E15.4. The next crest happens when $kx_2 = 5\pi/2 = 2\pi + \pi/2$. The distance between these crests is thus

$$\lambda \equiv x_2 - x_1 = \frac{1}{k}\left(\frac{5\pi}{2} - \frac{\pi}{2}\right) = \frac{2\pi}{k} \tag{E15.2}$$

You can think of the quantity k as expressing the number of radians worth of oscillation the wave goes through in a unit distance:

$$k = \frac{2\pi}{\lambda} = \frac{\text{radians/cycle}}{\text{distance/cycle}} = \frac{\text{radians}}{\text{distance}} \tag{E15.3}$$

This quantity is called the **wavenumber** of the wave.

The wave's behavior at a fixed position defines its period, frequency, and angular frequency

Now let us consider what happens to the wave in time as we watch it from a particular *place*, say, $x = 0$. A graph of the sinusoidal wave as a function of time at $x = 0$ is shown in figure E15.5. Note that we see the wave move up and down between $+A$ and $-A$ as time passes.

The **period** T of the wave is defined to be the time between adjacent crests. By analogy to how we determined the wavelength, you can show that the period is related to ω as follows:

$$T = \frac{2\pi}{\omega} \tag{E15.4}$$

Exercise E15X.3

Verify equation E15.4.

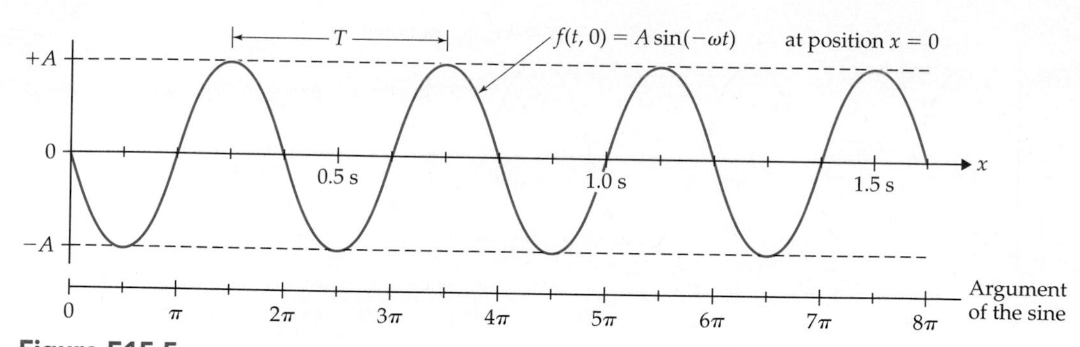

Figure E15.5
A graph of a sinusoidal wave as a function of t at position $x = 0$. In this case, $\omega = 2\pi/(0.5\text{ s}) \approx$ 12.6 (radians) per second.

The quantity ω can be thought of as expressing the number of radians worth of oscillation that the wave moves through per unit time:

$$\omega = \frac{2\pi}{T} = \frac{\text{radians/cycle}}{\text{time/cycle}} = \frac{\text{radians}}{\text{time}} \tag{E15.5}$$

The constant ω is called the **phase rate** (see chapter N11) or more commonly the **angular frequency** of the oscillation.

The ordinary **frequency** f of the oscillation (in cycles per second, or Hz) is defined to be equal to $1/T$:

$$f = \frac{\text{cycles}}{\text{second}} = \frac{1}{\text{seconds/cycle}}$$
$$= \frac{1}{T} = \frac{\omega}{2\pi} \tag{E15.6}$$

[Note that the f here is not related to the function $f(t, x)$ considered earlier.]

Exercise E15X.4

If a sinusoidal water wave has a wavelength of 2.0 cm and a frequency of 2.0 Hz, what are the values (with appropriate units) of the wave's wavenumber k and angular frequency ω?

So in summary, here is the constellation of equations that describe a one-dimensional sinusoidal wave:

A summary of the one-dimensional sinusoidal wave formula and associated quantities

$$f(t, x) = A \sin(kx - \omega t) \tag{E15.7}$$

$$\text{where} \quad k \equiv \frac{2\pi}{\lambda} \qquad \omega \equiv \frac{2\pi}{T} \qquad f = \frac{\omega}{2\pi} = \frac{1}{T} \tag{E15.8}$$

Purpose: These equations describe an idealized one-dimensional sinusoidal wave that varies with time t and position x along the x axis.

Symbols: $f(t, x)$ quantifies the "disturbance" the wave represents at point x at time t, A is the wave's **amplitude,** k (not the Coulomb k!) its **wavenumber,** λ its **wavelength,** ω its **angular frequency,** T its **period,** and f its **frequency** (don't confuse this with the disturbance function).

Limitations: This is an idealization of a real wave.

E15.3 The Phase Velocity of a Wave

Features (such as crests) of a traveling wave move at a specific rate

The wave $f(t, x) = A\sin(kx - \omega t)$ has one other important feature: *it moves* as time progresses. Figure E15.6 shows successive snapshots of such a wave at various different times. You can see in this diagram that a given crest of the wave progresses to the right as time passes. We call a wave whose basic spatial shape is translated in space like this as time passes a **traveling wave:** most of the waves we encounter in nature are traveling waves.

Why does the sinusoidal wave given by $f(t, x) = A\sin(kx - \omega t)$ move like this? Consider a given crest of the wave, say, the first crest to the right

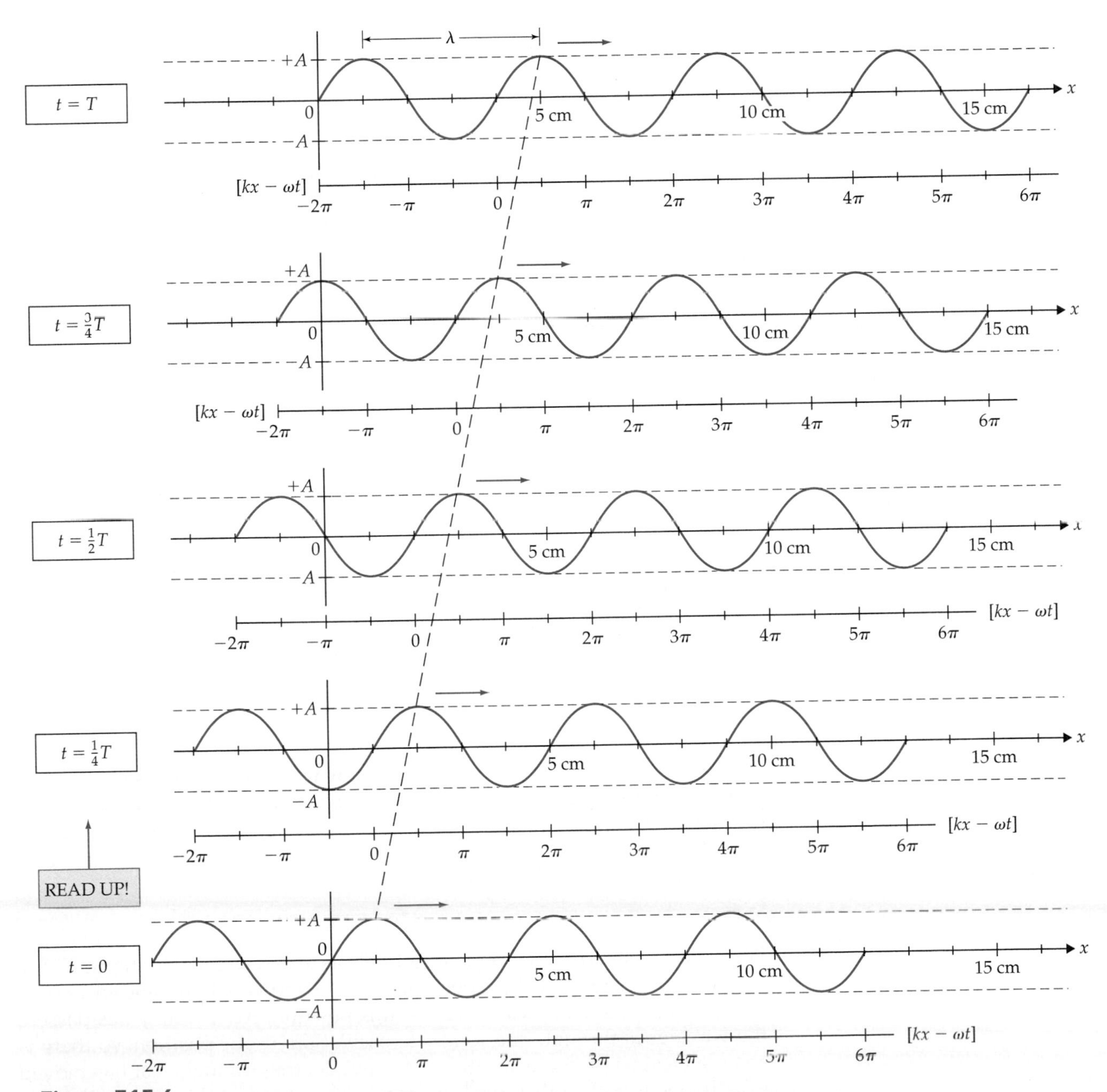

Figure E15.6
A successive series of snapshots of a sinusoidal wave (read from the bottom up!). Note how the crest that was originally at $x = 1$ cm moves to the right as time passes.

of $x = 0$ at time $t = 0$. This particular crest is the place where the argument of the sine (the quantity in the parentheses that the sine function operates on) has the value $\pi/2$ (the first positive angle where sine becomes $+1$). So for all time, the location x_{crest} of *this* particular crest is specified by the condition that

$$\frac{\pi}{2} = kx_{\text{crest}} - \omega t \tag{E15.9}$$

Now, at time $t = 0$, this crest is located where $kx_{\text{crest}} = \pi/2$, that is at $x_{\text{crest}} = \pi/2k$. But as t increases, x_{crest} must also increase in proportion to keep the *difference* in equation E15.9 fixed. In fact, if we take the time derivative of both sides of equation E15.9, we find that

$$0 = k\frac{dx_{\text{crest}}}{dt} - \omega \quad \Rightarrow \quad \frac{dx_{\text{crest}}}{dt} = +\frac{\omega}{k} \tag{E15.10}$$

This crest thus moves in the $+x$ direction with speed ω/k.

Exercise E15X.5

Show that the crest corresponding to the place where the argument of the sine is $5\pi/2$ moves with the same velocity.

Definition of the phase velocity of a sinusoidal wave

The velocity of a given feature (such as a given crest) of a traveling wave is called the wave's **phase velocity** (don't confuse this with the wave's *phase rate* ω). We see that our sinusoidal traveling wave has a phase velocity in the $+x$ direction whose magnitude (the wave's **phase speed**) is

$$v = |v_x| = +\frac{\omega}{k} \quad \text{(phase speed of our sinusoidal wave)} \tag{E15.11}$$

This equation, in combination with equations E15.8, implies the following relationships between the phase speed and the wave's wavelength, period, and frequency:

$$v = \frac{\omega}{k} = \frac{\lambda}{T} = \lambda f \tag{E15.12}$$

Purpose: This equation describes how we can calculate a sinusoidal wave's phase speed v from information about its angular frequency ω, its wavenumber k, its wavelength λ, its period T, and/or its frequency f.

Limitations: This expression applies only to sinusoidal traveling waves.

We can understand $v = \lambda f$ more intuitively as follows. Consider figure E15.6, and imagine that we sit at the position $x = \pi/2$ and watch the sine wave pass by. At time $t = 0$, there was a crest at this position. After time T has passed, the wave goes through one complete oscillation at our position, so there is again a crest passing our position. Meanwhile, the original crest has moved exactly one wavelength λ ahead in space. The speed of this crest is thus indeed λ/T, as claimed by equation E15.12.

Example E15.1

Problem The sound wave from a flute playing the A above middle C has a frequency of 440 Hz. If sound waves move at a speed of 340 m/s in air at 20°C, what is the approximate wavelength of this wave (assuming it is sinusoidal)?

Solution According to equation E15.12, we have

$$\lambda = \frac{v}{f} = \frac{340\text{ m/}\cancel{\text{s}}}{440\,\cancel{\text{Hz}}}\left(\frac{1\,\cancel{\text{Hz}}}{1\text{ cycle/}\cancel{\text{s}}}\right) = 0.77\text{ m} = 77\text{ cm} \tag{E15.13}$$

Exercise E15X.6

Seismic P waves radiating from an earthquake travel at a speed of very roughly 6 km/s near the earth's surface. If such waves for a given earthquake have a period of 0.2 s, what is their wavelength?

E15.4 The Wave Equation

One of the most important equations in physics is the **wave equation:**

$$0 = \frac{1}{v^2}\frac{d^2f}{dt^2} - \frac{d^2f}{dx^2} \tag{E15.14}$$

The wave equation

Purpose: If this equation (where v is a constant independent of t and x) accurately describes the behavior of a disturbance $f(t, x)$ in a medium, that disturbance will travel through the medium as a traveling wave moving in the $\pm x$ direction with speed v.

Limitations: This equation applies only to cases where the disturbance depends only on one spatial coordinate x.

Note: This equation assumes that when we evaluate the derivative of $f(t, x)$ with respect to one of the variables t or x, we treat the other variable as if it were a constant.[†]

This equation appears again and again in all areas of physics: it accurately describes mechanical disturbances of a stretched string or spring; pressure or density disturbances in solids, liquids, and gases; plasma oscillations in the ionosphere; electrical disturbances in a coaxial cable; surface oscillations on a body of water; some kinds of quantum-mechanical wave

[†]The double derivatives in this equation are technically *partial* derivatives for those of you who know what I mean. It seemed to me, however, that defining carefully what a partial derivative is and explaining the traditional partial derivative notation would be an unnecessary distraction. The note specifies everything that we need to know to evaluate these derivatives correctly.

functions; and so on. Physicists rapidly learn to recognize this equation as the basic indicator that traveling sinusoidal disturbance waves are possible in the medium in question.

A proof that our sinusoidal traveling wave is a solution of the wave equation

Let us show that our sinusoidal traveling wave $f(t, x) = A \sin(kx - \omega t)$ is indeed a possible solution of this equation. If we take the derivative of $f(t, x)$ with respect to x (while treating t as constant), we find that, according to the chain rule,

$$\begin{aligned}\frac{df}{dx} &= A\frac{d}{dx}\sin(kx - \omega t)\\ &= A\cos(kx - \omega t)\frac{d}{dx}(kx - \omega t)\\ &= A\cos(kx - \omega t)(k)\\ &= kA\cos(kx - \omega t)\end{aligned} \tag{E15.15}$$

If we take the derivative again, we get

$$\frac{d^2f}{dx^2} = kA\frac{d}{dx}\cos(kx - \omega t) = -k^2A\sin(kx - \omega t) \tag{E15.16}$$

In a similar way, you can show that

$$\frac{d^2f}{dt^2} = -\omega^2A\sin(kx - \omega t) \tag{E15.17}$$

Exercise E15X.7

Verify that equation E15.17 is correct.

Plugging these results into the right side of equation E15.14, we get

$$\begin{aligned}\frac{1}{v^2}\frac{d^2f}{dt^2} - \frac{d^2f}{dx^2} &= -\frac{\omega^2}{v^2}A\sin(kx - \omega t) + k^2A\sin(kx - \omega t)\\ &= \left(k^2 - \frac{\omega^2}{v^2}\right)A\sin(kx - \omega t)\\ &= \frac{k^2}{v^2}\left(v^2 - \frac{\omega^2}{k^2}\right)A\sin(kx - \omega t)\end{aligned} \tag{E15.18}$$

But according to equation E15.12, the phase speed v of our traveling wave solution is $v = \omega/k$, so the factor in parentheses in the last term above is zero:

$$\frac{1}{v^2}\frac{d^2f}{dt^2} - \frac{d^2f}{dx^2} = \frac{k^2}{v^2}\left(\frac{\omega^2}{k^2} - \frac{\omega^2}{k^2}\right)A\sin(kx - \omega t) = 0 \tag{E15.19}$$

as required by the wave equation. We see, therefore, that our traveling sinusoidal wave is indeed one possible solution of the wave equation. One can in fact show (see problem E15S.4) that any *sum* of sinusoidal traveling waves also satisfies this equation. Since the Fourier theorem tells us that any traveling wave can be written as a sum of sinusoidal traveling waves, this means that any arbitrary traveling wave will satisfy the wave equation. Therefore, if this equation accurately describes the behavior of a disturbance in a medium, then a whole range of traveling wave solutions exist for this medium.

How the wave equation emerges from basic physics in the context of a taut string

How would we know whether this equation "accurately describes the behavior of a disturbance" in a given medium? Let's see how this works by considering the special case of a transverse wave on a stretched string.

Example E15.2 Waves on a Stretched String

Problem Imagine that we place a string under tension by exerting a tension force of magnitude F_T on its ends. Assume that the string has a mass per unit length of μ. Show that small transverse disturbances on this string obey the wave equation and determine the phase speed v of traveling waves on this string.

Translation and Model We will model the string as being a series of particles of mass m connected by identical springs, as shown in figure E15.7a. (We can eventually take the limit that the distance Δx between the masses goes to zero to better model a continuous string.) In our model, saying that the tension on the string has a magnitude of F_T means that each spring is stretched sufficiently that each of its ends exerts a force of magnitude F_T on the mass to which that end is connected.

Figure E15.7a shows the string in its "undisturbed" configuration, where the string is straight and each mass is at rest on the x axis. We consider the string's ith mass (the one at position x_i) to be "disturbed" if it is displaced vertically away from the x axis to a nonzero y coordinate y_i. A listing of the y coordinates $y_i(t, x_i)$ for all the masses on the string at a given time t completely describes the wave on that string at that time. In this case, therefore, $y_i(t, x_i)$ corresponds to the disturbance function we more generally described earlier as being $f(t, x)$.

How will the masses respond to being disturbed? Figure E15.7b shows some of the masses in a disturbed configuration. The forces acting on the ith mass in this case are the leftward and rightward tension forces $\vec{F}_{TL}$ and $\vec{F}_{TR}$, respectively, shown in figure E15.7c. Newton's second law for that mass therefore reads

$$m\vec{a}_i = \vec{F}_{\text{net},i} = \begin{bmatrix} F_{TL,x} + F_{TR,x} \\ F_{TL,y} + F_{TR,y} \\ F_{TL,z} + F_{TR,z} \end{bmatrix} = \begin{bmatrix} -F_{TL}\cos\theta_L + F_{TR}\cos\theta_R \\ -F_{TL}\sin\theta_L + F_{TR}\sin\theta_R \\ 0 \end{bmatrix} \quad \text{(E15.20)}$$

At this point, I am going to make an approximation. During a realistic oscillation, the angle that any part of the string makes with the horizontal direction is going to be very small (imagine, for example, a vibrating guitar string: it remains almost straight even as it vibrates, right?). Indeed, I have greatly

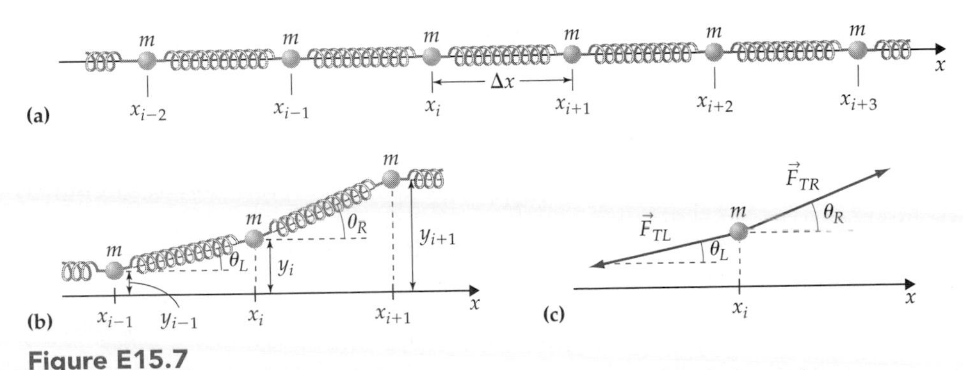

Figure E15.7
(a) We can model a stretched spring as being a sequence of particles with mass m connected by springs. This diagram shows the string at rest. (b) This diagram shows a possible set of disturbed vertical positions for the ith particle and its nearest neighbors. (c) This diagram shows the forces that are exerted on the ith particle by the springs connecting it to its nearest neighbors.

exaggerated the string's curvature in figure E15.7 just to make the angles visible at all. In this "small oscillation" limit, no individual spring will be stretched much more or less than the general stretching that gives the string its tension F_T. Therefore, the magnitude of the force that each individual spring exerts will be essentially equal to F_T. Moreover, in this small oscillation limit, the angles θ_L and θ_R are small. This means that $\cos\theta_L \approx \cos\theta_R \approx 1$, so the x component of the net force on the ith mass is $(F_{\text{net},i})_x = F_{TL}\cos\theta_L - F_{TR}\cos\theta_R \approx F_T - F_T \approx 0$. It is therefore a good approximation in this limit to assume that the x-position of any mass on this string is essentially fixed.

In the small-angle limit, we also have $\sin\theta_L \approx \tan\theta_L$ and $\sin\theta_R \approx \tan\theta_R$. If we take the angles to be positive when measured counterclockwise from the x direction, then

$$\sin\theta_L \approx \tan\theta_L = \frac{y_i - y_{i-1}}{\Delta x} \qquad \text{(E15.21}a\text{)}$$

$$\sin\theta_R \approx \tan\theta_R = \frac{y_{i+1} - y_i}{\Delta x} \qquad \text{(E15.21}b\text{)}$$

where y_i is the vertical position of the ith mass, y_{i-1} is the same for the adjacent mass to the left, y_{i+1} is the same for the adjacent mass to the right, and Δx is the horizontal distance between masses. If we plug this back into equation E15.20, we see that the only significant component of the acceleration of the ith mass is the y component, whose value is

$$a_{i,y} = \frac{(F_{\text{net},i})_y}{m} \approx \frac{F_T}{m}(\sin\theta_R - \sin\theta_L) \approx \frac{F_T}{m}\left(\frac{y_{i+1} - y_i}{\Delta x} - \frac{y_i - y_{i-1}}{\Delta x}\right)$$
$$= \frac{F_T \Delta x}{m}\left[\frac{1}{\Delta x}\left(\frac{y_{i+1} - y_i}{\Delta x} - \frac{y_i - y_{i-1}}{\Delta x}\right)\right] \qquad \text{(E15.22)}$$

Now, let us look at the quantity in square brackets in this expression in the limit that the spacing Δx between masses becomes very small. In that limit, $y_i(t, x_i)$ will become a continuous function $y(t, x)$. Now, the ratio $(y_{i+1} - y_i)/\Delta x$ describes the string's slope just to the right of the ith mass: this ratio best approximates the derivative dy/dx of the continuous function at a point halfway between the ith and $(i+1)$th mass, that is, at $x = x_i + \frac{1}{2}\Delta x$. Similarly, the ratio $(y_i - y_{i-1})/\Delta x$ describes the string's slope just to the left of the ith mass and best approximates dy/dx at $x = x_i - \frac{1}{2}\Delta x$. Therefore, in the limit that $\Delta x \to 0$,

$$\lim_{\Delta x \to 0}\left[\frac{1}{\Delta x}\left(\frac{y_{i+1} - y_i}{\Delta x} - \frac{y_i - y_{i-1}}{\Delta x}\right)\right]$$
$$= \lim_{\Delta x \to 0}\left[\frac{(dy/dx)_{x_i + \Delta x/2} - (dy/dx)_{x_i - \Delta x/2}}{\Delta x}\right]$$
$$\equiv \frac{d^2 y}{dx^2} \quad \text{(evaluated at position } x_i \text{ and time } t\text{)} \qquad \text{(E15.23}a\text{)}$$

since this amounts to the definition of the derivative of y. Similarly,

$$a_{i,y} \equiv \frac{d^2 y}{dt^2} \quad \text{(evaluated at position } x_i \text{ and time } t\text{)} \qquad \text{(E15.23}b\text{)}$$

Finally, in this same limit

$$\lim_{\Delta x \to 0} \frac{m}{\Delta x} \equiv \mu \equiv \text{mass per unit length on string} \qquad \text{(E15.24)}$$

Solution Plugging these results back into equation E15.22, we therefore have, in the limit that $\Delta x \to 0$,

$$\frac{\mu}{F_T}\frac{d^2y}{dt^2} = \frac{d^2y}{dx^2} \quad \Rightarrow \quad 0 = \frac{1}{v^2}\frac{d^2y}{dt^2} - \frac{d^2y}{dx^2} \quad \text{where } v \equiv \sqrt{\frac{F_T}{\mu}} \tag{E15.25}$$

This has the same mathematical form of the wave equation, considering that our disturbance function is $y(x, t)$ instead of $f(x, t)$ and noting that F_T/μ is indeed a constant independent of t and x. Therefore, there do exist traveling wave solutions for transverse disturbances on a stretched string, and they travel with speed $v = (F_T/\mu)^{1/2}$.

Evaluation This result makes good intuitive sense: experience with stretched strings suggests that the speed of waves on the string would increase if we increased the string tension and/or decreased the string's density.

Exercise E15X.8

Show that F_T/μ has the units of a squared speed, and calculate the speed of transverse waves on a string whose mass per unit length is 2.0 g/m and whose tension force is 100 N (roughly 22 lb).

A more intuitive approach to identifying media that support waves

Example E15.2 illustrates how basic physical principles applied to a segment of a stretched string lead directly to the wave equation (in the small-oscillation limit, at least). One finds that the same kind of thing happens in a variety of media (particularly in the small-oscillation limit): this equation is an extraordinarily common outcome of such analyses! (See the problems for other examples.)

The wave equation is a powerful and useful mathematical tool, but it is somewhat abstract. How can we recognize more *intuitively* when a medium can support a wave? Let's go back for a moment to equation E15.22, which we can write in the more general form

$$\frac{d^2f_i}{dt^2} \approx \frac{v^2}{\Delta x}\left(\frac{f_{i+1} - f_i}{\Delta x} - \frac{f_i - f_{i-1}}{\Delta x}\right) \tag{E15.26}$$

where v^2 is some medium-dependent constant and f_{i-1}, f_i, and f_{i+1} are values of the disturbance function $f(t, x)$ of interest evaluated at three points separated by a very small displacement Δx. As we have seen, this equation reduces to the wave equation in the limit that $\Delta x \to 0$.

Now, assume for the sake of argument that the medium is initially undisturbed at a certain time for $x \geq x_i$, so that $f_i = f_{i+1} = 0$, but at that time we suddenly displace the medium at $x = x_{i-1}$ so that $f_{i-1} \neq 0$. Under these circumstances, equation E15.26 reduces to

$$\frac{d^2f_i}{dt^2} \approx \frac{v^2}{\Delta x^2} f_{i-1} \tag{E15.27}$$

The following **wave rule** expresses this relationship in words.

The *wave rule*

> A disturbance in a homogeneous medium obeys the wave equation if a sudden disturbance at a given point in the otherwise undisturbed medium causes the disturbance value at a point a distance Δx away to *accelerate in direct proportion to the original disturbance divided by* Δx^2 in

the limit that we take $\Delta x \to 0$ while preserving the medium's macroscopic properties (such as total mass per unit length in the case of a string). The constant of proportionality is the squared wave speed.

For example, in the string model we considered in section E15.2, if the string is initially at rest and I suddenly displace one of its masses, it will exert forces on the two neighboring masses that cause them to accelerate. Once these masses have displaced significantly, they will cause the *next* farther masses to accelerate, and so on. We can easily see that this will cause a wave to move away from the original point of disturbance. Moreover, it is plausible that the speed at which these waves will move should increase as the constant of proportionality between the displacement and the acceleration: the more rapidly the mass at point adjacent to the disturbance accelerates, the more rapidly it will affect the next point and so move the disturbance forward.

Any medium in which displacing one point causes a nearby point to accelerate might support *some* kind of traveling disturbance pattern, but the medium obeys the wave equation only if (1) the acceleration is *directly proportional* to the disturbance, (2) the acceleration is *inversely* proportional to Δx^2 (at least in the limit that $\Delta x \to 0$), and (3) the constant of proportionality v^2 is independent of position and time. The special media that obey the wave equation have a number of nice properties: such media support sinusoidal traveling waves (as we have seen), obey the superposition principle (see problem E15S.4), and preserve the shape of arbitrary waves traveling through them (see problem E15A.1).

The wave rule greatly reduces the work required to identify a medium that satisfies the wave equation, as example E15.3 illustrates.

Example E15.3 Applying the Wave Rule

Problem Use the wave rule to show that the mass-and-spring model of a string obeys the wave equation in the limit of small disturbances if the string has a well-defined mass per unit length μ.

Translation Figure E15.8 shows the string when the masses at x_i and x_{i+1} are at their equilibrium positions but the mass at x_{i-1} has been displaced a certain small distance y_{i-1}.

Model If the displacement y_{i-1} is small, then the angle θ is small, and the spring is not stretched much from its equilibrium length. Therefore, the magnitude of the force that it exerts on the ith mass is still very nearly its equilibrium value F_T. According to the y component of Newton's second law, the acceleration of the disturbance (that is, the y displacement) of the ith mass is

$$\frac{d^2 y_i}{dt^2} = a_{i,y} = \frac{(F_{\text{net},i})_y}{m} = \frac{F_T \sin\theta}{m} \approx \frac{F_T \tan\theta}{m} = \frac{F_T}{m}\frac{y_{i-1}}{\Delta x} \quad \text{(E15.28)}$$

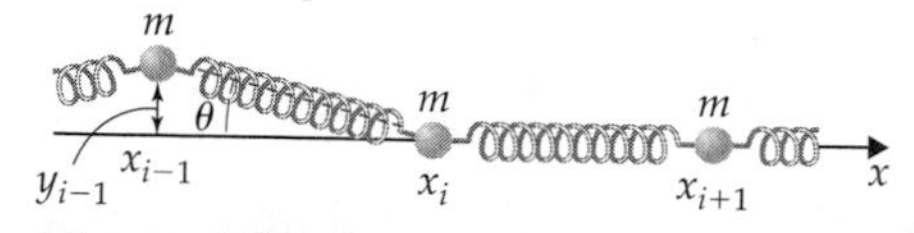

Figure E15.8
This diagram illustrates the situation when we disturb only the particle to the left of the ith particle. This will exert a net vertical force on the ith particle that will cause it to accelerate upward.

since $\sin\theta \approx \tan\theta$ for small angles. This acceleration *is* directly proportional to the disturbance y_{i-1}, but it looks at first glance as if it is inversely proportional to Δx, not Δx^2. However, if we are preserve the "macroscopic properties" of the medium as we take $\Delta x \to 0$, we have to decrease m as we decrease Δx to keep the string's mass per unit length $\mu = m/\Delta x$ fixed.

Solution Substituting $m = \mu\,\Delta x$ into equation E15.28, we get

$$\frac{d^2 y_i}{dt^2} = \frac{F_T}{\mu}\frac{y_{i-1}}{\Delta x^2} \tag{E15.29}$$

This does satisfy the wave equation. The constant of proportionality between the acceleration and $y_{i-1}/\Delta x^2$ is the squared wave speed, so $v = \sqrt{F_T/\mu}$.

Evaluation This is consistent with what we found in example E15.2.

In chapter E16, we will use the wave rule to show that disturbances in an electromagnetic field obey the wave equation.

The three-dimensional wave equation

For the record, a wave in a fully three-dimensional medium is described by a disturbance function $f(t, x, y, z)$ that is a function of all three position coordinates as well as time. The wave equation in such a medium is

$$0 = \frac{1}{v^2}\frac{d^2 f}{dt^2} - \frac{d^2 f}{dx^2} - \frac{d^2 f}{dy^2} - \frac{d^2 f}{dz^2} \tag{E15.30}$$

(You might note some similarity to the metric equation of special relativity, which is not entirely accidental.) We will not do anything in this course with the full wave equation: as I said at the beginning of this chapter, we can use one-dimensional waves to illustrate what we really need to know about waves. I only include the equation here for the sake of completeness.

TWO-MINUTE PROBLEMS

E15T.1 The speed of a transverse wave traveling on a spring is necessarily equal to the actual speed of a given part of the spring as it moves in response to the passing wave, true (T) or false (F)?

E15T.2 Sound waves move through air at about 340 m/s. The sound waves produced by bats doing echolocation can have frequencies in excess of 60 kHz. What would be the wavelength of such a sound wave?

A. 2×10^7 m
B. 180 m
C. 28 m
D. 1.3 m
E. 6 mm
F. Some other wavelength (specify)

E15T.3 A sinusoidal sound wave with a frequency of 510 Hz moves through water. The wave is observed to have wavelength of 2.94 m. What is the phase speed of sound waves in water according to this information?

A. 1.5 km/s
B. 173 m/s
C. 0.6 mm/s
D. Some other phase speed (specify)

E15T.4 A sinusoidal water surface wave has a wavelength of 10 cm and a frequency of 0.318 Hz. What is k?

A. 62.8 cm
B. 31.4 cm
C. 3.18 cm
D. 0.628 cm^{-1}
E. Some other value (specify)

E15T.5 A sinusoidal water surface wave has a wavelength of 10 cm and a frequency of 0.318 Hz. What is ω?

A. 0.32 s^{-1}
B. 1.0 s^{-1}
C. 2.0 s^{-1}

(more)

D. 3.2 cm/s
E. Some other value (specify)

E15.6 Imagine that sinusoidal waves of frequency f_0 on a certain stretched string are observed to have a wavelength λ_0. Imagine now that we stretch the string more tightly, so that the phase speed of disturbance waves on the string increases. How will the wavelength λ_{new} of waves with the same frequency f_0 on the tighter string compare to the original wavelength λ_0?
A. $\lambda_{new} > \lambda_0$
B. $\lambda_{new} = \lambda_0$
C. $\lambda_{new} < \lambda_0$
D. We need more information to determine the relationship.

HOMEWORK PROBLEMS

Basic Skills

E15B.1 Sound waves move through air at a speed of about 340 m/s. Compute the wavelength of the following sound waves:
(a) An organ pipe playing middle C (260 Hz)
(b) The highest audible pitch (≈20,000 Hz)
(c) The lowest audible pitch (≈15 Hz)

E15B.2 What are the wavelengths of the following kinds of electromagnetic waves?
(a) Radio waves on the AM band (≈1000 kHz)
(b) Radio waves on the FM band (≈100 MHz)
(c) EM waves in a microwave oven (≈30 GHz)

E15B.3 Sound waves move through air at a speed of about 340 m/s. What would be the frequency of a sound wave that had a wavelength of 1 m? 1 in.? 1 mm?

E15B.4 Visible light has wavelengths between 700 nm and about 400 nm. If light really is an electromagnetic wave, then what are the corresponding frequencies of these waves?

E15B.5 A sinusoidal traveling water wave has a wavelength of 25 cm and a frequency of 0.60 Hz. What are k and ω for this wave? What is the phase speed of this wave?

E15B.6 A sinusoidal wave moving down a taut rope has a wavelength of 2.0 m and a period of about 0.5 s. What are k and ω for this wave? What is the phase speed of the wave?

E15B.7 Consider the sinusoidal traveling wave shown (this is a snapshot at a certain instant of time). Assume the wave travels at 1.0 m/s.

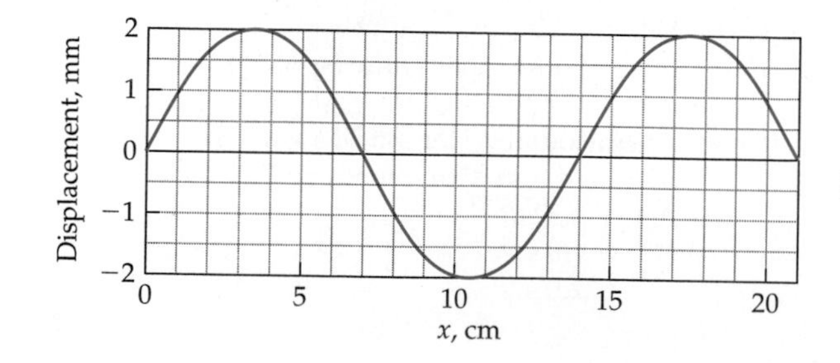

(a) What is the wave's amplitude?
(b) What is its wavenumber k?
(c) What is its angular velocity ω?
(d) What is its period?
(e) What is its frequency?

Synthetic

E15S.1 Sinusoidal water waves are created 120 km offshore by an earthquake near a small island. Observers in helicopters above the island report that the waves have an amplitude of about 2.0 m, a wavelength of 15 m, and a frequency of about 0.5 Hz. How long do lifeguards on the mainland have to evacuate beaches before the waves arrive?

E15S.2 Imagine that a geologist is measuring the waves produced by small earthquakes by using two seismographs, one 12 km from the volcano and another 17 km from the volcano. During one earthquake, the waves feel like the gentle rocking of a boat at a frequency of about 1.5 Hz and an amplitude of about 1 cm. The geologist later notices that the closer seismograph registered the waves about 0.85 s sooner than the other. What was the approximate wavelength of the waves during this episode?

E15S.3 Consider the function $f(x, t) = A\sin(kx + \omega t)$. Does this function describe a *traveling* sinusoidal wave? If not, why not? If so, what is the speed (in terms of ω and k) and the direction of motion of this wave? Does this wave satisfy the wave equation? Explain your responses carefully.

E15S.4 Argue that if $f(t, x)$ and $g(t, x)$ separately satisfy the wave equation, then $h(t, x) = f(t, x) + g(t, x)$ satisfies the wave equation. (By extension, any sum of sinusoidal waves will satisfy the wave equation.)

E15S.5 By rocking a boat, a person produces water waves on a previously undisturbed lake. This person observes that the boat oscillates 12 times in 20 s, each oscillation producing a wave crest 5 cm above the undisturbed level of the lake, and that the waves reach the shore (12 m away) in about 6 s. At any given instant of time, about how many wave crests are there between the boat and the shore?

E15S.6 Consider a series of identical masses m arranged along the x axis that are connected by identical springs with spring constant k_s. The masses are all free to slide in the $\pm x$ direction on a frictionless surface. When all the masses are in their equilibrium positions, their centers are equal distances Δx apart and the springs are all relaxed. Let's define these positions x_i to be the "home" positions of the masses. We can then define the disturbance function $s(x_i, t)$ for the ith mass to be its horizontal displacement from its home position at time t (s is positive if the mass is displaced in the $+x$ direction from home and negative if it is displaced in the $-x$ direction).

(a) Make a careful drawing of the mass at an arbitrary position x and its two adjacent neighbors, and argue that the x component of the net force on the mass at a given instant of time is given by

$$F_x = k_s[s(x + \Delta x) - s(x)] - k_s[s(x) - s(x - \Delta x)] \tag{E15.31}$$

where $s(x)$ is the displacement of the mass at home position x away from home, $s(x + \Delta x)$ is the displacement of the mass at home position $x + \Delta x$ away from home, etc.

(b) Argue (using the definition of the double derivative) that in the limit that Δx is very small,

$$\frac{1}{\Delta x}\left[\frac{s(x + \Delta x) - s(x)}{\Delta x} - \frac{s(x) - s(x + \Delta x)}{\Delta x}\right] \approx \frac{d^2 s}{dx^2} \tag{E15.32}$$

(c) Use this to show that

$$F_x \approx k_s \Delta x^2 \frac{d^2 s}{dx^2} \tag{E15.33}$$

(d) Show that Newton's second law and the previous results together imply that longitudinal disturbances in this set of interconnected masses obey the wave equation

$$0 = \frac{1}{v^2}\frac{d^2 s}{dt^2} - \frac{d^2 s}{dx^2} \tag{E15.34}$$

and find the wave speed v in terms of k_s, m, and Δx. (This means that disturbances in this system will move as traveling waves do up and down the x axis. If we consider the masses to be atoms and the springs to be interatomic bonds, this could represent a simplified model of a one-dimensional elemental solid.)

E15S.7 Consider the system discussed in problem E15S.6.

(a) Draw a diagram of the masses at x_{i-1}, x_i, and x_{i+1}, where the masses at the last two positions are undisturbed but the first mass is disturbed away from its home position.

(b) Use this picture and the wave rule to prove that longitudinal disturbances in this system obey the wave equation and determine the wave speed. (Assume that when we reduce Δx, we keep the mass per unit length of the system fixed.)

E15S.8 Figure E15.9 shows a "wave machine" of a type commonly used for classroom demonstrations of traveling waves. The wave machine consists of rods of length L and mass m separated by distance Δx along a wire spine. The disturbance function

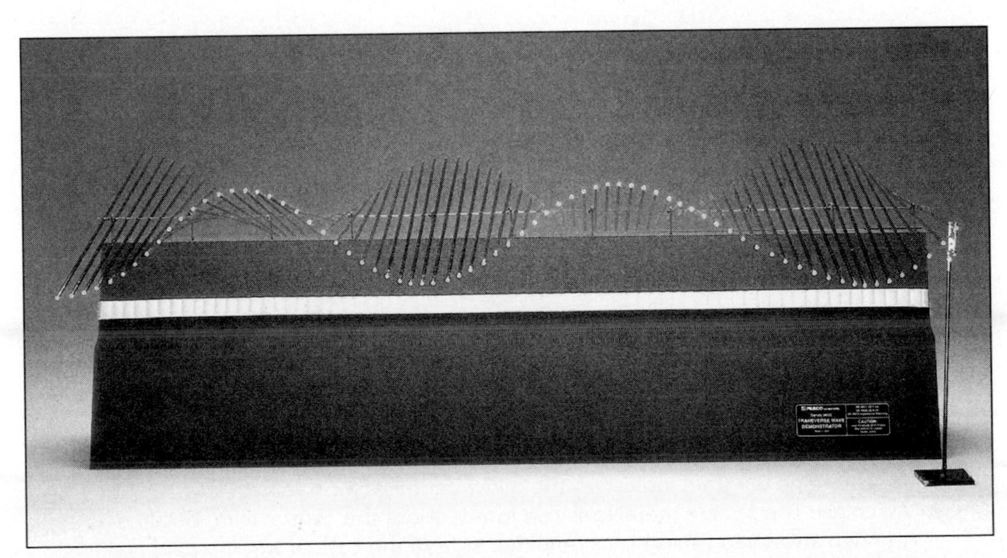

Figure E15.9
How fast do waves travel on this torsion-rod wave machine? (See problem E15S.8.)

in this case is the angle $\theta(t, x)$ that the rod at position x makes with the horizontal plane at time t. When a segment of the spine of length Δx is twisted through a small angle $\Delta\theta$, the segment exerts a torque on the rod at each end whose magnitude is

$$\tau = k_t \frac{\Delta\theta}{\Delta x} \qquad \text{(E15.35)}$$

where k_t is a constant expressing the spine's stiffness (a kind of a spring constant for twisting). Use the definitions of torque and the angular momentum of a symmetric object (see chapter C13) and the wave rule to show that an angular disturbance on this medium obeys the wave equation; determine the speed of the wave in terms of m, L, and k_t. (*Hint:* The moment of inertia of a rod rotating around its center is $I = \frac{1}{12}mL^2$, as discussed in chapter C9. When you consider Δx dependence, make sure that you hold fixed the total rod mass per unit length along the x axis.)

Rich-Context

E15R.1 Waves do not have to be sinusoidal! Consider the following disturbance function, which might be a good model for a disturbance consisting of a single pulse:

$$f(t, x) = Ae^{-(kx+\omega t)^2} \qquad \text{(E15.36)}$$

(a) Sketch a graph of what this wave looks like at $t = 0$, assuming that $k = 1.0/\text{cm}$. (Label the vertical axis in units of A.)

(b) Argue that this is a traveling wave, and find the speed of the peak. Does this wave move in the $+x$ direction or the $-x$ direction?

(c) Does this wave satisfy the wave equation? Defend your answer. (*Hint:* You can save yourself a lot of writing if you define $u \equiv kx + \omega t$ and then show that $du/dx = k$ and $du/dt = \omega$. Then you won't have to ever write out $kx + \omega t$ when you compute the derivatives.)

E15R.2 Consider a sinusoidal wave moving down a taut elastic string. Find a formula for the ratio of the maximum speed of motion of a point on the string to the phase speed of the wave. Express the ratio in terms of A, k, and ω for the wave and anything else that you need.

Advanced

E15A.1 Consider a completely arbitrary function $f(u)$ where $u \equiv kx - \omega t$. (a) Argue that no matter what the shape of the function is, its features will move in the $+x$ direction with speed $v = \omega/k$. (b) Show that no matter what $f(u)$ might be (as long as it has at least two derivatives), this disturbance function satisfies the wave equation.

E15A.2 Figure E15.10 shows a drawing of a "wave machine" that consists of a series of square aluminum rods spaced along a set of two parallel strands of fishing line flanking a stiff central wire. The rods have length L and a mass per unit rod length of μ. The distance between adjacent rods is Δx. The rods pivot about the central wire. The two side strands, which are separated by a distance a, are put under the same tension F_T. The disturbance in this case is the angle $\theta(t, x)$ that the rod at position x makes with the horizontal plane at time t. Show that for small angles, such a disturbance obeys the wave equation, and determine the wave speed. (*Hint:* Instead of Newton's second law, you will have to use the definitions of torque and the angular momentum of a symmetric object: see chapter C13).

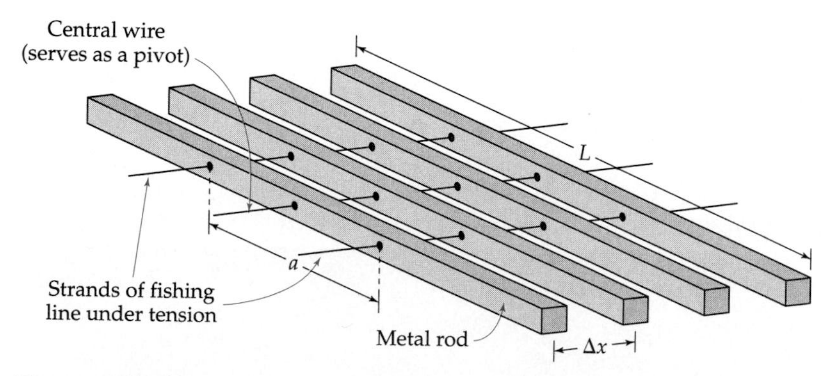

Figure E15.10
A schematic diagram indicating the construction of a certain wave machine. The square metal rods can rotate about the central wire, but the two stretched fishing lines on either side of the central wire try to keep the rods level. How fast do waves move on this wave machine? (See problem E15A.2.)

ANSWERS TO EXERCISES

E15X.1 P waves are longitudinal, S waves are transverse.

E15X.2 Earthquake waves shake objects, which means that the waves must have given the objects kinetic energy. This energy ultimately comes from the sudden relaxation of strains in rock due to the slippage along a fault at the epicenter. This energy may be carried many miles from the epicenter by the wave.

E15X.3 The first crest to pass $x = 0$ after $t = 0$ occurs when $-\omega t_1 = -3\pi/2$ [since $\sin(-3\pi/2) = +1$]. The next crest passes when $-\omega t_2 = -7\pi/2$. Thus $T \equiv t_2 - t_1 = -2\pi/(-\omega) = 2\pi/\omega$, as claimed.

E15X.4 $k = 3.14$ per centimeter and $\omega = 12.57$ per second.

E15X.5 (The calculation is essentially identical with what we did before except that we now substitute $5\pi/2$ everywhere that we had $\pi/2$ before.)

E15X.6 Solving equation E15.12 for λ and plugging in the numbers, we get $\lambda = 1.2$ km.

E15X.7 The first derivative of $f = A\sin(kx - \omega t)$ is

$$\begin{aligned}\frac{df}{dt} &= A\frac{d}{dt}\sin(kx - \omega t)\\ &= A\cos(kx - \omega t)\frac{d}{dt}(kx - \omega t)\\ &= -\omega A\cos(kx - \omega t)\end{aligned} \tag{E15.37}$$

Taking the derivative again, we get

$$\begin{aligned}\frac{d^2f}{dt^2} &= -\omega A\frac{d}{dt}\cos(kx - \omega t)\\ &= +\omega A\sin(kx - \omega t)\frac{d}{dt}(kx - \omega t)\\ &= -\omega^2 A\sin(kx - \omega t)\end{aligned} \tag{E15.38}$$

E15X.8 We can do both parts at once here:

$$\begin{aligned}v = \sqrt{\frac{F_T}{\mu}} &= \sqrt{\frac{100\ \text{N}}{2.0\ \text{g/m}}\left(\frac{1\ \text{kg}\cdot\text{m/s}^2}{1\ \text{N}}\right)\left(\frac{1000\ \text{g}}{1\ \text{kg}}\right)}\\ &= \sqrt{50{,}000\ \text{m}^2/\text{s}^2} = 220\ \text{m/s}\end{aligned} \tag{E15.39}$$

Note that the units do work out correctly!

E16 Electromagnetic Waves

▷ Electric Field Fundamentals
▷ Controlling Currents
▷ Magnetic Field Fundamentals
▷ Calculating Static Fields
▽ Dynamic Fields
The Electromagnetic Field
Maxwell's Equations
Induction
Introduction to Waves
Electromagnetic Waves

Chapter Overview

Introduction

Maxwell's greatest triumphs were his prediction of the existence of electromagnetic waves that could move through empty space at the speed of light and his identification of light as electromagnetic waves. We close this unit by exploring what Maxwell's equations can tell us about the existence and characteristics of electromagnetic waves.

Section E16.1: Electromagnetic Waves

For simplicity's sake, we will mostly consider **plane waves** of the form $\vec{E}(t, x)$, which are independent of y and z. This section shows that a sudden disturbance of E_z on the $x = x_0$ plane in empty space (according to Faraday's law) creates a magnetic field that grows in the $-y$ direction, which in turn (according to the Ampere-Maxwell law) causes the value of E_z at $x = x_1 = x_0 + \Delta x$ to *accelerate:* $d^2E_{z1}/dt^2 = c^2(E_{z0}/\Delta x^2)$. As we saw in chapter E15, this behavior tells us that a disturbance in E_z will obey the wave equation and travel with speed c. Maxwell's discovery of this behavior led him to assert that light was an electromagnetic wave.

Section E16.2: Characteristics of Electromagnetic Waves

In this section, we will see that Maxwell's equations also imply that electromagnetic waves that move in a well-defined direction have the following characteristics.

1. Their $\vec{E}$ and $\vec{\mathbb{B}}$ vectors have equal magnitudes at a given point and time.
2. Their $\vec{E}$ and $\vec{\mathbb{B}}$ vectors are perpendicular to each other and to the direction of motion.

We can write a general sinusoidal electromagnetic plane wave that moves in the $+x$ direction as the sum of **horizontally polarized** and **vertically polarized** waves:

$$\vec{E}_H(x,t) = \begin{bmatrix} 0 \\ A_H \\ 0 \end{bmatrix} \sin(k_w x - \omega t) \qquad \vec{\mathbb{B}}_H(t,x) = \begin{bmatrix} 0 \\ 0 \\ A_H \end{bmatrix} \sin(k_w x - \omega t) \tag{E16.8a}$$

$$\vec{E}_V(x,t) = \begin{bmatrix} 0 \\ 0 \\ A_V \end{bmatrix} \sin(k_w x - \omega t) \qquad \vec{\mathbb{B}}_V(t,x) = \begin{bmatrix} 0 \\ -A_V \\ 0 \end{bmatrix} \sin(k_w x - \omega t) \tag{E16.8b}$$

Purpose: These equations describe the fields $\vec{E}_H$ and $\vec{\mathbb{B}}_H$ associated with a horizontally polarized sinusoidal plane wave moving in the $+x$ direction and the fields $\vec{E}_V$ and $\vec{\mathbb{B}}_V$ associated with a vertically polarized wave.

Symbols: A_H and A_V are constants, ω is the wave's angular frequency, and k_w is its wavenumber (the subscript distinguishes this quantity from the Coulomb constant k). Once ω is chosen, k_w is determined: $k_w = \omega/c$.

Limitations: This equation applies only to plane waves moving in the $+x$ direction, but you can easily adapt the equations to other directions if you remember that $\vec{E} \times \vec{\mathbb{B}}$ points in the direction in which the wave moves.

Note: Of course, the directions we describe as being "horizontal" and "vertical" depend on our arbitrary choice of coordinates.

The Fourier theorem implies that any arbitrary wave can be constructed from a sum of sinusoidal plane waves having these characteristics.

Section E16.3: The Energy in an Electromagnetic Wave

We can use the formula $u_E = (E^2 + \mathbb{B}^2)/8\pi k$ for energy density in an electromagnetic field to calculate the intensity (energy delivered per unit area per unit time) by a sinusoidal electromagnetic (EM) wave in terms of its electric field amplitude:

$$\text{Average intensity of an EM wave} = \frac{c}{4\pi k}(E^2)_{\text{avg}} = \frac{cE_0^2}{8\pi k} \qquad \text{(E16.13)}$$

Purpose: This equation describes the energy per unit time per unit area deposited on a surface that absorbs a sinusoidal electromagnetic wave.

Symbols: E_0 is the amplitude of the wave's oscillating electric field, c is the speed of light, and k is the Coulomb constant.

Limitations: This equation applies only to sinusoidal waves.

Section E16.4: The Power Radiated by a Charge

A charge moving at a constant velocity cannot create an electromagnetic wave, because whether or not such a wave exists is a frame-independent idea, and in the rest frame of the charge, there clearly is no electromagnetic wave. Accelerating charges, however, do create electromagnetic waves; the formula linking the power (energy per unit time) radiated by an accelerating charge in the form of such waves is

$$P = \frac{2}{3}\frac{kq^2a^2}{c^3} \qquad \text{(E16.19)}$$

Purpose: This is the **Larmor formula** for the power P that a point charge q radiates in the form of electromagnetic waves when it is accelerated.

Symbols: c is the speed of light, and a is the particle's acceleration.

Limitations: This equation applies only to nonrelativistic point particles.

Section E16.5: Why the Sky Is Blue

Sunlight is scattered by air molecules in a process known as **Raleigh scattering**: the oscillating electric field in the light from the sun causes the electrons in a molecule to oscillate sympathetically, which causes it to reradiate electromagnetic waves of the same angular frequency ω. The power of the light reradiated by this oscillating electron, by the Larmor formula, ends up being proportional to ω^4, which means that colors of light corresponding to high frequencies (blue and violet) are much more effectively reradiated than those corresponding to low frequencies (red and orange). This is what gives the light scattered by the sky its characteristic blue color.

Section E16.6: Maxwell's Rainbow

This section provides an overview of the spectrum of electromagnetic waves as we now understand it, naming the various frequency ranges involved. (Almost none of this spectrum was known when Maxwell published his theory in 1865!)

E16.1 Electromagnetic Waves

We are now finally in a position to see that Maxwell's equations imply the existence of electromagnetic waves. We begin by visualizing empty space with no electric or magnetic fields as being analogous to an undisturbed medium and an electromagnetic field as a disturbance of this medium. This is just an *analogy:* empty space is *empty* and is thus not really a "medium" for anything (though it took the physics community about 40 years after Maxwell to understand this fully!). Nonetheless, we will find it a useful analogy.

Electromagnetic plane waves

To simplify matters, we will at first consider an electromagnetic disturbance where the electric field has the form $\vec{E} = [0, 0, E_z(t, x)]$. We call such a disturbance a **plane wave,** because while E_z might vary in the x direction, it is independent of y and z and so has the same value anywhere on a plane where x has a fixed value. For example, if $E_z(t, x)$ were a sinusoidal function, the set of all points corresponding to a crest at a given instant of time would lie on the plane $x = x_{\text{crest}}$, which is parallel to the $x = 0$ plane. Successive crests of this wave would correspond to a stack of parallel planes, as shown in figure E16.1.

A hypothetical disturbance in the electric field

Consider now a specific disturbance of this type in empty space where we somehow suddenly "deflect" the value of E_z everywhere on the plane $x = x_0$ to some value E_{z0} while leaving the electric field at other points in the $+x$ direction undisturbed (i.e., zero). In example E15.3, we used Newton's second law to show that an analogous deflection of a point on a string caused a point a distance Δx away to accelerate in the manner specified by the wave rule, which then implied that disturbances on that string obey the wave equation. In a similar way, let us see how this electric field deflection affects the value of the electric field at a point a distance Δx away. The dynamic behavior of an electromagnetic field is governed not by Newton's second law but rather by Maxwell's equations. The relevant equations in this case are Faraday's law and the Ampere-Maxwell law for empty space (where $i_{\text{enc}} = 0$):

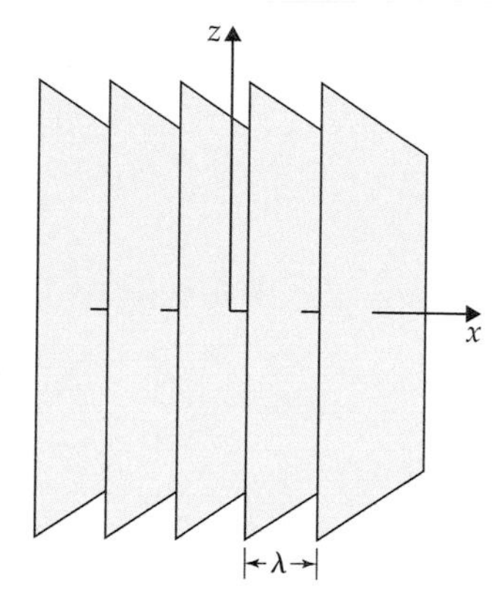

Figure E16.1
If $\vec{E}(t, x)$ is a sinusoidal plane wave, then at a given instant of time, the set of all points where this wave has a crest would correspond to successive planes parallel to the $x = 0$ plane. (The y axis points away from us in this drawing.)

$$\oint \vec{E} \cdot d\vec{S} + \frac{1}{c}\frac{d\Phi_{\mathbb{B}}}{dt} = 0 \qquad \text{(Faraday's law)} \qquad \text{(E16.1}a\text{)}$$

$$\oint \vec{\mathbb{B}} \cdot d\vec{S} - \frac{1}{c}\frac{d\Phi_E}{dt} = 0 \qquad \text{(the Ampere-Maxwell law in empty space)} \qquad \text{(E16.1}b\text{)}$$

Figure E16.2 shows two rectangular amperian loops. Loop A lies in the xz plane with left and right legs at x_0 and $x_1 \equiv x_0 + \Delta x$, respectively. This loop's midpoint is at $x_m = x_0 + \frac{1}{2}\Delta x$. Loop B lies in the xy plane, with its left and right legs at x_m and $x_m + \Delta x$, respectively. This loop's midpoint is at x_1. We will apply Faraday's law and the Ampere-Maxwell law to loops A and B, respectively.

What Faraday's law tells us

At the instant that the disturbance reaches x_0, the electric field is still zero on the right leg of loop A and is perpendicular to the top and bottom legs, so these legs contribute nothing to the circulation. Since the segment vectors on the loop's left leg point in the $-z$ direction, the component version of the dot product tells us that $\vec{E} \cdot d\vec{S} = E_{z0}\, dS_z = E_{z0}(-dS)$ on this leg. Since E_{z0} does not depend on z, this leg's contribution, and thus the net circulation around the loop, is $-E_{z0}\,\Delta z$. The loop rule tells us that tile vectors on the surface bounded by this loop point in the $-y$ direction, so the component version of the dot product tells us that $\vec{\mathbb{B}} \cdot d\vec{A} = -\mathbb{B}_y\, dA$ for all tiles on the surface. The total flux will be the loop's area $\Delta x\,\Delta z$ times the average value of $-\mathbb{B}_y$ over

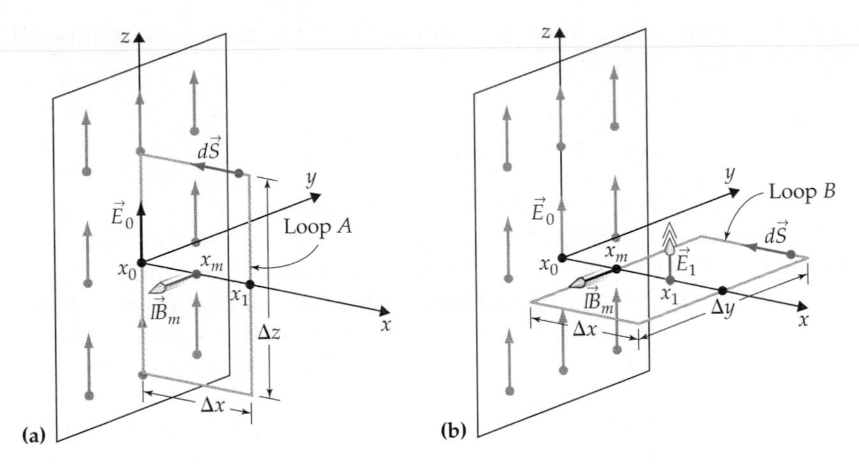

Figure E16.2
These drawings display amperian loops we can use to extract the implications of (a) Faraday's law and (b) the Ampere-Maxwell law for what happens after a sudden disturbance in the value of E_z at $x = x_0$ (the vectors on the plane illustrate the disturbed field). Note that both the electric field and the magnetic field are initially undisturbed (zero) in front of the plane. We find that these laws imply that the magnetic field at x_m starts to grow steadily in the $-y$ direction and the electric field at x_1 starts to accelerate in the $+z$ direction in response to this disturbance.

the loop. If the loop is small enough, this average value will be very nearly equal to $-\mathbb{B}_{ym}$, the value at the loop's midpoint. Therefore, Faraday's law in this case becomes

$$-E_{z0}\,\Delta z - \frac{1}{c}\frac{d\mathbb{B}_{ym}}{dt}\Delta x\,\Delta z = 0 \quad\Rightarrow\quad \frac{d\mathbb{B}_{ym}}{dt} = -\frac{c}{\Delta x}E_{z0} \qquad \text{(E16.2)}$$

This tells us that a sudden electric field disturbance in the $+z$ direction at x_0 causes the magnetic field at loop A's midpoint to grow in the $-y$ direction.

Exercise E16X.1

By considering loops at the same location but in the yz and xy planes, show that Faraday's law implies that our hypothetical electric field disturbance will *not* cause any other component of the magnetic field to change.

What the Ampere-Maxwell law tells us

Now consider loop B. After the magnetic field has grown a bit, it is nonzero on this loop's left leg but still zero on the right leg. Using arguments very similar to the ones in the previous paragraph, you can show that the net magnetic circulation around this loop is $-\mathbb{B}_{ym}\,\Delta y$ and the net electric flux through the surface bounded by that loop is $+E_{z1}\Delta x\,\Delta y$. The Ampere-Maxwell law therefore implies that

$$-\mathbb{B}_{ym}\,\Delta y - \frac{1}{c}\frac{dE_{z1}}{dt}\Delta x\,\Delta y = 0 \quad\Rightarrow\quad \frac{dE_{z1}}{dt} = -\frac{c}{\Delta x}\mathbb{B}_{ym} \qquad \text{(E16.3)}$$

Exercise E16X.2

Verify that equation E16.3 is correct.

If we take the time derivative of equation E16.3 and then use equation E16.2 to eliminate $d\mathbb{B}_{ym}/dt$, we get

$$\frac{d^2E_{z1}}{dt^2} = -\frac{c}{\Delta x}\frac{d\mathbb{B}_{ym}}{dt} = -\frac{c}{\Delta x}\left(-\frac{c}{\Delta x}E_{z0}\right) = +\frac{c^2}{\Delta x^2}E_{z0} \qquad \text{(E16.4)}$$

We see, therefore, that a disturbance in E_z at x-position x_0 causes the value of E_z at x-position x_1 to *accelerate* in proportion to $c^2/\Delta x^2$ in the limit that the distance between the points Δx is very small. This exactly fulfills the conditions of the wave rule, so that rule implies that such a disturbance obeys the wave equation, and waves in this "medium" will have speed c. Therefore, *Maxwell's equations allow for the existence of wavelike electromagnetic disturbances that move through empty space at the speed of light.*

Our electromagnetic field theory is consistent with the cosmic speed limit

This provides the final answer to the action-at-a-distance issue that we first discussed in chapter E2. Imagine that I wiggle a charged particle so that it creates a disturbance in the local electric field. Equation E16.4 tells us that this disturbance will *not* be instantly transmitted to any distant particle, but rather (in exactly the same way as wiggling one end of a stretched string will cause a wave to move to the other end) the disturbance will flow to that distant particle in the form of an electromagnetic wave traveling at the speed of light. Thus we finally have an electromagnetic field theory that is fully consistent with relativity's cosmic speed limit!

Maxwell's assertion that light was an electromagnetic wave

Of course, when Maxwell saw the wave equation emerging from his equations in 1865, he was not thinking about consistency with relativity (which had not yet been invented). Instead he saw compelling evidence supporting the conjecture that *light itself was an electromagnetic wave*. Some decades before, W. Weber and R. Kohlrausch noticed that the product of the electrostatic permittivity ε_0 ($= 1/4\pi k$ in our notation) and the magnetic permeability constant μ_0 ($= 4\pi k/c^2$), which had been independently determined by electrostatic and magnetostatic experiments, had the units of an inverse squared speed and a magnitude very similar to the recently measured speed of light. This led them to suspect that light might be somehow linked to electricity and magnetism. If you write Maxwell's equations entirely in traditional units, the constant of proportionality in equation E16.4 is $1/\varepsilon_0\mu_0$ (see problem E16S.4). Therefore, Maxwell could see directly from the wave equation emerging from his equations that electromagnetic waves moving at a speed equal to the measured speed of light must exist. Moreover, experiments done early in the 1800s had firmly established that light exhibited wavelike behavior. Therefore, Maxwell was able to assert with confidence that visible light consisted of electromagnetic waves.

E16.2 Characteristics of Electromagnetic Waves

Even more importantly, Maxwell was able to predict that one could *create* electromagnetic waves (essentially by wiggling charges), and that such waves (including light waves) should display certain previously unsuspected characteristics. In this section, we will explore some of the defining characteristics of electromagnetic waves in empty space.

Electromagnetic waves are *transverse*

Equation E16.4 implies that disturbances in the electric field's z component obey the wave equation. If we simply rotate the coordinate system shown in figure E16.2 counterclockwise 90° around the x axis, we can see that the same line of reasoning tells us that a disturbance in the value of E_y also must obey the wave equation (see figure E16.3).

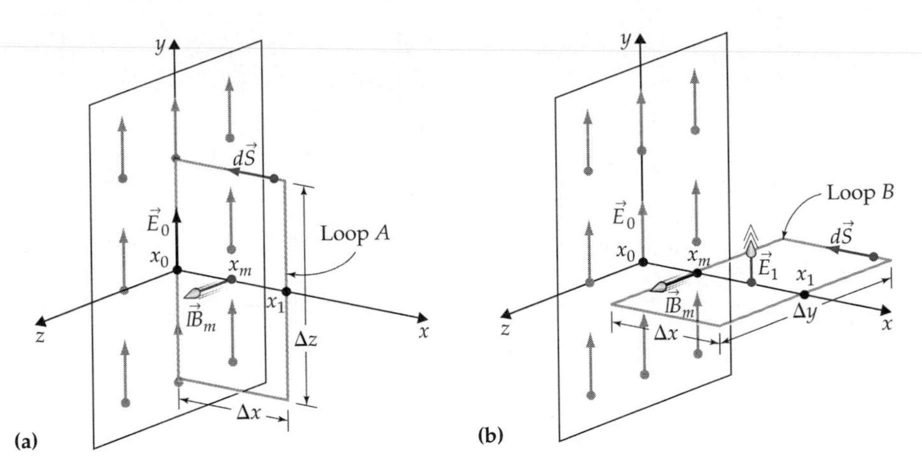

Figure E16.3
This drawing is the same as figure E16.2, except that the y and z axes have been rotated 90° counterclockwise around the x axis. It illustrates that the same logic we used in section E16.1 to show that disturbances in E_z obey the wave equation implies that disturbances in E_y do as well.

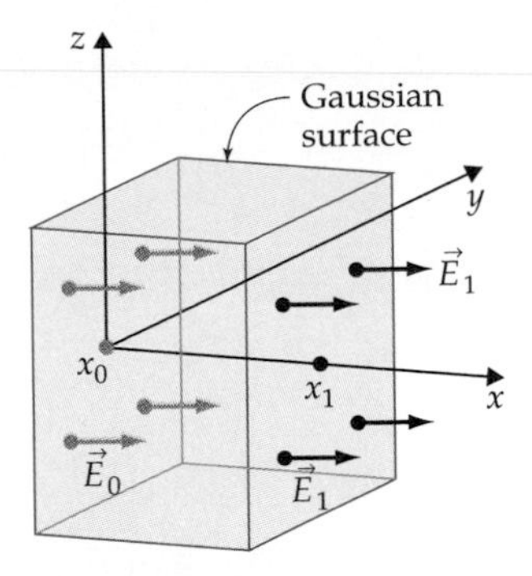

Figure E16.4
An electric field disturbance moving in the x direction in empty space that has a nonzero value of the E_x component would violate Gauss's law, because the value of E_x would be different on the right and left faces of the surface shown, implying that there will be a net flux even though the surface encloses no charge.

However, we *cannot* have a disturbance in E_x that moves through empty space along the x axis. Since E_x depends on x in such a wave, *any* wave of this form would generally have a different value of E_x at the left and right faces of the gaussian surface shown in figure E16.4. The net flux through the other faces of this surface must be zero, because $\vec{E}$ is parallel to those faces. Therefore, there would generally be a nonzero flux through the surface, even though that surface encloses no charge in empty space. This is not consistent with Gauss's law, so such a wave is impossible. We see, therefore, that *Maxwell's equations imply that electromagnetic waves are necessarily transverse:* the disturbance in the electric field can *only* have components in the directions perpendicular to the direction in which the wave is moving.

Maxwell's equations require that an EM wave have both electric and magnetic components

Another essential feature of an electromagnetic wave is that a disturbance in the electric field is necessarily accompanied by a disturbance in the magnetic field. We saw in section E16.1 that a disturbance in the electric field's z component at x_0 first creates (according to Faraday's law) a changing magnetic field in the $-y$ direction at position $x_0 + \frac{1}{2}\Delta x$, and it is this changing magnetic field that (according to the Ampere-Maxwell law) makes the value of the electric field at position $x_1 = x_0 + \Delta x$ accelerate. The accompanying magnetic disturbance is therefore an essential part of the process that allows the disturbance to move forward.

The relationship between E and $\mathbb{B}$

We can see how $\vec{E}$ and $\vec{\mathbb{B}}$ have to be related by considering a simple disturbance where E_z has a constant value behind a planar wave front that is perpendicular to the x axis and that moves in the $+x$ direction. We will assume that both $\vec{E}$ and $\vec{\mathbb{B}}$ are zero in front of the wave front. According to what we have already found, such a disturbance must move at the speed of light.

Figure E16.5 shows an amperian loop whose left leg is behind the plane and whose right leg is in front of the plane. Since the electric field is zero on the right leg and perpendicular to the top and bottom legs, the net electric circulation is equal to the contribution from the left leg, which is $-E_{z0}\,\Delta z$. Note that this value is *constant* during the entire time that it takes the wave front to move from the left leg to the right leg.

$$\frac{1}{c}\frac{d\Phi_{\mathbb{B}}}{dt} = -\oint \vec{E}\cdot d\vec{S} = E_{z0}\,\Delta z = \text{constant} \qquad \text{(E16.5)}$$

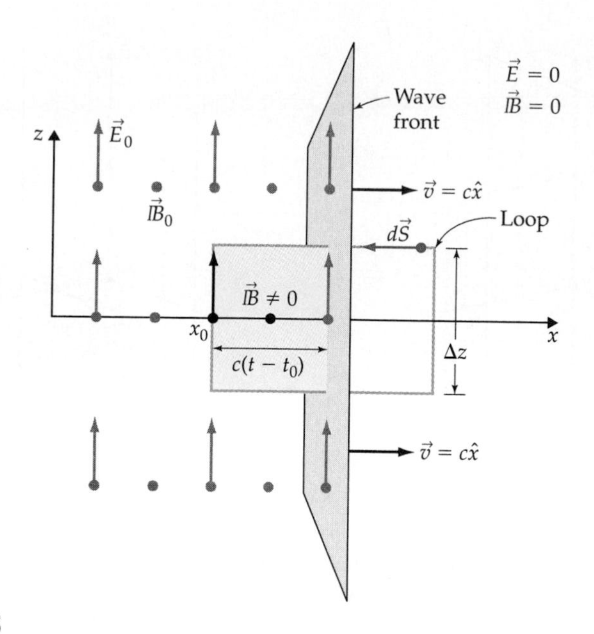

Figure E16.5

This picture shows a "step" wave moving in the $+x$ direction at the speed of light. Behind the wave front, $\vec{E}$ has a constant value of $\vec{E}_0$ and points in the z direction. In front of the wave front, both $\vec{E}$ and $\vec{B}$ are zero. Faraday's law applied to the amperian loop shown implies that the y component of the magnetic field must also be constant behind the wave front and have magnitude $B_0 = E_0$. (Note that the shaded region is the only region of the surface bounded by the loop where the magnetic field is not zero. The dots indicate that the magnetic field points in the $-y$ direction, that is, toward the viewer.)

Now, since tile vectors on the surface bounded by the loop point in the $-y$ direction, $\vec{B} \cdot d\vec{A} = -B_y\, dA$ at all points on this surface. But B_y is only nonzero at points behind the wave front, an area that has width $c(t - t_0)$, where t_0 is the time that the front passed the left leg (this is the shaded area in figure E16.5). Since the area of this region increases at a constant rate as the wave front moves forward, the flux through that surface will increase at a constant rate only if B_y has some fixed value, call it B_{y0}, behind the wave front. Under such circumstances,

$$\frac{1}{c}\frac{d\Phi_B}{dt} = \frac{1}{c}\frac{d}{dt}[-B_{y0}c(t - t_0)\Delta z] = -B_{y0}\,\Delta z \tag{E16.6}$$

Plugging this into equation E16.5, we find that behind the wave front, $-B_{y0} = E_{z0}$. Since the electric field points purely in the z direction by hypothesis, and since (as we saw in section E16.1) only the y component of $\vec{B}$ is affected by the wave, this means that

The electric and magnetic fields have the same magnitude in an electromagnetic wave

$$B = E \tag{E16.7}$$

This is actually a general result that applies to all points in *all* electromagnetic waves. Moreover, we have seen in *all* our examples (see figures E16.2, E16.3, and E16.5) that the associated magnetic field is perpendicular to the electric field. Figure E16.6 illustrates the electric and magnetic fields associated with a sinusoidal plane wave, evaluated at various points along the x axis at a given instant of time (a full picture of the field would show these vectors duplicated along all lines parallel to the x axis). The hand illustrates that an easy way to remember the relative directions of $\vec{E}$ and $\vec{B}$ is to note that $\vec{E} \times \vec{B}$ always points in the direction in which the wave moves.

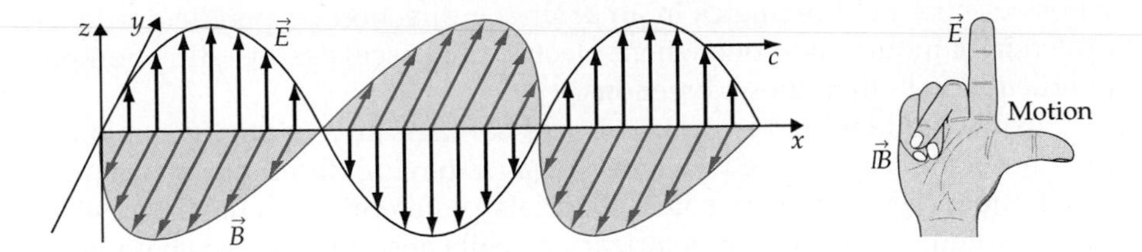

Figure E16.6
A snapshot at given time of a vertically polarized electromagnetic wave. The electric and magnetic field vectors shown have been evaluated at various points along the x axis (they will look the same for a given x at all values of y and z). The hand illustrates that $\vec{E} \times \vec{\mathbb{B}}$ points in the direction in which the wave is moving.

Because any $\vec{E}$ perpendicular to the x direction can be written as $\vec{E} = E_y\hat{y} + E_z\hat{z}$, any plane wave whose electric field is perpendicular to the x axis can be written as a superposition of two fundamental plane waves, one involving a disturbance in E_y and the other a disturbance in E_z. We call these two fundamental waves different **polarizations** of an electromagnetic wave: the first is *horizontally polarized* while the second is *vertically polarized* (the names assume that the z axis is vertical).

So the results of this section imply that electromagnetic fields for horizontally and vertically polarized sinusoidal waves moving in the $+x$ direction must have the following form:

The electric and magnetic fields for horizontally and vertically polarized plane waves

$$\vec{E}_H(x,t) = \begin{bmatrix} 0 \\ A_H \\ 0 \end{bmatrix} \sin(k_w x - \omega t) \qquad \vec{\mathbb{B}}_H(t,x) = \begin{bmatrix} 0 \\ 0 \\ A_H \end{bmatrix} \sin(k_w x - \omega t) \tag{E16.8a}$$

$$\vec{E}_V(x,t) = \begin{bmatrix} 0 \\ 0 \\ A_V \end{bmatrix} \sin(k_w x - \omega t) \qquad \vec{\mathbb{B}}_V(t,x) = \begin{bmatrix} 0 \\ -A_V \\ 0 \end{bmatrix} \sin(k_w x - \omega t) \tag{E16.8b}$$

Purpose: These equations describe the electromagnetic fields $\vec{E}_H$ and $\vec{\mathbb{B}}_H$ associated with a horizontally polarized sinusoidal plane wave moving in the $+x$ direction and the fields $\vec{E}_V$ and $\vec{\mathbb{B}}_V$ associated with the analogous vertically polarized wave.

Symbols: A_H and A_V are constants, ω is the wave's angular frequency, and k_w is its wavenumber (the subscript distinguishes this quantity from the Coulomb constant k). The value of ω is arbitrary, but once it is chosen, k_w is determined: $k_w = \omega/c$.

Limitations: This equation applies only to plane waves moving in the $+x$ direction, but you can easily adapt the equations to other directions if you remember that $\vec{E} \times \vec{\mathbb{B}}$ points in the direction in which the wave moves.

The Fourier theorem implies that *any* electromagnetic wave moving in the $+x$ direction can be constructed of a superposition of such waves at various frequencies.

Of course, the directions that we describe as being "horizontal" and "vertical" depend on our arbitrary choice of coordinates. The point here is that we can describe any electromagnetic wave as the sum of two waves, one whose electric field oscillates in an arbitrary direction perpendicular to the direction of motion, and one whose electric field oscillates in a direction perpendicular to both of these directions.

Visible light emitted by most sources contains an essentially incoherent mixture of sinusoidal electromagnetic waves having different polarizations and frequencies. However, some materials preferentially reflect, scatter, or transmit light waves of one polarization while absorbing waves having the other polarization. For example, the lenses of Polaroid sunglasses transmit vertically polarized light waves while absorbing the horizontally polarized light that is preferentially reflected from horizontal surfaces. This reduces the glare from such reflections.

In summary, we have seen that electromagnetic waves that move in a well-defined direction have the following characteristics:

1. The waves move at the speed of light.
2. Their $\vec{E}$ and $\vec{\mathbb{B}}$ vectors have equal magnitudes at a given point and time.
3. Their $\vec{E}$ and $\vec{\mathbb{B}}$ vectors are perpendicular to each other and to the direction of motion, which is indicated by the direction of $\vec{E} \times \vec{\mathbb{B}}$.

In 1887 and 1888, Heinrich Hertz was finally able to vindicate Maxwell's predictions by producing the first artificial electromagnetic waves (what we would now call radio waves) and showing that they had exactly the characteristics described in this section. By 1901, Guglielmo Marconi had used such waves to send a message across the Atlantic Ocean, and the wireless age had begun. Thus the musings of a middle-class Scottish college professor began to completely change the world.

E16.3 The Energy in an Electromagnetic Wave

Electromagnetic waves must carry energy

A microwave oven operates by generating electromagnetic waves having a wavelength of roughly 1 cm. The fact that food placed in a microwave oven gets warm is clear evidence that such waves can carry energy. In this section we will use the results of section E16.2 to quantify the energy carried by an electromagnetic sinusoidal wave.

In section E14.5 we found that the density of energy in an electromagnetic field is given by

$$u_{\text{EM}} = \frac{1}{8\pi k}(E^2 + \mathbb{B}^2) \tag{E16.9}$$

Since $E = \mathbb{B}$ everywhere in an electromagnetic wave, a sinusoidal wave's total energy density at any specific point in space is

$$u_{\text{EM}} = \frac{E^2 + \mathbb{B}^2}{8\pi k} = \frac{E^2}{4\pi k} = \frac{E_0^2}{4\pi k}\sin^2(k_w x - \omega t) \tag{E16.10}$$

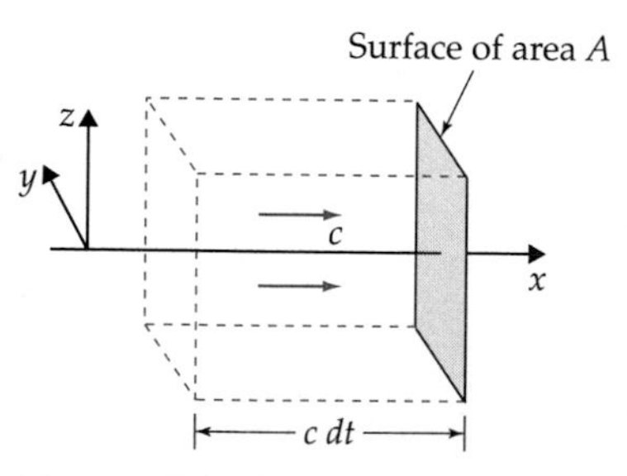

Figure E16.7
All the electromagnetic energy within the box shown will hit the plate within time dt.

where E_0 is the amplitude of the electric field wave. Now imagine that we have a flat surface of area A oriented perpendicular to the direction in which the wave is moving, as shown in figure E16.7. All the energy contained within the box of volume $Ac\,dt$ will hit the plate within time dt, so the rate at which energy is delivered to the surface per unit time must be

$$\frac{dU}{dt} = \frac{u_{\text{EM}} Ac\,dt}{dt} = Ac\frac{E_0^2}{4\pi k}\sin^2(k_w x - \omega t) \tag{E16.11}$$

if dt is small enough that $\sin^2(k_w x - \omega t)$ doesn't vary much within the box.

Definition of the intensity of an electromagnetic wave

The **intensity** of an electromagnetic wave is defined to be the total power per unit area that it would deposit on a surface held perpendicular to the wave. According to equation E16.11, the intensity of an electromagnetic sinusoidal wave is

$$\text{Intensity of EM wave} = \frac{1}{A}\frac{dU}{dt} = \frac{cE_0^2}{4\pi k}|\sin(k_w x - \omega t)|^2 \qquad \text{(E16.12)}$$

Note that the intensity is proportional to the *square* of the electric field amplitude. Now, the average of $\sin^2\theta$ (which oscillates between 0 and 1) over many oscillations is simply $\frac{1}{2}$, so the average intensity is

$$\text{Average intensity of an EM wave} = \frac{c}{4\pi k}(E^2)_{\text{avg}} = \frac{cE_0^2}{8\pi k} \qquad \text{(E16.13)}$$

Purpose: This equation describes the energy per unit time per unit area deposited on a surface that absorbs a sinusoidal electromagnetic wave.
Symbols: E_0 is the amplitude of the wave's oscillating electric field, c is the speed of light, and k is the Coulomb constant.
Limitations: This equation applies only to sinusoidal waves.

A formula for the average intensity of an EM wave

Exercise E16X.3

Show that the units of the final expression are watts per square meter (the appropriate SI units for a power per unit area). Show also that the constant $4\pi k/c$ has a value of about 377 Ω.

Exercise E16X.4

Sunlight has an intensity of roughly 1 kW/m^2. Estimate the amplitude of the electric field involved in such light.

E16.4 The Power Radiated by a Charge

Our goal here is to find the power radiated by an accelerating charge

We have seen that we can find wavelike solutions to Maxwell's equations, and we have seen that these waves carry energy. But how can we *create* electromagnetic waves? In this section, I will argue that *accelerating* charges create electromagnetic waves and estimate the energy radiated by an accelerating charge.

We might think that any *moving* charge would create an electromagnetic wake like the wake of a boat moving through water. But the theory of relativity precludes this. Consider a charge moving at a constant velocity. We can find an inertial frame where this charge is at rest, and we *know* that a charge at rest does not radiate electromagnetic waves. But electromagnetic waves, since they carry energy and have distinctive physical effects, cannot exist in one inertial frame and not in others. So a charge cannot generate an electromagnetic wave in a frame where it moves at a constant velocity: it must *accelerate*.

We will use the method of *dimensional analysis*

We could work through a complicated derivation of the power radiated by an accelerating charge, but we can use unit consistency to arrive at essentially the same result much more quickly. What could this power depend on? It must depend on the charge q of the particle being accelerated, and as we've seen, it is plausible that it depends on the magnitude of the particle's acceleration a. The Coulomb constant k appears in nearly every formula having to do with electricity and magnetism. Finally, our formula should plausibly involve the speed of light c, since we are talking about electromagnetic waves. The particle's mass, however, is likely to be irrelevant for this electromagnetic problem, and though the formula might depend on the particle's velocity, we will consider this later and find that it leads to physically unreasonable results. There are no other quantities associated with the particle that could reasonably be involved. So let us assume that our formula depends on q, a, k, and c, and nothing else.

Doing the analysis

Note that if either the charge q or the particle's acceleration a is zero, then the radiated power will be zero: a neutral particle will not radiate no matter how violently it is accelerated, and a charged particle moving at a constant velocity will not radiate, no matter how large its charge is. This suggests that the power must depend on the product of these quantities raised to positive powers. Let's see if we can find a reasonable equation of the form:

$$P = bk^i c^j q^m a^n \tag{E16.14}$$

where b is an unknown unitless constant of proportionality (like 8π or $\frac{4}{3}$ or something) and i, j, m, and n are unknown numbers (note that m and n should be positive).

Now, in order for a formula like this to work, its *units* must be consistent. The units of power are $\text{W} = \text{J/s} = \text{kg}\cdot\text{m}^2/\text{s}^3$. The units of q are coulombs, and the units of a are meters per second squared. The units of the Coulomb constant k are

$$\frac{\text{N}\cdot\text{m}^2}{\text{C}^2} = \frac{(\text{kg}\cdot\text{m/s}^2)\text{m}^2}{\text{C}^2} = \frac{\text{kg}\cdot\text{m}^3}{\text{C}^2\cdot\text{s}^2} \tag{E16.15}$$

So if equation E16.14 is to have self-consistent units, we must have

$$\frac{\text{kg}\cdot\text{m}^2}{\text{s}^3} = \left(\frac{\text{kg}\cdot\text{m}^3}{\text{C}^2\cdot\text{s}^2}\right)^i \left(\frac{\text{m}}{\text{s}}\right)^j \text{C}^m \left(\frac{\text{m}}{\text{s}^2}\right)^n = \frac{\text{kg}^i\cdot\text{m}^{3i+j+n}\cdot\text{C}^{m-2}}{\text{s}^{2i+j+2n}} \tag{E16.16}$$

There is only one power of kilograms on the left, so there can be only one power of kilograms on the right: thus $i = 1$. There are no units of coulombs on the left, so we can't have any on the right, so $m = 2$. If we use $i = 1$, we also find that consistency requires

$$2 = 3 + j + n \quad \text{(to make the distance units consistent)} \tag{E16.17a}$$

$$3 = 2 + j + 2n \quad \text{(to make the time units consistent)} \tag{E16.17b}$$

If you subtract equation E16.17a from E16.17b, you can solve for n and then plug this result back into either equation to find j. If you then plug the resulting values back into equation E16.14, you should find that

$$P = \frac{bkq^2a^2}{c^3} \tag{E16.18}$$

Exercise E16X.5

Verify equation E16.18.

Note that the powers of q and a did come out positive, as hoped. We see that this technique (which is called **dimensional analysis**) allows us to determine virtually everything about the formula without a detailed derivation. It does not really matter what a "correct" derivation would look like: if the formula has the form given in equation E16.14 (which was *plausible* in this case) and depends only on q, a, k, and c, then it *must* be as given by equation E16.18.

Unfortunately, this method does not determine the value of b, the potential "unitless constant" in this equation. Our only comfort is that in many of the equations that we have seen before of this type, such constants are rarely an order of magnitude bigger or smaller than 1, so substituting $b = 1$ will likely be correct to within an order of magnitude. A full derivation in this case shows that $b = \frac{2}{3}$, implying that

The Larmor formula

$$P = \frac{2}{3}\frac{kq^2a^2}{c^3} \qquad \text{(E16.19)}$$

Purpose: This is the **Larmor formula** for the power P that an accelerating charge q radiates in the form of electromagnetic waves.
Symbols: c is the speed of light, and a is the particle's acceleration.
Limitations: This applies only to nonrelativistic point particles.

Exercise E16X.6

Rework the argument to include the possibility that P depends on the particle's speed v, and argue that the result is physically absurd.

Exercise E16X.7

Imagine that you whirl a ball with a charge of 10 nC at the end of a 1.0-m string around your head at a speed of 5 m/s. At roughly what rate would the ball radiate energy in the form of electromagnetic waves?

The method of *dimensional analysis* used in this section is one of the standard tools of the practicing scientist! It is *very* useful when you know enough physics to choose the right *variables* but not enough to do the detailed derivation. Here is an outline of the method:

An outline of the method of dimensional analysis

1. Determine the variables on which the quantity of interest depends.
2. Invent a plausible equation for the quantity of interest that involves products of powers of those variables.
3. Determine unknown powers in the equation by unit consistency.
4. Assume any unitless constants to be of order of magnitude of 1.

While it is by no means foolproof, the method can provide a useful first guess if thoughtfully applied. Even when a formula is actually a more complicated function of the variables than the simple products of powers assumed here, the product of powers is often a good approximation. Unitless constants rarely differ from 1 by more than a factor of 10.

E16.5 Why the Sky Is Blue

It turns out that the Larmor formula provides the key to understanding a simple, everyday mystery: why is the sky blue? The answer has nothing much to do with any of the particulars about our atmosphere, but rather with some very basic physics.

A description of the Rayleigh scattering process

We have seen that light is an electromagnetic wave that causes the electric and magnetic field vectors at a given location to oscillate as the wave passes; so when sunlight hits an atom in the air, it causes the electric field vector at the atom's location to oscillate back and forth. Any electric field at the atom's location will push the atom's positively charged nucleus one way and its negatively charged electron cloud in the other way (as we discussed in chapter E2). Light will therefore cause an atom's nucleus and electron clouds to oscillate in opposite directions (if the nucleus goes right, the electron cloud goes left, and vice versa) at the same frequency as the light.

Now, the charged nucleus and electron cloud accelerate as they oscillate, so they will radiate in all directions electromagnetic waves that carry away some of the energy received from the initial light. So the light you see from the sky is simply sunlight that has been reradiated by oscillating air atoms. This physical process is called **Rayleigh scattering.**

You may already know that the *color* of light depends on its frequency. Light from the sun is a mixture of waves with various frequencies. It might seem that since the light scattered by an atom must have the same frequency as that striking it, the light we see scattered from the sky should have the same color mix as sunlight. So why is the sky blue, not yellow like the sun?

Why some colors are more effectively radiated than others

It turns out that some frequencies get reradiated much more *effectively* than others. The position (in some suitable reference frame) of the center of an electron cloud oscillating in response to a light wave will be given by something like

$$x(t) = x_0 \sin \omega t \tag{E16.20}$$

where ω is the angular frequency of the incident light and x_0 is the amplitude of the oscillation. If so, the magnitude of the acceleration of the electron cloud's center is

$$a = \left| \frac{d^2x}{dt^2} \right| = \omega^2 x_0 \left| \sin \omega t \right| \tag{E16.21}$$

Exercise E16X.8

Verify equation E16.21.

The sky is blue because blue light and violet light are preferentially scattered

According to the Larmor formula, the energy per unit time that an atom reradiates in the form of scattered light waves is proportional to a^2 and thus to ω^4! Since light at the violet end of the visible spectrum has nearly twice the frequency of light at the red end, violet light is scattered very roughly 16 times more effectively than red light! Thus colors at the violet end of the spectrum are strongly emphasized over those at the red end in the scattered light.

So why doesn't the sky look violet instead of blue? One reason is that sunlight itself contains much less violet light than blue light, so even though violet light is scattered very effectively, there isn't as much to begin with. Another reason is that human eyes are much less sensitive to violet light than they are to blue light, so we do not see the violet light that is there well.

This also explains the red of the setting sun

This also explains why the sun looks red while setting. When the sun is near the horizon, its light has to travel through much more air to get to our eyes than when it is overhead. This air scatters away most of the energy associated with the higher-frequency portions of the sunlight as it passes. Only the lower-frequency colors (such as red and orange) in the sunlight therefore survive the journey to reach our eyes.

E16.6 Maxwell's Rainbow

While Maxwell's equations require that electromagnetic waves move with the speed of light, they do not put any constraints on the *frequency* that such waves can have, which is usually determined by the frequency of the oscillating charges that create it. There are many kinds of physical situations where charged particles accelerate or oscillate for different reasons. Hertz created the first artificial electromagnetic waves by using a circuit that was the electric analogy of a mass on a spring to accelerate electrons back and forth about a billion times per second: the electromagnetic waves he created thus had a frequency of roughly 1 GHz (corresponding to a wavelength $\lambda = c/f \approx 33$ cm).

Different physical processes produce waves with characteristic frequencies

In general, the electromagnetic waves created by the bulk motion of many charges have frequencies that are classified as being either in the **radio wave** range (very roughly 300 MHz or lower) or the **microwave** range (300 MHz to about 10^{12} Hz). For example, the electromagnetic waves that carry information to your FM radio have frequencies of about 100 MHz and are created by electrons sloshing back and forth in the radio station's antenna (driven by the transmitter).

There are other kinds of physical systems where charges accelerate even more rapidly. For example, the atoms in any object that has internal thermal energy vibrate and/or move randomly about. The resulting atomic oscillations and collisions cause electron clouds to accelerate and radiate electromagnetic waves. This becomes obvious when the object is *very* hot: your hand gets warm near a hot coal because of the energy it receives from electromagnetic waves radiated by the coal. The electromagnetic waves radiated by thermal processes typically have frequencies of roughly 10^{12} to 10^{14} Hz ($\lambda \approx 100\ \mu$m to $1\ \mu$m), which we call the **infrared** range.

Electromagnetic waves in the **ultraviolet** range ($f \approx 10^{16}$ Hz, $\lambda \approx 100$ nm) are usually created by atomic electrons changing energy levels. When electrons are accelerated using an electron gun and then crashed into a metal plate, their sudden deceleration produces **X-rays** with frequencies of very roughly 10^{18} Hz ($\lambda \approx 1$ nm). Transformations inside an atomic nucleus can cause charged protons to shift position suddenly and energetically, radiating **gamma rays** with frequencies of very roughly 10^{22} Hz ($\lambda \approx 0.1$ pm).

The range of possible frequencies is, of course, continuous, but physicists have chosen to break this continuous distribution of frequencies into ranges that roughly reflect the different physical processes that produce such waves (see figure E16.8).

Visible light is only a small part of this range

You can see that visible light, which has frequencies ranging from about 4.3×10^{14} Hz ($\lambda = 700$ nm) to about 7.5×10^{14} Hz ($\lambda = 400$ nm), constitutes only a very narrow part of this range (although, not coincidentally, it is a range where the sun radiates a significant fraction of its energy). Figure E16.9 shows various everyday objects that emit nonvisible electromagnetic waves in various frequency ranges.

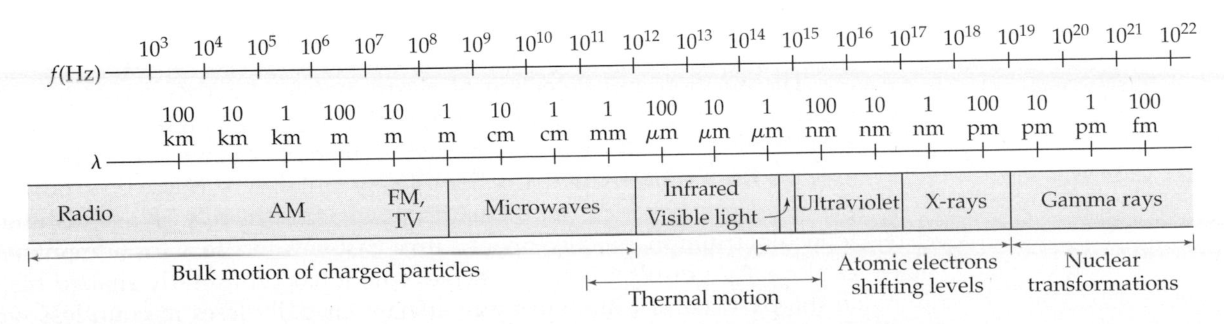

Figure E16.8
Maxwell's rainbow.

(a)

(b)

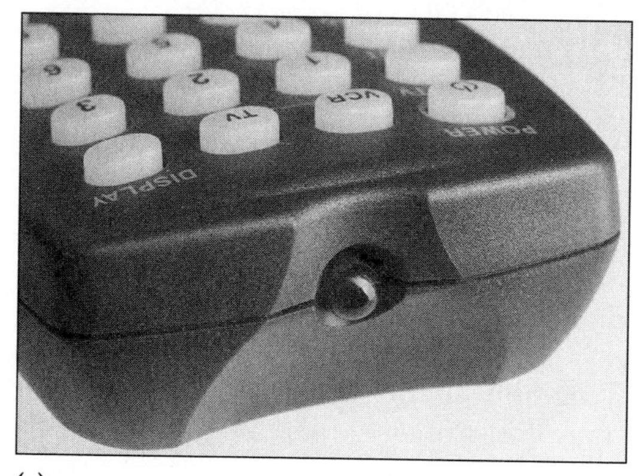

(c)

(d)

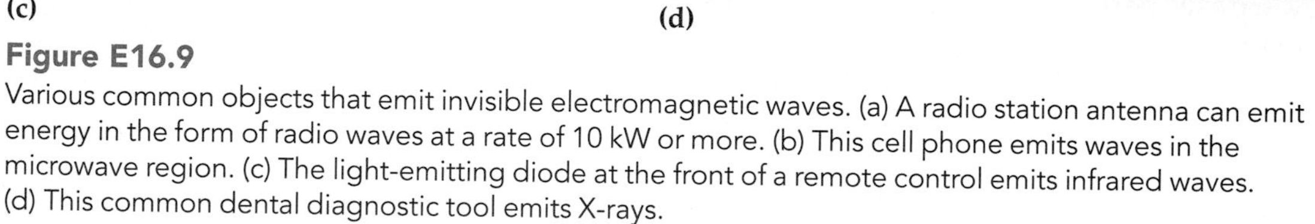

Figure E16.9

Various common objects that emit invisible electromagnetic waves. (a) A radio station antenna can emit energy in the form of radio waves at a rate of 10 kW or more. (b) This cell phone emits waves in the microwave region. (c) The light-emitting diode at the front of a remote control emits infrared waves. (d) This common dental diagnostic tool emits X-rays.

The creation and detection of electromagnetic waves across this spectrum are so much a part of 21st-century technology that it is hard to imagine that before Maxwell's work in the 1860s, most of "Maxwell's rainbow" outside of the visible range was completely unknown. Maxwell's efforts to express Faraday's ideas mathematically were not merely an astonishing intellectual triumph, but also opened up a vast new territory for humankind to explore. As a result, Maxwell's achievement has profoundly shaped history in the past century, and currently affects our daily lives in countless ways. Maxwell's equations are thus among the most important of the great ideas that shaped physics.

TWO-MINUTE PROBLEMS

E16T.1 An electromagnetic wave moving in the $+x$ direction is best described as being
A. A transverse wave
B. A longitudinal wave
C. Either a transverse or a longitudinal wave (depending on how it was created)
D. A wave that fits neither category well

E16T.2 It is possible in principle to create a traveling electromagnetic wave in empty space that is purely electric ($\vec{\mathbb{B}} = 0$), true (T) or false (F)?

E16T.3 In both possible polarizations of a sinusoidal electromagnetic wave, $\vec{E} \times \vec{\mathbb{B}}$ points in the direction in which the wave is moving, T or F?

E16T.4 Imagine that the electric field part of an electromagnetic wave moving in the $+z$ direction is given by $E_x = A\sin(kz - \omega t)$, $E_y = E_z = 0$. The x and z components of the magnetic field part are both zero, T or F? The y component of the magnetic field part is
A. $\mathbb{B}_y = +A\sin(kx - \omega t)$
B. $\mathbb{B}_y = -A\sin(kx - \omega t)$
C. $\mathbb{B}_y = \mathbb{B}_0\sin(kx - \omega t)$, with $\mathbb{B}_0$ not necessarily equal to A
D. $\mathbb{B}_y = 0$
E. Given by some other expression (specify)

E16T.5 Without Maxwell's correction to Ampere's law, electromagnetic disturbances could not move in empty space, T or F?

E16T.6 The energy carried by a sinusoidal electromagnetic wave is nonnegative everywhere at all times, T or F?

E16T.7 Imagine that when you switch on your study lamp, you increase the intensity of light shining on your textbook by a factor of 16. By what factor does the average electric field strength in this light increase?
A. 256
B. 16
C. 8
D. 4
E. 2
F. 1 (the field remains unchanged)
T. <1 (the electric field strength gets smaller)

E16T.8 A charge moving at a constant speed cannot radiate electromagnetic waves, T or F?

E16T.9 A charged atom oscillating in a certain molecule has a period of oscillation of about 1 ps. What kind of electromagnetic radiation will it produce?
A. Radio
B. Microwave
C. Infrared
D. Visible
E. Ultraviolet
F. Some other kind (specify)

HOMEWORK PROBLEMS

Basic Skills

E16B.1 Imagine that the amperian loop shown in figure E16.10 has a length equal to one-half of the wavelength of a vertically polarized sinusoidal wave. Assume that the electromagnetic field in this wave is as given by equation E16.8*b*. Consider the instant that the electric field on the loop's left leg is zero. Argue that Faraday's law will be satisfied at this instant. (*Hint:* The average value of $-\mathbb{B}_y$ over the loop is a maximum at this instant. Why?)

E16B.2 Argue that sinusoidal waves of the type described by equations E16.8 satisfy Gauss's law for the magnetic field.

E16B.3 Use equations E12.4 to argue that if the ratio $E/\mathbb{B} = 1$ for an electromagnetic wave in any inertial reference frame, it has the same ratio in all reference frames moving parallel to the wave. (*Hint:* Orient

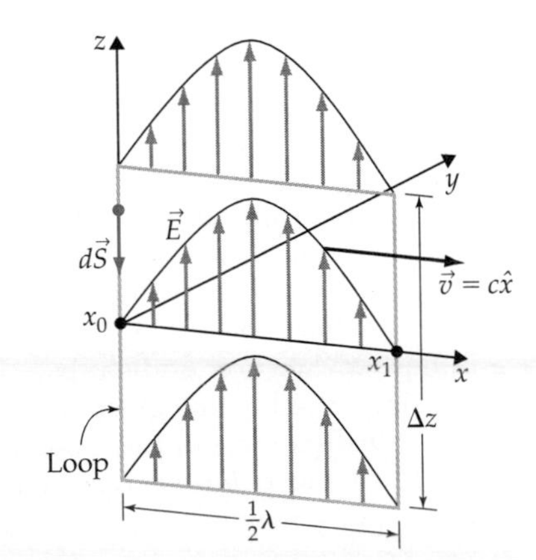

Figure E16.10
An amperian loop that spans an entire half-wavelength of a sinusoidal electromagnetic wave. The gray vectors show the wave's electric field at points along the x axis and along the loop's top and bottom legs. (See problem E16B.1.)

your axes so that the x direction is the direction in which the wave is moving and the electric and magnetic fields point in the $+y$ and $+z$ directions, respectively.)

E16B.4 Imagine that a sinusoidal electromagnetic wave has an electric field that has an amplitude of 300 N/C and points parallel to the yz plane along a line 30° from the z axis toward the y axis. Imagine that we write this wave as a sum of horizontally and vertically polarized waves.

(a) What are the values of A_H and A_V? (*Hint:* One of these values could be negative.)

(b) At an instant when E_z is positive, what are the magnitude and direction of $\vec{\mathbb{B}}$?

E16B.5 In a fairly brightly-lit room, the intensity of light is about 200 W/m^2. What is the amplitude of the electric field of the light waves in this room? If the room is 5 m long by 4 m wide by 2.5 m high and the light intensity is fairly uniform throughout the room, about how much energy is stored in the room in the form of light waves?

E16B.6 Near a certain source of light, the amplitude of the electric field in the light waves is about 25 N/C. What is the intensity of this light?

E16B.7 Imagine that a certain light beam has an energy density of 1.0 μJ per cubic meter. What is the intensity of this light? How does this compare to the intensity of sunlight?

E16B.8 Electric field strengths in excess of 10^9 N/C can tear electrons out of atoms. What intensity of light will do this? How does this compare to the intensity of sunlight?

E16B.9 Imagine that an object with a charge of 0.1 μC traveling at 30 m/s hits a wall and comes to rest within about 10 ms. How much total energy is radiated in the form of electromagnetic waves in this situation?

Synthetic

E16S.1 Imagine disturbing an electric field by somehow suddenly "deflecting" the value of E_y everywhere on the $x = x_0$ plane to some value E_{y0} while leaving the electric field at other points in nearby empty space in the $+x$ direction undisturbed. Present an argument similar to the one given in section E16.1 to show that disturbances of this component of the electric field should obey the wave equation and move at the speed of light.

E16S.2 There is nothing special about the electric field in sections E16.1 and E16.2: we could have just as easily showed that Maxwell's equations in empty space lead to wave equations for the *magnetic* field, found the two possible polarizations for such magnetic wave solutions, and calculated the electric field that goes along with each polarization. Let us just take the first along this path. Imagine disturbing the magnetic field by somehow suddenly "deflecting" the value of $\mathbb{B}_z$ from zero to some value $\mathbb{B}_{z0}$ everywhere on the $x = x_0$ plane while leaving all points in the $+x$ direction in empty space unchanged. (*Hint:* All the parts of this problem are closely analogous to what we did with the electric field in section E16.1 and exercise E16X.1.)

(a) Consider a small loop in the xz plane with one leg at $x = x_0$ and another at $x = x_1 = x_0 + \Delta x$. Draw a picture of this loop, and show that the Ampere-Maxwell law implies that the value of E_y at the middle of this loop (at x-position x_m) must start growing at a rate proportional to $\mathbb{B}_{z0}$.

(b) Consider loops in the xy plane and yz plane, and argue that the Ampere-Maxwell law does not require either E_x or E_z to change.

(c) Now consider a small loop in the xy plane with one leg at x_m and the other at $x_m + \Delta x$. Argue that the time derivative of Faraday's law for this loop, combined with the result of part (a), implies that a disturbance in $\mathbb{B}_z$ will obey the wave equation and move with speed c.

E16S.3 Consider a step wave, where we assume that $\vec{E} = E_0\hat{y}$ and $\vec{\mathbb{B}} = \mathbb{B}_0\hat{z}$ behind and $\vec{E} = 0$ and $\vec{\mathbb{B}} = 0$ in front of a planar wave front that is perpendicular to the x direction and moves in the $+x$ direction with speed v through empty space. Consider two amperian loops, each of which has a leg at an x position we will define to be $x = 0$ and another at $x = L$, but which lie in the xy and xz planes, respectively. Do *not* assume that L is small or that the value of the flux over a surface bounded by a loop is equal to the field's value in the middle multiplied by the surface's area; instead, calculate the time derivatives of these fluxes *exactly* at an instant when the wave front's x-position is between 0 and L. Assume that we have defined $t = 0$ to be the instant when the wave front passes $x = 0$: then the position of the wave front will be $x = vt$.

(a) Show that Faraday's law implies in this case that $E_{y0} = (v/c)\mathbb{B}_{z0}$.

(b) Show that the Ampere-Maxwell law for empty space implies that $\mathbb{B}_{z0} = (v/c)E_{y0}$.

(c) Prove that this means that this step wave has to move at the speed of light and that $E_0 = \mathbb{B}_0$.

E16S.4 Express Maxwell's equations in empty space in terms of traditional units (that is, use $\vec{B}$, ε_0, and μ_0 instead of $\vec{\mathbb{B}}$, k, and c). Then go through the process illustrated in section E16.1 and show that you end up with wave equations where $1/\varepsilon_0\mu_0$ appears instead of c^2.

E16S.5 Use equation E12.4 to prove that the value of $\vec{E} \cdot \vec{\mathbb{B}}$ is frame-independent, and use this result to argue that if the electric and magnetic fields in an electromagnetic traveling wave are perpendicular in one inertial reference frame, they are perpendicular in *all* inertial reference frames (even frames not moving parallel to the electromagnetic wave).

E16S.6 What is the approximate electric field strength of the electromagnetic waves radiated by a 100-W lightbulb, measured 3.0 m from the bulb?

E16S.7 The *Poynting vector* $\vec{S}$ for an electromagnetic wave has a direction equal to the wave's direction of motion and a magnitude equal to the wave's intensity in watts per square meter. Show that for both polarizations described by equations E16.8*a* and E16.8*b*

$$\vec{S} = \frac{c}{4\pi k}(\vec{E} \times \vec{\mathbb{B}}) = \frac{1}{\mu_0}(\vec{E} \times \vec{B}) \tag{E16.22}$$

E16S.8 An electron moves at a speed of $0.01c$ in a magnetic field of $\mathbb{B} = 600$ MN/C perpendicular to its direction of motion. (Remember that this magnetic field will cause the electron to move in a circular orbit: see chapter E7.) How much energy does the electron lose each second? Express your answer in electronvolts (1 eV $= 1.602 \times 10^{-19}$ J).

E16S.9 What electric charge would the moon have to have so that it would radiate away 1 percent of its current orbital kinetic energy in the form of electromagnetic waves over the course of 1 billion years?

E16S.10 Imagine that I place a charge of 10 nC on one prong of a tuning fork that vibrates at 440 Hz with an amplitude of about 0.4 mm. What is the average power radiated by such a charge in the form of electromagnetic waves?

E16S.11 In the Rutherford model of the hydrogen atom (proposed by Ernst Rutherford in roughly 1910), an electron is imagined to orbit the proton in a circle whose radius is about $r_0 \approx 0.053$ nm.

(a) Use Newton's second law, Coulomb's law, and what you know about acceleration in circular motion to show that the electron's acceleration in a circular orbit of radius r is $a = ke^2/mr^2$ and its kinetic energy is $K = ke^2/2r$, where $e =$ charge on a proton $= |$charge on an electron$|$ and m is the mass of the electron.

(b) Show that the total orbital energy of the electron in such an orbit is $E = -ke^2/2r$.

(c) Because the electron is accelerating, it will radiate energy in the form of electromagnetic waves. Assuming that it does so slowly enough that its orbit remains essentially circular, show that the Larmor formula predicts that the rate at which it radiates energy is

$$P = \frac{2ke^2c}{3r^4}\left(\frac{ke^2}{mc^2}\right)^2 \tag{E16.23}$$

(d) This has to come at the expense of the electron's orbital energy E, so (since E decreases with time) $P = -dE/dt$. Show, using the result of part (b), that we independently have

$$\frac{dE}{dt} = \frac{ke^2}{2r^2}\frac{dr}{dt} \tag{E16.24}$$

(e) Set P given by equation E16.23 equal to $-dE/dt$ given by equation E16.24, and rearrange to show that

$$r^2\,dr = -\frac{4}{3}\left(\frac{ke^2}{mc^2}\right)^2 c\,dt \tag{E16.25}$$

(f) Define $t = 0$ to occur when $r = r_0$. As it radiates energy, the electron will spiral inward until it reaches $r = 0$ at a time we will define to be $t = T$. By integrating the left side from $r = r_0$ to $r = 0$ and the right side from $t = 0$ to $t = T$ and solving for T, show that

$$T = \frac{r_0^3}{4c}\left(\frac{mc^2}{ke^2}\right)^2 \tag{E16.26}$$

(g) Show that $ke^2/mc^2 = 2.8 \times 10^{-15}$ m, and go on to calculate T. Is the Rutherford model plausible?

Rich-Context

E16R.1 How long would it take an object like the sun to collapse to a black hole if the forces holding it up against its own gravity size were suddenly to disappear? Estimate the time roughly, using dimensional analysis. (*Answer:* roughly 1 h.)

E16R.2 Imagine that some electrons are moving at a speed of $0.001c$ in a magnetic field that keeps them in a circular orbit 10 cm in radius. Assume that the electrons are moving in a vacuum. About how much time will pass before the radius of the orbit of the electrons has increased to 20 cm?

Advanced

E16A.1 A radio transmitter creates a time-dependent current $I(t) = I_0 \sin\omega t$ in the radio antenna, where $I_0 = 20$ A and $\omega = 6.3 \times 10^6$ per second (corresponding to about the middle of the AM band). What is the average power radiated by this antenna if it is 160 m long? [*Hint:* The EM radiation is produced by charge sloshing back and forth in the antenna. Argue that at any given instant $I(t) = Qv/L$, where Q is the total charge involved in the sloshing, v is the drift speed of the charges, and L is the antenna's length.]

ANSWERS TO EXERCISES

E16X.1 Consider first a loop in the yz plane, as shown in figure E16.11. The disturbed electric field is perpendicular to the top and bottom legs, so these legs contribute nothing to the net circulation. Since the disturbed electric field is independent of y, it will have the same value at all points along the left and right legs. However, since the disturbed electric field is parallel to the segment vectors on the right leg but antiparallel to those on the left leg, the right leg's contribution to the net circulation will be exactly canceled by the contribution from the left leg. Therefore Faraday's law simply says that

$$\oint \vec{E} \cdot d\vec{S} + \frac{1}{c}\frac{d\Phi_{B}}{dt} = 0 \quad \Rightarrow \quad \frac{d\Phi_{B}}{dt} = 0 \tag{E16.27}$$

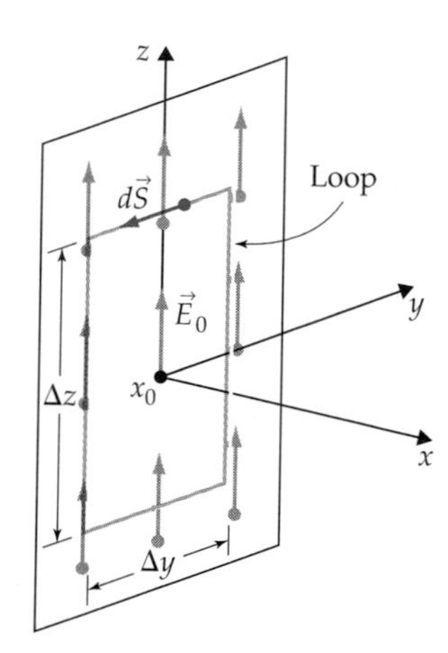

Figure E16.11
The electric circulation around the amperian loop shown is zero, because the contributions of the vertical legs cancel. Faraday's law then implies that the electric field disturbance at $x = x_0$ will not affect the value of B_x.

Since the tile vectors on this loop point in the $+x$ direction, the net flux through this loop must be

$$\Phi_{B} = \int \vec{B} \cdot d\vec{A} = \int B_x \, dA \tag{E16.28}$$

Therefore, saying that $d\Phi_{B}/dt = 0$ amounts to saying that $B_{x,\text{avg}}$ over the surface does not change. The disturbance in the electric field therefore has no effect on this component of the magnetic field.

Now consider a loop in the xy plane. The disturbed electric field is perpendicular to all points on the loop, so the net circulation is again zero and Faraday's law again implies that $d\Phi_{B}/dt = 0$. In this case, the tile vectors point in the $+z$ direction, so saying that $d\Phi_{B}/dt = 0$ amounts to saying that $B_{z,\text{avg}}$ over the surface does not change. Again, the disturbance in the electric field has no effect on this component of the magnetic field.

E16X.2 The value of $\vec{B} \cdot d\vec{S}$ for loop B is zero on the right leg of this loop because the magnetic field is still not disturbed there. It is zero on the top and bottom legs because $\vec{B}$ points entirely in the z direction and so is perpendicular to segment vectors on these legs. The segment vectors in the left leg have components $[0, -dS, 0]$, so the component version of the dot product means that $\vec{B} \cdot d\vec{S} = B_{ym}\, dS_y = -B_{ym}\, dS$. Now because the initial disturbance E_{z0} is independent of z, the growing value of $-B_{ym}$ will also be independent of z. This means that the total magnetic circulation around the loop is therefore (as claimed)

$$\oint \vec{B} \cdot d\vec{S} = \int (-B_{ym}\, dS) = -B_{ym} \int_{\text{left leg}} dS = -B_{ym}\,\Delta y \tag{E16.29}$$

as claimed. Now, the tile vectors on the surface bounded by this loop all point in the $+z$ direction, so the component definition of the dot product implies that $\vec{E} \cdot d\vec{A} = E_z\, dA_z = +E_z\, dA$. The total flux through the loop will therefore be the average value of E_z over the loop's face multiplied by the loop's area $\Delta x \Delta y$. Since the loop's center is at $x = x_1$, the value of E_z at the loop's center is E_{z1}. If the loop is small enough that $E_{z1} \approx$ the average value of E_z over the loop's face, then the flux through the loop is $\Phi_E \approx E_{z1}\Delta x \Delta y$. Therefore the Ampere-Maxwell law in this case reads

$$\oint \vec{B} \cdot d\vec{S} - \frac{1}{c}\frac{d\Phi_E}{dt} = 0$$

$$\Rightarrow \quad -B_{ym}\,\Delta y - \frac{1}{c}\frac{dE_{z1}}{dt}\Delta x \Delta y = 0 \tag{E16.30}$$

This is the first equation in equation E16.3. To get the second, multiply both sides by $c\,(\Delta x \Delta y)$ and add dE_{z1}/dt to both sides.

E16X.3 k/c has units of $(\text{N}\cdot\text{m}^2/\text{C}^2)(\text{s/m}) = \text{N}\cdot\text{m}\cdot\text{s}/\text{C}^2$. If you multiply the reciprocal of this by the square of the electric field units N/C, you get $(\text{N/C})^2 \times (\text{C}^2\cdot\text{N}^{-1}\cdot\text{m}^{-1}\cdot\text{s}^{-1}) = \text{N}\cdot\text{m}^{-1}\cdot\text{s}^{-1} = \text{N}\cdot\text{m}\cdot\text{m}^{-2}\cdot\text{s}^{-1} = \text{J}/(\text{m}^2\cdot\text{s}) = \text{W/m}^2$. We can convert the units of k/c to ohms as follows: $\text{N}\cdot\text{m}\cdot\text{s}/\text{C}^2 = \text{J}\cdot\text{s}/\text{C}^2 = (\text{J/C})/(\text{C/s}) = \text{V/A} = \Omega$. Finding the numerical result is simply a matter of plugging in numbers.

E16X.4 Solving equation E16.13 for E_0, we get

$$E_0^2 = \frac{8\pi k}{c}(\text{intensity})$$

$$= 2\left(377\,\frac{\text{N}\cdot\text{m}\cdot\text{s}}{\text{C}^2}\right)\left(1000\,\frac{\text{J}}{\text{m}^2\cdot\text{s}}\right)$$

$$= 754{,}000\left(\frac{\text{N}\cdot\cancel{\text{m}}\cdot\text{s}}{\text{C}^2}\right)\left(\frac{\cancel{\text{J}}}{\cancel{\text{m}^2}\cdot\text{s}}\right)\left(\frac{1\,\text{N}\cdot\cancel{\text{m}}}{1\cancel{\text{J}}}\right)$$

$$= 754{,}000\ \text{N}^2/\text{C}^2$$

$$\Rightarrow \quad E_0 = \sqrt{754{,}000\ \text{N}^2/\text{C}^2} = 870\ \text{N/C} \tag{E16.31}$$

E16X.5 Subtracting equation E16.17*a* from equation E16.17*b*, we find that

$$1 = -1 + 0 + n \quad \Rightarrow \quad n = 2 \tag{E16.32}$$

Plugging this result back into equation E16.17*a*, we get

$$2 = 3 + j + 2 \quad \Rightarrow \quad j = -3 \tag{E16.33}$$

Plugging these exponents (and $i = 1$) back into equation E16.14 yields equation E16.18.

E16X.6 If we put a factor of speed in the numerator, the power would be zero when the speed was zero, even though the charge might be accelerating violently from positive to negative velocity or vice versa. This contradicts relativity, because in a different frame, the particle's speed would not be zero, and so we would have the relativistically impossible situation where a charge radiates energy in one inertial frame but not in another. If we put a factor of speed in the denominator, radiated power would go to infinity when the particle came to rest. The bottom line is that neither of these predictions seems very realistic.

E16X.7 This is simply a matter of plugging in the numbers (using $a = v^2/r$). The result is about 1.4×10^{-29} W.

E16X.8 Taking the first derivative, we get

$$\frac{dx}{dt} = \frac{d}{dt}(x_0 \sin \omega t) = \omega x_0 \cos \omega t \tag{E16.34}$$

Taking the second derivative, we get

$$\frac{d^2x}{dt^2} = \frac{d}{dt}(\omega x_0 \cos \omega t) = -\omega^2 x_0 \sin \omega t \tag{E16.35}$$

Glossary

action-at-a-distance model: a model for the interaction between particles that assumes that each particle acts directly and instantaneously on the other across the distance between them. (Section E2.1)

alternating current (AC): a flow of electric current that varies sinusoidally with time: $I(t) = I_0 \sin(\omega t)$, where ω is the angular frequency of the variation in radians per second and I_0 is the maximum current that flows during the cycle. (Section E14.6)

ampere (A): the SI unit of current, corresponding to 1 coulomb per second (C/s) of charge flowing through a specified surface oriented perpendicular to the moving charge. (Section E4.1)

Ampere-Maxwell law: the equation

$$\oint \vec{\mathbb{B}} \cdot d\vec{S} - \frac{1}{c}\frac{d\Phi_E}{dt} = 4\pi k \frac{i_{\text{enc}}}{c} \qquad \text{(E13.26}b\text{)}$$

This is the differential form of Ampere's law with an additional time derivative term that makes it possible for the law to work for time-dependent fields. (Section E13.2)

Ampere's law: a law of physics declaring that for a static magnetic field $\vec{\mathbb{B}}$, the sum of $\vec{\mathbb{B}} \cdot d\vec{S}$ for all segments on a closed loop is equal to $4\pi k/c$ times the current flowing through the surface bounded by the loop (technically, the flux of the current density through that surface). (Section E11.1)

Ampere's law for the electric field: a law of physics declaring that for a static electric field $\vec{E}$, the sum of $\vec{E} \cdot d\vec{S}$ for all segments on a closed loop is equal to zero. (Section E11.6)

amperian loop: a closed loop chosen so that it is easy to calculate the sum of $\vec{\mathbb{B}} \cdot d\vec{S}$ or $\vec{E} \cdot d\vec{S}$ for that loop (thus making it easy to apply Ampere's law). (Section E11.2)

amplitude: the magnitude of the maximum displacement of the medium in a sinusoidal wave. (Section E15.2)

angular frequency or **phase rate** ω: the number of radians of oscillation that a sinusoidal traveling wave goes through per unit time at a fixed point in space. (Section E15.2)

anode: see **electron gun.**

aurora: also called the *northern lights;* a phenomenon that looks like glowing curtains in the air, caused by energetic charged particles from the sun that, because they must spiral along the direction of the earth's magnetic field, are directed by the field into the upper atmosphere near the earth's magnetic poles. (Section E7.4)

(infinite) **axial current distribution:** a current distribution where the current everywhere flows parallel to a central axis and which is unchanged by rotations around that axis. (Section E9.4)

battery: one or more cells wired together. For example, a standard flashlight battery is a *cell*, but a 12-V car battery is a *battery* consisting of six cells wired in series. (Section E5.1)

Biot-Savart law (pronounced "Bee-o Sahvahr"): the law that describes how to compute the magnetic field of an arbitrarily shaped wire by summing the fields created by short wire segments (see equation E8.10). (Section E8.3)

bounded (surface): an element of surface that is completely enclosed by a boundary, like a field is enclosed by a fence. This is in contrast to a **closed** surface (like a sphere), which has no boundary. (Section E9.5)

brushes: sliding contacts that connect the rotating coil of an electric motor or generator to the outside world. (Section E8.1)

capacitance C: the ratio of the charge on a capacitor's positive plate to the magnitude of the potential difference between its plates. A capacitor's capacitance depends only on the size, shape, and arrangement of its plates and the material in which the plates are embedded. (Section E4.4)

capacitor: a pair of conductors that always have opposite charges and that interact electrostatically with each other much more strongly than with anything else. We call the two conductors the capacitor's **plates.** (Section E4.4)

cathode: see **electron gun.**

cathode-ray tube (CRT): a vacuum tube having an electron gun at one end and a phosphor-coated screen at the other, and (usually) magnetic coils or electric plates that modify the path of the electron beam created by the gun. The screen glows where the electron beam hits it. CRTs are used in TV sets, computer monitors, etc. (Section E7.3)

cell: an electrical device consisting of two conducting plates (**electrodes**) and some kind of charge transport mechanism (commonly chemical) that gives each charge carrier transported a characteristic energy. (Section E5.3)

central plane: The plane midway between the parallel surface planes of an infinite slab. (Section E9.2)

charge carriers: the particles that actually transport the charge in an electric current. In metals, the charge carriers are electrons, but in batteries and in conducting fluids, the charge carriers are often ionized atoms. (Section E4.1)

charge distribution: a huge number of charged particles (usually electrons or positive ions) distributed over the surface or throughout the interior of a macroscopic object. We can calculate the electric field of such a distribution by dividing it into infinitesimal pointlike bits. (Section E2.6)

circuit diagram: a schematic diagram that helps us describe and visualize the structure of an electric circuit, using stylized symbols to represent circuit elements and straight lines to represent connections between elements. (Section E6.3)

circuit element: any object or device (such as wires, lightbulbs, batteries, and electric motors) that appears in an electric circuit. (Section E6.2)

circulation: the dot product of a field vector with a segment vector that represents a small part of a closed loop, summed over all segment vectors in the loop. The standard notation for the circulation of $\vec{\mathbb{B}}$, for example, around a loop whose segment vectors are $d\vec{S}$ is $\oint \vec{\mathbb{B}} \cdot d\vec{S}$, where the integral symbol tells us to sum over small quantities and the circle tells us that we are calculating this for a closed loop. (Section E11.1)

circulation direction: the arbitrarily-chosen direction along the curve that surrounds a bounded surface that we use to define the direction of tile vectors on the bounded surface. (Section E9.5)

closed surface: a surface that completely encloses a certain volume in space. (Section E9.5)

commutator: an arrangement of contacts on a motor's axle that, as the commutator turns under the brushes, feeds an alternating current to that motor's coil. (Section E8.1)

compass: (see **magnetic compass.**)

conduction electrons: the electrons in a metal that are free to move and thus carry charge through the metal. Usually, each atom in a metal contributes one electron to the pool of conduction electrons. (Section E4.2)

conductivity σ_c: a scalar quantity (having SI units of $C^2 \cdot s \cdot kg^{-1} \cdot m^{-3}$) that represents the constant of proportionality between the current density in a substance and the electric field driving that current: $\vec{J} = \sigma_c \vec{E}$. Under ordinary conditions, the conductivity depends only on the characteristics of the substance: if Drude's model accurately describes the substance, $\sigma_c = nq^2\tau/m$, where n is the number density of charge carriers, q is their charge, m is their mass, and τ is the mean time between collisions. (Section E4.3)

conductor: a material through which charges can easily move. Salty water and most metals are good conductors. In metals, it is electrons that are free to move, but this is not always the case: in salty water, charge is carried by ionized atoms. (Section E1.6)

conventional current: a long-established convention that considers the direction of any current flow to be the direction in which *positive* particles would flow if they were the charge carriers in the current. If negative particles (such as electrons) actually carry the charge, the direction of the conventional current is opposite to the direction of the actual motion of the negative carriers. (Section E4.1)

coulomb (C): the SI unit of electric charge, defined so that $1\ C = 6.242 \times 10^{18}$ proton charges. (Section E1.3)

Coulomb constant k: a constant whose numerical value is $8.99 \times 10^9\ N \cdot m^2/C^2$ that essentially specifies the strength of the electromagnetic interaction. (Section E1.5)

Coulomb's law: the law that expresses the force that one point charge exerts on another in terms of the values q_1 and q_2 of the charges, the distance r between them, and the **Coulomb constant** k. The law states that the magnitude of the force on either charge is given by $F_e = |kq_1q_2/r^2|$ and that its direction is toward the other charge if the charges have unlike signs and away from the other charge if the charges have like signs. (Section E1.5)

crests and **troughs:** the extreme values of the displacement of the medium in a sinusoidal wave. A crest exists where the medium is displaced the maximum amount in the positive direction, while a trough exists where the medium is displaced maximally in the negative direction (however these directions are defined). (Section E15.2)

cross-product rule: the right-hand rule that specifies that the direction of the cross product $\vec{u} \times \vec{w}$ is given by your right thumb if you point your index finger in the direction of $\vec{u}$ and your second finger in the direction of $\vec{w}$. (Section E8.5)

curl: a vector quantity whose components represent differential versions of the circulation around infinitesimal loops whose tile vectors point in the x, y, and z directions. For example, the x component of the curl of $\vec{\mathbb{B}}$ is defined to be

$$\text{curl}(\vec{\mathbb{B}})_x \equiv \lim_{dA \to 0} \frac{\oint \vec{\mathbb{B}} \cdot d\vec{S}}{dA} \quad \begin{array}{l}\text{around loop of area } dA \\ \text{whose tile vectors all point} \\ \text{in } +x \text{ direction}\end{array} \tag{E13.31}$$

The curl for other components and other fields is defined analogously. (Section E13.6)

current density $\vec{J}$: a vector (having SI units of amperes per square meter) that quantifies the local flow of current near a point in a conductor: $\vec{J} = nq\vec{v}_d = \rho\vec{v}_d$, where n is the number density of charge carriers, q is the charge of the charge carrier, ρ is the density of charge carriers, and $\vec{v}_d$ is the local drift velocity of carriers at the point in question. The magnitude of the current density can be thought of as the limit as $A \to 0$ of the *current per unit area I/A* flowing through a surface of area A perpendicular to the local flow. (Section E4.3)

cyclotron frequency: the number of times per second that an otherwise free charged particle in a magnetic field moves around in its circular orbit. This frequency is interesting because it is independent of the particle's velocity (as long as $v \ll c$). (Section E7.4) (infinite) **cylindrically symmetric charge distribution:** a charge or current distribution that is unchanged by (1) rotation around some central axis and (2) sliding along that axis. Examples include infinitely long wires, cylinders, pipes, and so on. (Section E10.4)

dielectric constant $\varepsilon/\varepsilon_0$: the ratio of a substance's permittivity to the permittivity of free space. This substance-dependent constant (which is always greater than 1) expresses how much weaker the electric field inside the substance is than it would be if the substance were replaced by a vacuum. (Problem E4S.12)

differential form of Maxwell's equations: the set of equations (see equations E13.30) that express the behavior of the electromagnetic field in the form of local field equations. These are to be contrasted with the *integral* form of Maxwell's equations given by equation E13.26. (Section E13.6)

dimensional analysis: an approach to guessing the formula for a quantity by deciding what variables are relevant and then making the units come out correctly. (Section E16.4)

dipole: see **electric dipole.**

direct current (DC): a flow of electric current that is essentially constant with time, like the current flowing in a flashlight powered by batteries. (Section E14.6)

divergence: a quantity that represents the differential expression of the flux through a surface. The divergence of $\vec{E}$ at a point P is defined to be

$$\operatorname{div}(\vec{E}) \equiv \lim_{dV \to 0} \frac{\oint \vec{E} \cdot d\vec{A}}{dV} \quad \text{for gaussian surface of volume } dV \text{ surrounding } P \tag{E13.29}$$

The divergence of the magnetic field is defined analogously. (Section E13.6)

domain: a (typically) tiny region inside a ferromagnetic material where atomic current loops are all aligned. When an external magnetic field is applied, the domains aligned with the field grow by "recruiting" at their edges, causing the material to develop its own magnetic field, reinforcing the external field. (Section E8.6)

drift velocity $\vec{v}_d$: the average velocity of charge carriers in a conductor. (Section E4.2)

Drude model: a model that explains how electrons flow through metals by imagining electrons to be particles that move freely between collisions with metal atoms. The thermal velocities of these particles are imagined to be large and random. If an electric field is applied to the metal, electrons pick up a small amount of momentum opposite to the direction of the electric field between collisions. This causes the electrons (on average) to slowly drift in the direction opposite to the applied field. (Section E4.2)

dynamic (field): A field that changes with time. (Section E1.1)

e: the magnitude of the charge of a proton or electron: $e = 1.602 \times 10^{-19}$ C. Unfortunately, this symbol is easy to confuse with the base of the natural logarithms $e = 2.718$: the meaning of an italic e must be determined from context. (Section E1.3)

electric charge: a fundamental property of matter that acts as the source of an electromagnetic field. There are two kinds of charge, which are arbitrarily designated "positive" and "negative." (Section E1.2)

electric circuit: a collection of devices or objects through which an electric current can flow. (Section E5.1)

electric current: any flow of charged particles from one place to another. Current is quantified by determining the rate at which charge flows through a specified surface oriented perpendicular to the flow. One *ampere* (1 A) of current corresponds to a flow rate through the surface of one coulomb of charge per second (1 C/s). (Section E4.1)

electric dipole: an object consisting of point charges with equal absolute values but opposite signs separated by a distance. (Section E1.7)

electric field: the entire electric field of a charge distribution at a given time t that is described by a vector function $\vec{E}(x, y, z)$ that specifies the electric field vector at every point in space at that instant of time. (Section E2.2)

electric field vector $\vec{E}$: the vector that describes the quantitative value of an electric field at a certain point. This vector is defined to be the *electric force per unit charge* experienced by a small positive test charge at rest at the point in question. $\vec{E}$ is measured in units of newtons per coulomb. (Section E2.2)

electric motor: a device that converts electric power to mechanical power (most often first in the form of rotational kinetic energy). (Section E8.1)

electric potential (or just **potential**) $\phi(x, y, z)$: the electrostatic potential energy per unit charge that a test particle at $[x, y, z]$ would have compared to what it has at infinity. The potential provides an alternative to representing the field around a charge distribution using the electric field $\vec{E}(x, y, z)$ that has some useful advantages. (Section E3.1)

electrodes: the two terminals of a battery or cell. Chemical reactions in the battery or cell transport charge from one electrode to the other. (Section E5.3)

electrolyte: the material through which ions flow in a battery or cell. (Section E5.3)

electromagnetic field: the concept that physicists use to describe the unity that lies behind electric and magnetic phenomena. An electromagnetic field is created by an object involving unbalanced, separated, and/or moving charges, and it mediates the electromagnetic interaction between two such objects. The electromagnetic field generally has an electric part $\vec{E}$ and a magnetic part $\vec{\mathbb{B}}$ (the total electromagnetic field in a given reference frame is therefore represented by *six* components). The electromagnetic field has a frame-independent reality, but the way that it divides into electric and magnetic parts depends on one's choice of inertial reference frame. (Section E12.1)

electromagnetic wave: a traveling wave disturbance in the electromagnetic field. In empty space, such waves travel with speed c and have perpendicular electric and magnetic field components such that $E = \mathbb{B}$ and $\vec{E} \times \vec{\mathbb{B}}$ points in the direction the wave is moving. (Section E16.1).

electron: a negatively charged subatomic particle that is an important constituent of all atoms. Fairly mobile electrons are the typical charge carriers in metals. See also **proton, neutron,** and **electron.** (Section E1.3)

electron cloud: the probability distribution of electrons orbiting an atomic nucleus. Quantum mechanics asserts that the positions of individual electrons orbiting an atom are not well defined: treating the electrons as a cloudlike dispersed charge is a better model. (Section E1.3)

electron gun: a device that creates a beam of electrons with a fairly well-defined velocity. Electrons boil off a heated, negatively charged plate (the **cathode**) and accelerate toward a positively charged plate (the **anode**). A small hole in the positive plate allows some electrons to escape to form a narrow beam. (Section E7.3)

electrostatic interaction: that aspect of the electromagnetic interaction that operates between charges at rest. (Section E1.1)

electrostatic polarization: the phenomenon whereby the electrostatic forces exerted by an external charge distort atoms in a neutral substance, making them electric dipoles. This phenomenon makes it possible for charged objects to attract electrically neutral but polarizable objects. (Section E1.7)

emf $\mathscr{E}$**:** the energy per unit charge transferred by *non*-electrical forces *to* electrons or ions as they are transported against the electric field in a battery. A battery generally has a characteristic and fixed emf. It is also called the **electromotive force** (even though it is not a force). (Section E5.3)

equipotential curve (or just **equipotential**)**:** a curve on a two-dimensional diagram that connects points having the same potential. (Section E3.1)

equipotential diagram: a diagram showing the equipotential curves on a two-dimensional slice of three-dimensional space. The equipotential curves actually show where the equipotential surfaces intersect the plane of the drawing. (Section E3.1)

equipotential surface: a surface that connects points in a three-dimensional space that have the same potential. (Section E3.1)

farad (F)**:** The SI unit of capacitance: $1\text{ F} = 1\text{ C/V}$. (Section E4.4)

Faraday's law: the equation

$$\oint \vec{E} \cdot d\vec{S} + \frac{1}{c}\frac{d\Phi_{\mathbb{B}}}{dt} = 0 \qquad \text{(E13.26}d\text{)}$$

This is a corrected version of **Ampere's law for the electric field** with an additional time derivative term that makes it possible for the law to work for time-dependent fields. (Section E13.3)

Faraday's law of induction: the equation

$$\mathscr{E}_{\text{loop}} = -\frac{1}{c}\frac{d\Phi_{\mathbb{B}}}{dt} \quad \text{or} \quad \mathscr{E}_{\text{loop}} = -\frac{d\Phi_B}{dt} \qquad \text{(E14.2)}$$

This equation describes how the induced emf in a conducting loop depends on the rate of change of magnetic flux through the loop. This equation turns out to be quite general, useful in all nonrelativistic frames and situations. (Section E14.1)

ferromagnetic: any type of material that (when initially unmagnetized) is attracted strongly and equally by both poles of a magnet. Only a few kinds of metals and alloys (notably iron and certain iron oxides) are ferromagnetic: most materials respond only weakly (if at all) to magnets. (Section E7.1). A type of material in which it is energetically favorable for atoms to *align* their magnetic moments. Iron, cobalt, and nickel are common ferromagnetic materials. (Section E8.6)

field diagram: a depiction of the total field created by an object that shows the field vectors at a large number of positions in the space surrounding the object. (Section E2.3)

field model: a model for the interaction between particles that assumes that each particle creates a *field* at all points in the space surrounding it. The other particle interacts with the field created by the first particle, not directly with the particle itself. (Section E2.1)

flux (of a field through a tile)**:** the dot product of the field vector and the tile's tile vector. (Section E9.5)

(net) **flux** Φ (of a field through a surface)**:** the dot product of a field vector and a tile vector for all tiles on a bounded or closed surface. (We conventionally indicate the field whose flux we are calculating by attaching a subscript to Φ.) If the surface is closed, we conventionally assume that the tile vectors point outward; if the surface is bounded and the boundary loop has a circulation direction, we assume that the tiles point as indicated by the loop rule. Note that if $\vec{u}$ is a field indicating the local velocity of a moving fluid such as water, then Φ_u (in cubic meters per second) expresses the rate at which fluid moves through a surface. (Section E9.5)

Fourier's theorem: a mathematical theorem asserting that *any* wave can be treated as a superposition of sinusoidal waves. Thus if one understands the behavior of sinusoidal waves, one really understands the behavior of all waves. (Section E15.2)

frequency f: the number of complete oscillations that an observer at a fixed position would measure a sinusoidal wave to go through per unit time. (Section E15.2)

gamma rays: electromagnetic waves with wavelengths shorter than about 30 pm (frequencies in excess of 10^{19} Hz). Such waves are generated primarily by protons and/or neutrons shifting energy levels in an atomic nucleus. (Section E16.6)

gauss: a historical unit of magnetic field strength, defined to be equal 10^{-4} tesla.

gaussian surface: an imaginary closed surface whose shape we choose in a given situation so that it is easy to calculate the flux of the electric field through it. This helps us apply Gauss's law to the situation in question. (Section E10.2)

Gauss's law: a law of electromagnetism that states that the net flux of the electric field through an arbitrary closed surface is $4\pi k$ times the charge enclosed by that surface. (Section E10.1)

Gauss's law for the magnetic field: a law of electromagnetism that states that the net flux of *any* magnetic field through a closed surface is identically zero. This expresses an intrinsic characteristic of all magnetic fields. (Section E10.6)

generator: a device that converts energy (usually mechanical energy of rotation) to electric energy. (Section E8.1)

gravitational field vector $\vec{g}$: the vector that (in newtonian gravity) represents the gravitational field at a point in space in the same way that the vector $\vec{E}$ represents the electric field. This vector is defined to be the *gravitational force per unit mass* experienced by an infinitesimal test mass placed at the given point: $\vec{g} \equiv \vec{F}_g / m_{\text{test}}$. (Section E2.2)

gravitational potential ϕ_g: a scalar field that can be defined at a given point by placing a test particle at that point, determining its gravitational potential energy, and dividing that potential energy by the particle's mass. The gravitational potential is thus *gravitational potential energy per unit mass*. It is directly analogous to electric potential (which is electrostatic potential energy per unit charge). (Section E3.1)

henry (H): the SI unit of inductance. 1 henry $= 1\ \Omega \cdot$ s. (Section E14.3)

horizontally polarized: see **polorization** (of an electromagnetic wave).

ideal battery approximation: the assumption that energy losses to thermal energy during charge transport in a battery are zero. If this is true, then *all* the energy given to the transported charge is converted to electrostatic potential energy, and the potential difference between the battery's electrodes will be equal to the battery's emf, irrespective of the current flowing in the battery. (Section E5.3)

inductance L: a quantity that expresses the degree to which a coil or loop induces an emf in *itself* in response to a given change in the current flowing through the coil. The inductance of a given coil is defined implicitly by the equation $|\mathcal{E}| = L\,|dI/dt|$. The SI unit of inductance is the *henry*, where 1 H $= 1\ \Omega \cdot$ s. (Section E14.3)

induction: the process by which an emf in a loop or coil is created as a result of its interaction with a magnetic field. (Section E8.2). The phenomenon whereby a *changing* magnetic flux through a conducting loop induces a current to flow in that loop. (Section E14.1)

infrared (light): electromagnetic waves with frequencies ranging from about 1×10^{12} Hz to a bit over 3×10^{14} Hz (wavelengths from 0.3 mm to 0.7 μm). Such waves are created mostly by thermal vibrations and processes. (Section E16.6)

insulator: in electrical contexts, a material through which charges cannot easily move. Air, rubber, plastic, glass, and amber are good insulators. (Section E1.6)

intensity I: the energy per time that an electromagnetic wave delivers to a unit surface area directly perpendicular to the wave. (Section E16.3)

kilowatt-hour: a unit of electrical energy equal to the amount of electrical energy that is transformed in a hour to other forms of energy (or vice versa) if the energy is being transformed at a rate of 1000 W. (Section E5.5)

Larmor formula: the formula $P = \frac{2}{3}kq^2a^2/c^3$ that expresses the power radiated in the form of electromagnetic radiation by an accelerating charge. (Section E16.4)

Lenz's law: the emf and/or current induced in a loop or coil seeks to *oppose* the change in magnetic field that induces it. (Section E14.2)

local field equation: a field equation that sets some characteristic of the field in question evaluated at some specific point in space and time equal to a quantity describing the density of charge and/or current at that *same* point in space and time, while *completely ignoring* the effects of charges and/or currents at distant points in space and time. This avoids the instantaneous communication problem usually encountered by equations that connect a field at one point with charges and/or currents at distant points. (Section E13.6)

longitudinal wave: a mechanical wave in which the passing wave displaces the medium in a direction parallel to the wave's direction of motion. (Section E15.1)

loop-to-magnet rule: a rule stating that if you curl your right fingers in the direction of the current flowing in a loop, then your right thumb indicates (1) the direction of the magnetic field created at the loop's center and (2) the south-to-north direction of the magnet equivalent to the loop. When applied to an amperian loop, this rule says that if you curl your right fingers in the direction of the circulation direction defined by the loop's segment vectors, your thumb indicates the direction of tile vectors on the surface bounded by the loop. (Sections E7.1, E8.5, and E9.5)

Lorentz force law: the equation $\vec{F}_{em} = q[\vec{E} + (\vec{v}/c) \times \vec{\mathbb{B}}]$, which specifies the total electromagnetic force on a particle with charge q moving with speed v in a frame where the electric and magnetic fields are $\vec{E}$ and $\vec{\mathbb{B}}$. Even though the relative mix of $\vec{E}$ and $\vec{\mathbb{B}}$ varies from frame to frame, this equation gives the correct physical force measured in any given reference frame. (This equation also implicitly *defines* the electric and magnetic fields.) (Section E12.1)

LR circuit: a circuit consisting of a coil with significant inductance, a resistor, and possibly a power supply or battery. (Section E14.4)

magnet: any object able to exert magnetic forces on another magnet *and* on a ferromagnetic substance. (Section E7.1)

magnetic compass: a magnet suspended so that it can rotate freely in response to an external magnetic field (usually the earth's magnetic field). (Section E7.1)

magnetic field $\vec{B}$ or $\vec{\mathbb{B}}$: a vector field whose direction at a given point in space indicates the direction that a magnetic compass would indicate is "north," and whose magnitude is proportional to the ratio of the velocity-dependent magnetic force $\vec{F}_m$ a charge particle at that point would feel to the magnitude of the particle's charge q, the particle's speed v, and the sine of the angle θ between $\vec{v}$ and $\vec{F}_m$: $B = F_m/v \sin\theta$, $\mathbb{B} = cB$. (Sections E7.2 and E7.3)

magnetic force law: $\vec{F}_m = q(\vec{v}/c) \times \vec{\mathbb{B}}$. This law summarizes empirical observations of the effect of magnetic fields on moving charged particles and defines the magnitude and units of $\vec{\mathbb{B}}$. (Section E7.3)

magnetic moment μ: a quantity that characterizes how strongly a current-carrying loop or a dipole magnet responds to a magnetic field. For example, the energy needed to turn from having its axis aligned with the field to having it anti-aligned with the field is $2\mu B$. (Problem E8R.3)

magnetic monopole: a hypothetical particle or object having a single, isolated, magnetic pole. To the best of our current knowledge, magnetic monopoles do not exist. (Section E7.2)

magnetic permeability μ_0: a constant equal to $4\pi k/c^2$, where k is the Coulomb constant and c is the speed of light. This constant is conventionally used in place of the Coulomb constant in equations describing the magnetic field created by a current. (Section E8.2)

magnetic poles: the parts of a magnet that seem to be the source of the magnetic forces, analogous to the charges on an electric dipole. A magnet always has exactly two unlike poles (breaking a magnet in two yields not two isolated poles but two magnets each with two poles). (Section E7.1)

Maxwell's equations: the set of four equations that describe the behavior of dynamic electromagnetic fields. These equations are Gauss's law, Gauss's law for the magnetic field, the Ampere-Maxwell law, and Faraday's law. See the inside front cover for a list of the equations. Together, these four equations provide a complete and relativistically self-consistent description of the electromagnetic field in all known (macroscopic) circumstances. (Section E13.4)

mechanical battery model: an idealized model of a cell that visualizes it as two charged conducting plates connected by a conveyor belt that transports positive charge from the negative plate to the positive plate against the electric field of the plates. If a wire is connected to the plates, the belt runs at whatever speed will maintain the charges on the plates (and thus the potential difference between the plates). (Section E5.3)

mechanical wave: any kind of wave where the medium carrying the wave is physically displaced by the wave. (Section E15.1)

microwave: electromagnetic wave with frequencies above about 330 MHz and below about 1×10^{12} Hz (wavelengths from 1 m to 0.3 mm). (Section E16.6)

net charge: the sum of an object's total positive charge and its total negative charge, expressing the latter as a negative number. Thus an object having equal magnitudes of positive and negative charge has zero net charge. (Section E1.3)

net flux: see **flux** (of a field through a surface).

neutron: see **proton, neutron,** and **electron.**

node rule: the rule that says that when two or more wires are connected at a point in an electric circuit, the total current entering that point is equal to the total current leaving that point. This rule is basic consequence of the conservation of charge. (Problem E6A.1)

north and south (magnetic poles)**:** the conventional names for the two kinds of magnetic poles. A magnet's north pole is the pole that ends up pointing geographically northward when the magnet is suspended so that it can rotate freely in the earth's magnetic field. (Section E7.1)

number density n: the number of a specified type of particle per cubic meter in a given medium. In chapter E4, n usually describes the number density of charge carriers. (Section E4.3)

ohm (Ω)**:** the SI unit of resistance: $1\,\Omega = 1$ V/A. (Section E5.4)

ohmic: a conducting object or device that has an essentially constant resistance and thus obeys Ohm's law. (Section E5.4)

Ohm's law: the assertion that the resistance of a conductor is generally independent of the current it conducts (for reasonably small currents). (Section E5.4)

one-dimensional wave: a wave that depends on only x and t (as opposed to x, y, z, and t). (Section E15.2)

(circuit elements in) **parallel:** a set of circuit elements as connected in parallel (1) if a given electron flowing through the set will flow through exactly one element and (2) if the elements are connected so that the voltage across each element is the same as the voltage across the set. (Section E6.4)

parallel-plate capacitor: a capacitor consisting of two parallel flat conducting plates whose separation is very small compared to the size of each plate. If its plates have area A and are separated by a distance s, the capacitance of such a capacitor is $C = A/4\pi ks = \varepsilon_0 A/s$. This is useful as a simple model for many kinds of real capacitors. (Section E4.4)

period T**:** the time that an observer at a fixed position measures between successive crests of a sinusoidal wave. (Section E15.2)

permeability: (see **magnetic permeability**).

permittivity ε**:** expresses the ratio of the electric field that exists in a polarized substance to the electric field that would exist if the substance were replaced by a vacuum. (Problem E4S.12)

permittivity of empty space ε_0**:** a constant equal to $1/4\pi k$, where k is the Coulomb constant. This constant has been historically used in place of k in most electrostatic equations. (Section E2.3)

phase velocity: the velocity of a given feature (such as the crest) of a sinusoidal wave. The magnitude of a wave's phase velocity is its **phase speed.** (Section E15.3)

plane wave: a wave in a three-dimensional space whose value depends on only one of the three spatial coordinates, meaning that the disturbance has the same value on the plane spanned by the other two coordinates. (Section E16.1)

polarization (of an electromagnetic wave): a special case of an electromagnetic wave where the wave's electric field changes with time in a specific, well-defined manner. In this text, we will only be interested in polarizations where the electric field oscillates back and forth along a certain specific direction. If a wave moving in the $+x$ oscillates along whatever direction we have defined to be the y axis, we say that the wave is **horizontally polarized;** if it oscillates along the direction we have defined to be the z axis, we say that it is **vertically polarized.** (Section E16.2)

potential: (see **electric potential**).

potential difference V**:** the difference in electric potential between two points in an electric circuit (usually the two terminals of a battery or the two ends of a circuit element). (Section E5.4)

principle of relativity: the principle that the laws of physics are the same in all inertial reference frames. (Section E12.1)

proton, neutron, and electron: the three subatomic particles out of which all atoms are constructed. An atom consists of a small but massive nucleus of protons and neutrons surrounded by a cloud of electrons. The *proton* is a positively charged particle with a mass such that Avogadro's number of protons has a mass of about 1 g. The *neutron* is an electronically neutral particle with about the same mass. The *electron* is a negatively charged particle with a much smaller mass (more than 1800 times smaller). (Section E1.3)

radio waves: electromagnetic waves with frequencies below about 330 MHz. Such waves are primarily created by the bulk motion of charges. (Section E16.6)

Rayleigh scattering: a process where incident light causes the charged parts of an atom or molecule to oscillate and thus reradiate the light in all directions with the same frequency. Higher-frequency light is more effectively scattered than lower-frequency light by a factor of ω^4. (Section E16.5)

resistance R**:** the ratio of the potential difference V between a conductor's ends to the current I driven through the conductor by that potential difference: $R \equiv V/I$. The resistance so defined usually depends on temperature and may also be a function of current, but for many conductors, R is essentially independent of I. (Section E5.4)

resistor: technically, a circuit element specifically manufactured to have a certain fixed resistance. More generally, we can use this word to describe any circuit element that is at least approximately ohmic. (Section E6.3)

(magnetic) **right-hand rule:** a rule stating that when a particle with charge q moves with velocity $\vec{v}$ in a magnetic field $\vec{\mathbb{B}}$, pointing your right index finger in the direction of $q\vec{v}$ and your longest finger in the direction of $\vec{\mathbb{B}}$ aligns your thumb with the direction of the magnetic force $\vec{F}_m$ on the particle. The same rule yields the direction of the cross product of two vectors: if you point your right index finger in the direction of $\vec{u}$ and your second finger in the direction of $\vec{w}$, your thumb indicates the direction of $\vec{u} \times \vec{w}$ (see the **cross-product rule**). (Section E7.3)

segment vector $d\vec{S}$**:** a vector describing an infinitesimal displacement along a curve. (Section E11.1)

seismic wave: a wave created by an earthquake that causes the earth's crust to displace as the wave passes. (Section E15.1)

self-energy: the energy required to assemble a charge distribution from point charges essentially at infinity. (Section E3.5) (circuit elements in) **series:** two batteries are *in series* if the positive terminal of one is connected to the negative terminal of the other. The voltage difference across two batteries in series is the sum of the voltage differences across each battery alone. (Section E5.3) We say that a general set of circuit elements is *in series* if every charge carrier that goes through one element goes through each other element in the set in sequence. (Section E6.2)

shell theorem: a theorem stating that (1) the potential or electric field outside a spherical shell with a uniformly distributed charge is exactly as if that charge had been concentrated at a point at the shell's center, and (2) the potential in the empty space inside the shell is a constant, implying that the electric field there is zero. (Section E3.4)

sinusoidal wave: a wave in which the disturbance of the medium as a function of time can be described by the function $f(t, x) = A \sin(kx - \omega t)$. (Section E15.2)

solenoid: a coil of wire consisting of a number of turns of wire wrapped around some cylindrical form. (Section E11.5)

(infinite) **solenoidal current distribution:** a current distribution which is unchanged by rotation around or sliding along some straight axis, and where the current flows in a circle about that axis. (Section E9.4)

sound wave: a wave traveling through a medium that compresses or stretches the medium as the wave passes. (Section E15.1)

spherically symmetric charge distribution: a distribution that is unchanged by rotation around any axis. (Section E9.2)

static (field): a field that does not change with time. (Section E1.1)

superposition principle (for electric fields): the statement that the electric field created at a given point by a collection of charged particles is the vector sum of the fields that each individual particle produces at that point individually. (Section E2.4) Also applies to magnetic fields. (Section E8.3)

symmetry argument: a technique for determining useful characteristics of an electromagnetic field created by a charge or current distribution that is not changed by an operation such as a rotation around an axis, translation along an axis, or mirror reflection across a plane. (Section E9.1)

tension wave: a wave traveling along a flexible line under tension (such as a string or rope) that causes displacements perpendicular to the line as the wave passes. (Section E15.1)

tesla T: the SI unit of magnetic field strength, defined to be one newton-second per coulomb-meter $[\text{N} \cdot \text{S}/(\text{m} \cdot \text{C})]$. (Section E7.3)

test charge: a point particle with a tiny charge that we use (at least in our imagination) to measure the electric field around a charge distribution. The electric field at a point in space at a given instant of time is defined to be the force experienced by such a test charge at rest divided by the charge of the test charge: $\vec{E} \equiv \vec{F}_e / q_{\text{test}}$. (The test charge is assumed to be tiny so that it doesn't disturb the charge distribution whose field is being measured.) (Section E2.2)

tile: a piece of area on a surface that is small enough to be considered flat. (Section E9.5)

tile vector $d\vec{A}$: a vector whose magnitude is the tile's area and whose direction points perpendicular to the tile in the direction away from a closed surface's interior or in the direction indicated by the loop rule for a surface bounded by a closed loop with a well-defined circulation direction. This vector efficiently encodes useful information about a tile. (Section E9.5)

transformer: a device that uses induction to transfer electrical power from one circuit to another while changing the emf at which that power is delivered. (Section E14.6)

transverse wave: a mechanical wave in which the passing wave displaces the medium in a direction perpendicular to the wave's direction of motion. (Section E15.1)

traveling wave: a wave whose basic shape is preserved in time and whose shape moves through the medium as time progresses. (Section E15.3)

triboelectric series: a list of materials arranged in such an order that if two materials on the list are rubbed, the one higher on the list loses electrons to the one lower on the list. (Section E1.3)

ultraviolet (light): electromagnetic waves whose wavelengths range from about 400 nm to 3 nm. Such waves are primarily produced by electrons changing energy levels in atoms. (Section E16.6)

uniform (field): a field whose field vectors have the same magnitude and direction at all points. (Section E7.4)

vector function $\vec{f}(x, y, z)$: a function that implicitly assigns a vector to every point $[x, y, z]$ in space by specifying how we can calculate the vector $\vec{f}$, given the coordinate values x, y, and z of that point. (Section E2.2)

velocity selector: a device that uses crossed electric and magnetic fields to permit only those charged particles having a certain specific velocity to pass through it. (Section E12.4)

vertically polarized: see **polarization** (of an electromagnetic wave).

volt (V): the SI unit of electric potential. $1 \text{ V} = 1 \text{ J/C}$. (Section E3.1)

voltage (at a point in a circuit): another name for potential (*voltage* is commonly used in circuit contexts). (Section E6.1)

water wave: a wave traveling along the surface of a body of water that causes water near the surface to displace both horizontally and vertically as the wave passes. (Section E15.1)

wave: a disturbance that moves through a medium, while the medium itself remains basically at rest. (Section E15.1)

(one-dimensional) **wave equation:** an equation of the form shown in equation E15.14. If such an equation follows from the basic physics of a certain medium, then disturbances in the form of one-dimensional traveling waves can exist in the medium. (Section E15.4)

wavelength λ: the distance between successive crests in a sinusoidal wave at any given instant of time. (Section E15.2)

wavenumber k: the number of radians of oscillation that a sinusoidal wave goes through per unit length. (Section E15.2)

wave rule: the statement that if a sudden disturbance at a point in a homogeneous medium causes the disturbance value at a point nearby to *accelerate* in proportion to the original disturbance divided by the square of the distance between the points Δx^2 (in the limit that Δx is very small), then such a disturbance will obey the wave equation. The constant of proportionality is the squared speed of the wave. (Section E15.4)

wire rule: the rule stating that if you point your thumb in the direction of the conventional current carried by a wire, the magnetic field will curl around the wire in the same direction as your right fingers. (Section E8.4)

X-rays: electromagnetic waves with frequencies ranging from roughly 10^{17} Hz to 10^{19} Hz (wavelengths from 3 nm down to 30 pm). Such waves are created technologically primarily by crashing electron beams into metal targets. (Section E16.6)

Index

Note: Page numbers followed by *f* indicate figures.

acceleration, of charge, 329–331
action-at-a-distance model, 24, 324, 341
alternating current, definition of, 292, 341
amber, electrical charge in, 5
Ampère, Andre Marie, 155
ampere, definition of, 66, 341
Ampere-Maxwell law, 256, 262, 322–323, 341
Ampere's law
for axial current distribution, 220–221
correcting (*See* Ampere-Maxwell law; Faraday's law)
definition of, 341
for electric field, 215, 226–228 (*See also* Faraday's law)
for infinite planar slab, 222–223, 222*f*, 223*f*
for infinite solenoid, 224–225
for magnetic field, 214, 216–217 (*See also* Ampere-Maxwell law)
for moving infinite planar slab, 246–247, 246*f*, 264–267, 265*f*
problem with, 258–262, 258*f*
using, 214, 218–219
amperian loop
for axial current distribution, 220, 220*f*
choosing, 218–219, 219*f*
for cylindrical charge distribution, 226
definition of, 341
electric field disturbance and, 322–323, 323*f*
for infinite planar slab, 221, 222*f*, 265
for infinite solenoid, 223, 224*f*
magnetic field disturbance and, 325–326, 326*f*
amplitude, definition of, 305, 341
angular frequency, definition of, 301, 306
anode, definition of, 130–131, 131*f*
antineutrino, in conservation of charge, 9
atom, 15, 15*f*. *See also* electron(s); neutrons; protons
atomic model, electrical charge and, 7–8, 8*f*
aurora, 138, 139*f*, 341
Avogadro's number, mass represented by, 7
axial current distribution
Ampere's law for, 215*t*, 219–221, 220*f*
definition of, 341
magnetic field vector of, 173*t*, 179–180, 180*f*

$\vec{B}$. *See* magnetic field vector
bar magnet
dipoles and, 162
and induced current, 284–285
batteries
capacity of, 92–93
chemical (*See* chemical battery)
in circuit, 109, 109*f*
in circuit diagram, 112, 112*f*
definition of, 84, 87, 341
ideal battery approximation, 92, 119, 119*f*, 345
magnetic (*See* magnetic battery)
mechanical model of, 86–87, 87*f*
plates of (*See* electrodes)
realistic, 119, 119*f*
safety of, 120
batteries in series, 90–91, 90*f*, 347
Biot, Jean, 155
Biot-Savart law, 159, 341
blue light, 332
bounded surface
definition of, 172, 183, 183*f*, 341
net flux through, 185
tile vector for, 184, 184*f*
brush
in electric motor, 141, 141*f*, 341

capacitance, 65, 77–78, 77*t*, 341
capacitor
definition of, 65, 77, 341
discharging, 85, 97–99
in electronic circuits, 99, 99*f*
parallel-plate (*See* parallel-plate capacitor)
capacity, of batteries, 92–93
car battery. *See* lead-acid battery
cathode, definition of, 130–131, 131*f*
cathode-ray tube (CRT), definition of, 131, 131*f*, 341
cell
in circuit diagram, 112, 112*f*
definition of, 91, 341
cell batteries, safety of, 120
central plane
of infinite planar slab, 177, 177*f*, 341
charge. *See* electric charge; point charge
charge carriers
in circuits, 109, 109*f*
definition of, 66, 341
charge distribution. *See also* equilibrium charge distribution; *specific charge distributions*
definition of, 342
electric field of, 23
electric field vector for, 51
electric potential for, 51
on macroscopic objects, 33–36
theorems for, 53–55
charged object
neutral object attracted by, 16–17, 16*f*
water and, 17, 17*f*
charged particle. *See also* electron(s); protons
in aurora, 138, 139*f*
electric force on, computing, 26–27
electromagnetic force on, 241
moving (*See* moving charged particle)

charged particle—*Cont.*
near current-carrying wire, and electromagnetic field, 238–241, 239*f*
point, 10–11, 10*f*
space around (*See* electric field)
in static magnetic field, 135
test, charge of (*See* test charge)
charged point particle, force exerted by, 10–11, 10*f*
chemical battery, 90–92, 91*f*
circuit diagram, 112–113, 112*f*, 113*f*, 114*f*
definition of, 107, 112, 342
circuit elements
in circuit diagram, 107
definition of, 110, 342
in parallel (*See* parallel circuit)
in series (*See* series circuit)
circular loop. *See also* ring
magnetic field of, 151, 161–163, 161*f*
circulation
for axial current distribution, 220
calculating (*See* Ampere's law)
definition of, 216, 342
electromagnetic waves and, 323
for infinite planar slab, 221–222
for infinite solenoid, 224
notation for, 216*f*
circulation direction, definition of, 172, 183, 183*f*, 342
clock, capacitor as, 99, 99*f*
closed curve, definition of, 183, 183*f*
closed surface
definition of, 172, 183, 183*f*, 342
for gaussian surface (*See* gaussian surface)
net flux through, 185
tile vector for, 183–184, 184*f*
commutator
definition of, 342
in electric motor, 141, 141*f*
complex circuit, analyzing, 116–118, 117*f*, 118*f*
conduction electrons, definition of, 69, 342
conductivity, 65, 73, 73*t*, 342
conductor
charging, 14*f*, 15
conductivity of, 65, 73, 73*t*, 342
definition of, 3, 14, 342
electrons in, 14, 66
electrostatic potential energy converted in, 85, 95–96
equilibrium charge distributions on, 75–76
hollow, charge distribution on, 76
human body as, 120
isolated, as capacitor, 78
moving (*See* moving conductor)
ohmic, 85, 95, 346
potential difference across, notation for, 94
power dissipated in, 95–96
static charges on, 74–76
surface of, charge on, 195, 195*f*
thermal energy in, 85, 95–96
connections, in circuit diagram, 107, 112–114, 112*f*, 114*f*
conservation of electrical charge, 9
conservation of energy
Ampere's law and, 226–227
in battery, 92
in inductor, 289–291
in Lenz's law, 285
potential difference and, 49
constant of proportionality, for magnetic force, 133
conventional current, 67, 140, 342
copper wire, drift velocity in, 70–71
coulomb, definition of, 2, 8, 342
Coulomb constant, 28, 330, 342
Coulomb's law
definition of, 10–11, 342
electric field of point charge and, 27
equation for, 3, 10
Gauss's law and, 194–195, 199
limitations of, 11
magnetic field and, 150, 156–157, 240–241
crests
definition of, 305, 342
of traveling waves, 307–308, 307*f*
cross-product rule, 155, 162*f*, 342
curl, 270–272, 342
current. *See* electric current
current density, 71–74
current and, 72
definition of, 64–65, 72, 342
electric field vector and, 73
flux of, 217, 217*f*
current distribution. *See specific current distributions*
current-carrying coil, self-induction of, 285–286
current-carrying wire
charged particle near, and electromagnetic field, 238–241, 239*f*
electric field in, 89
magnetic field of, 151, 155, 155*f*, 158–161, 158*f*, 160*f*
cyclotron frequency, definition of, 136, 342
cylindrical charge distribution
Ampere's law for, 226
definition of, 343
electric field of, 53*t*
electric field vector of, 173*t*, 176–177, 176*f*
Gauss's law for, 193, 199–201, 200*f*
nested, 243–245, 244*f*

dielectric constant, definition of, 343
dimensional analysis, 330–331, 343
dipole. *See* electric dipole
direct current, 88, 292, 343
disk, equilibrium charge distribution on, 75, 75*f*
dispersed energy, electric field as, 55–57
divergence, 270, 343
domains, 164, 164*f*, 343
drift, of electrons, electric field and, 68, 68*f*
drift velocity
in circuit, 87
in copper wire, 70–71
definition of, 64, 343
local (*See* current density)
magnitude of, 68–69
thermal speed and, 71
Drude model
conductor characteristics in, 73
definition of, 64, 343
electric field in, 68, 68*f*
purpose of, 67
dynamic fields
definition of, 343
in electromagnetic model, 4, 258
equations for, 262, 268–269
for infinite planar slab, 258*f*, 259–260, 259*f*

$\vec{E}$. *See* electric field vector
electric charge. *See also* negative charge; positive charge
in amber, 5
and atomic model, 7–8, 8*f*
behavior of, 6
conservation of, 9
creation of, 7–9
definition of, 343
electrical field of, describing (*See* electric field vector)
magnetic poles and, 130
mathematical description of, 10–11, 11*f*
nature of, 5–7
point (*See* point charge)
power radiated by, 329–331
tendency to accept, 9, 9*t*
unit (*See* point charge)
electric circuit. *See also* batteries
complex (*See* complex circuit)
current in, 87
definition of, 87, 343
drift velocity in, 87
household, 120
parallel (*See* parallel circuit)
position *vs.* potential in, 110, 110*f*
potential in, defining, 109–110, 110*f*
series (*See* series circuit)
simple, 87
electric current. *See also* drift
in chemical battery, 91–92, 91*f*
in circuit, 87
in complex circuit, 117–118, 117*f*, 118*f*
current density and, 72
definition of, 64, 66, 67*f*, 343
direction of, 66–67, 67*f*
of discharging capacitor, 98
in electric motor, 140–141
electric potential and, 94
emf and, in electric power, 292
energy transfer and, 69
flow of (*See* current density)
induced (*See* induced current)
magnetic battery and, 152–153, 153*f*
magnetic field created by, 154
from magnetic force, 153–154
magnetic force and, 139
magnitude of, 66, 67*f*
in moving conductors, 152–154, 262–267, 263*f*
oscillating, 292
in parallel circuit, 114–115
through parallel-plate capacitor, 86
safety issues with, 119–120
in series circuit, 110–111
steady flow of, and net electric field, 88
time-dependence of, 288, 288*f*, 289*f*
in transformer, 292–293, 292*f*
in wires in series, 108, 108*f*
electric dipole
creation of, 16
definition of, 15, 15*f*, 16, 343
field diagram of, 31*f*
field of, 23, 30*f*, 31–33, 32*f*
magnets and, comparison of, 126, 129–130, 130*f*, 162
natural, 17, 17*f*
neutral object as, 16–17
electric field. *See also* electromagnetic field
Ampere's law for, 215, 226–228, 262–267, 263*f*
of charge distribution, 23 (*See also* Gauss's law)
in circuits, 109, 109*f*
combining (*See* superposition principle)
created by moving conductor, 152
in current-carrying wire, 89
defining, with vectors (*See* electric field vector)
definition of, 24, 343
of dipole, 23, 30*f*, 31–33, 32*f*
as dispersed energy, 55–57
disturbance in (*See* electromagnetic waves)
divergence of, 270
in Drude model, 68
dynamic (*See* dynamic fields)
in electromagnetic field, 241
electron motion and, 68, 68*f*
energy density of, 43, 57
field model for, 24–25
frame-dependence of, 238
Gauss's law for (*See* Gauss's law)
magnetic field and, in electromagnetic wave, 325–326
magnitude of (*See* electric field magnitude)
of parallel-plate capacitor, 86, 86*f*
between parallel plates, 54, 55*f*
physically possible *vs.* impossible, 227–228, 228*f*
of point charge (*See* point charge field)
for polarized plane waves, 327–328
potential at any point in, 47–48
of ring, 33–34
in silver wire, 74
of spherical surface, 56
in static equilibrium, 75
in steady current flow, 88
surface elements and, 196
of symmetric objects, 174–178
time derivative of, 260–261
transformation and, 174
electric field magnitude. *See also* electric field vector
benchmarks for, 27
and drift velocity, 69
and wire length, 93–94
electric field vector
calculating, from electric potential, 43, 49–50, 51
curl of, 270–272
current density and, 73
of cylindrical charge distribution, 173*t*, 176–177, 176*f*, 200
definition of, 22, 25–26, 343
direction of, finding, 49–50
electric force computed from, 26–27
electric potential calculated from, 43, 47–79
electric potential *vs.*, advantages of, 44
on equipotential diagram, 50, 51*f*
equipotentials and, 50
explicit *vs.* implicit, 29
Gauss's law for, 193, 197–201, 200*f*
gravitational field vector and, 27
for infinite planar slab, 193, 201–203
and magnetic field vector, transformation to, 237, 242–243
magnetic field vector and, 129–130
near finite plate, 54, 55*f*
oscillation of, 331–332
in parallel-plate capacitor, 86, 86*f*
of point charge, 28
for spherical charge distribution, 173*t*, 174–175, 175*f*
for surface charge distribution, 53–55, 53*t*
vs. total field, 28–29
of wire, 34–36

electric force, 26–27, 29
electric hand-drier, power used by, 97, 97*f*
electric motors
definition of, 343
energy per revolution in, 142–143, 142*f*
magnetic fields in, 140–143, 140*f*, 141*f*
electric potential
analogy to gravitational potential, 109
calculating, from electric field vector, 43, 47–79
change in, along infinitesimal displacement, 47
in circuit, defining, 109–110, 110*f*
current and, 94
definition of, 42, 44, 343
vs. electric field vector, advantages of, 44
electric field vector calculated from, 43, 49–50, 51
vs. emf, 84
in household circuits, 120
vs. position, in circuit, 110, 110*f*
of spherical surface, 56
in static equilibrium, 75
electric potential difference
between battery electrodes, 84
in capacitance, 77
between capacitor plates, 85
across complex circuit, 117–118, 117*f*, 118*f*
definition of, 347
across discharging capacitor, 98–99, 99*f*
across electrolyte, 92
emf and, 90, 92
across lead-acid battery, 92
and lethal current, 119–120
maintenance of, 92
across moving conductor, 152
notation for, 94
across parallel circuit, 114, 116, 116*f*
across realistic battery, 119
across series circuit, 110–111
across static field, 49
across Van de Graaff generator, 120
between wire ends, 85, 93, 93*f*
across wires in series, 108, 108*f*
electric potential field, definition of, 42
electric power, 292, 293
electricity, 4, 154–155
electrodes
in chemical battery, 91
definition of, 84, 90, 343
vs. surface charge distribution, in current flow, 89
electrolyte, 91, 92, 343
electromagnetic field. *See also* electric field; magnetic field
charged particles and, 236
definition of, 343–344
dynamic, 258
energy in, 291
frame-independence of, 240
Lorentz force law for, 236–237
of moving conductor, 248–249, 248*f*
of moving infinite slab, 246–247, 246*f*, 263–268, 264*f*, 267*f*
of nested cylindrical charge distribution, 243–245
transformation equations for, 242–243
electromagnetic force, 241, 249
electromagnetic model, dynamic fields in, 4, 258
electromagnetic plane wave, definition of, 322, 322*f*
electromagnetic waves
Ampere-Maxwell law for empty space and, 322–323
characteristics of, 320, 324–328
definition of, 344
energy in, 321, 328–329
from everyday objects, 333–334, 334*f*
Faraday's law and, 322–323
frequencies of, 332–333, 333*f*
intensity of, 328–329, 328*f*
magnetic disturbance in, 325
polarizations of, 320–321, 327–328
speed of, 324
wave rule and, 324
electromotive force. *See* emf
electron(s)
charge of, 7–8
in conductor, 14, 66
in conservation of charge, 9
as current loops, 163
definition of, 344
in insulators, 13, 66
mass of, 7
in metal, 67–68, 70
motion of, and electric fields, 68, 68*f*
near current-carrying wire, and electromagnetic field, 238–241, 239*f*
transfer of, 8–9
electron beam, 130–131, 132*f*
electron cloud, 7–8, 8*f*, 344
electron gun, 130–131, 131*f*, 344
electronic circuits, capacitors in, 99, 99*f*
electrostatic, definition of, 5, 344
electrostatic force
of charged particles, equation for, 22–23
of macroscopic objects, 11, 12–13
of point particles, magnitude of, 10, 11–12
electrostatic interaction, action-at-a-distance model for, 24, 324, 341
electrostatic polarization, 16–17, 344
electrostatic potential energy
in conductor, 85, 95–96
vs. electric potential, 44
emf
of batteries in series, 90–91, 90*f*
current and, in electric power, 292
definition of, 84, 90, 344
vs. electric potential, 84
induction of (*See* induction)
of lead-acid battery, 92
from magnetic force, 153
potential difference and, 90, 92
in transformers, 291–293, 292*f*
energy
in batteries, 90
conservation of (*See* conservation of energy)
in electric motor, per revolution, 142–143, 142*f*
in electromagnetic field, 291
in electromagnetic waves, 321, 328–329
in magnetic field, 281, 289–291
in mechanical waves, 303–304
sources of, behavior of, 93
transfer of, and electric current, 69
energy density
of electric field, 43, 57
in infinite solenoid magnetic field, 290–291
equilibrium charge distribution, 75–76, 75*f*, 76*f*
equipotential. *See* equipotential curves
equipotential curves, 46, 46*f*, 50, 344
equipotential diagram, 46–47, 46*f*, 50, 51*f*, 344
equipotential surface, definition of, 46–47, 344

ϕ. *See* electric potential
$\vec{F}$. *See* electromagnetic force
farad, definition of, 344
Faraday, Michael, 155, 269, 282
Faraday's law
  definition of, 256, 266–267, 344
  and electric field disturbance, 322–323
Faraday's law of induction, definition of, 280, 282, 344
feedback mechanism
  for current flow control, 88
  in wires in series, 108
ferromagnetic substances
  atom alignment in, 163–164
  definition of, 344
field. *See* dynamic fields; electric field; electromagnetic field; magnetic field
field diagram
  definition of, 344
  of dipole, 31*f*
  of total electric field, 28–29, 28*f*
field model, 22, 24–25, 344
finite plate, 54, 55*f*
flat surface model, infinite, 54, 55*f*
flux
  calculating (*See* Gauss's law; symmetry argument)
  of current density, 217, 217*f*
  of cylindrical charge distribution, 199–200
  definition of, 183, 344
  of infinite planar slab, 202, 203
  net (*See* net flux)
  notation for, 216*f*
  sign of, 185, 217, 217*f*
  of spherical charge distribution, 197, 198
  through tile, 172–173, 184–185
Fourier's theorem, 300, 304, 327, 344–345
frame-independence, of electromagnetic fields, 240–241
free particle
  helical path of, 137, 137*f*
  in magnetic field, 127, 135–138
  in particle physics, 137, 138*f*
frequency
  definition of, 301, 345
  of electromagnetic waves, 332–333, 333*f*
  phase speed and, 308
  ranges of, 333, 333*f*
  reradiated, 332

gamma rays, 333, 333*f*, 345
gauss, definition of, 133, 345
gaussian surface
  choosing, 196
  for cylindrical charge distribution, 199, 200*f*
  definition of, 345
  for infinite planar slab, 201–202, 202*f*, 203*f*
  for moving infinite slab, 268, 268*f*
  for moving point particle, 204*f*
  for spherical charge distribution, 197, 197*f*, 198, 198*f*
Gauss's law
  Coulomb's law and, 194–195, 199
  definition of, 192, 194, 345
  for electric field, 192, 195–197, 267–268
  for infinite planar slab, 201–203, 267–268, 267*f*
  local field equations and, 270
  for magnetic fields, 204–206, 345
  for spherical charge distribution, 197–199
  superposition principle and, 194
  using, 192, 195–196
generators, 153–154, 154*f*, 345
gravitational field vector, 27, 345
gravitational potential, 44–45, 109, 345
gravitational potential energy, 45
gravitomagnetic field, 241–242
ground, current conducted into, 120

hand-drier, electric, power used by, 97, 97*f*
helical motion, of free particle in magnetic field, 137, 137*f*
henry, definition of, 345
Hertz, Heinrich, 328
horizontally polarized waves, 320–321, 327–328
household circuits, potential in, 120
human body, 119–120

ideal battery approximation, 92, 119, 119*f*, 345
induced current, direction of, 284–285, 284*f*
inductance
  definition of, 280–281, 286, 345
  of solenoid, 286–287
  in transformers, 293, 293*f*
induction. *See also* self-induction
  conservation of energy and, 285
  definition of, 4, 280, 282, 283*f*, 345
  from emf, 153
  in transformers, 291
inductor, 281, 287–291, 289*f*
inertial reference frames, in principle of relativity, 238
infinite flat surface model, 54, 55*f*
infinite planar slab
  Ampere's law and, 258–260, 258*f*, 259*f*, 260*f*
  central plane of, 177, 177*f*
  electric field of, and surface charge distribution, 53*t*
  electric field vector of, 173*t*, 177–178, 177*f*
  Gauss's law for, 193, 201–203, 203*f*
  infinite planar (*See* infinite planar slab)
  magnetic field of, 215*t*, 221–223
  moving (*See* moving infinite slab)
infinite solenoid
  inductance of, 286–287
  magnetic field of, 223–226, 290–291
infinite symmetric objects. *See also specific objects*
  charge distributions of, 183
  current distributions of, 183
infrared range, 333, 333*f*, 345
insulator
  atom near, charge of, 15, 15*f*
  charge distribution by, 53
  charging, 14
  definition of, 3, 13–14, 345
  electrons in, 13, 66
intensity, 328–329, 328*f*, 345
ion, in velocity selector, 249, 249*f*
iron, magnetized, domains in, 164, 164*f*

kilowatt-hours, definition of, 96, 345
kinetic energy, transfer of, and electric current, 69

Larmor formula, definition of, 321, 331, 345
lead-acid battery, 91–93, 91*f*
Lenz's law, definition of, 280, 284, 345
light
electromagnetism of, 2, 4, 324
frequencies of, 332, 333
mystery of, 4
waves in, 324, 328
lightbulb, 96, 112, 112*f*
local field equations, 257, 270, 345
longitudinal waves, definition of, 300, 303, 303*f*, 345
loop-to-magnet rule, definition of, 345–346
for magnet direction, 127, 134, 134*f*, 141
for magnetic field vector direction, 151, 162–163, 162*f*
Lorentz contraction, 240
Lorentz force law, 236–237, 241, 346
LR circuit, 288, 346

macroscopic objects, 11, 33–36
magnet(s)
definition of, 346
dipoles and, comparison of, 126, 129–130, 130*f*, 162
in particle accelerators, 137
permanent (*See* permanent magnet)
properties of, 126, 128
magnetic battery, current driven by, 152–153, 153*f*
magnetic charge, existence of, 205
magnetic compass, definition of, 128, 346
magnetic field. *See also* electromagnetic field
Ampere's law for, 214, 216–217, 258–262, 258*f*
charged particle in, 135
of circular loop, 151, 161–163, 161*f*
circulation of, 216, 216*f*
created by point charge, 150
of current-carrying wire, 151, 155, 155*f*, 158–161, 158*f*, 160*f*
current created by (*See* induction)
of current distribution (*See* Ampere's law)
definition of, 346
describing, 126
direction of, 135
displaying, 129, 129*f*
disturbance in (*See* electromagnetic waves)
dynamic (*See* dynamic fields)
from electric current, 154, 155
electric field and, in electromagnetic wave, 325–326
in electric motor, 140–141, 140*f*
in electromagnetic field, 241
electron beam affected by, 131, 132*f*
energy in, 281, 289–291
flux through (*See* flux)
frame-dependence of, 238
free particle in, 127, 135–138
Gauss's law for, 193, 204–206
of inductor, 289–290
investigation of, 131
magnitude of (*See* magnetic field magnitude)
mirror rule for, 178, 179*f*
and moving charged particle, 156, 157*f*, 204, 204*f*
in particle detector, 137–138
of permanent magnet, 163
physically possible *vs.* impossible, 205–206, 205*f*, 206*f*
for polarized plane waves, 327–328
residual, 164
in solenoid, 283–284, 290–291
source of, 164–165
from static charge, 156
superposition principle for, 150–151, 158
surface elements and, 196
of symmetric current distributions, 173*t*, 179–183
symmetry arguments for, 178
transformation and, 174
in transformers, 291–292
of wire, 151, 155, 155*f*, 158–161, 158*f*, 160*f*
magnetic field magnitude. *See also* magnetic field vector
benchmarks for, 133, 134*t*
magnetic field vector
alternate version of, 133
Ampere's law for, 215*t*, 219–226, 220*f*
of axial current distribution, 173*t*, 179–180, 180*f*
curl of, 270–271
definition of, 128, 129*f*
direction of, 163
and electric field vector, transformation to, 237, 242–243
electric field vector and, 129–130
of infinite solenoid, 223–226
magnitude of, 133–134
oscillation of, 331–332
of solenoidal current distribution, 173*t*, 181, 181*f*
magnetic force
constant of proportionality for, 133
and conventional current, 140
current created by, 153–154
direction of, 135
emf from, 153
formula for, 131
on moving charged particle, 126–127, 133–134
on moving conductor, 152, 152*f*
on wire, 127
magnetic force law, 133–134, 346
magnetic induction, discovery of, 155
magnetic moment, definition of, 346
magnetic monopole, definition of, 130, 346
magnetic permeability, definition of, 346
magnetic poles. *See* pole, magnetic
magnetic resonance imaging (MRI), 226, 226*f*
magnetism
electricity linked to, 154–155
electromagnetism of, 4
mystery of, 4
phenomenon of, 127
magnetized iron, domains in, 164, 164*f*
Marconi, Guglielmo, 328
Maxwell, James Clerk, 2, 155, 269, 334
Maxwell's equations, 2
definition of, 257, 268–269, 346
differential forms of, 270–271, 343
and dynamic fields, 4
and electromagnetic waves, 324
Maxwell's rainbow, 332–334
mechanical battery, 86–87, 87*f*, 346
mechanical waves, 300, 303–304, 346
media, for waves, 313–315
metal, electrons in, 67–68

metal bar. *See* conductor
microamperes, definition of, 66
microwaves, 333, 333*f*, 346
milliamperes, definition of, 66
mirror rule
in axial current distribution, 179–180, 180*f*
definition of, 172, 178–179, 179*f*
in planar current distribution, 182–183
in solenoidal current distribution, 181, 181*f*
moving charged particle
direction of, 135–136, 136*f*
frame-dependence and, 238
magnetic field from, 156, 157*f*, 164–165, 204, 204*f*
magnetic force on, 126–127, 133–134
net flux through, 204
period and speed of, 136
radius and momentum of, 136
relativistic, 136–137
moving conductor
Ampere's law and, 262–263, 263*f*
creating currents in, 152–154
electromagnetic field of, 248–249, 248*f*
moving infinite slab, charged
Ampere's law and, 259–260, 259*f*, 260*f*
electromagnetic field of, 246–247, 246*f*
Gauss's law and, 267–268, 267*f*
moving infinite slab, current-carrying
Ampere's law and, 263–267, 264*f*, 265*f*
Gauss's law and, 268, 268*f*
MRI (magnetic resonance imaging), 226, 226*f*

natural dipole, 17, 17*f*
negative charge, behavior of, 6
nested cylindrical charge distribution, 243–245
net charge, 9, 346
net flux
definition of, 173, 185–186, 344
electromagnetic waves and, 323
through moving charged particle, 204
neutral object
as dipole, 16–17, 16*f*
force on, by point charge, 32
neutrons, 7–8
node rule, definition of, 346
number density, 70, 346

Oersted, Hans Christian, 154–155
ohm, definition of, 95, 346
ohmic, definition of, 85, 95, 346
Ohm's law
in complex circuit analysis, 116–117
definition of, 85, 95, 347
one-dimensional waves, 300, 304–305, 347. *See also* sinusoidal wave

parallel circuit, 107, 114–116, 116*f*, 347
parallel-plate capacitor, 54, 55*f*, 78, 86–87, 86*f*, 347. *See also* batteries
particle. *See* antineutrino; charged particle; free particle; neutrons; particle charge; point particles
particle accelerators, magnets in, 137
particle charge, 9. *See also* point charge
particle detector, magnetic fields in, 137–138
particle physics, motion of free particles in, 137, 138*f*
period, 301, 305, 308, 347
permanent magnet, 163, 164
permittivity constant, 28, 347
phase rate. *See* angular frequency
phase speed, definition of, 301, 308
phase velocity, definition of, 301, 308, 347
planar charge distribution. *See also* infinite planar slab
central plane of, 177, 177*f*
electric field vector of, 173*t*, 177–178, 177*f*
electromagnetic field of, 246–247, 246*f*
planar current distribution. *See also* infinite planar slab
magnetic field vector of, 173*t*, 182–183, 182*f*
mirror rule in, 182–183, 182*f*
planar slab, infinite. *See* infinite planar slab
plane wave, definition of, 322, 347
plates, battery. *See* electrodes
plates, parallel. *See* parallel-plate capacitor
point charge. *See also* electric dipole
behavior of, in field model, 24–25
force of, 30, 32
magnetic field created by, 150
potential field of, 45
spherical surface charge and, 54
superposition principle for (*See* superposition principle)
point charge field, 23, 27–29, 28*f*, 51
and equilibrium charge distribution, 75–76, 76*f*
point particles
charged, 10–11, 10*f*
charge of (*See* point charge)
magnetic field of, 157
spherical object modeled as, 54
"sufficiently close," 54
polarization, 16–17, 32–33
definition of, 347
electromagnetic field and, 248–249
in light waves, 328
polarized waves, 320–321, 327–328
pole, magnetic
behavior of, 128, 130
definition of, 346
in electric motor, 140–141, 140*f*
electrical charges and, 130
naming, 128*f*
in permanent magnets, 163, 163*f*
positive charge, 6, 76–77, 77*f*. *See also* protons
potential. *See* electric potential
power
in electrostatic potential energy conversion, 96
radiated by charge, 329–331
units of, 330
principle of relativity, 238, 347
protons
charge of, 7–8
in conservation of charge, 9
mass of, 7

q_{test}. *See* test charge

radio waves
definition of, 347
first, 328
frequencies of, 333, 333*f*
Rayleigh scattering process, 331–332, 347
realistic battery, 119, 119*f*
red light, 332
relativity
definition of, 347
electromagnetism and, 4
field theory and, 25, 240–241, 269
magnetic interactions and, 128
principle of, 238
resistance
definition of, 85, 95, 347
of discharging capacitor, 97
of human body, 119–120
of parallel circuit, 115–116, 116*f*
of series circuit, 110–111
resistor
in circuit diagram, 107, 112, 112*f*
in complex circuit analysis, 116–118, 117*f*, 118*f*
definition of, 112, 112*f*, 347
in parallel, 114, 115*f*
right-hand rule. *See* loop-to-magnet rule; wire rule
ring
charge distribution of, 33–34, 34*f*
electric field vector for, 51–52, 52*f*
electric potential of, 51–52, 52*f*
roller-coaster analogy, for circuits, 109, 109*f*

safety issues, 119–120
Savart, Felix, 155
Scotch tape, electrical charge in, 5–6, 5*f*, 6*f*
segment vector, definition of, 347
seismic wave, definition of, 347
self-energy, 56, 347
self-induction
of current-carrying coil, 285–286
definition of, 280–281
in discharging inductor, 287–289
series. *See* batteries in series; series circuit; wires in series
series circuit, 106, 110–111, 347
shell theorem, definition of, 348
silver wire, electric field in, 74
sinusoidal plane wave, 322, 322*f*, 328
sinusoidal wave. *See also* one-dimensional waves
definition of, 304, 348
equations for, 306
mathematical form of, 304–305, 305*f*, 306*f*
wave equation and, 310
skin, resistance of, 119–120
sky, color of, 321, 331–332
slab. *See* infinite planar slab
small oscillation limit, 312–313
solenoid
definition of, 181, 348
inductance of, 286–287
infinite (*See* infinite solenoid)
magnetic field inside, 283–284, 290–291
solenoidal current distribution
Ampere's law for, 215*t*, 223–226
definition of, 348
magnetic field vector of, 173*t*, 181, 181*f*
sound wave, 300, 302, 302*f*, 309, 348
speed of light, electromagnetic waves and, 324
spherical charge distribution
definition of, 348
electric field of, 53*t*
electric field vector for, 173*t*, 174–175, 175*f*
Gauss's law for, 193, 197–199
spherical surface
compressed, energy field of, 56
electric potential of, at surface, 56
energy field of, 56
equilibrium charge distribution on, 75, 75*f*
positive charge on, 76–77, 77*f*
static charges, 74–76, 156
static equilibrium, definition of, 74
static field, definition of, 348
string, transverse wave on, 310–315, 311*f*, 314*f*
"sufficiently close" point, definition of, 54
superposition principle
definition of, 11–13, 348
equation for, 23, 29
Gauss's law and, 194
for magnetic fields, 150–151, 158, 204
for point charge, 29–30, 30*f*
for potentials, 42–43, 45–46
surface charge
current control with, in circuit, 88, 88*f*
in parallel circuits, 115
self-energy of, 56
surface charge distribution
electric fields produced by, 53–55, 53*t*
vs. electrodes, in current flow, 89
surface elements, 196–197
symmetric charge distributions, electric fields of, 173*t*, 174–178. *See also specific charge distributions*
symmetric current distributions, magnetic fields of, 173*t*, 179–183, 215*t*. *See also specific current distributions*
symmetry argument
Ampere's law and, 218–219
for axial current distribution, 220
for cylindrical charge distribution, 197
definition of, 172, 174, 348
Gauss's law and, 195–196, 201
for infinite planar slab, 197, 221
for infinite solenoid, 223
for magnetic fields, 178
for spherical charge distribution, 197
structure of, 175–176
symmetry principle, definition of, 174

tension wave, definition of, 348
tesla, definition of, 133, 348
test charge
definition of, 348
in electric field vector definition, 26
net electric force on, 29
thermal energy
in chemical reaction, 92
in conductor, 85, 95–96
in discharging capacitor, 99
thermal speed, 71
three-dimensional wave equation, 315
tile(s), 172–173, 183–185, 348

tile vector, definition of, 172, 183–184, 348
time between collisions, 69, 74
transformation, effect of, on fields, 174
transformer
current in, 292–293, 292*f*
definition of, 281, 292, 348
emf in, 292–293, 292*f*
idealized, 291
inductance in, 293, 293*f*
use of, 292
transparent tape, electrical charge in, 5–6, 5*f*, 6*f*
transverse waves
definition of, 300, 303, 303*f*, 348
electromagnetic, 324–325, 325*f*
on string, 310–315, 311*f*, 314*f*
traveling waves
definition of, 307–308, 307*f*, 348
wave equation and, 310
triboelectric series, 9, 9*t*, 348
troughs, definition of, 305, 342

ultraviolet range, 333, 333*f*, 348
uniform field, definition of, 348
unit charge. *See* point charge

Van de Graaff generator, 14*f*, 15, 87, 87*f*, 120
vector function
definition of, 348
for electric field (*See* electric field vector)
for magnetic field (*See* magnetic field vector)
velocity selector, 249–250, 249*f*, 348
vertically polarized waves, 320–321, 327–328
violet light, 332
voltage. *See* electric potential
voltage drop. *See* electric potential difference

water, natural dipole in, 17, 17*f*
water waves, 303, 304*f*, 348
watts, definition of, 96
wave(s)
definition of, 300, 302, 348
electromagnetic (*See* electromagnetic waves)
longitudinal (*See* longitudinal waves)
mathematical representation of, 304
media for, 313–315
transverse (*See* transverse waves)
traveling (*See* traveling waves)
types of, 300, 302–304, 302*f*
wave equation
applications of, 309–310
definition of, 301, 309, 348–349
sinusoidal traveling waves and, 310
solution to, 310
three-dimensional, 315
transverse wave on string and, 310–313, 311*f*
wave rule and, 313–315
wavelength
definition of, 301, 305, 349
phase speed and, 308
of sound wave, 309
wavenumber, definition of, 301, 305, 349
wave rule, 301, 313–315, 324, 349
wire(s)
capacitor discharged through, 97–99
charge distribution of, 34–36, 35*f* (*See also* cylindrical charge distribution)
in circuit diagram (*See* connections, in circuit diagram)
current-carrying (*See* current-carrying wire)
current in, in magnetic force, 139
in electric motor, 140–142, 140*f*, 141*f*
electric potential difference across, 85, 93, 93*f*
length of, in electric field magnitude, 93–94
length of, in magnetic force, 139
magnetic field of, 151, 155, 155*f*
magnetic force on, 127, 138–143
resistance of, 95
segment, magnetic field of, 158–161, 158*f*, 160*f*
wire rule, 155, 155*f*, 162*f*, 349
wires in series, 108, 108*f*

X-rays, 333, 333*f*, 349

Periodic Table of the Elements

1 Atomic number
H Symbol
1.008 Atomic mass

1 1A	2 2A	3 3B	4 4B	5 5B	6 6B	7 7B	8 8B	9 	10 	11 1B	12 2B	13 3A	14 4A	15 5A	16 6A	17 7A	18 8A
1 **H** 1.008																	2 **He** 4.003
3 **Li** 6.941	4 **Be** 9.012											5 **B** 10.81	6 **C** 12.01	7 **N** 14.01	8 **O** 16.00	9 **F** 19.00	10 **Ne** 20.18
11 **Na** 22.99	12 **Mg** 24.31											13 **Al** 26.98	14 **Si** 28.09	15 **P** 30.97	16 **S** 32.07	17 **Cl** 35.45	18 **Ar** 39.95
19 **K** 39.10	20 **Ca** 40.08	21 **Sc** 44.96	22 **Ti** 47.88	23 **V** 50.94	24 **Cr** 52.00	25 **Mn** 54.94	26 **Fe** 55.85	27 **Co** 58.93	28 **Ni** 58.69	29 **Cu** 63.55	30 **Zn** 65.39	31 **Ga** 69.72	32 **Ge** 72.59	33 **As** 74.92	34 **Se** 78.96	35 **Br** 79.90	36 **Kr** 83.80
37 **Rb** 85.47	38 **Sr** 87.62	39 **Y** 88.91	40 **Zr** 91 .22	41 **Nb** 92.91	42 **Mo** 95.94	43 **Tc** (98)	44 **Ru** 101.1	45 **Rh** 102.9	46 **Pd** 106.4	47 **Ag** 107.9	48 **Cd** 112.4	49 **In** 114.8	50 **Sn** 118.7	51 **Sb** 121.8	52 **Te** 127.6	53 **I** 126.9	54 **Xe** 131.3
55 **Cs** 132.9	56 **Ba** 137.3	57 **La** 138.9	72 **Hf** 178.5	73 **Ta** 180.9	74 **W** 183.9	75 **Re** 186.2	76 **Os** 190.2	77 **Ir** 192.2	78 **Pt** 195.1	79 **Au** 197.0	80 **Hg** 200.6	81 **Tl** 204.4	82 **Pb** 207.2	83 **Bi** 209.0	84 **Po** (210)	85 **At** (210)	86 **Rn** (222)
87 **Fr** (223)	88 **Ra** (226)	89 **Ac** (227)	104 **Rf** (257)	105 **Db** (260)	106 **Sg** (263)	107 **Bh** (262)	108 **Hs** (265)	109 **Mt** (266)	110	111	112	(113)	114	(115)	116	(117)	

58 **Ce** 140.1	59 **Pr** 140.9	60 **Nd** 144.2	61 **Pm** (147)	62 **Sm** 150.4	63 **Eu** 152.0	64 **Gd** 157.3	65 **Tb** 158.9	66 **Dy** 162.5	67 **Ho** 164.9	68 **Er** 167.3	69 **Tm** 168.9	70 **Yb** 173.0	71 **Lu** 175.0
90 **Th** 232.0	91 **Pa** (231)	92 **U** 238.0	93 **Np** (237)	94 **Pu** (242)	95 **Am** (243)	96 **Cm** (247)	97 **Bk** (247)	98 **Cf** (249)	99 **Es** (254)	100 **Fm** (253)	101 **Md** (256)	102 **No** (254)	103 **Lr** (257)